noncommutative geometry and physics

Proceedings of the COE International Workshop

edited by

Yoshiaki Maeda Nobuyuki Tose Naoya Miyazaki *Keio University, Japan*

Satoshi Watamura *Tohoku University, Japan*

Daniel Sternheimer *Bourgogne University, France*

noncommutative geometry and physics

Proceedings of the COE International Workshop

Yokohama, Japan 26-28 February, 1-3 March 2004

World Scientific

NEW JERSEY • LONDON • SINGAPORE • SHANGHAI • HONG KONG • TAIPEI • BANGALORE

Published by

World Scientific Publishing Co. Pte. Ltd.
5 Toh Tuck Link, Singapore 596224
USA office: 27 Warren Street, Suite 401-402, Hackensack, NJ 07601
UK office: 57 Shelton Street, Covent Garden, London WC2H 9HE

British Library Cataloguing-in-Publication Data
A catalogue record for this book is available from the British Library.

NONCOMMUTATIVE GEOMETRY AND PHYSICS
Proceedings of the COE International Workshop

ISBN 981-256-492-6

Printed in Singapore by World Scientific Printers (S) Pte Ltd

Preface

A workshop on *Noncommutative Geometry and Physics 2004* was held at Keio University, Yokohama, Japan from February 26 through March 3, 2004. The aim of the workshop was to enhance international cooperation in various aspects of this field between mathematicians and physcists.

Noncommutative differential geometry is a novel approach to geometry, aimed in part at applications in physics. It was founded in the early eighties by 1982 Fields Medalist Alain Connes on the basis of his fundamental works in operator algebras. It is now a very active branch of mathematics with present and potential applications to a number of domains in physics, from solid state to quantization of gravity. The strategy is to formulate usual differential geometry in a somewhat unusual manner, using in particular operator algebras and related concepts, so as to be able to "plug in" noncommutativity in a natural way. Algebraic tools such as K-theory and (cyclic) cohomology and homology play an important role.

These Proceedings contain papers presented at the workshop. All papers were submitted by conference speakers and participants, and were duly refereed. The papers contain presentations of new results which have not appeared previously in professional journals, or comprehensive reviews including an original part of the present developments in those topics. Noncommutativity in a geometric setting, and possible physical applications thereof, are present (explicitly or as a watermark) in all contributions. The domains go beyond noncommutative geometry *stricto sensu*, as any reader can discover from looking at the table of contents. But a closer look at the presentations shows that these deal with complementary aspects. Together they contribute to the symphony, each in its own way.

The volume is accessible to researchers and graduate students interested in a variety of mathematical areas related to noncommutative geometry, and in the interface with modern theoretical physics.

The workshop was held in the framework and with the support of the 21st century Center of Excellence (COE) program at Keio, *Integrative Mathematical Sciences: Progress in Mathematics motivated by Natural and Social Phenomena.*

The editors and workshop organizers are grateful for the generous financial support of the COE and for its help and encouragement in the planning phase.

The World Scientific Publishing company has been very helpful in the production of this volume; special tribute is due to Ms. Zhang Ji for her editorial guidance throughout the production of the volume. But above all we wish to thank all authors for their important contributions and the referees for their valuable comments and suggestions.

The Editors:
Yoshiaki MAEDA,
Naoya MIYAZAKI,
Daniel STERNHEIMER,
Nobuyuki TOSE,
Satoshi WATAMURA

Contents

NONANTICOMMUTATIVE HARMONIC SUPERSPACE AND $N = 2$ SUPERSYMMETRIC $U(1)$ GAUGE THEORY

TAKEO ARAKI

Department of Physic
Tohoku University
Sendai, 980-8578, Japan
araki@tuhep.phys.tohoku.ac.jp

KATSUSHI ITO AND AKIHISA OHTSUKA

Department of Physics
Tokyo Institute of Technology
Tokyo, 152-8551, Japan
ito@th.phys.titech.ac.jp
ohtsuka@th.phys.titech.ac.jp

We study $N = 2$ supersymmetric $U(1)$ gauge theory in nonanticommutative harmonic superspace. We calculate the Lagrangian in the Wess-Zumino gauge up to the first order in the deformation parameter. The deformed gauge and supersymmetry transformations are also studied.

1. Introduction

Supersymmetric gauge theories in nonanticommutative superspace[1] have been taken much attentions recently. They are realized as the field theories on the D-branes in the graviphoton background in the superstring theory compactified on a Calabi-Yau threefold. Nonanticommutativity is introduced by the Ramond-Ramond fields in the hybrid formalism of superstrings [2,3,4].

From the field theoretical viewpoint, nonanticommutative superspace provides various interesting problems. Compare to the field theories in noncommutative spacetime, only a finite number of the deformation terms are necessary in the case of nonanticommutative $N = 1$ superspace. This property makes the deformed Lagrangian simple, so that the noncommutative effects can be studied explicitly [5,6,7,8,9].

It is known that there are two types of deformations in $N = 1$ superspace. One is called the Q-deformation[5], whose Poisson structure is constructed by chiral supersymmetry generators. This deformation preserves the chirality of a theory but breaks half of $N = 1$ supersymmetry. This deformed superspace is also called $N = 1/2$ superspace. The other is called the D-deformation[10], whose Poisson structure is constructed by the chiral supercovariant derivatives. This deformation preserves the full $N = 1$ supersymmetry but breaks the (anti-)chiral structure of the theory.

In a previous paper [9], we studied the deformed Lagrangian of $N = 2$ supersymmetric $U(N)$ gauge theory in Q-deformed $N = 1$ superspace by adding adjoint chiral superfields to $N = 1$ super Yang-Mills theory. In this formalism, only the deformed $N = 1/2$ supersymmetry is formulated manifestly.

Noncommutative $N = 2$ superspace provides another approach to study the deformation of $N = 2$ supersymmetric gauge theories, where extended supersymmetry is realized manifestly. $N = 2$ rigid superspace is convenient to construct $N = 2$ supersymmetric Yang-Mills theory[13]. The Lagrangian is constructed from $N = 2$ chiral superfields, which obey certain constrains connecting the chiral part and anti-chiral parts. One may consider the noncommutative deformation of rigid superspace. It turns out that the constraint equations are highly nontrivial to solve. Moreover, in order to introduce matter hypermultiplets and study supersymmetry off-shell, we need infinitely many auxiliary fields [18].

Harmonic superspace [14] is known to provide a good off-shell formulation of quantum field theory with extended supersymmetry including $N = 2$ matter hypermultiplet. This superspace is obtained by adding the harmonic coordinates of S^2 to the rigid $N = 2$ superspace. In this formalism, analytic superfields play an important role instead of $N = 2$ chiral superfields. These superfields are unconstrained and include infinitely many auxiliary fields in their components arising from the harmonic expansions.

In refs. [11,12], the nonanticommutative harmonic superspace has been introduced. Ferrara and Sokatchev [11] studied $N = 2$ $U(1)$ gauge theories based on the singlet deformation whose Poisson structure includes only the supercovariant derivatives (the singlet D deformation). The Lagrangian depends on certain function of anti-holomorphic scalar field. Ivanov *et al.* have studied the Q-deformation of $N = 2$ harmonic superspace and its Poisson structure[12].

In this paper, we study $N = 2$ supersymmetric $U(1)$ gauge theory in Q-deformed noncommutative $N = 2$ harmonic superspace. In this case, one can decompose the deformation parameters into the non-singlet part and the singlet part with respect to the $SU(2)_R$ R-symmetry group. We shall discuss the deformed action with general deformation parameters in the component formalism and their symmetry. To write down the component action in the simplest way, it is convenient to take the Wess-Zumino(WZ) gauge. It is shown that the action has a infinite power series in the deformation parameters. We calculate the first order correction with respect to the deformation parameter. But for the singlet deformation, we are able to write down the fully deformed action.

The undeformed gauge and supersymmetry transformations do not keep the WZ gauge. So we need to perform additional C-dependent gauge transformation to retain the WZ gauge. For the generic deformation case, we shall discuss the first order correction to the undeformed symmetry. In the case of the singlet deformation, we can obtain the exact gauge and supersymmetry transformations.

This paper is based on the papers [19,20,21], where the work [20] was done by the first two authors of the present paper.

2. Noncommutative Harmonic Superspace

We begin with reviewing the $N=2$ harmonic superspace [14] and its Q-deformation. Let $(x^\mu, \theta^\alpha_i, \bar\theta^{\dot\alpha i})$ be the coordinates of the $N=2$ rigid superspace, where $\mu = 0,1,2,3$ are indices of spacetime coordinates, $\alpha, \dot\alpha = 1,2$ spinor indices and $i = 1,2$ labels the $SU(2)_R$ doublet. The signature of spacetime is Euclidean. Spinors with upper and lower indices are related through the ε-tensor with $\varepsilon^{12} = -\varepsilon_{12} = 1$[15]. On the other hand, raising and lowering $SU(2)$ indices are done with the help of antisymmetric tensor ϵ_{ij} with $\epsilon^{12} = -\epsilon_{12} = -1$, $\theta^i = \epsilon^{ij}\theta_j$ and $\theta_i = \epsilon_{ij}\theta^j$.

The supersymmetry generators Q^i_α, $\bar Q_{\dot\alpha i}$ and the supercovariant derivatives D^i_α, $\bar D_{\dot\alpha i}$ are defined by

$$Q^i_\alpha = \frac{\partial}{\partial\theta^\alpha_i} - i(\sigma^\mu)_{\alpha\dot\alpha}\bar\theta^{\dot\alpha i}\frac{\partial}{\partial x^\mu}, \quad \bar Q_{\dot\alpha i} = -\frac{\partial}{\partial\bar\theta^{\dot\alpha i}} + i\theta^\alpha_i(\sigma^\mu)_{\alpha\dot\alpha}\frac{\partial}{\partial x^\mu}$$
$$D^i_\alpha = \frac{\partial}{\partial\theta^\alpha_i} + i(\sigma^\mu)_{\alpha\dot\alpha}\bar\theta^{\dot\alpha i}\frac{\partial}{\partial x^\mu}, \quad \bar D_{\dot\alpha i} = -\frac{\partial}{\partial\bar\theta^{\dot\alpha i}} - i\theta^\alpha_i(\sigma^\mu)_{\alpha\dot\alpha}\frac{\partial}{\partial x^\mu}. \quad (1)$$

They satisfy the anticommutation relations

$$\{D^i_\alpha, D^j_\beta\} = \{\bar D_{\dot\alpha i}, \bar D_{\dot\beta j}\} = 0, \quad \{D^i_\alpha, \bar D_{\dot\beta j}\} = -2i\delta^i_j(\sigma^\mu)_{\alpha\dot\beta}\frac{\partial}{\partial x^\mu},$$
$$\{Q^i_\alpha, Q^j_\beta\} = \{\bar Q_{\dot\alpha i}, \bar Q_{\dot\beta j}\} = 0, \quad \{Q^i_\alpha, \bar Q_{\dot\beta j}\} = 2i\delta^i_j(\sigma^\mu)_{\alpha\dot\beta}\frac{\partial}{\partial x^\mu},$$
$$\{D^i_\alpha, Q^j_\beta\} = \{D^i_\alpha, \bar Q_{\dot\beta j}\} = \{\bar D_{\dot\alpha i}, Q^j_\beta\} = \{\bar D_{\dot\alpha i}, \bar Q_{\dot\beta j}\} = 0. \quad (2)$$

We introduce the left-handed chiral basis $(x_L^\mu, \theta_i^\alpha, \bar\theta^{\dot\alpha i})$ in superspace, where

$$x_L^\mu = x^\mu + i\theta_i\sigma^\mu\bar\theta^i. \tag{3}$$

In this basis, supercharges and supercovariant derivatives take the form

$$\begin{aligned} Q_\alpha^i &= \frac{\partial}{\partial\theta_i^\alpha}, \quad \bar Q_{i\dot\alpha} = -\frac{\partial}{\partial\bar\theta^{i\dot\alpha}} + 2i\theta_i^\alpha\sigma^\mu{}_{\alpha\dot\alpha}\frac{\partial}{\partial x_L^\mu} \\ D_\alpha^i &= \frac{\partial}{\partial\theta_i^\alpha} + 2i\sigma^\mu{}_{\alpha\dot\alpha}\bar\theta^{\dot\alpha i}\frac{\partial}{\partial x_L^\mu} \quad \bar D_{\dot\alpha i} = -\frac{\partial}{\partial\bar\theta^{\dot\alpha i}}. \end{aligned} \tag{4}$$

In the $N=2$ rigid superspace, we introduce nonanticommutative deformation:

$$\{\theta_i^\alpha, \theta_j^\beta\}_* = C_{ij}^{\alpha\beta}, \tag{5}$$

Here the anticommutation relation is calculated by using the $*$-product, which is defined by

$$f * g(x,\theta,\bar\theta) = f(x,\theta,\bar\theta)\exp(P)g(x,\theta,\bar\theta), \quad P = -\frac{1}{2}\overleftarrow{Q_\alpha^i}C_{ij}^{\alpha\beta}\overrightarrow{Q_\beta^j}. \tag{6}$$

for functions f and g on superspace. P is called the Poisson structure. With this $*$-product, x_L^μ and $\bar\theta_{\dot\alpha i}$ (anti-)commute with other coordinates

$$\begin{aligned} &[x_L^\mu, x_L^\nu]_* = [x_L^\mu, \theta_i^\alpha]_* = [x_L^\mu, \bar\theta^{\dot\alpha i}]_* = 0, \\ &\{\bar\theta^{\dot\alpha i}, \bar\theta^{\dot\beta j}\}_* = \{\bar\theta^{\dot\alpha i}, \theta_j^\alpha\}_* = 0, \end{aligned} \tag{7}$$

which is a generalization of $N=1$ noncommutative superspace[2,5]. $C_{ij}^{\alpha\beta}$ is symmetric under the exchange of pairs of indices (αi),(βj): $C_{ij}^{\alpha\beta} = C_{ji}^{\beta\alpha}$. We decompose the nonanticommutative parameter $C_{ij}^{\alpha\beta}$ into symmetric and antisymmetric parts with respect to $SU(2)_R$ indices, such as

$$C_{ij}^{\alpha\beta} = C_{(ij)}^{\alpha\beta} + \frac{1}{4}\epsilon_{ij}\varepsilon^{\alpha\beta}C_s. \tag{8}$$

Here C_s corresponds to the singlet deformation parameter[11]. $C_{(ij)}^{\alpha\beta}$ is symmetric with respect to i and j, and also α and β. Since x_L^μ are commuting variables, spacetime coordinates x^μ have nontrivial (anti-)commutations:

$$\begin{aligned} [x^\mu, x^\nu]_* &= C_{(ij)}^{\mu\nu}\bar\theta^i\bar\theta^j + \frac{1}{2}C_s\varepsilon_{ij}\bar\theta^i\bar\sigma^{\mu\nu}\bar\theta^j \\ [x^\mu, \theta_i^\alpha]_* &= iC_{ij}^{\alpha\beta}(\sigma^\mu)_{\beta\dot\beta}\bar\theta^{\dot\beta j} \end{aligned} \tag{9}$$

where

$$C_{(ij)}^{\mu\nu} = C_{(ij)}^{\alpha\beta}(\sigma^{\mu\nu})_\alpha{}^\gamma\varepsilon_{\beta\gamma} \tag{10}$$

and $\sigma^{\mu\nu} = \frac{1}{4}(\sigma^\mu\bar{\sigma}^\nu - \sigma^\nu\bar{\sigma}^\mu)$, $\bar{\sigma}^{\mu\nu} = \frac{1}{4}(\bar{\sigma}^\mu\sigma^\nu - \bar{\sigma}^\nu\sigma^\mu)$.

We next discuss $N = 2$ harmonic superspace. It is formulated by introducing harmonic variables $u_i^\pm$ which form a 2×2 $SU(2)$ matrix. The harmonic variables satisfy the condition

$$u^{+i}u_i^- = 1, \quad \overline{u^{+i}} = u_i^-. \tag{11}$$

The coordinates of the harmonic superspace are $(x^\mu, \theta_i^\alpha, \bar{\theta}_{\dot{\alpha}i}, u_i^\pm)$. Using $u_i^\pm$, $SU(2)$ indices can be projected into two parts which have ± 1 $U(1)$ charges,

$$D_\alpha^\pm = u_i^\pm D_\alpha^i, \quad \bar{D}_{\dot{\alpha}}^\pm = u_i^\pm \bar{D}_{\dot{\alpha}}^i. \tag{12}$$

D_α^i is solved by $D_\alpha^\pm$ such as $D_{\alpha i} = u_i^+ D_\alpha^- - u_i^- D_\alpha^+$ with the help of the completeness condition $u_i^+u_j^- - u_j^+u_i^- = \epsilon_{ij}$. In the harmonic superspace, an important notion is an analytic superfield rather than chiral superfield satisfying $\bar{D}_{\dot{\alpha}}^i\Phi = 0$. An analytic superfield $\Phi(x, \theta, \bar{\theta}, u)$ is defined by

$$D_\alpha^+\Phi = \bar{D}_{\dot{\alpha}}^+\Phi = 0, \tag{13}$$

which can be conveniently written in terms of analytic coordinates:

$$x_A^\mu = x^\mu - i(\theta^i\sigma^\mu\bar{\theta}^j + \theta^j\sigma^\mu\bar{\theta}^i)u_i^+u_j^- = x^\mu - i(\theta^+\sigma^\mu\bar{\theta}^- + \theta^-\sigma^\mu\bar{\theta}^+), \tag{14}$$

$$\theta_\alpha^\pm = u_i^\pm\theta_\alpha^i, \quad \bar{\theta}_{\dot{\alpha}}^\pm = u_i^\pm\bar{\theta}_{\dot{\alpha}}^i. \tag{15}$$

The supercharges and supercovariant derivatives take the form

$$\begin{aligned}
Q_\alpha^+ &= \frac{\partial}{\partial\theta^{-\alpha}} - 2i\sigma^\mu_{\alpha\dot{\alpha}}\bar{\theta}^{+\dot{\alpha}}\frac{\partial}{\partial x_A^\mu}, \quad Q_\alpha^- = -\frac{\partial}{\partial\theta^{+\alpha}} \\
\bar{Q}_{\dot{\alpha}}^+ &= \frac{\partial}{\partial\bar{\theta}^{-\dot{\alpha}}} + 2i\theta^{+\alpha}\sigma^\mu_{\alpha\dot{\alpha}}\frac{\partial}{\partial x_A^\mu}, \quad \bar{Q}_{\dot{\alpha}}^- = -\frac{\partial}{\partial\bar{\theta}^{+\dot{\alpha}}} \\
D_\alpha^+ &= \frac{\partial}{\partial\theta^{-\alpha}}, \quad D_\alpha^- = -\frac{\partial}{\partial\theta^{+\alpha}} + 2i(\sigma^\mu)_{\alpha\dot{\alpha}}\bar{\theta}^{-\dot{\alpha}}\frac{\partial}{\partial x_A^\mu}, \\
\bar{D}_{\dot{\alpha}}^+ &= \frac{\partial}{\partial\bar{\theta}^{-\dot{\alpha}}}, \quad \bar{D}_{\dot{\alpha}}^- = -\frac{\partial}{\partial\bar{\theta}^{+\dot{\alpha}}} - 2i\theta^{-\alpha}\sigma^\mu_{\alpha\dot{\alpha}}\frac{\partial}{\partial x_A^\mu}.
\end{aligned} \tag{16}$$

Their nontrivial anticommutation relations are

$$\begin{aligned}
\{Q_\alpha^+, \bar{Q}_{\dot{\alpha}}^-\} &= 2i\sigma^\mu_{\alpha\dot{\alpha}}\frac{\partial}{\partial x_A^\mu}, \quad \{Q_\alpha^-, \bar{Q}_{\dot{\alpha}}^+\} = -2i\sigma^\mu_{\alpha\dot{\alpha}}\frac{\partial}{\partial x_A^\mu}, \\
\{D_\alpha^+, \bar{D}_{\dot{\alpha}}^-\} &= -2i\sigma^\mu_{\alpha\dot{\alpha}}\frac{\partial}{\partial x_A^\mu}, \quad \{\bar{D}_{\dot{\alpha}}^+, D_\alpha^-\} = 2i\sigma^\mu_{\alpha\dot{\alpha}}\frac{\partial}{\partial x_A^\mu}.
\end{aligned} \tag{17}$$

An analytic superfield Φ is functions of $(x_A^\mu, \theta^+, \bar{\theta}^+, u)$,

$$\Phi = \Phi(x_A^\mu, \theta^+, \bar{\theta}^+, u). \tag{18}$$

Expanding in θ, Φ takes the form

$$\Phi = \phi(x_A,u) + \theta^+\psi(x_A,u) + \bar\theta^+\bar\chi(x_A,u) \\ +(\theta^+)^2 M(x_A,u) + (\bar\theta^+)^2 N(x_A,u) + \theta^+\sigma^\mu\bar\theta^+ A_\mu(x_A,u) \\ +(\bar\theta^+)^2\theta^+\lambda(x_A,u) + (\theta^+)^2\bar\theta^+\bar\kappa(x_A,u) + (\theta^+)^2(\bar\theta^+)^2 D(x_A,u) \quad (19)$$

and each θ-components are expanded in the harmonic variables, e.g.,

$$\phi(x_A,u) = \sum_{n=0}^{\infty} \phi^{(i_1\cdots i_{q+n} j_1 \cdots j_n)}(x_A) u^+_{(i_1}\cdots u^+_{i_{q+n}} u^-_{j_1}\cdots u^-_{j_n)} \quad (20)$$

for the field with $U(1)$ charge $q \geq 0$. Similar formula holds for $q < 0$. Here we define $A_{(i_1\cdots i_n)}$ as the symmetrized sum over the permutation of indices $i_1, \cdots i_n$:

$$A_{(i_1\cdots i_n)} = \frac{1}{n!}\sum_{\sigma\in S_n} A_{i_{\sigma(1)}\cdots i_{\sigma(n)}} \quad (21)$$

and S_n denotes the permutation group of degree n. We define the harmonic derivatives $D^{\pm\pm}$ and D^0 by

$$D^{\pm\pm} = u^{+i}\frac{\partial}{\partial u^{-i}}, \quad D^0 = u^{+i}\frac{\partial}{\partial u^{+i}} - u^{-i}\frac{\partial}{\partial u^{-i}}. \quad (22)$$

They form an $SU(2)$ algebra. In the analytic basis, they are expressed as

$$D^{\pm\pm} = \partial^{++} - 2i\theta^{\pm}\sigma^\mu\bar\theta^{\pm}\frac{\partial}{\partial x^\mu_A} + \theta^{\pm\alpha}\frac{\partial}{\partial\theta^{\mp\alpha}} + \bar\theta^{\pm\dot\alpha}\frac{\partial}{\partial\bar\theta^{\mp\dot\alpha}},$$

$$D^0 = \partial^0 + \theta^{+\alpha}\frac{\partial}{\partial\theta^{+\alpha}} - \theta^{-\alpha}\frac{\partial}{\partial\theta^{-\alpha}} + \bar\theta^{+\dot\alpha}\frac{\partial}{\partial\bar\theta^{+\dot\alpha}} - \bar\theta^{-\dot\alpha}\frac{\partial}{\partial\bar\theta^{-\dot\alpha}}, \quad (23)$$

where $\partial^{\pm\pm}$ and ∂^0 denote the harmonic derivatives with fixed x_A, $\theta^\pm$ and $\bar\theta^\pm$.

In the analytic basis, one can compute the $*$-product by using $Q^i_\alpha = u^{+i}Q^-_\alpha - u^{-i}Q^+_\alpha$. For example we have

$$\{\theta^{\eta\alpha}, \theta^{\eta'\beta}\}_* = C^{\eta\eta'\alpha\beta}, \quad (\eta,\eta' = \pm)$$
$$[x^\mu_A, x^\nu_A]_* = 4C^{--\mu\nu}(\bar\theta^+)^2$$
$$[x^\mu_A, \theta^{\eta\alpha}]_* = -2iC^{-\eta\beta\alpha}(\sigma^\mu\bar\theta^+)_\beta \quad (24)$$

where $C^{\eta\eta'\mu\nu} = u^{\eta i}u^{\eta' j}C^{\mu\nu}_{ij}$. For superfields A and B, the $*$-product takes the form

$$A*B = AB + APB + \frac{1}{2}AP^2B + \frac{1}{6}AP^3B + \frac{1}{24}AP^4B, \quad P^5 = 0. \quad (25)$$

We note that the Poisson structure P commutes with the supercovariant derivatives. Therefore the analytic structure is preserved when we introduce noncommutativity in harmonic superspace. Supersymmetry generated by Q^i_α only is conserved for generic deformation. Field theory in this noncommutative superspace has chiral $N = 1$ (or $N = (1,0)$ in the sense of [12]) supersymmetry.

3. $N = 2$ Supersymmetric $U(1)$ gauge theory

We now construct the action of $N = 2$ supersymmetric $U(1)$ gauge theory in this non(anti)commutative superspace. We introduce an analytic superfield $V^{++}(\zeta, u)$ with $\zeta = (x^\mu_A, \theta^+, \bar\theta^+)$ by covariantizing the harmonic derivative $D^{++} \to \nabla^{++} = D^{++} + iV^{++}$. Generalizing the construction in [16,17], the action is given by

$$S_* = \frac{1}{2}\sum_{n=2}^{\infty} \int d^4x d^8\theta du_1 \ldots du_n \frac{(-i)^n}{n} \frac{V^{++}(1) * \cdots * V^{++}(n)}{(u_1^+ u_2^+) \cdots (u_n^+ u_1^+)} \tag{26}$$

where $V^{++}(i) = V^{++}(\zeta_i, u_i)$, $\zeta_i = (x_A, \theta_i^+, \bar\theta_i^+)$ and $d^8\theta = d^4\theta^+ d^4\theta^-$ with $d^4\theta^\pm = d^2\theta^\pm d^2\bar\theta^\pm$. The harmonic integral is defined by the rules:
(i)

$$\int du f(u) = 0$$

for $f(u)$ with non-zero $U(1)$ charge.
(ii)

$$\int du 1 = 1.$$

(iii)

$$\int du u^+_{(i_1} \cdots u^+_{i_n} u^-_{j_1} \cdots u^-_{j_n)} = 0, \ (n \geq 1).$$

The action (26) is invariant under the gauge transformation

$$\delta^*_\Lambda V^{++} = -D^{++}\Lambda + i[\Lambda, V^{++}]_*, \tag{27}$$

with an analytic superfield Λ. The generic superfield $V^{++}(\zeta, u)$ includes infinitely many auxiliary fields. Most of these fields are gauged away except the lowest component fields in the harmonic expansion. One can take the WZ gauge

$$\begin{aligned} V^{++}_{WZ}(x_A, \theta^+, \bar\theta^+, u) = -i\sqrt{2}(\theta^+)^2\bar\phi(x_A) + i\sqrt{2}(\bar\theta^+)^2\phi(x_A) \\ - 2i\theta^+\sigma^\mu\bar\theta^+ A_\mu(x_A) \\ +4(\bar\theta^+)^2\theta^+\psi^i(x_A)u_i^- - 4(\theta^+)^2\bar\theta^+\bar\psi^i(x_A)u_i^- \\ +3(\theta^+)^2(\bar\theta^+)^2 D^{ij}(x_A)u_i^- u_j^-, \end{aligned} \tag{28}$$

which is convenient to study the theory in the component formalism.

The component action S_* in the WZ gauge can be expanded in a power series of the deformation parameter C. In [19], we have computed the $O(C)$ action explicitly. In this section, we will consider the non-singlet deformation case. The quadratic part $S_{*,2}$ in S_* is the same as the commutative one:

$$S_{*,2} = \int d^4x \left\{ -\frac{1}{4}F_{\mu\nu}F^{\mu\nu} - \frac{1}{4}F_{\mu\nu}\tilde{F}^{\mu\nu} + \phi\partial^2\bar{\phi} - i\psi^i\sigma^\mu\partial_\mu\bar{\psi}_i + \frac{1}{4}D^{ij}D_{ij} \right\}. \tag{29}$$

The cubic part $S_{*,3}$ in S_* is of order $O(C)$ and given by

$$\begin{aligned} S_{*,3} = \int d^4x \Big[& -\frac{2\sqrt{2}}{3} iC^{\alpha\beta}_{(ij)}\psi^i_\alpha(\sigma^\nu\partial_\nu\bar{\psi}^j)_\beta\bar{\phi} - 2\sqrt{2}iC^{\alpha\beta}_{(ij)}\psi^i_\alpha(\sigma^\nu\bar{\psi}^j)_\beta\partial_\nu\bar{\phi} \\ & + \frac{2}{3}iC^{\alpha\beta}_{(ij)}A_\mu(\sigma^\mu\bar{\psi}^i)_\alpha(\sigma^\nu\partial_\nu\bar{\psi}^j)_\beta - iC^{\mu\nu}_{(ij)}\bar{\psi}^i\bar{\psi}^j F_{\mu\nu} \\ & + \sqrt{2}C^{\mu\nu}_{(ij)}D^{ij}A_\mu\partial_\nu\bar{\phi} + \frac{1}{\sqrt{2}}C^{\mu\nu}_{(ij)}D^{ij}F_{\mu\nu}\bar{\phi} \Big]. \end{aligned} \tag{30}$$

Note that here we have already dropped the C_s dependent terms. We will refer $S_{*,2} + S_{*,3}$ as the $O(C)$ action.

In the commutative case, the gauge parameter $\Lambda = \lambda(x_A)$ preserves the WZ gauge and gives rise to the gauge transformation for component fields. In the non(anti)commutative case, however, the gauge transformation (27) with the same gauge parameter does not preserve the WZ gauge because of the C-dependent terms arising from the commutator. In order to preserve the WZ gauge, one must include the C-dependent terms. The gauge parameter is shown to take the form

$$\begin{aligned} \lambda_C(\zeta, u) = & \lambda(x_A) + \theta^+\sigma^\mu\bar{\theta}^+\lambda^{(-2)}_\mu(x_A, u; C) + (\bar{\theta}^+)^2\lambda^{(-2)}(x_A, u; C) \\ & + (\bar{\theta}^+)^2\theta^{+\alpha}\lambda^{(-3)}_\alpha(x_A, u; C) + (\theta^+)^2(\bar{\theta}^+)^2\lambda^{(-4)}(x_A, u; C) \end{aligned} \tag{31}$$

which has been determined by solving the WZ gauge preserving conditions expanded in harmonic modes [19]. The gauge transformation is also fully determined[19], which reads at $O(C)$

$$\begin{aligned} \delta^*_{\lambda_C}A_\mu &= -\partial_\mu\lambda + O(C^2), \\ \delta^*_{\lambda_C}\phi &= O(C^2), \\ \delta^*_{\lambda_C}\psi_{\alpha i} &= \frac{2}{3}(\varepsilon C_{(ij)}\sigma^\mu\bar{\psi}^j)_\alpha\,\partial_\mu\lambda + O(C^2), \\ \delta^*_{\lambda_C}D_{ij} &= 2\sqrt{2}C^{\mu\nu}_{(ij)}\partial_\mu\lambda\partial_\nu\bar{\phi} + O(C^2) \\ \delta^*_{\lambda_C}(\text{others}) &= 0. \end{aligned} \tag{32}$$

The $O(C)$ action is invariant under the $O(C)$ gauge transformation (32).

These gauge transformations are not canonical, in particular for fermions ψ^i_α. But if we redefine the component fields such as

$$\begin{aligned}
\hat{A}_\mu &= A_\mu + O(C^2),\\
\hat{\phi} &= \phi + O(C^2), \quad \hat{\bar{\phi}} = \bar{\phi},\\
\hat{\psi}_{\alpha i} &= \psi_{\alpha i} + \frac{2}{3}(\varepsilon C_{(ij)}\sigma^\mu \bar{\psi}^j)_\alpha A_\mu + O(C^2), \quad \hat{\bar{\psi}}^{\dot{\alpha}} = \bar{\psi}^{\dot{\alpha}}\\
\hat{D}_{ij} &= D_{ij} + 2\sqrt{2}C^{\mu\nu}_{(ij)} A_\mu \partial_\nu \bar{\phi} + O(C^2),
\end{aligned} \tag{33}$$

the newly defined fields are shown to transform canonically: $\delta^*_{\lambda_C}\hat{A}_\mu = -\partial_\mu \lambda$, $\delta^*_{\lambda_C}(\text{others}) = 0$. In terms of redefined fields, the $O(C)$ action can be written as

$$\begin{aligned}
S_{*,2} + S_{*,3} = \int d^4x \Big[&-\frac{1}{4}\hat{F}_{\mu\nu}(\hat{F}^{\mu\nu} + \tilde{\hat{F}}^{\mu\nu}) + \hat{\phi}\partial^2\hat{\bar{\phi}} - i\hat{\psi}^i\sigma^\mu\partial_\mu\hat{\bar{\psi}}_i + \frac{1}{4}\hat{D}^{ij}\hat{D}_{ij}\\
&- 2\sqrt{2}iC^{\alpha\beta}_{(ij)}\hat{\psi}^i_\alpha(\sigma^\mu\hat{\bar{\psi}}^j)_\beta\partial_\mu\hat{\bar{\phi}} - \frac{2\sqrt{2}}{3}iC^{\alpha\beta}_{(ij)}\hat{\psi}^i_\alpha(\sigma^\mu\partial_\mu\hat{\bar{\psi}}^j)_\beta\hat{\bar{\phi}}\\
&- iC^{\mu\nu}_{(ij)}\hat{\bar{\psi}}^i\hat{\bar{\psi}}^j\hat{F}_{\mu\nu} + \frac{1}{\sqrt{2}}C^{\mu\nu}_{(ij)}\hat{D}^{ij}\hat{F}_{\mu\nu}\hat{\bar{\phi}} + O(C^2)\Big],
\end{aligned} \tag{34}$$

where $\hat{F}_{\mu\nu} = \partial_\mu\hat{A}_\nu - \partial_\nu\hat{A}_\mu$.

Now we study the supersymmetry transformation that is generated by the chiral part of the supersymmetry generators: the $N = (1,0)$ supersymmetry generated by Q^i_α. The deformed supersymmetry transformation of the gauge multiplet,

$$\begin{aligned}
\delta^*_\xi V^{++}_{WZ} = &-i\sqrt{2}(\theta^+)^2\delta^*_\xi\bar{\phi}(x_A) + i\sqrt{2}(\bar{\theta}^+)^2\delta^*_\xi\phi(x_A) - 2i\theta^+\sigma^\mu\bar{\theta}^+\delta^*_\xi A_\mu(x_A)\\
&+ 4(\bar{\theta}^+)^2\theta^+\delta^*_\xi\psi^i(x_A)u^-_i - 4(\theta^+)^2\bar{\theta}^+\delta^*_\xi\bar{\psi}^i(x_A)u^-_i\\
&+ 3(\theta^+)^2(\bar{\theta}^+)^2\delta^*_\xi D^{ij}(x_A)u^-_i u^-_j,
\end{aligned} \tag{35}$$

is given by

$$\delta^*_\xi V^{++}_{WZ} = \tilde{\delta}_\xi V^{++}_{WZ} + \delta^*_\Lambda V^{++}_{WZ}, \tag{36}$$

where

$$\tilde{\delta}_\xi V^{++}_{WZ} = \left(-\xi^{+\alpha}Q^-_\alpha + \xi^{-\alpha}Q^+_\alpha\right)V^{++}_{WZ} \tag{37}$$

and $\delta^*_\Lambda V^{++}_{WZ}$ is a deformed gauge transformation of V^{++}_{WZ} with an appropriate analytic gauge parameter $\Lambda(\zeta, u)$ to retain the WZ gauge:

$$\delta^*_\Lambda V^{++}_{WZ}(\zeta, u) = -D^{++}\Lambda(\zeta, u) + i[\Lambda, V^{++}_{WZ}]_*(\zeta, u). \tag{38}$$

We will denote the analytic gauge parameter as

$$
\begin{aligned}
\Lambda(\zeta, u) = & \lambda^{(0,0)}(x_A, u) + \bar{\theta}^{+}_{\dot{\alpha}} \lambda^{(0,1)\dot{\alpha}}(x_A, u) + \theta^{+\alpha} \lambda^{(1,0)}_{\alpha}(x_A, u) \\
& + (\bar{\theta}^{+})^2 \lambda^{(0,2)}(x_A, u) + (\theta^{+})^2 \lambda^{(2,0)}(x_A, u) + \theta^{+} \sigma^{\mu} \bar{\theta}^{+} \lambda^{(1,1)}_{\mu}(x_A, u) \\
& + (\bar{\theta}^{+})^2 \theta^{+\alpha} \lambda^{(1,2)}_{\alpha}(x_A, u) + (\theta^{+})^2 \bar{\theta}^{+}_{\dot{\alpha}} \lambda^{(2,1)\dot{\alpha}}(x_A, u) \\
& + (\theta^{+})^2 (\bar{\theta}^{+})^2 \lambda^{(2,2)}(x_A, u),
\end{aligned} \tag{39}
$$

where $\lambda^{(n,m)}(x_A, u)$ is the $(\theta^+)^n(\bar{\theta}^+)^m$-component. Then we find the deformed supersymmetry transformation laws in the WZ gauge:

$$
\begin{aligned}
\delta^*_\xi \phi &= -\sqrt{2} i \xi^i \psi_i - \frac{8}{3} i (\xi^j \varepsilon C_{(jk)} \psi^k) \bar{\phi} - \frac{2\sqrt{2}}{3} i (\xi^j \varepsilon C_{(jk)} \sigma^\nu \bar{\psi}^k) A_\nu + O(C^2), \\
\delta^*_\xi \bar{\phi} &= 0, \\
\delta^*_\xi A_\mu &= i \xi^i \sigma_\mu \bar{\psi}_i + 2\sqrt{2} i (\xi^j \varepsilon C_{(jk)} \sigma_\mu \bar{\psi}^k) \bar{\phi} + O(C^2), \\
\delta^*_\xi \psi^{\alpha i} &= -(\xi^i \sigma^{\mu\nu})^\alpha F_{\mu\nu} - D^{ij} \xi^\alpha_j \\
& \quad + 2\sqrt{2} D^{(ij} (\xi^{k)} \varepsilon C_{(jk)})^\alpha \bar{\phi} - 2i (\bar{\psi}^{(i} \bar{\psi}^{j}) (\xi^{k)} \varepsilon C_{(jk)})^\alpha \\
& \quad - \left\{ 2\sqrt{2} (\xi^j \varepsilon C_{(jk)} \sigma^{\mu\nu})^\alpha + \frac{2\sqrt{2}}{3} (\xi^j \sigma^{\mu\nu} \varepsilon C_{(jk)})^\alpha + \sqrt{2} C^{\mu\nu}_{(jk)} \xi^{\alpha j} \right\} \epsilon^{ki} \bar{\phi} F_{\mu\nu} \\
& \quad + \left\{ \frac{4\sqrt{2}}{3} (\xi^j \sigma^{\mu\nu} \varepsilon C_{(jk)})^\alpha + 2\sqrt{2} C^{\mu\nu}_{(jk)} \xi^{\alpha j} \right\} \epsilon^{ki} \partial_\mu \bar{\phi} A_\nu \\
& \quad - \frac{2\sqrt{2}}{3} (\xi^j \varepsilon C_{(jk)})^\alpha \epsilon^{ki} \partial^\mu \bar{\phi} A_\mu + O(C^2), \\
\delta^*_\xi \bar{\psi}^i_{\dot{\alpha}} &= +\sqrt{2} (\xi^i \sigma^\nu)_{\dot{\alpha}} \partial_\nu \bar{\phi} + 2 (\xi^j \varepsilon C_{(jk)} \sigma^\nu)_{\dot{\alpha}} \partial_\nu (\bar{\phi}^2) \epsilon^{ki} + O(C^2), \\
\delta^*_\xi D^{ij} &= -2i \xi^{(i} \sigma^\nu \partial_\nu \bar{\psi}^{j)} - 6\sqrt{2} i \epsilon^{k(l} \partial_\nu \{ (\xi^i \varepsilon C_{(kl)} \sigma^\nu \bar{\psi}^{j)}) \bar{\phi} \} + O(C^2).
\end{aligned} \tag{40}
$$

We can check that the $O(C)$ action $S_{*2} + S_{*3}$ is indeed invariant under the deformed supersymmetry transformation (40). Denoting the deformed supersymmetry transformation δ^*_ξ as $\delta^*_\xi = \delta^{*(0)}_\xi + \delta^{*(1)}_\xi + \cdots$ where $\delta^{*(n)}_\xi$ represents the $O(C^n)$ variations, we can see that $\delta^{*(1)}_\xi S_{*,2} + \delta^{*(0)}_\xi S_{*,3} = 0$. Note that $\delta^{*(1)}_\xi S_{*,3}$ is non zero for generic deformation parameters. Therefore we need higher order terms in C in order to obtain fully supersymmetric action.

After the redefinition (33), the deformed supersymmetry transformation

(40) becomes

$$
\begin{aligned}
\delta^*_\xi \hat{\phi} &= -\sqrt{2} i \xi^i \hat{\psi}_i - \frac{8}{3} i (\xi^j \varepsilon C_{(jk)} \hat{\psi}^k) \hat{\bar{\phi}} + O(C^2), \\
\delta^*_\xi \hat{\bar{\phi}} &= 0, \\
\delta^*_\xi \hat{A}_\mu &= i \xi^i \sigma_\mu \hat{\bar{\psi}}_i + 2\sqrt{2} i (\xi^j \varepsilon C_{(jk)} \sigma_\mu \hat{\bar{\psi}}^k) \hat{\bar{\phi}} + O(C^2), \\
\delta^*_\xi \hat{\psi}^{\alpha i} &= -(\xi^i \sigma^{\mu\nu})^\alpha \hat{F}_{\mu\nu} - \hat{D}^{ij} \xi^\alpha_j \\
&\quad - i (\xi^i \sigma_{\mu\nu})^\alpha C^{\mu\nu}_{(jk)} (\hat{\bar{\psi}}^j \hat{\bar{\psi}}^k) + 2\sqrt{2} \hat{D}^{(ij} (\xi^{k)} \varepsilon C_{(jk)})^\alpha \hat{\bar{\phi}} \\
&\quad - \left\{ 2\sqrt{2} (\xi^j \varepsilon C_{(jk)} \sigma^{\mu\nu})^\alpha + \frac{2\sqrt{2}}{3} (\xi^j \sigma^{\mu\nu} \varepsilon C_{(jk)})^\alpha + \sqrt{2} C^{\mu\nu}_{(jk)} \xi^{\alpha j} \right\} \epsilon^{ki} \hat{\bar{\phi}} \hat{F}_{\mu\nu} \\
&\quad + O(C^2), \\
\delta^*_\xi \hat{\bar{\psi}}^i_{\dot{\alpha}} &= +\sqrt{2} (\xi^i \sigma^\nu)_{\dot{\alpha}} \partial_\nu \hat{\bar{\phi}} + 2 (\xi^j \varepsilon C_{(jk)} \sigma^\nu)_{\dot{\alpha}} \partial_\nu (\hat{\bar{\phi}}^2) \epsilon^{ki} + O(C^2), \\
\delta^*_\xi \hat{D}^{ij} &= -2i \xi^{(i} \sigma^\nu \partial_\nu \hat{\bar{\psi}}^{j)} \\
&\quad - 6\sqrt{2} i \epsilon^{k(l} \partial_\nu \{ (\xi^i \varepsilon C_{(kl)} \sigma^\nu \hat{\bar{\psi}}^{j)}) \hat{\bar{\phi}} \} + 2\sqrt{2} i \epsilon^{il} \epsilon^{jm} (\xi^k \varepsilon C_{(lm)} \sigma^\nu \hat{\bar{\psi}}_k) \partial_\nu \hat{\bar{\phi}} \\
&\quad + O(C^2).
\end{aligned} \tag{41}
$$

For generic non-singlet deformations, it seems difficult to find the appropriate field redefinition such that both gauge and supersymmetry transformations become canonical. However, in the case of singlet deformation it is possible to peform such a field redefinition.

4. Singlet Deformation

In this section, we consider the singlet deformation case $C^{\alpha\beta}_{(ij)} = 0$. In the singlet case, the order $O(C_s)$ Lagrangian[19] reads $L = L^{(0)} + L^{(1)}$, where

$$
\begin{aligned}
L^{(0)} &= -\frac{1}{4} F_{\mu\nu} (F^{\mu\nu} + \tilde{F}^{\mu\nu}) - i \psi^i \sigma^\mu \partial_\mu \bar{\psi}_i - \partial^\mu \phi \partial_\mu \bar{\phi} + \frac{1}{4} D_{ij} D^{ij}, \\
L^{(1)} &= \frac{1}{\sqrt{2}} C_s A_\nu \partial_\mu \bar{\phi} (F^{\mu\nu} + \tilde{F}^{\mu\nu}) + \frac{i}{\sqrt{2}} C_s \bar{\phi} (\psi^k \sigma^\nu \partial_\nu \bar{\psi}_k) \\
&\quad + \frac{i}{\sqrt{2}} C_s (\psi^k \sigma^\nu \bar{\psi}_k) \partial_\nu \bar{\phi} - \frac{i}{2} C_s \varepsilon^{\alpha\beta} A_\mu (\sigma^\mu \bar{\psi}^k)_\alpha (\sigma^\nu \partial_\nu \bar{\psi}_k)_\beta \\
&\quad + \frac{i}{2} C_s \bar{\psi}^i \bar{\psi}^j D_{ij} + \frac{\sqrt{2}}{4} C_s A_\mu A^\mu \partial^2 \bar{\phi} - \frac{\sqrt{2}}{4} C_s \bar{\phi} D^{ij} D_{ij}.
\end{aligned} \tag{42}
$$

In this case, it is shown that the gauge parameter $\lambda(x_A)$ keeps the WZ gauge. The analytic gauge parameter is of the same form as in the commutative case: $\Lambda(\zeta, u) = \lambda(x_A)$. Then we immediately find the deformed gauge transformation laws for the component fields in the case of the singlet deformation:

$$\delta^*_\Lambda A_\mu = -\Big(1+\frac{1}{\sqrt{2}}C_s\bar{\phi}\Big)\partial_\mu\lambda, \quad \delta^*_\Lambda\phi = \frac{1}{\sqrt{2}}C_s A_\mu\partial^\mu\lambda,$$
$$\delta^*_\Lambda\psi^i_\alpha = -\frac{1}{2}C_s\partial_\mu\lambda\,(\sigma^\mu\bar{\psi}^i)_\alpha, \quad \delta^*_\Lambda\bar{\phi} = \delta^*_\Lambda\bar{\psi}^i_{\dot{\alpha}} = \delta^*_\Lambda D^{ij} = 0. \qquad (43)$$

Note that there is no higher order correction in C_s.

In deformed supersymmetry transformation (36), it is shown that the gauge parameter to keep the WZ gauge is the same one as in the case of $C_s = 0$. Then we determine the deformed supersymmetry transformations as

$$\delta_\xi A_\mu = i\xi^i\sigma_\mu\bar{\psi}_i, \quad \delta_\xi\phi = -\sqrt{2}i\xi^i\psi_i, \quad \delta_\xi\bar{\phi} = 0,$$
$$\delta_\xi\psi^i_\alpha = \Big(1+\frac{1}{\sqrt{2}}C_s\bar{\phi}\Big)(\sigma^{\mu\nu}\xi^i)_\alpha F_{\mu\nu} - D^{ij}\xi_{\alpha j} + \frac{1}{\sqrt{2}}C_s\xi^i_\alpha\partial_\mu\bar{\phi}A^\mu,$$
$$\delta_\xi\bar{\psi}^{\dot{\alpha}i} = -\sqrt{2}(\bar{\sigma}^\mu\xi^i)^{\dot{\alpha}}\Big(1+\frac{1}{\sqrt{2}}C_s\bar{\phi}\Big)\partial_\mu\bar{\phi},$$
$$\delta_\xi D^{kl} = -i\xi^k\sigma^\mu\partial_\mu\Big\{\bar{\psi}^l\Big(1+\frac{1}{\sqrt{2}}C_s\bar{\phi}\Big)\Big\} - i\xi^l\sigma^\mu\partial_\mu\Big\{\bar{\psi}^k\Big(1+\frac{1}{\sqrt{2}}C_s\bar{\phi}\Big)\Big\}, \qquad (44)$$

The $N = 2$ vector multiplet $(D^{ij}, \hat{A}_\mu, \hat{\psi}^i, \bar{\psi}_i, \hat{\phi}, \bar{\phi})$, however, does not transform canonically under the supersymmetry transformation. But, if we instead perform field redefinitions as

$$a_\mu = F(\bar{\phi})A_\mu, \quad \varphi = F(\bar{\phi})^2\left(\phi + G(\bar{\phi})A_\mu A^\mu\right), \quad \bar{\varphi} = \bar{\phi},$$
$$\lambda^i_\alpha = F(\bar{\phi})^2\left(\psi^i_\alpha + H(\bar{\phi})A_\mu(\sigma^\mu\bar{\psi}^i)_\alpha\right), \quad \bar{\lambda}^{\dot{\alpha}i} = F(\bar{\phi})\bar{\psi}^{\dot{\alpha}i},$$
$$\tilde{D}^{ij} = F(\bar{\phi})^2\left(D^{ij} - 2iH(\bar{\phi})\bar{\psi}^i\bar{\psi}^j\right), \qquad (45)$$

where

$$F(\bar{\phi}) = \frac{1}{1+\frac{1}{\sqrt{2}}C_s\bar{\phi}}, \quad G(\bar{\phi}) = \frac{\frac{1}{2\sqrt{2}}C_s}{1+\frac{1}{\sqrt{2}}C_s\bar{\phi}}, \quad H(\bar{\phi}) = -\frac{\frac{1}{2}C_s}{1+\frac{1}{\sqrt{2}}C_s\bar{\phi}}. \qquad (46)$$

We can show that the multiplet $(\tilde{D}^{ij}, a_\mu, \lambda^i, \bar{\lambda}_i, \varphi, \bar{\varphi})$ now transforms canonically under the supersymmetry transformation as well as the gauge transformation:

$$\delta_\xi a_\mu = i\xi^i\sigma_\mu\bar{\lambda}_i, \quad \delta_\xi\varphi = -i\sqrt{2}\xi^i\lambda_i, \quad \delta_\xi\bar{\varphi} = 0,$$
$$\delta_\xi\lambda^i_\alpha = (\sigma^{\mu\nu}\xi^i)_\alpha f_{\mu\nu} - \tilde{D}^{ij}\xi_{j\alpha}, \quad \delta_\xi\bar{\lambda}^{\dot{\alpha}i} = -\sqrt{2}(\bar{\sigma}^\mu\xi^i)^{\dot{\alpha}}\partial_\mu\bar{\varphi},$$
$$\delta_\xi\tilde{D}^{ij} = -i\left(\xi^i\sigma^\mu\partial_\mu\bar{\lambda}^j + \xi^j\sigma^\mu\partial_\mu\bar{\lambda}^i\right), \qquad (47)$$

where $f_{\mu\nu} \equiv \partial_\mu a_\nu - \partial_\nu a_\mu$. We can construct the Lagrangian which is invariant under (43) and (44) and reduces to the Lagrangian (42) at the order $O(C_s)$. In terms of these newly defined fields, it becomes

$$L = if_2(\bar{\varphi})\Big(1 + \frac{1}{\sqrt{2}}C_s\bar{\varphi}\Big)^3\Big\{-\frac{1}{4}f_{\mu\nu}(f^{\mu\nu} + \tilde{f}^{\mu\nu}) - i\lambda^i\sigma^\mu\partial_\mu\bar{\lambda}_i + \varphi\partial^2\bar{\varphi} + \frac{1}{4}\tilde{D}_{ij}\tilde{D}^{ij}\Big\}, \quad (48)$$

up to a function $f_2(\bar{\varphi})$. In [22], the function $f_2(\bar{\varphi})$ is found to be $\Big(1 + \frac{1}{\sqrt{2}}C_s\bar{\varphi}\Big)^{-1}$.

5. Discussion

In this paper, we have studied $N = 2$ supersymmetric $U(1)$ gauge theory in noncommutative harmonic superspace with both singlet and non-singlet deformations. We have obtained the deformed action and discussed gauge and $N = (1, 0)$ supersymmetry of the theory

It is interesting to study the reduction of deformation parameters such that only $N = 1$ subspace becomes noncommutative. In this case $N = (1, 1/2)$ supersymmetry generated by Q^1, Q^2, and $\bar{Q}_2$ commute with the Poisson structure[12]. Hence, we expect that $N = (1, 0)$ supersymmetry enhanced to $N = (1, 1/2)$ by this reduction. A detailed analysis will appear in a forthcoming paper[25].

Another interesting problem is to realize the present deformed theory as the field theory on the D-brane. Recently, relationship between pure spinor formalism[23] and harmonic superspace is pointed out[24]. Deformation of harmonic superspace would be obtained by introducing appropriate R-R background in this formalism.

References

1. J. H. Schwarz and P. Van Nieuwenhuizen, *Speculations concerning a fermionic substructure of space-time.* Lett. Nuovo Cim. **34** (1982), 21.
2. H. Ooguri and C. Vafa, *The C-deformation of gluino and non-planar diagrams.* Adv. Theor. Math. Phys. **7** (2003), 53, **hep-th/0302109**. *Gravity induced C-deformation.* Adv. Theor. Math. Phys. **7** (2004), 405, **hep-th/0303063**.
3. N. Berkovits and N. Seiberg, *Superstrings in graviphoton background and N = 1/2 + 3/2 supersymmetry.* JHEP **0307** (2003), 010, **hep-th/0306226**.
4. J. de Boer, P. A. Grassi and P. van Nieuwenhuizen, *Non-commutative superspace from string theory.* Phys. Lett. **B574** (2003), 98, **hep-th/0302078**.

5. N. Seiberg, *Noncommutative superspace, $N = 1/2$ supersymmetry, field theory and string theory.* JHEP **0306** (2003), 010, **hep-th/0305248**.
6. R. Britto, B. Feng and S. J. Rey, *Deformed superspace, $N = 1/2$ supersymmetry and (non)renormalization theorems.* JHEP **0307** (2003) 067, **hep-th/0306215**.
   S. Terashima and J. T. Yee, *Comments on noncommutative superspace.* JHEP **0312** (2003), 053, *hep-th/0306237.*
   M. T. Grisaru, S. Penati and A. Romagnoni, *Two-loop renormalization for nonanticommutative $N = 1/2$ supersymmetric WZ model.* JHEP **0308** (2003), 003, **hep-th/0307099**,
   R. Britto and B. Feng, *$N = 1/2$ Wess-Zumino model is renormalizable.* Phys. Rev. Lett. **91** (2003), 201601, **hep-th/0307165**.
   A. Romagnoni, *Renormalizability of $N = 1/2$ Wess-Zumino model in superspace.* JHEP **0310** (2003), 016, **hep-th/0307209**.
   O. Lunin and S. J. Rey, *Renormalizablity of non(anti)commutative gauge theories with $N = 1/2$ supersymmetry.* JHEP **0309** (2003), 045, **hep-th/0307275**
   D. Berenstein and S. J. Rey, *Wilsonian proof for renormalizability of $N = 1/2$ supersymmetric field theories.* Phys. Rev. **D68** (2003), 121701, hep-th/0308049.
   M. Alishahiha, A. Ghodsi and N. Sadooghi, *One-loop perturbative corrections to non(anti)commutativity parameter of $N = 1/2$ supersymmetric $U(N)$ gauge theory.* Nucl. Phys. **B691** (2004), 111, **hep-th/0309037**.
   A.T. Banin, I.L. Buchbinder and N.G. Pletnev, JHEP **0407** (2004), 011, **hep-th/0405063**.
   I. Jack, D.R.T. Jones and L.A. Worthy, *One-loop renormalisation of $N = 1/2$ supersymmetric gauge theory.* **hep-th/0412009**.
   S. Penati and A. Romagnoni, *Covariant quantization of $N = 1/2$ SYM theories and supergauge invariance.* **hep-th/0412041**.
7. A. Imaanpur, *On instantons and zero modes of $N = 1/2$ SYM theory.* JHEP **0309** (2003), 077, **hep-th/0308171**; *Comments on Gluino Condensates in $N = 1/2$ SYM theory.* JHEP **0312** (2003), 009, **hep-th/0311137**,
   P. A. Grassi, R. Ricci and D. Robles-Llana, *Instanton Calculations for $N = \frac{1}{2}$ super Yang-Mills theory.* JHEP **0407** (2004), 065, **hep-th/0311155**.
   R. Britto, B. Feng, O. Lunin and S. J. Rey, *$U(N)$ instantons on $N = \frac{1}{2}$ superspace- exact solution & geometry of moduli space.* Phys. Rev. **D69** (2004), 126004, **hep-th/0311275**.
   M. Billo, M. Frau, I. Pensando and A. Lerda, *$N = 1/2$ gauge theory and its instanton moduli space from open strings in R-R background.* JHEP **0405** (2004), 023, **hep-th/0402160**.
8. B. Chandrasekhar and A. Kumar, *$D = 2$, $N = 2$, supersymmetric theories on non(anti)commutative superspace.* JHEP **0403** (2004), 013, **hep-th/0310137**.
   T. Inami and H. Nakajima, *Supersymmetric $CP(N)$ sigma model on noncommutative superspace.* Prog. Theor. Phys. **111** (2004), 961, **hep-th/0402137**.
   B. Chandrasekhar, *$D = 2$, $N = 2$ supersymmetric sigma models on*

non(anti)commutative superspace. Phys. Rev. **D70** (2004), 125003, **hep-th/0408184**.

9. T. Araki, K. Ito and A. Ohtsuka, *Supersymmetric gauge theories on noncommutative superspace.* Phys. Lett. **B573** (2003), 209, **hep-th/0307076**.
10. D. Klemm, S. Penati and L. Tamassia, *Non(anti)commutative superspace.* Class. Quant. Grav. **20** (2003), 2905, **hep-th/0104190**.
    S. Ferrara and M. A. Lledo, *Some aspects of deformations of supersymmetric field theories.* JHEP **0005** (2000), 008, **hep-th/0002084**.
    M. Hatsuda, S. Iso and H. Umetsu, *Noncommutative superspace, supermatrix and lowest Landau level.* Nucl. Phys. **B671** (2003), 217, **hep-th/0306251**.
    S. Ferrara, M. A. Lledo and O. Macia, *Supersymmetry in noncommutative superspaces.* JHEP **0309** (2003), 068, **hep-th/0307039**.
11. S. Ferrara and E. Sokatchev, *Non-anticommutative $N = 2$ super-Yang-Mills theory with singlet deformation.* Phys. Lett. **B579** (2004), 226, **hep-th/0308021**.
12. E. Ivanov, O. Lechtenfeld and B. Zupnik, *Nilpotent Deformations of $N = 2$ Superspace.* JHEP **0402** (2004), 012,**hep-th/0308012**.
13. R. Grimm, M. Sohnius and J. Wess, *Extended supersymmetry and gauge theories.* Nucl. Phys. B **133** (1978), 275.
14. A. Galperin, E. Ivanov, V. Ogievetsky and E. Sokatchev, *Harmonic Superspace.* Cambridge University Press , 2001
15. J. Wess and J. Bagger, *Supersymmetry and Supergravity.* Princeton University Press, 1992.
16. B.M. Zupnik, *The action of the supersymmetric $N = 2$ gauge theory in harmonic superspace.* Phys. Lett. **B183** (1987), 175.
17. I.L. Buchbinder and I.B. Samsonov, *Noncommutative $N = 2$ supersymmetric theories in harmonic superspace* Grav. Cosmol. **8** (2002), 17, **hep-th/0109130**.
18. P.S. Howe, K.S. Stelle and P.C. West, *$N = 1$ $D = 6$ harmonic superspace.* Class. Quant. Grav. **2** (1985), 815.
    K.S. Stelle, preprint NSF-ITP-85-001
19. T. Araki, K. Ito and A. Ohtsuka, *$N = 2$ supersymmetric $U(1)$ gauge theory in noncommutative harmonic superspace.* JHEP **0401** (2004), 046, **hep-th/0401012**.
20. T. Araki and K. Ito, *Singlet deformation and non(anti)commutative $N = 2$ supersymmetric $U(1)$ gauge theory.* Phys. Lett. **B595** (2004), 513, **hep-th/0404250**.
21. T. Araki, K. Ito and A. Ohtsuka, *Deformed supersymmetry in non(anti)commutative $N = 2$ supersymmetric $U(1)$ gauge theory.* Phys. Lett. **B606** (2005), 202, **hep-th/0410203**.
22. S. Ferrara, E. Ivanov, O. Lechtenfeld, E. Sokatchev and B. Zupnik, *Non-anticommutative chiral singlet deformation of $N =$ (1,1) gauge theory.* Nucl. Phys. **B704** (2004), 154, **hep-th/0405049**.
23. N. Berkovits, *Super-Poincare covariant quantization of the superstring.* JHEP **04** (2000), 018, **hep-th/0001035**.
24. P.A. Grassi and P. van Nieuwenhuizen, *Harmonic superspaces from super-*

strings. Phys. Lett. **B593** (2004), 271, **hep-th/0402189**.
25. T. Araki, K. Ito and A. Ohtsuka, work in progress

NONCOMMUTATIVE LOCALLY ANTI-DE SITTER BLACK HOLES

P. BIELIAVSKY

Département de Mathématique
Université Catholique de Louvain
Chemin du cyclotron, 2,
1348 Louvain-La-Neuve, Belgium
bieliavsky@math.ucl.ac.be

S. DETOURNAY * AND PH. SPINDEL

Mécanique et Gravitation
Université de Mons-Hainaut, 20
Place du Parc
7000 Mons, Belgium
stephane.detournay@umh.ac.be, spindel@umh.ac.be

M. ROOMAN[†]

Service de Physique théorique
Université Libre de Bruxelles
Campus Plaine, C.P.225
Boulevard du Triomphe, B-1050
Bruxelles, Belgium
mrooman@ulb.ac.be

This note is based on a talk given by one of us at the workshop 'Noncommutative Geometry and Physics 2004' (Feb. 2004, Keio University, Japan). We give a review of our joint work on strict deformation of BHTZ 2+1 black holes.

This note is based on a talk given by one of us at the workshop 'Noncommutative Geometry and Physics 2004' (Feb. 2004, Keio University, Japan). We give a review of our joint work on strict deformation of BHTZ 2+1 black holes [4,5]. However some results presented here are not published

* "chercheur fria", belgium
† fnrs research director

elsewhere, and an effort is made for enlightening the instrinsical aspect of the constructions. This shows in particular that the three dimensional case treated here could be generalized to an anti-de Sitter space of arbitrary dimension provided one disposes of a universal deformation formula for the actions of a parabolic subgroup of its isometry group.

1. Motivations

The universal covering space of a generic BHTZ space-time [3]—realized as some open domain in AdS_3— is, in a canonical way, the total space of a principal fibration over $\mathbb{R}$ with structure group a minimal parabolic subgroup $\mathfrak{R}$ of $G := \widetilde{SL(2,\mathbb{R})}$ [4]. The action of the structure group is isometric w.r.t. the AdS_3 metric on the total space. For spinless BHTZ black holes, the fibers are dense open subsets of twisted conjugacy classes in G associated to an exterior automorphism of G. These twisted conjugacy classes are known to be WZW branes in $G = \mathrm{AdS}_3$ [2] i.e. extremal for the DBI brane action associated to a specific 2-form B on AdS_3 (referred hereafter as the 'B-field'). In the present work, we are interested in studying deformations of BHTZ spaces which are supported on (or tangential to) the fibration fibers. We will require our deformations to be non-formal ('strict' in the sense of Rieffel, see below) as well as compatible with the action of the structure group. Our motivation for this is threefold. First, a deformation of the brane in the direction of the B-field is generally understood as the (non-commutative) geometrical framework for studying interactions of strings with endpoints attached to the brane [22]. Despite the fact that curved situations have been extensively studied in various non-commutative contexts, in the framework of strict deformation theory i.e. in a purely operator algebraic framework, non-commutative spaces emerging from string theory have mainly been studied in the case of constant B-fields in flat (Minkowski) backgrounds. Second, the maximal fibration preserving isometry group of (the universal cover of) a BHTZ space being the above mentioned minimal parabolic group $\mathfrak{R}$, it is natural to ask for a deformation which is invariant under the action of $\mathfrak{R}$. But there is also a deeper reason for requiring the invariance. From considerations on black hole entropy, or just by geometrical interest, one may be interested in defining higher genus locally AdS_3 black holes (every BHTZ black hole is topologically $S^1 \times \mathbb{R}^2$). This would of course imply implementing the action of a Fuchsian group in the BHTZ picture. At the classical level, this question was at the center of important works by Brill et al. [1]. Our point here is that one may also attend this question at the deformed level. Indeed, the symmetry group $\mathfrak{R}$ is certainly too small to contain large Fuchsian groups. However, if the deformation is already $\mathfrak{R}$-invariant one may ask wether the (classical) action of $\mathfrak{R}$ would extend to a (deformed) action of the entire G by automorphisms of the deformed algebra. If this is the case (and it is), one obtains an action of every Fuchsian group at the deformed functional level [9,11]. The third motivation relies on Connes-Landi's and Connes-Dubois-Violette's work [16,17] where they use Rieffel's strict defomation method for actions of tori to define spectral triples for non-commutative spherical manifolds. The main point of their construction is that the data of an *isometric* action of a torus on a spin manifold yields not only a strict deformation of its function algebra but a compatible deformation of the Dirac operator as well. Therefore, disposing of a strict deformation formula for actions of $\mathfrak{R}$ would produce, in the present situation of BHTZ spaces, examples of spectral triples for non-commutative non-compact Lorentzian manifolds with constant curvature with the additional feature that the deformation would be supported on the $\mathfrak{R}$-orbits. The difficulty here is of course that these orbits cannot be obtained as orbits of an isometric action of $\mathbb{R}^d$.

2. Generic BHTZ spaces revisited

A BHTZ black hole is defined as the quotient space of (an open subset of) AdS_3 by an isometric action of the integers $\mathbb{Z}$. Namely, if Ξ denotes a Killing vector field on AdS_3 and if $\{\phi_t^\Xi\}_{t\in\mathbb{R}}$ denotes its flow, the $\mathbb{Z}$-action will be of the form:

$$\mathbb{Z}\times\mathrm{AdS}_3 \longrightarrow \mathrm{AdS}_3 : (n,x)\mapsto n.x := \phi_n^\Xi(x). \tag{1}$$

Not all Killing fields Ξ give rise to black hole solutions. The relevant ones can be defined as follows. As a Lorentzian manifold, the space AdS_3 is identified with the universal cover, G, of the group $SL(2,\mathbb{R})$ of 2×2 real matrices of determinant equal to 1 endowed with its Killing metric. The Lie algebra of the isometry group of AdS_3 (i.e. the algebra of Killing fields) is then canonically isomorphic to the semisimple Lie algebra $\mathfrak{g}\oplus\mathfrak{g}$ where $\mathfrak{g} := \mathfrak{sl}(2,\mathbb{R})$ denotes the Lie algebra of G. An element $\Xi\in\mathfrak{g}\oplus\mathfrak{g}$ defines a generic (i.e. non-extremal) BHTZ black hole if and only if its adjoint orbit only intercepts split Cartan subalgebras of $\mathfrak{g}\oplus\mathfrak{g}$. The squares of the (Killing) norms of the projections of Ξ onto each ideals of $\mathfrak{g}\oplus\mathfrak{g}$ are therefore both positive. Their sum —the Killing norm of Ξ— is called the *mass* and is denoted by M. Their difference —which measures how far Ξ is from being conjugated to a diagonal element in $\mathfrak{g}\oplus\mathfrak{g}$— is called the *angular momentum* and is denoted by J. Generic spaces, as opposed to extremal spaces, correspond to non-zero values of M and J. In the present letter, we will mainly be concerned with the two following geometrical properties of generic BHTZ spaces:

(i) In a generic BHTZ space, the horizons consist in a finite union of (projected) lateral classes of minimal parabolic subgroups of G.
(ii) Every generic BHTZ space is canonically endowed with a regular Poisson structure whose characteristic distribution is (locally) generated by Killing vector fields.

We now give a detailed description of the relevant geometry in the spinless as well as in the rotating cases.

2.1. *Spinless BHTZ spaces* $(J = 0 \neq M)$

The global geometry in this case can be derived from the two following observations. First, the action (1) appears as the restriction to a fixed split (connected) Cartan subgroup A of G of the action of G on itself by twisted conjugation:

$$\tau : G \times G \longrightarrow G : (g, x) \mapsto gx\sigma(g^{-1}) =: \tau_g(x), \tag{2}$$

where σ denotes the unique involutive exterior automorphism of G fixing pointwise the Cartan subgroup A. Indeed, if $\mathfrak{a}$ denotes the Lie subalgebra of A and if $H \in \mathfrak{a}$ is a vector of Killing length equal to 1, one has:

$$\phi_n^{\Xi} = \tau_{\exp(n\sqrt{M}H)} \qquad n \in \mathbb{Z}. \tag{3}$$

Second, the action τ yields the following global decomposition of G. The map

$$\phi : A \times G/A \longrightarrow G : (a, gA) \mapsto \tau_g(a) =: \phi(a, gA) \tag{4}$$

is well-defined as a global diffeomorphism [a].

As a consequence, the space $G = \mathrm{AdS}_3$ appears as the total space of a trivial fibration over $A = SO(1,1) \simeq \mathbb{R}$ whose fibers are the τ_G-orbits i.e.

[a] We refer to the above decomposition (4) as the *σ-twisted Iwasawa* decomposition of G with repsect to A. Indeed, if $G = KNA$ denotes an Iwasawa decomposition of G associated to the Cartan subgroup A, the homogeneous space G/A can be identified with the submanifold KN. In this setting, the decomposition (4) then reads:

$$K \times N \times A \longrightarrow G : (k, n, a) \mapsto kna\sigma(kn)^{-1},$$

which motivates our terminology.

the σ-twisted conjugacy classes[b]. As a homogeneous G-space every fiber is isomorphic to the affine symmetric space $G/A = \mathrm{AdS}_2$. Moreover, the $\mathbb{Z}$-action (1) which will define the BHTZ space is fiberwise [c]. The Killing metric on $G = \mathrm{AdS}_3$ turns out to be globally diagonal with respect to the twisted Iwasawa decomposition:

$$\mathrm{d}s^2_{\mathrm{AdS}_3} = \mathrm{d}a^2_A - \frac{1}{4}\cosh^2(a)\,\mathrm{d}s^2_{G/A}, \tag{6}$$

where $\mathrm{d}s^2_{G/A}$ denotes the canonical projected AdS_2-metric on G/A. The study of the quotient space $\mathbb{Z}\backslash G$ therefore reduces to the study of $\mathbb{Z}\backslash(G/A)$.

For a pictural visualization, we realize the space G/A as the G-equivariant universal covering space of the adjoint orbit $\mathcal{O} := \mathtt{Ad}(G)(H)$ of the element $H \in \mathfrak{g}$. Sitting in the Minkowski space $\mathfrak{g}$ endowed with its Killing form B, the orbit $\mathcal{O}$ is the sphere of radius 1—a one sheet hyperboloid. The orbits of A in $\mathcal{O}$—along which the $\mathbb{Z}$-identifications will be made— are planar: every intersection $\mathcal{O} \cap (\mathfrak{a}^\perp + \xi) \quad \xi \in \mathcal{O}$ is constituted by a union of respectively two or five A-orbits. The two planar sections $\mathcal{O} \cap (\mathfrak{a}^\perp \pm H) =: \mathcal{S}_\pm$ divide $\mathcal{O} - \mathcal{S}_\pm$ into six connected open components. In only two of them, $\mathcal{M}_\pm$, the A-orbits are timelike curves with respect to the metric on $\mathcal{O}$ induced by the Killing form B. The preimage $\mathcal{M}$ of $\mathcal{M}_+ \cup \mathcal{M}_-$ in $A \times G/A = G$ is constituted by a countable union of connected open components. Denoting by $\mathcal{M}^x$ the component of $\mathcal{M}$ containing $x \in \mathcal{M}$, one has

$$\mathcal{M} = \coprod_{z \in Z(G)} \mathcal{M}^{zJ} \qquad \text{(disjoint union)}, \tag{7}$$

where $Z(G)$ denotes the center of G and where J is determined (up to sign) by

$$J \in K, \qquad J^2 = -I. \tag{8}$$

Due to the minus sign in the expression of the Killing metric (6), the open set $\mathcal{M} \subset \mathrm{AdS}_3$ is constituted by all the points $x \in \mathrm{AdS}_3$ where the Killing vector Ξ_x is spacelike. The $\mathbb{Z}$-action (1) is proper and totally discontinuous on $\mathcal{M}$, one therefore has that the quotient $\mathcal{M}^x \to \mathbb{Z}\backslash\mathcal{M}^x$ defines a metric

---

[b]Remark that the foliation of $G = \mathrm{AdS}_3$ in τ_G-orbits (4) is the unique (up to conjugation) foliation tangent to Ξ and generated by Killing vector fields.

[c]Indeed, through the twisted Iwasawa decomposition (4) it reads:

$$n.(a, gA) = (a, \exp(n\sqrt{M}H)gA) \qquad n \in \mathbb{Z}. \tag{5}$$

covering of (non-complete) connected Lorentzian manifolds. These quotient spaces are all isometric to one another and homeomorphic to $S^1 \times \mathbb{R}^2$. In this picture, the black hole singularity appears at the level of G as the preimage $\mathcal{S}$ of $\mathcal{S}_+ \cup \mathcal{S}_-$ in $A \times G/A$ i.e. the boundary of $\mathcal{M}$ (on $\mathcal{S}$ the Killing field Ξ is null while it is timelike on the complement of the closure of $\mathcal{M}$, therefore yielding closed timelike curves in the quotient).

At the level of G, the horizons may be characterized as follows: a point $x \in \mathcal{M}$ belongs to an horizon if and only if the null directions from x which intersect the singularity $\mathcal{S}$ constitute a discrete set among all null directions from x. The union $\mathcal{H}$ of future and past horizons are then given by the union of the lateral classes through the element J of both minimal parabolic subgroups of G associated to A. Formally, if N and $\overline{N}$ denote the nilpotent subgroups of G normalized by A, one has

$$\mathcal{H} = (J\,Z(G)AN)\bigcup(J\,Z(G)A\overline{N}), \tag{9}$$

while the singularity is the union of the subgroups:

$$\mathcal{S} = Z(G)AN\bigcup Z(G)A\overline{N}. \tag{10}$$

Note that since the conjugation by J defines the K-adapted Cartan involution, the union $\mathcal{H}$ does not depend on the choice of right or left classes. At the level of the black hole (e.g. $\mathbb{Z}\backslash\mathcal{M}^J$), the horizons are obtained by projecting $\mathcal{H}$.

2.2. *Rotating BHTZ black holes ($J \neq 0 \neq M$)*

Like in the spinless case, we look for a foliation of G which is tangent to Ξ and generated by Killing vector fields. Here again there is no choice (up to conjugation) for such a foliation and it appears to underly an Iwasawa type decomposition of G. We adopt the same notations as in the preceding section and we consider $\alpha \in \mathfrak{a}^\star$ be such that $|\alpha| < 1$. Without loss of generality, one may then write the $\mathbb{Z}$-action (1) defining the rotating BHTZ space as

$$\mathbb{Z} \times G \longrightarrow G : (n, x) \mapsto n.x := \exp(nH)x\exp(n < \alpha, H > H). \tag{11}$$

We now observe that the map

$$A \times N \times K \longrightarrow G : (a, n, k) \mapsto anka^\alpha \tag{12}$$

is a global diffeomorphism [d].

---

[d]for $a \in A$, we set $a^\alpha := \exp(< \alpha, \log(a) > \log(a))$.

We call the decomposition (12) a *modified Iwasawa* decomposition. Note that this defines a trivial fibration of G onto K whose fibers are the orbits of the following action $\overline{\tau}$ of the solvable group $\mathfrak{R} := AN$:

$$\overline{\tau} : \mathfrak{R} \times G \longrightarrow G : (an, x) \mapsto anxa^{\alpha}. \tag{13}$$

Consequently, every $\overline{\tau}_{\mathfrak{R}}$-orbit is stable under the $\mathbb{Z}$-action (1) and through the modified Iwasawa decomposition (12) it simply reads:

$$n.(r, k) = \overline{\tau}_{\exp(nH)}(r, k) = (\exp(nH)r, k) \qquad r \in \mathfrak{R}. \tag{14}$$

There are three main differences with the $J = 0$ case. First, the action $\overline{\tau}$ of $\mathfrak{R}$ does not extend to an action of G by isometries. Second, there is no global transversal which is everywhere orthogonal to the foliation. The $\mathbb{Z}$-action is everywhere proper and discontinuous so that the quotient space $\mathbb{Z}\backslash AdS_3$ is globally a complete Lorentzian manifold.

The same techniques as in the spinless case nevertheless apply to obtain the causal structure in the rotating case. Again, one finds that the horizons are (projected) lateral classes of minimal parabolic subgroups.

We summarize this section by observing that to every generic BHTZ space are canonically associated two objects. First, a trivial fibration of AdS_3 whose fibers are orbits under an isometric action of a subgroup of G of dimension at least two. We call it *the action associated to the BHTZ space.* Second, a pair of conjugated minimal parabolic subgroups whose lateral classes define the horizons. We call them *the parabolic subgroups associated to the BHTZ space.*

3. BHTZ domains and extensions

We consider a generic BHTZ space with corresponding $\mathbb{Z}$-action (1) on AdS_3. We define a *BHTZ domain* of AdS_3 as a $\mathbb{Z}$-stable connected simply connected open subset $\mathcal{U}$ of AdS_3 maximal for the following two properties.

(i) the restricted action $\mathbb{Z}\times\mathcal{U} \to \mathcal{U}$ is proper and totally discontinuous;
(ii) the Killing vector field Ξ is everywhere spacelike on $\mathcal{U}$.

Roughly speaking, a BHTZ domain is thus a realization in AdS_3 of the universal covering space of a generic BHTZ space-time.
Similarly, we define a *BHTZ extension domain* as a $\mathbb{Z}$-stable connected simply connected open subset $\mathcal{E}$of AdS_3 maximal only for property (i) above.

We call the corresponding quotient manifold $\mathbb{Z}\backslash\mathcal{E}$ a *BHTZ extension* of the BHTZ space-time at hand. The latter therefore lies in every of its maximal extensions.
Remark that for rotating generic BHTZ spaces, the whole AdS_3 constitutes an extension domain. While for spinless BHTZ spaces, we have seen that any extension domain never entirely contains a τ_G-orbit.
The rest of this section will be devoted to the following assertion: every BHTZ domain admits a unique (up to conjugation) extension domain foliated by the orbits of the restriction of its associated action to the neutral component of one of its associated minimal parabolic subgroup.
The above statement is obvious in the rotating case. While for the spinless case, it is just worth observing that at the level of the adjoint orbit $\mathcal{O} \subset \mathfrak{g}$ (cf. Subsection 2.1), the subgroup AN (resp. $A\overline{N}$) has exactly two open orbits: the connected components of the complement of the planar intersection $\mathcal{O} \cap \mathfrak{n}^\perp$ (resp. $\overline{\mathfrak{n}}^\perp$) where $\mathfrak{n}$ (resp. $\overline{\mathfrak{n}}$) denotes the Lie algebra of N (resp. $\overline{N}$). Each of them being identified with the subgroup itself (the stabilizer groups are all trivial), the restricted action of $\exp(\mathbb{Z}\sqrt{M}H)$ is proper and discontinuous. One then readily verifies that any connected component of the preimage in $A \times G/A = G$ of one of these open orbits defines a BHTZ extension domain.

4. Poisson structures

We have seen that the orbits for the associated action of a non-spinning BHTZ space are the σ-twisted conjugacy classes in $G = AdS_3$ (σ is an involutive exterior automorphism of G). This remark naturally leads to consider a particular Poisson structure on AdS_3 whose symplectic leaves are the τ_G-orbits. Indeed, on the one hand, these submanifolds are known to be WZW 1- branes. That is, extremals of the DBI action associated with the data of a 2-form B on AdS_3 such that its exterior derivative dB is, up to a constant factor, equal to the metric volume ν on AdS_3. On the second hand, each orbit is, as a G-homogeneous space, isomorphic to the symmetric space G/A which is endowed with a unique (up to a constant factor) G-invariant symplectic structure $\omega^{G/A}$ [12]. It is then natural to ask wether the restriction of the B-field to the tangent distribution of the orbit foliation yields on each orbit (a multiple of) its canonical G-invariant symplectic structure. It actually does it, and in a canonical way. Indeed, the twisted Iwasawa decomposition (4) $\phi : A\times G/A \to G$ yields a global transversal (A) to the orbit foliation. By extending by 0 on A the 2-form $\omega^{G/A}$ to $A \times G/A$

one gets a closed 2-form on $G = \mathrm{AdS}_3$: $\omega := (\phi^{-1})^\star(0 \oplus \omega^{G/A})$. Every τ-invariant 2-form on G is then of the form $f\omega$ where f is leafwise constant i.e. $f = f(a)$ $a \in A$. Such a form therefore defines an admissible B-field if and only if $\nu = f'\, da \wedge \omega$. This last condition determines a unique (up to an additive constant) function f. Indeed, both ν and ω being τ-invariant, it is enough to check the condition on the transversal, which defines f'[e].

Now for a given generic BHTZ space, the above remark leads us to consider the particular class of Poisson structures on AdS_3 constituted by leafwise symplectic structures which are invariant under the associated action. Since the group $\mathfrak{R}$ has a unique (up to a constant factor) left-invariant symplectic structure, such Poisson structures exist in the rotating case as well.

5. Strict deformations for solvable Lie group actions

5.1. *Connes spectral triples for Abelian Lie group actions*

In this subsection we very roughly recall Rieffel strict deformation theory for actions of $\mathbb{R}^d$ [21] and how Connes-Landi and Connes-Dubois-Violette used it to define non-commutative spaces [16,17].

5.1.1. *Rieffel's machinery*

Given an action τ of the Abelian group $V := \mathbb{R}^d$ on a manifold $\mathcal{X}$ Rieffel's machinery produces a one parameter deformation $\{\mathbb{A}_\theta\}_{\theta\in\mathbb{R}}$ of the commutative algebra $\mathbb{A}_0 := C(\mathcal{X})$ of complex valued functions on X vanishing at infinity [f]. The algebras $\mathbb{A}_\theta$'s involved in the deformation are in fact $C^\star$-algebras [g] and the family $\{\mathbb{A}_\theta\}_{\theta\in\mathbb{R}}$ is a continuous field of $C^\star$-algebras. One starts with a fixed anti-symmetric matrix $\mathcal{J} \in \mathfrak{so}(\mathbb{R}^d)$. Note that, via the action τ, it determines a Poisson structure on $C^\infty(X)$: $\{u, v\} := \mathcal{J}^{ij} X_i^\star.u\, X_j^\star.v$ where $u, v \in C^\infty(\mathcal{X})$ and where $\{X_i^\star\}$ are the fundamental vector fields w.r.t. the action τ associated to a basis $\{X_i\}$ of (the Lie algebra of) $\mathbb{R}^d$. Denote by $\alpha : V \times C(\mathcal{X}) \to C(\mathcal{X})$ the action on functions induced by τ and by $\mathbb{A}_0^\infty$ the space of smooth vectors of α. One

[e] a computation at the level of the fundamental vector fields yields

$$f(a) = \tanh(\frac{a}{2}) + \text{const.}$$

[f] if X is compact, $C(\mathcal{X})$ is the whole algebra of continuous functions.
[g] the $C^\star$-norm on $C(\mathcal{X})$ is the sup-norm.

then defines the deformed product at the level of the smooth vectors by the following oscillatory integral formula:

$$a \star_\theta b := \int_{V\times V} e^{ix.y} \alpha_x(a) \alpha_{\theta \mathcal{J} y}(b) \, \mathrm{d}x \mathrm{d}y \tag{15}$$

where $a, b \in \mathbb{A}_0^\infty$ and where $x.y$ denotes the Euclidean dot product on $V = \mathbb{R}^d$. The pair $(A_0^\infty, \star_\theta)$ then turns out to be a pre-$C^\star$-algebra for a suitable choice of involution and norm. Its $C^\star$-completion is denoted by $\mathbb{A}_\theta$. Note that for $\theta = 0$, the RHS of Formula (15) reduces to the usual commutative product of functions ab. Also, the first order expansion term in θ of the oscillatory integral (15) is $\frac{\theta}{2i}\{a, b\}$. The product $a \star_\theta b$ is thus an associative deformation of ab in the direction of the Poisson bracket coming from the action τ.

An important example obtained from Rieffel's machinery is the so-called *quantum torus* $C(\mathbb{T}^d)_\theta$. It is defined as the deformation of the d-torus $\mathbb{T}^d$ obtained from the above procedure when applied to the natural action of $\mathbb{R}^d$ on $\mathbb{T}^d$.

5.1.2. *Deformed spectral triples for compact spin $\mathbb{T}^d$-manifolds*

Assume M is a compact spin manifold which the torus $\mathbb{T}^d$ acts on by isometries. The point is that Rieffel's machinery applied to M only produces a deformation of the *topological* space M. One would want to deform the metric structure as well. For doing this, Connes introduced the notion of spectral triple. The typical commutative example of a spectral triple is the one associated to a compact spin manifold; it roughly goes as follows. The basic idea is that the metric aspect of the Riemannian manifold M is encoded, in the operator algebra framework, by the Dirac operator D[15]. The spectral triple associated to M is then defined as the triple $(\mathbb{A}_0, \mathcal{H}, D)$ where $\mathbb{A}_0 = C(M)$ and where $\mathcal{H}$ is the Hilbert space of L^2-spinor fields on which the Dirac operator D acts as well as the $C^\star$-algebra $\mathbb{A}_0$ (simply by multiplication). Generally, a (not necessarily commutative) spectral triple is defined as a triple $(\mathbb{A}, \mathcal{H}, D)$ where $\mathbb{A}$ is a $C^\star$-algebra, $\mathcal{H}$ is a Hilbert space representation of $\mathbb{A}$ by bounded operators and D is a self-adjoint unbounded operator on $\mathcal{H}$ such that, among other things, $[D, a]$ is bounded for all a in $\mathbb{A}$[h]. To produce a deformed spectral triple from an action of $\mathbb{T}^d$ on M, one first observes that the deformation $C^\infty(M)_\theta$ obtained from Rieffel's machinery can equivalently be described as the closed subalgebra

[h] this is the essential property needed to define the metric distance in operator terms.

$(C^\infty(M)\otimes C(\mathbb{T}^d)_\theta)^{\tau\times L^{-1}}$ of $C^\infty(M)\otimes C(\mathbb{T}^d)_\theta$ constituted by the elements invariant under the natural action of $\mathbb{T}^d$ on the latter tensor product algebra (L denotes the regular action). Essentially, this is the co-action $C^\infty(M)\to C^\infty(M)\otimes C(\mathbb{T}^d)$ which yields the identification with $C^\infty(M)_\theta$. Second, one uses the previous identification to deform the $C^\infty(M)$-module of sections $\Gamma(M,S)$ of the spinor bundle $S\to M$. Since the action τ is isometric, up to a double covering of $\mathbb{T}^d$, the spinor bundle is $\mathbb{T}^d$-equivariant. The space of invariants $\Gamma(M,S)_\theta := (\Gamma(M,S)\otimes C(\mathbb{T}^d)_\theta)^{\tau\times L^{-1}}$ is therefore a $C^\infty(M)_\theta$-module stable by the operator $D\otimes I$. The restriction of the latter on $\Gamma(M,S)_\theta$ then defines the deformed Dirac operator D_θ.

We end this subsection by stressing the fact that the way Connes and Dubois-Violette define a deformation of the Dirac operator via an action of $\mathbb{T}^d$ essentially relies on two crucial properties. First, the invariance of Rieffel's deformed product on $C(\mathbb{T}^d)_\theta$ under the (left or right) regular representation (this allows to define the algebra (or module) structure on the space of invariant elements in the tensor product). Second, the fact that the action τ of $\mathbb{T}^d$ on M is isometric.

5.2. *Symmetric spaces and universal deformation formulae*

In this section we recall a consruction of a universal (strict) deformation formula for the actions of a non-Abelian Lie group. This formula was obtained via geometric methods based on symplectic symmetric spaces. We begin by recalling the notion of symplectic symmetric space, we then pass to the specific example which will be relevant here.

A **symplectic symmetric space** [7,8] is triple (M,ω,s) where (M,ω) a symplectic manifold and $s: M\times M\to M$ is a smooth map such that for each point $x\in M$ the partial map $s_x: M\to M: y\mapsto s_x(y):=s(x,y)$ is an involutive symplectic diffeomorphism which admits x as isolated fixed point. Under these hypotheses, the following formula defines a symplectic torsionfree affine connection ∇ on M:

$$\omega_x(\nabla_X Y, Z) := \frac{1}{2}X_x.\omega(Y + s_{x\star}Y, Z); \tag{16}$$

where X,Y and Z are smooth vector fields on M. This connection is the only one being invariant under the *symmetries* $\{s_x\}_{x\in M}$ which turns out to be the geodesic symmetries. The group $G(M)$ generated by the products

$s_x \circ s_y$ is a Lie group of transformations of M which acts transitively on M. It is called the **transvection** group of M.

The particular example we will be concerned with is the one where the transvection group is the Poincaré group $P_{1,1} := SO(1,1)\times\mathbb{R}^2$. It is a three dimensional solvable Lie group whose generic (two dimensional) coadjoint orbits are symplectic symmetric spaces admitting $P_{1,1}$ as tranvection group. They are all equivalent to oneanother ($P_{1,1}/\mathbb{R}$ as homogeneous spaces) and topologically $\mathbb{R}^2$.

Let (M,ω) be such an orbit and let ∇ be tis canonical connection. Such a space turns out to be stricly geodesically convex in the sense that between two points in M there is a unique geodesic. It has moreover the property that given any three points x, y and z in M, there's a unique fourth point t in M satisfying

$$s_x s_y s_z t = t. \tag{17}$$

This allows to define the **three point phase** of M as the three point function $S \in C^\infty(M\times M\times M, \mathbb{R})$ [23] given by

$$S(x,y,z) := \int_{\Delta(t, s_z t, s_y s_z t)} \omega \tag{18}$$

where $\Delta(x,y,z)$ denotes the geodesic triangle with vertices x, y and z. The three point **amplitude** $A \in C^\infty(M\times M\times M,\mathbb{R})$ [20,6] is defined as the ratio

$$A(x,y,z) := \left(\frac{\int_{\Delta(t,s_z t, s_y s_z t)}\omega}{\int_{\Delta(x,y,z)}\omega}\right)^{\frac{1}{2}}. \tag{19}$$

Both S and A are invariant under the diagonal action of the symmetries. For compactly supported $u, v \in C_c^\infty(M)$, one sets

$$u \star_\theta v(x) := \frac{1}{\theta^2}\int_{M\times M} A(x,y,z)\, e^{\frac{i}{\theta}S(x,y,z)} u(y)\, v(z)\, dy\, dz, \tag{20}$$

$$\text{or shortly } u\star_\theta v = \frac{1}{\theta^2}\int_{M\times M} A\, e^{\frac{i}{\theta}S}\, u\otimes v; \tag{21}$$

where dx denotes the symplectic measure on (M,ω). This formula defines a symmetry invariant associative (non-formal) deformation of M led by the Poisson structure associated to ω on some topological function space $\mathcal{E}_\theta(M) \subset C(M)$. For every real value of θ the space $\mathcal{E}_\theta(M)$ contains the test functions. Moreover, the deformed product is *strongly closed* in the sense that, one has the trace property:

$$\int_M u\star_\theta v = \int_M uv \tag{22}$$

for all compactly supported smooth functions u and v.

We now explain how this is relevant to our situation of BHTZ spaces. The transvection group $P_{1,1}$ contains a copy of the group $\mathfrak{R}$ which acts on maximal extensions of BHTZ spaces as discussed above. Moreover the restricted action $\mathfrak{R} \times M \to M$ turns out to be strictly transitive, so that one has the identification $M \simeq \mathfrak{R}$. Transporting the above structure to the group manifold $\mathfrak{R}$, one gets a left-invariant deformation, again denoted by $\star_\theta$, on the symplectic group $(\mathfrak{R}, \omega^{\mathfrak{R}})$. Let us denote by $K_\theta^{\mathfrak{R}}$ the associated (oscillating) kernel:

$$u \star_\theta v = \int_{\mathfrak{R}\times\mathfrak{R}} K^{\mathfrak{R}}{}_\theta \, u \otimes v \quad u, v \in \mathcal{E}_\theta(\mathfrak{R}), \tag{23}$$

where the integration is taken w.r.t. a left invariant Haar measure on $\mathfrak{R}$. Now, in a similar spirit as Rieffel's machinery, this defines a universal deformation formula for actions of $\mathfrak{R}$. Indeed, if $\mathcal{X}$ is a space the group $\mathfrak{R}$ acts on via an action τ, one defines for $a \in C(\mathcal{X})$ and $x \in \mathcal{X}$ the function $\alpha^x : \mathfrak{R} \to \mathbb{C} : g \mapsto a(\tau_{g^{-1}}x)$. Setting

$$\mathcal{E}_\theta(\mathcal{X}) := \{a \in C(\mathcal{X}) : \alpha^x a \in \mathcal{E}_\theta(\mathfrak{R}) \; \forall x \in \mathcal{X}\}, \tag{24}$$

the following formula yields an associative product on the space $\mathcal{E}_\theta(\mathfrak{R})$:

$$(a \star_\theta b)(x) := \int_{\mathfrak{R}\times\mathfrak{R}} K^{\mathfrak{R}}{}_\theta(e, g, h) \, \alpha^x a(g) \, \alpha^x b(h) \, dg \, dh; \tag{25}$$

or shortly: $a \star_\theta b =: \int_{\mathfrak{R}\times\mathfrak{R}} K^{\mathfrak{R}}{}_\theta \, \alpha a \otimes \alpha b$. Of course, such a product enjoys trace properties similar to (22).

Analoguous universal strict deformation formulae for more general solvable groups have been given in .

5.3. *Spectral triples from non-Abelian Lie group actions*

We now give a noncommutative spectral triple deforming the commutative one canonically associated to a maximally extended non-rotating BHTZ-space. We start, more generally, with the data of a pseudo-Riemannian spin manifold M of signature (m, n). We set D for the Dirac operator on M acting on the $C^\infty(M)$-bi-module of smooth sections $\underline{S}$ of the spinor bundle $S \to M$ over M[i]. We note that the fact that, for all a in $C^\infty(M)$, $[D, a]$ is continuous with respect to any reasonable topology on the spinor bi-module

[i]More generally, given a vector bundle $E \to M$ over M, we shall denote its space of smooth sections by $\underline{E}$.

relies on the property that D is a derivation of the module structure. In the case of Connes and Landi's construction, this property of the (undeformed) Dirac operator remains through the deformation because the torus is an Abelian group. One difficulty in the case of a non-Abelian Lie group action is therefore to force the derivation property of the Dirac operator with respect to the deformed module structure.

5.3.1. *Deformations of the spinor module and module endomorphisms*

Assume our deformation at the level of functions on M comes from an action α of a symplectic solvable group $\mathfrak{R}$ in such a way that the spinor module is $\mathfrak{R}$-equivariant. This means that there exists an action $\alpha^S : \mathfrak{R} \times \underline{S} \to \underline{S}$ such that $\alpha_g(a.\Psi) = \alpha_g(a).\alpha^S_g(\Psi)$ for all $a \in C^\infty(M)$. Denoting by K_θ the convolution kernel which defines the deformation, the following formula yields, at least formally, a deformation of the (right) module structure:

$$\Psi \star_\theta a := \int_{\mathfrak{R}\times\mathfrak{R}} K_\theta(e, g, h)\, \alpha_h(a).\alpha^S_g(\Psi)\, \mathrm{d}g\, \mathrm{d}h\,; \tag{26}$$

where e denotes the identity in $\mathfrak{R}$. One writes a similar formula $a \star_\theta \Psi$ for a deformation of the left module structure. In the same way, one deforms the module structure attached to any vector bundle associated to the spin structure over M. Note that, for the particular case of the endomorphism bundle, one formally has, for all $\Psi \in \underline{S}$, $\gamma \in \underline{S^\star \otimes S}$ and $a \in C^\infty M$,

$$(\gamma \star_\theta \Psi) \star_\theta a = \gamma \star_\theta (\Psi \star_\theta a). \tag{27}$$

In other words, the left deformed module of endomorphisms acts on the right deformed spinor module.

5.3.2. *Deforming the Dirac operator: case of trivial fibrations*

First consider the case of a differential operator $\Theta : \underline{S} \to \underline{S}$ of the form

$$\Theta(\Psi) = \gamma\,(\nabla_X \Psi)\,, \tag{28}$$

where ∇ denotes the spin connection, where X is a vector field on M and where $\gamma \in \underline{S^\star \otimes S}$ is a section of the endomorphism bundle. Note that Θ acts as a derivation of the (undeformed) module structure.

If the vector field X commute with the actions α and α^S, i.e. $X.\alpha_g a = \alpha_g(X.a)$ and $\nabla_X \alpha^S_g \Psi = \alpha^S_g\,(\nabla_X \Psi)$, then it automatically acts as derivation of the deformed multiplications. Indeed, one formally has: $X.(a \star_\theta b) =$

$\int K_\theta \, X.(\alpha a \otimes \alpha b) = \int K_\theta \, (\alpha(X.a) \otimes \alpha b + \alpha a \otimes \alpha(X.b)) = (X.a) \star_\theta b + a \star_\theta (X.b)$, and similarly for $\nabla_X(\Psi \star_\theta a)$. In particular, in the case the element γ is constant, the operator Θ is then a derivation of the deformed module structure.

Now, we turn back to our (extended) BHTZ spaces. In the preceeding sections, we have seen that their maximal extension domains are total spaces of trivial principal fibrations over $\mathbb{R}$ with structure group $\mathfrak{R}$. In particular, this implies that the space $\mathcal{E}_\theta(\mathcal{U})$, resulting from the application of the deformation formula to the left action τ of $\mathfrak{R}$ by isometries on $\mathcal{U}$, is non-trivial. Moreover, the (non-isometric) *right* action of $\mathfrak{R}$ on itself—and hence on a maximal extension domain $\mathcal{U}$— defines for any element X in the Lie algebra of the group $\mathfrak{R}$ a fundamental vector field again denoted by X which commutes with the actions α and α^S of $\mathfrak{R}$. Observing that $\tau_\star X = X$, one immediately concludes in the case of the non-rotating case $J = 0$ [j] that the Dirac operator D on $\mathcal{U}$ acts as a derivation of the (right) deformed module structure. This yields the formula $[D, a].\Psi = \Psi \star_\theta (Da)$, showing that for every a the commutator $[D, a]$ is a bounded operator in the space $\mathcal{H}$ of L^2 spinors. The triple $(\mathcal{E}_\theta(\mathcal{U}), \mathcal{H}, D)$ is thus a spectral triple in the sense of Connes[k]. A distance formula for such Lorentzian spectral triples has been formulated by V. Moretti in [19]. In a future work, we plan to investigate the question of understanding how causality is encoded at the operator algebraic level.

Remark.

(i) In the more general case of a non-constant element γ, the deformed operator Θ_θ defined as

$$\Theta_\theta(\Psi) := (\nabla_X \Psi) \star_\theta \gamma, \tag{29}$$

acts as a derivation of the left deformed module structure (as a immediate consequence of the fact that the right deformed module of endomorphisms acts on the left deformed spinor module).

(ii) Another situation of interest is when the vector field X is Hamiltonian with respect to the Poisson structure induced by the action α. Assume moreover that the spinor bundle $S \to M$ is trivial—as it is

[j] In the rotating case, the lack of a transversal everywhere orthogonal to the foliation makes the argument more involved.

[k] More precisely, because of noncompactess, in the sense of .

the case for AdS_3 or BHTZ-spaces. In this case, one has a globally defined connection form $\Gamma \in \Omega^1(S^\star \otimes S)$ such that

$$\nabla_X \Psi = X.\Psi + \Gamma(X)\Psi. \tag{30}$$

Denoting by $\lambda_X \in C^\infty(M)$ the hamiltonian function associated to X, one deforms the covariant derivative in the direction of X as

$$(\nabla_X)^\theta \Psi := \lambda_X \star_\theta \Psi - \Psi \star_\theta \lambda_X + \Psi \star_\theta \Gamma(X). \tag{31}$$

Note that $(\nabla_X)^\theta$ provides a derivation of the left deformed module structure in the sense that $(\nabla_X)^\theta(a \star_\theta \Psi) = [\lambda_X, a]_\theta \star_\theta \Psi + a \star_\theta (\nabla_X)^\theta \Psi$ where $[a, b]_\theta := a \star_\theta b - b \star_\theta a$.

These remarks have been used in to define strict deformations of BHTZ spaces themselves rather than their extension domains.

References

1. S. Aminneborg, I. Bengtsson, D. Brill, S. Holst, P. Peldan, Class.Quant.Grav. **15** (1998), 627 ;D. Brill, Annalen Phys. **9** (2000), 217.
2. C. Bachas, M. Petropoulos, JHEP 0102 (2001) 025
3. M. Bañados, M. Henneaux, C. Teitelboim, J. Zanelli, Phys. Rev. **D48** (1993) 1506, gr-qc/9302012
4. P. Bieliavsky, M. Rooman, Ph. Spindel, Nucl.Phys. **B645** (2002), 349-364.
5. P. Bieliavsky, S. Detournay, M. Herquet, M. Rooman, Ph. Spindel, Phys.Lett. **B570** (2003), 231-236.
6. P. Bieliavsky, S. Detournay, M. Rooman, Ph. Spindel,JHEP06(2004)031.
7. P. Bieliavsky, *Espaces symétriques symplectiques.* Ph. D. thesis, Université Libre de Bruxelles, 1995.
8. P. Bieliavsky, M. Cahen and S. Gutt, Modern group theoretical methods in Physics, Math. Studies, **18** (1995),63–75.
9. P. Bieliavsky, Advances in geometry, 71–82, Progr. Math., 172, Birkhäuser Boston, Boston, MA, 1999.
10. P. Bieliavsky, Journal of Symplectic Geometry, Vol.**1** N.2 (2002), 269, **math.QA/0010004**
11. P. Bieliavsky; M. Pevzner, Noncommutative harmonic analysis, 61–77, Progr. Math., 220, Birkhäuser Boston, Boston, MA, 2004.
12. P. Bieliavsky, Geom. Dedicata **73** (1998), no. 3, 245–273.
13. P. Bieliavsky and Y. Maeda, Lett. Math. Phys. **62** (2002), no. 3, 233–243.
14. P. Bieliavsky and M. Massar, Lett. Math. Phys. **58** (2001), no. 2, 115–128.
15. A. Connes, *Noncommutative geometry*, Academic Press, San Diego (1994)

16. A. Connes and G. Landi, Comm. Math. Phys. **221** (2001), no. 1, 141–159.
17. A. Connes, M. Dubois-Violette, Comm. Math. Phys. **230** (2002), no. 3, 539–579.
18. V. Gayral, J.M. Gracia-Bondia, B. Iochum, T. Schucker, J.C. Varilly, Comm. Math. Phys. **246** (2004), no. 3, 569–623.
19. V. Moretti, Rev.Math.Phys. Vol.**15** (2003), 1-46, **gr-qc/0203095**.
20. Z. Qian, *Groupoids, Midpoints and Quantizations.* Ph.D. Thesis, University of California, Berkeley 1997.
21. M. A. Rieffel, Mem. Amer. Math. Soc. **106** (1993), no. 506.
22. N. Seiberg, E. Witten, JHEP 9909:032, 1999, **hep-th/9908142**
23. A. Weinstein, Symplectic geometry and quantization (Sanda and Yokohama, 1993), Contemp. Math. **179** (1994), 261–270.

DYNAMICS OF FUZZY SPACES

M. BURIĆ
Faculty of Physics, P.O. Box 368 11001 Belgrade
Maja.Buric@phy.bg.ac.yu

J. MADORE
Laboratoire de Physique Théorique Université de Paris-Sud,
Bâtiment 211, F-91405 Orsay
John.Madore@th.u-paud.fr

The noncommutative extension of a dynamical 2-dimensional space-time is given and some of its properties discussed. Wick rotation to euclidean signature yields a surface which has as commutative limit the donut but in a singular limit in which the radius of the hole tends to zero.

1. Introduction

There is a very simple argument due to Pauli that the quantum effects of a gravitational field will in general lead to an uncertainty in the measurement of space coordinates. It is based on the observation that two 'points' on a quantized curved manifold can never be considered as having a purely space-like separation. If indeed they had so in the limit for infinite values of the Planck mass, then at finite values they would acquire for 'short time intervals' a time-like separation because of the fluctuations of the light cone. Since the 'points' are in fact a set of four coordinates, that is scalar fields, they would not then commute as operators. This effect could be considered important at least at distances of the order of Planck length, and perhaps greater. This is one motivation to study noncommutative geometry. A second motivation, which is the one we consider ours, is the fact that it is possible to study noncommutative differential geometry, and there is no reason to assume that even classically coordinates commute at all length scales. One can consider for example coordinates as order parameters as in solid-state physics and suppose that singularities in the gravitational field become analogs of core regions; one must go beyond the classical approximation to describe them. A straightforward and conservative way to do this is to represent them by operators. The space-time manifold is thus replaced with an algebra $\mathcal{A}$ (noncommutative 'space'), generated by a set of noncommutative 'coordinates' x^i. We think of x^i as linear operators on some vector space, and therefore we assume that the multiplication in the algebra is associative. The essential element which allows us to interpret a noncommutative algebra as a space-time is the possibility [1,2] to introduce a differential structure on the former.

We use a noncommutative version of Cartan's frame formalism [3]; the differential structure has been also studied from other points of view [2,4,5]. In order to develop some intuition in the complete absence of experimental evidence, one is obliged to consider examples. Several of these have been found. A series of models in all dimensions has been found [6], as well as some models in dimensions two [3,7] and four [8,9]. We shall present here the 'fuzzy donut', an example which we believe to be new. We shall argue that the moving frame formalism is in this respect a natural way to implement gravity. It enables one to introduce a sort of correspondence principle as a guide of how to construct the frame from its commutative limit. The parameter $\hbar$ with the dimensions of length squared is introduced in to facilitate the discussion of the commutative limit; a mass parameter μ as well as Newton's constant, are also introduced.

The outline of the presentation is the following. In the following section we shall introduce differential calculi based on sets of derivations, which are the noncommutative version of vector fields. In Section 5 we show that the momenta close under the commutator to generate a quadratic algebra. Implicit in our calculations is the assumption that this algebra can be at least formally identified with the original algebra generated by the 'coordinates'. Although not explicitly mentioned, this algebra is present in most of the examples; Section 3 presents what could be considered a counterexample with an algebra which is not quadratic. We conclude by introducing connection, metric and curvature in the noncommutative case in Section 6 and discussing their relations in Section 7.

2. Differential calculi

In the following sections we shall recall some of the more elementary aspects of the noncommutative extension of the de Rham differential calculus. In ordinary geometry a topological manifold can have more than one differential structure. In the noncommutative case this is generically so: many non-equivalent differential structures on a given algebra exist. This means that there are several 'geometries' which one can associate to the underlying noncommutative space. Once the differential calculus is chosen, some more or less obvious assumptions as hermiticity and bilinearity fix the linear connection almost uniquely. This is to be contrasted with the commutative case, where for each differential structure linear connection can be chosen almost arbitrarily. The choice therefore of a differential is one of the more important steps in the 'quantization'. We stress that we do not think that all noncommutative geometries are suitable as noncommutative models of space-time, any more so than in the commutative limit.

A noncommutative 'space' is an associative $*$-algebra $\mathcal{A}$ generated by a set of hermitean 'coordinates' x^i which in some limit tend to the (real) coordinates $\tilde{x}^i$ of a manifold; the latter we identify as the classical limit of the geometry. That is, in the classical or weak-field limit we impose the condition

$$x^i \to Z^{-1}\tilde{x}^i. \tag{2.1}$$

The Z could be perhaps singular; in all examples considered here we can choose $Z = 1$. Elements of $\mathcal{A}$ will be denoted by x^i, f, g, p_a, and so forth. In general the coordinates satisfy a set of commutation relations. We shall consider the algebra as a formal algebra and not attempt to represent it as an algebra of operators.

The simplest relation which can be used to define the algebra is

$$[x^i, x^j] = iJ^{ij}, \tag{2.2}$$

where J^{ij} are real numbers defining a canonical or symplectic structure. Often in the literature the notation θ^{ij} is used instead of J^{ij}. Another associative algebra is defined using the commutation relations

$$[x^i, x^j] = iC^{ij}{}_k x^k. \tag{2.3}$$

For simplicity we shall assume that the center of the algebra, the set of elements which commute with all generators, consists of complex multiples of the identity. The third important special case is a quantum space, defined by a homogeneous quadratic relation:

$$[x^i, x^j] = iC^{ij}{}_{kl} x^k x^l. \tag{2.4}$$

A combination of these three commutation relations will be satisfied by a set of generators p_a of $\mathcal{A}$ which we shall refer to as momenta; the commutation relations obeyed by the 'coordinates' in general are not even necessarily polynomial.

In ordinary geometry a vector field can be defined as a derivation of the algebra of smooth functions. This definition can be used also when the algebra is noncommutative. A derivation, we recall, is a linear map $f \mapsto Xf$ which satisfies the Leibniz rule, $X(fg) = (Xf)g + fXg$. Derivations will be denoted by X, Y, e_a and so forth and the set of all derivations by $\mathrm{Der}(\mathcal{A})$. A simple example is the algebra of 2×2 complex matrices M_2 with (redundant) generators the Pauli matrices. The algebra is of dimension four, the center is of dimension one and $\mathrm{Der}(M_2)$ is of dimension three with basis consisting of three derivations $e_a = \mathrm{ad}\,\sigma_a$:

$$e_a f = [\sigma_a, f]. \tag{2.5}$$

We notice that the Leibniz rule is here the Jacobi identity. We see also that the left multiplication $\sigma_a e_b$ of the derivation e_b by the generator σ_a no longer satisfies the Jacobi identity: it is not a derivation. The vector space $\mathrm{Der}(M_2)$ is not a left M_2 module. This property is generic. If X is a derivation of an algebra $\mathcal{A}$ and h an element of $\mathcal{A}$, then hX is not necessarily a derivation:

$$hX(fg) = h(Xf)g + hfXg \neq h(Xf)g + fhXg \tag{2.6}$$

if $hf \neq fh$.

Although derivations do not form a left module, one can introduce associated elements known as differential forms which form a bimodule; they can be multiplied from the left and from the right. We shall therefore express as much as possible physical quantities using the latter. We define here a 1-form ω is a linear map $\omega : \mathrm{Der}(\mathcal{A}) \to \mathcal{A}$. The set of 1-forms $\Omega^1(\mathcal{A})$ has a bimodule structure, that is, if ω is a 1-form, $f\omega$ and ωf are also 1-forms. The elements of $\Omega^1(\mathcal{A})$ will be typically denoted by ω, θ, ξ, η.

The important step is the definition of a differential d; it is a linear map from functions to 1-forms, $d : \mathcal{A} \to \Omega^1(\mathcal{A})$ which obeys the Leibniz rule. In general $fdg \neq dgf$ but we shall introduce later special forms θ^a which commute with the algebra. The exterior product $\xi$and η of two 1-forms ξ and η is a 2-form. There is no reason to assume the exterior product antisymmetric. We mention also that one can deduce the structure of the algebra of all forms from that 'of the module of 1-forms. The map d can be extended to all forms if one require that $d^2 = 0$. We should stress that in general one can associate many differential calculi to a given algebra.

There is then a variety of possibilities to define a differential. One problem is how to determine or at least restrict it by imposing some physical requirements. We shall use here a modification of the moving frame formalism and show that so defined differential calculi over an algebra admit essentially a unique metric and linear connection. We shall fix therefore the differential calculus by requiring that the metric have the desired classical limit. The idea is to define an analogue of a parallelizable manifold, which therefore has a globally defined frame. The frame is defined either as a set of vector fields e_a or as a set of 1-forms θ^a dual to them. The metric components with respect to the frame are then constant.

We choose a set of n derivations e_a which we assume to be inner generated by 'momenta' p_a:

$$e_a f = [p_a, f]. \tag{2.7}$$

We suppose that the momenta generate also the whole algebra $\mathcal{A}$. Since the center is trivial, this means that an element which commutes with all momenta must be a complex number. An alternative way is to use the 1-forms θ^a dual to e_a such that relation

$$\theta^a(e_b) = \delta^a_b \tag{2.8}$$

holds. To define the left hand side of this equation we define first the differential, exactly as in the classical case, by the condition

$$df(e_a) = e_a f. \tag{2.9}$$

The left and right multiplication by elements of the algebra $\mathcal{A}$ are defined by

$$f dg = f e_a g \theta^a, \qquad dg f = e_a g f \theta^a. \tag{2.10}$$

Since every 1-form can be written as sum of such terms the definition is complete. In particular, since

$$f\theta^a(e_b) = f\delta^a_b = (\theta^a f)(e_b), \tag{2.11}$$

we conclude that the frame necessarily commutes with all the elements of the algebra $\mathcal{A}$; this is a characteristic feature. If one does not insist on using differential calculi defined by inner derivations this condition can be generalized to include frames which commute only modulo an algebra morphism. For a recent discussion of this possibility we refer to [10].

In the case of the algebra M_2 considered above, the module of 1-forms is generated by three elements $d\sigma_a$ defined as the maps

$$d\sigma_a(e_b) = e_b\sigma_a = [\sigma_b, \sigma_a]. \tag{2.12}$$

The maps $\sigma_c d\sigma_a$ and $d\sigma_a\sigma_c$ are defined respectively as

$$\sigma_c d\sigma_a(e_b) = \sigma_c[\sigma_b, \sigma_a], \qquad d\sigma_a\sigma_c(e_b) = [\sigma_b, \sigma_a]\sigma_c. \tag{2.13}$$

Obviously, $\sigma_c d\sigma_a \neq d\sigma_a \sigma_c$.

The 1-form θ defined as

$$\theta = -p_a \theta^a \tag{2.14}$$

can be considered as an analog of the Dirac operator in ordinary geometry. It implements the action of the exterior derivative on elements of the algebra. That is

$$df = -[\theta, f] = [p_a \theta^a, f] = [p_a, f]\,\theta^a. \tag{2.15}$$

The differential is real if $(df)^* = df^*$. This is assured if the derivations e_a are real: $e_a f^* = (e_a f)^*$, which is the case if the momenta p_a are antihermitean. From the definitions one has $\theta^{a*} = \theta^a$, $\theta^* = -\theta$. Furthermore, $(f\xi)^* = \xi^* f^*$, $(\xi f)^* = f^* \xi^*$, and $(\xi\eta)^* = -\eta^* \xi^*$. Note that whereas the product of two hermitean elements is hermitean only if they commute, the product of two hermitean 1-forms is hermitean only if they anticommute.

The implementation of the differential structure as we have given is just as arbitrary as before since it amounts to a choice of the momenta. In some cases, the construction of the frame is not difficult. In the example (2.2) one can choose the differential such that $\left[x^i, dx^j\right] = 0$; a frame is $\theta^a = \delta^a_i dx^i$ since dx^i commute with all elements of the algebra. The most general form is $\theta^a = \Lambda^a_i dx^i$ with Λ^a_i real numbers. The duality relations give the momenta

$$\delta^a_b = \theta^a(e_b) = \delta^a_i dx^i(e_b) = \delta^a_i e_b(x^i) = \delta^a_i [p_b, x^i], \tag{2.16}$$

that is,

$$p_b = -i\delta^k_b J^{-1}_{ki} x^i. \tag{2.17}$$

In order to discuss noncommutative limit, (2.2) should in fact be rewritten as

$$[x^i, x^j] = i\kbar J^{ij}, \tag{2.18}$$

introducing the parameter $\kbar$ to describe the fundamental area scale on which noncommutativity becomes important. The $\kbar$ is presumably of order of the Planck area $G\hbar$; the commutative limit is defined by $\kbar \to 0$. The momenta read then

$$p_a = \frac{1}{i\kbar} J^{-1}_{ai} x^i, \tag{2.19}$$

and they are singular in the limit $\kbar \to 0$.

Since the frame is given by $\theta^a = \delta^a_i dx^i$, this space can be naturally thought of as the noncommutative generalization of flat space. The momenta are linear in the coordinates and hence

$$[p_a, p_b] = \frac{1}{(i\bar{k})^2} J^{-1}_{ai} J^{-1}_{bj} \left[x^i, x^j\right] = K_{ab}, \qquad K_{ab} = -\frac{1}{i\bar{k}} J^{-1}_{ab}. \tag{2.20}$$

In general only by explicit construction can one show that the frame exists. In the case of the Lie algebra (2.3), for example, one sees that the 1-forms dx^i do not define a frame because they do not commute with the algebra. In the example of M_2 with Pauli matrices as momenta, the frame which is the solution to the equation (2.8) is seen to be

$$\theta^a = \tfrac{1}{4} \sigma_b \sigma^a d\sigma^b. \tag{2.21}$$

This construction can be repeated [11] for the algebra M_n of $n \times n$ complex matrices.

As a last example we consider the algebra generated by two hermitean elements x and y related by

$$[x, y] = -2i\bar{k}\mu y. \tag{2.22}$$

This is related to the Jordanian deformation [12] of $\mathbb{R}^2$ with deformation parameter $h = i\bar{k}\mu^2$. The μ is the gravitational mass scale; the associated length $G\mu$ vanishes with $\bar{k}$. To find the frame, we rewrite this as follows

$$(x + i\bar{k}\mu)y = y(x - i\bar{k}\mu). \tag{2.23}$$

The differential must satisfy

$$dx\, y + (x + i\bar{k}\mu)\, dy = dy\, (x - i\bar{k}\mu) + y\, dx. \tag{2.24}$$

We shall impose separately the conditions

$$dx\, y = y\, dx, \qquad (x + i\bar{k}\mu)\, dy = dy\, (x - i\bar{k}\mu). \tag{2.25}$$

The first of these relations suggests that dx can be taken as a frame element, in fact $f(y)dx$ as well as dx. We set $\theta^1 = f(y)dx$. Rewriting (2.25) as

$$(x + i\bar{k}\mu)yy^{-1}\, dy = yy^{-1}\, dy\, y^{-1}y(x - i\bar{k}\mu), \tag{2.26}$$

we see that we can take $\theta^2 = -(\mu y)^{-1}dy$. We assume that θ^a commute with x and y. The duality relations (2.8) determine $f(y)$. They read

$$\begin{aligned} f(y)[p_1, x] &= 1, & f(y)[p_2, x] &= 0, \\ (\mu y)^{-1}[p_1, y] &= 0, & (\mu y)^{-1}[p_2, y] &= -1, \end{aligned} \tag{2.27}$$

and reduce to (2.25) for

$$p_2 = \frac{1}{2i\bar{k}}x, \qquad p_1 = \frac{1}{2i\bar{k}}y, \qquad f(y) = (\mu y)^{-1}. \tag{2.28}$$

The frame therefore is given by

$$\theta^1 = (\mu y)^{-1}dx, \quad \theta^2 = -(\mu y)^{-1}dy. \tag{2.29}$$

The corresponding calculus is the covariant one [12]. The momenta are proportional to the coordinates so their commutation relation is:

$$[p_1, p_2] = \mu p_1. \tag{2.30}$$

A short calculation shows that the frame elements anticommute. The line element

$$ds^2 = \pm(\tilde{\theta}^1)^2 \pm (\tilde{\theta}^2)^2 \tag{2.31}$$

of the commutative limit is that of the Lobachevski plane or of (anti) de Sitter space depending on the choice of signature.

3. Rindler space: frame rotations

As an example of a frame for which such a noncommutative extension does not exist we consider the 2-dim Rindler frame which is defined in one-half of 2-dim Minkowski space. We shall use this example to illustrate the fact that not all moving frames are suitable for 'quantization'; some are more suitable than others.

Let μ be a parameter with dimensions of mass and proportional to the Rindler acceleration. The commutative Rindler frame is given by $\tilde{\theta}^0 = \mu\tilde{x}d\tilde{t}$ and $\tilde{\theta}^1 = d\tilde{x}$; the commutative Minkowski frame is $\tilde{\theta}'^0 = d\tilde{t}'$ and $\tilde{\theta}'^1 = d\tilde{x}'$. The local Lorentz rotation from the former to the latter is defined by

$$\tilde{\theta}'^a = \tilde{\Lambda}^{-1a}{}_b\tilde{\theta}^b \tag{3.1}$$

with

$$\tilde{\Lambda} = \begin{pmatrix} \cosh\mu\tilde{t} & \sinh\mu\tilde{t} \\ \sinh\mu\tilde{t} & \cosh\mu\tilde{t} \end{pmatrix}. \tag{3.2}$$

The classical coordinate transformation from the Rindler coordinates to the Minkowski coordinates is given, for $x > 0$ by

$$\tilde{x}' = \tilde{x}\cosh\mu\tilde{t}, \qquad \tilde{t}' = \tilde{x}\sinh\mu\tilde{t}. \tag{3.3}$$

It is of course not to be confused with the rotation.

We shall first show that the Rindler frame is not a suitable frame; there are no dual momenta. If momenta p_a did exist then they would necessarily satisfy the relations

$$\begin{aligned} [p_0, t] &= (\mu x)^{-1}, \; [p_0, x] = 0, \\ [p_1, t] &= 0, \qquad\quad\; [p_1, x] = 1. \end{aligned} \tag{3.4}$$

But one easily sees that the solution is not a quadratic algebra. In fact if one set $[p_0, p_1] = L_{01}$ and $[t, x] = i\hbar J^{01}$, one finds from the Jacobi identities that

$$[L_{01}, t] = [p_0, [p_1, t]] - [p_1, [p_0, t]] = \mu^{-1} x^{-2}, \tag{3.5}$$

$$[L_{01}, x] = [p_0, [p_1, x]] - [p_1, [p_0, x]] = 0. \tag{3.6}$$

The L_{01} commutes with x and therefore belongs to the algebra generated by x. From the commutation with t one finds

$$i\hbar\mu(\frac{d}{dx} L_{01}) J^{01} = -x^{-2}. \tag{3.7}$$

Similarly one has

$$i\hbar[p_0, J^{01}] = [[p_0, t], x] - [[p_0, x], t] = 0, \tag{3.8}$$

$$i\hbar[p_1, J^{01}] = [[p_1, t], x] - [[p_1, x], t] = 0 \tag{3.9}$$

from which one concludes that J^{01} is constant. We shall set $J^{01} = 1$. We deduce therefore, neglecting integration constants, that

$$L_{01} = \frac{1}{i\hbar\mu} x^{-1}. \tag{3.10}$$

But the duality relations (3.4) require

$$p_0 = -\frac{1}{i\hbar\mu} \log(\mu x), \qquad p_1 = \frac{1}{i\hbar} t \tag{3.11}$$

and thus one easily sees that

$$L_{01} = \frac{1}{i\hbar} e^{i\hbar\mu p_0} \tag{3.12}$$

which is not a quadratic expression in p_0 and p_1.

The expressions (3.11) for the momenta seem quite different from the corresponding commutative expressions for the derivations $\tilde{e}_i$ dual to the frame:

$$\tilde{e}_0 = (\mu\tilde{x})^{-1}\partial_0, \qquad \tilde{e}_1 = \partial_1. \tag{3.13}$$

However in both cases one obtains the same action on the generators of the algebra. In particular

$$\tilde{e}_0\tilde{t} = (\mu\tilde{x})^{-1}, \qquad [p_0, t] = (\mu x)^{-1}. \tag{3.14}$$

Here one appreciates the importance of the space-time commutation relation.

Although the momenta p_a dual to the frame which we have used do not satisfy a quadratic relation it is easy to introduce another set $\bar{p}_a$ which do. We define the new momenta by the equations

$$\bar{p}_0 = -\frac{1}{i\bar{k}\mu}e^{-i\bar{k}\mu p_0}, \qquad \bar{p}_1 = p_1. \tag{3.15}$$

They obey the commutation relation

$$[\bar{p}_0, \bar{p}_1] = \frac{1}{i\bar{k}}. \tag{3.16}$$

From (3.11) one see also that $\bar{p}_a$ are related to the coordinate generators by the transformations

$$x = -i\bar{k}\bar{p}_0, \qquad t = i\bar{k}\bar{p}_1. \tag{3.17}$$

The frame defined by the new momenta is given by $\theta^a = \delta^a_i \bar{d}x^i$; it is a Minkowski-like frame in Rindler coordinates. We put here a bar on the differential to emphasize that the calculus is different. In spite of the apparent nonlocality in the transformation (3.15) the action of both $\bar{p}_0$ and p_0

$$[p_0, f] \to (\mu\tilde{x})^{-1}\tilde{\partial}_0 f, \qquad [\bar{p}_0, f] \to \tilde{\partial}_0 f \tag{3.18}$$

is local.

We now compare the differential calculi defined by two different frames, Rindler and Minkowski, and related by a noncommutative frame rotation of the type (3.1). Both frames can be used to define a differential calculus; each differential calculus has at most one basis as frame. Let $\tilde{\theta}^a$ be a global moving frame for some 2-dimensional commutative geometry and let $\{\tilde{\theta}'^a\}$ be the set of all moving frames $\tilde{\theta}'^a$ such that

$$\tilde{\theta}'^a = \tilde{\Lambda}^{-1}{}^a_b \tilde{\theta}^b \tag{3.19}$$

for some local Lorentz rotation $\tilde{\Lambda}$. The set of noncommutative versions will be described then each by a frame θ'^a which we shall suppose related by

$$\theta'^a = \Lambda^{-1}{}^a_b \theta^b \tag{3.20}$$

for the corresponding 'noncommutative' local Lorentz rotation. In the special cases we have been considering one can restrict the matrices Λ to the subset the elements of which depend on but one generator so they are well-defined. In more general situations the definition would require elaboration. It is clear that if $[f, \theta'^a] = 0$ then

$$[f, \theta^a] = [f, \Lambda^a_b]\Lambda^{-1b}{}_c\theta^c. \tag{3.21}$$

This can also be written as a rule

$$\theta^a f = \Lambda^a_b f \Lambda^{-1b}{}_c \theta^c \tag{3.22}$$

for relating the left- and right-module structures. In general then each local rotation defines a different calculus. The equivalence class of (commutative) moving frames gives rise to a set of inequivalent frames which have the same classical limit. If one wishes to consider one calculus, defined, say, by the condition $[f, \theta^a] = 0$ then each of the bases θ'^a satisfies the relation

$$\theta'^a f = \Lambda^{-1a}{}_b f \Lambda^b_c \theta'^c. \tag{3.23}$$

That is, θ'^a is not the frame for the same calculus unless Λ is a global rotation (a constant matrix).

Note that the Rindler metric can be considered equivalent to the 2-dim Kasner metric. The Kasner metric in dimension-4, for a special value of the parameters, is flat and a moving frame can be chosen which for these values become the ordinary flat frame:

$$\tilde\theta^0 = d\tilde t, \quad \tilde\theta^1 = d\tilde x - \tilde t^{-1}\tilde x d\tilde t, \quad \theta^2 = d\tilde y, \quad \theta^3 = d\tilde z. \tag{3.24}$$

We refer here to $\{\tilde\theta^0, \tilde\theta^1\}$ as the 2-dim Kasner moving frame. By a change of variables

$$\tilde x \to \tilde t, \qquad \tilde t \to \tilde t^{-1}\tilde x. \tag{3.25}$$

the Rindler frame can be brought to this form.

4. Kasner: noncommutative corrections

We shall here study the noncommutative corrections of the Kasner metric as defined in (3.24). For convenience instead of $\tilde x$ and $\tilde t$ we choose as classical variables $\tilde t$ and

$$\tilde\phi = \tilde t^{-1}\tilde x. \tag{4.1}$$

As we have already learned, not all moving frames attached to a metric are suitable for quantization; in this case the appropriate differential calculus is that determined by the flat Minkowski frame. The classical frame rotation from the Minkowski moving frame $\tilde{\theta}^a$ to the Kasner moving frame $\tilde{\eta}^a$ is given by

$$\begin{aligned}\tilde{\eta}^0 &= \cosh\tilde{\phi}\,\tilde{\theta}^0 - \sinh\tilde{\phi}\,\tilde{\theta}^1,\\ \tilde{\eta}^1 &= -\sinh\tilde{\phi}\,\tilde{\theta}^0 + \cosh\tilde{\phi}\,\tilde{\theta}^1.\end{aligned} \tag{4.2}$$

The relation between the two coordinate systems is given by

$$\tilde{x}' = \tilde{t}\sinh\tilde{\phi}, \qquad \tilde{t}' = \tilde{t}\cosh\tilde{\phi}. \tag{4.3}$$

It follows that $\tilde{t}^2 = \tilde{t}'^2 - \tilde{x}'^2$; the origin of the Kasner time coordinate, exactly at the flat-space values of the parameters and because of the singular nature of the transformation, becomes a null surface.

In the noncommutative case we choose the symmetric ordering; therefore the change of generators (4.3) becomes

$$\begin{aligned}x' &= t\sinh\phi - \tfrac{1}{2}[t, \sinh\phi],\\ t' &= t\cosh\phi - \tfrac{1}{2}[t, \cosh\phi].\end{aligned} \tag{4.4}$$

We normalize the Minkowski coordinates so that the commutator is given by $[t', x'] = i\bar{k}$. One finds that the corresponding Kasner commutator is

$$[t, x] = i\bar{k}(1 + o((i\bar{k})^2)). \tag{4.5}$$

We shall use rather the form

$$[t, \phi] = i\bar{k}t^{-1}(1 + o((i\bar{k})^2)). \tag{4.6}$$

The frame is given by

$$\theta^0 = dt', \qquad \theta^1 = dx'. \tag{4.7}$$

We can rewrite it in terms of t and x using the change of variables. It is of interest however to express also the calculus in terms of the Kasner moving frame; as we have already noticed in Equation (3.21) it is not a noncommutative frame since the frame rotation is local. We designate it therefore η^a and define as

$$\begin{aligned}\eta^0 &= \cosh\phi\,\theta^0 - \sinh\phi\,\theta^1,\\ \eta^1 &= -\sinh\phi\,\theta^0 + \cosh\phi\,\theta^1.\end{aligned} \tag{4.8}$$

Since from $[\phi, \theta^a] = 0$ we obtain $[\phi, \eta^a] = 0$, the equation (4.8) can be easily inverted. To complete the definition of the differential calculus we need the commutation relations $[t, \eta^a]$. These can only be calculated perturbatively. To lowest order one finds

$$[t, \eta^0] = -i\bar{k}t^{-1}\eta^1 + \tfrac{3}{2}(i\bar{k})^2 t^{-3}\eta^0, \tag{4.9}$$

$$[t, \eta^1] = -i\bar{k}t^{-1}\eta^0 + \tfrac{3}{2}(i\bar{k})^2 t^{-3}\eta^1. \tag{4.10}$$

To the same order, the transformation (4.4) of the generators reads

$$\begin{aligned} x' &= \sinh\phi\, t + \tfrac{1}{2} i k\!\!\!^{-} \cosh\phi\, t^{-1} + \tfrac{1}{4}(i k\!\!\!^{-})^2 \sinh\phi\, t^{-3}, \\ t' &= \cosh\phi\, t + \tfrac{1}{2} i k\!\!\!^{-} \sinh\phi\, t^{-1} + \tfrac{1}{4}(i k\!\!\!^{-})^2 \cosh\phi\, t^{-3}. \end{aligned} \tag{4.11}$$

These commutation relations determine a noncommutative geometry which is a natural extension of the flat Kasner geometry.

5. Fuzzy donut: momentum relations

Let us examine now further properties of the module structure defined by the differential (2.9). The exterior product is a map from the tensor product of two copies of the module of 1-forms into the module of 2-forms. We shall identify the latter as a subset of the former and write the product as

$$\theta^a\theta^b = P^{ab}{}_{cd}\theta^c \otimes \theta^d. \tag{5.1}$$

The $P^{ab}{}_{cd}$ are complex numbers which satisfy the projector condition and a hermiticity condition [6]. The basis 1-forms anticommute for $P^{ab}{}_{cd} = \frac{1}{2}(\delta^a_c\delta^b_d - \delta^b_c\delta^a_d)$. The exterior derivative of θ^a is a 2-form, so it can be written as

$$d\theta^a = -\frac{1}{2}C^a{}_{bc}\,\theta^b\theta^c. \tag{5.2}$$

The $C^a{}_{bc}$ are called the structure elements. They can be chosen to satisfy $C^a{}_{bc} = C^a{}_{de}P^{de}{}_{bc}$.

Impose now the condition $d^2 = 0$. It gives

$$0 = d(df) = d(-[\theta, f]) = -\left[d\theta + \theta^2, f\right], \tag{5.3}$$

so it implies that $d\theta + \theta^2$ commutes with all elements of the algebra. Since $d\theta + \theta^2$ is a 2-form, in the frame basis it can be written as

$$d\theta + \theta^2 = -\frac{1}{2}K_{ab}\theta^a\theta^b \tag{5.4}$$

where the elements K_{ab} are complex numbers. One can impose $K_{ab}P^{ab}{}_{cd} = K_{cd}$. A straightforward calculation shows that

$$\begin{aligned} d\theta &= -dp_a\theta^a - p_a d\theta^a = [p_b, p_a]\,\theta^b\theta^a + \frac{1}{2}p_a C^a{}_{bc}\theta^b\theta^c, \\ \theta^2 &= p_a p_b\theta^a\theta^b, \end{aligned} \tag{5.5}$$

and hence (5.4) reduces to

$$(p_c p_b + \frac{1}{2}C^a{}_{bc}p_a + \frac{1}{2}K_{bc})\theta^b\theta^c = 0. \tag{5.6}$$

This can be written as

$$2p_b p_a P^{ab}{}_{cd} + p_a C^a{}_{cd} + K_{cd} = 0. \tag{5.7}$$

The relation $d(f\theta^a - \theta^a f) = 0$ written in terms of the momenta gives further restrictions. It reads

$$[p_b \delta^a_c + p_c \delta^a_b + \frac{1}{2} C^a{}_{bc}, f]\theta^b\theta^c = 0 \tag{5.8}$$

which means

$$(C^a{}_{bc} + 2p_b\delta^a_c + 2p_c\delta^a_b - F^a{}_{bc})\,\theta^b\theta^c = 0, \tag{5.9}$$

where $F^a{}_{bc}$ are complex numbers. Thus the structure elements, defined in Equation (5.2) are linear in the momenta. It follows immediately that

$$e_a C^a{}_{bc} = 0. \tag{5.10}$$

This relation must be also satisfied in the commutative limit and constitutes a constraint on the frame. The example of Section 3 shows that this condition is not necessarily sufficient. A frame has four degrees of freedom in two dimensions. The constraint subtracts one therefrom. On the other hand having chosen a calculus, the choice of frame is equivalent to a gauge condition. This can be made more transparent if the momenta exist in which case the gauge condition can be expressed as the condition (5.10). The commutation relation (2.11) can be thought of also as a gauge condition since it is necessary for the existence of the momenta; there remain hence $4 - 1 - 1 = 2$ degrees of freedom. Combining (5.7) and (5.9) we obtain the relation

$$2p_c p_d P^{cd}{}_{ab} - p_c F^c{}_{ab} - K_{ab} = 0. \tag{5.11}$$

The coefficients in (5.11) are complex numbers. We see that the momentum generators p_a satisfy a quadratic relation.

One can readily find the conjugate momenta for a family of 2-dim metrics with one Killing vector. We shall exhibit all possible choices which yield differential calculi based on inner derivations. As frame we choose

$$\theta^0 = f(x)dt, \quad f > 0, \qquad \theta^1 = dx. \tag{5.12}$$

and we suppose that $J^{01} = J^{01}(x)$. The frame relations can be written as

$$\begin{aligned} dx\, x &= x\, dx, \quad & dx\, t &= t\, dx, \\ dt\, x &= x\, dt, \quad & dt\, t &= (t + i\bar{k}F)dt. \end{aligned} \tag{5.13}$$

We have set, for convenience

$$F = J^{01}\frac{d}{dx}\log f. \tag{5.14}$$

The differential structure of the algebra can be written as

$$(dx)^2 = 0, \quad dx\,dt = -dt\,dx, \quad (dt)^2 = -\tfrac{1}{2}i\bar{k}F'dx\,dt \tag{5.15}$$

or as the relations

$$(\theta^1)^2 = 0, \qquad \theta^0\theta^1 = -\theta^1\theta^0, \tag{5.16}$$

$$(\theta^0)^2 = \tfrac{1}{2}i\bar{k}fF'\theta^0\theta^1 = 2i\epsilon\theta^0\theta^1 \tag{5.17}$$

with

$$\epsilon = 4\bar{k}fF'. \tag{5.18}$$

It follows from the frame properties that ϵ is a constant.

Suppose now that the dual momenta exist. The duality relations are

$$\begin{aligned} &[p_0, t] = f^{-1}, \; [p_0, x] = 0, \\ &[p_1, t] = 0, \qquad [p_1, x] = 1. \end{aligned} \tag{5.19}$$

These relations allow us to identify p_1 with the partial derivative with respect to x. If $\phi = \phi(x)$ then

$$[p_1, \phi] = [p_1, x]\partial_x\phi = \partial_x\phi. \tag{5.20}$$

On the other hand, for $\phi = \phi(t, x)$ we can write to first order

$$[p_0, \phi] = [p_0, t]\partial_t\phi = f^{-1}\partial_t\phi. \tag{5.21}$$

If we denote as before $[p_0, p_1] = L_{01}$, the Jacobi identities imply the relations

$$\begin{aligned} &[p_0, J^{01}] = 0, \qquad &[p_1, J^{01}] = 0, \\ &[t, L_{01}] = -f'f^{-2}, \qquad &[x, L_{01}] = 0. \end{aligned} \tag{5.22}$$

One can conclude again that J^{01} is constant and also that L_{01} is a function of x alone. We set $J^{01} = 1$. It follows that, neglecting the integration constants, the 'Fourier transformation' between the position and momentum generators is given by

$$p_0 = -\frac{1}{i\bar{k}}\int f^{-1}, \qquad p_1 = -\frac{1}{i\bar{k}}t. \tag{5.23}$$

Each of the pairs (t, x) and (p_0, p_1) generates the algebra.

The array $P^{ab}{}_{cd}$ we write as

$$P^{ab}{}_{cd} = \tfrac{1}{2}\delta_c^{[a}\delta_d^{b]} + i\epsilon Q^{ab}{}_{cd} \tag{5.24}$$

In dimension two, if we assume that metric depends on x that is on p_0 only, we find that

$$P^{ab}{}_{cd}p_a p_b = \frac{1}{2}[p_c, p_d] + i\epsilon Q^{00}{}_{cd}p_0^2 \tag{5.25}$$

and therefore L_{01} is given by

$$L_{01} = K_{01} + p_0 F^0{}_{01} - 2i\epsilon p_0^2 Q^{00}{}_{01}. \tag{5.26}$$

The structure elements are given by

$$C^0{}_{01} = F^0{}_{01} - 4i\epsilon p_0 Q^{00}{}_{01}. \tag{5.27}$$

Symmetry and reality of the product imply that $Q^{ab}{}_{cd}$ has non-vanishing elements:

$$Q^{10}{}_{00} = -Q^{01}{}_{00} = 1, \qquad Q^{00}{}_{01} = -Q^{00}{}_{10} = 1. \tag{5.28}$$

We set also

$$K_{01} = \frac{1}{i\bar{k}J^{01}} = \frac{1}{i\bar{k}}, \qquad F^0{}_{01} = -ib\mu, \tag{5.29}$$

while $C^0{}_{10}$ is determined by the constraint

$$C^0{}_{ab}P^{ab}{}_{01} = C^0{}_{01}, \qquad C^0{}_{01} + C^0{}_{10} = -2i\epsilon C^0{}_{00}. \tag{5.30}$$

We have then finally the expressions

$$L_{01} = (i\bar{k})^{-1}(1 - b\mu^{-1}(i\epsilon p_0) - 2\mu^{-2}(i\epsilon p_0)^2), \tag{5.31}$$

$$C^0{}_{01} = -ib\mu - 4i\epsilon p_0, \tag{5.32}$$

and a differential equation

$$-i\epsilon\frac{dp_0}{dx} = \mu^2 - i\epsilon b\mu p_0 - 2(i\epsilon p_0)^2 \tag{5.33}$$

for p_0. There are three cases to be considered.

The simplest is the case with $\mu^2 \to \infty$. The equation reduces to

$$-i\bar{k}\frac{dp_0}{dx} = 1. \tag{5.34}$$

One finds the relations

$$i\bar{k}p_0 = -x, \qquad f(x) = 1. \tag{5.35}$$

This is noncommutative Minkowski space.

An equally degenerate case is the case with $\mu^2 \to \infty$ and with $\epsilon b = c\mu$. Equation (5.33) can be written in the form

$$-i\hbar \frac{dp_0}{dx} = 1 - icp_0. \tag{5.36}$$

One finds the solution

$$ip_0 = c^{-1}(e^{-\hbar^{-1}cx} - 1), \qquad f(x) = e^{\hbar^{-1}cx}. \tag{5.37}$$

The change of variables

$$t' = 2t, \qquad \mu x' = 2c^{-1}(e^{-cx} - 1), \tag{5.38}$$

transforms the algebra into the algebra of de Sitter space analyzed in Section 2.

The case which interests us the most here is that with μ finite. With $b = 0$ the equation becomes

$$-i\epsilon \frac{dp_0}{dx} = \mu^2 - 2(i\epsilon p_0)^2. \tag{5.39}$$

If we introduce the notation

$$\beta^2 = 2\mu^2 > 0 \tag{5.40}$$

the equation becomes

$$\frac{1}{\beta}\frac{d}{dx}\left(-2i\epsilon\beta^{-1}p_0\right) = 1 - \left(-2i\epsilon\beta^{-1}p_0\right)^2. \tag{5.41}$$

The solution is given by

$$i\hbar p_0 = -\beta^{-1}\tanh(\beta x), \qquad f(x) = \cosh^2(\beta x). \tag{5.42}$$

The function F is

$$F = -2i\beta^2\hbar p_0 = 2\beta\tanh(\beta x). \tag{5.43}$$

We find therefore the identity

$$F' + F^2 = f^{-1}f'' = 2\beta^2(1 + \tanh^2(\beta x)). \tag{5.44}$$

The frame (5.12) is given by

$$\theta^0 = \cosh^2(\beta x)dt = \tfrac{1}{2}(1 + \cosh(2\beta x))dt, \qquad \theta^1 = dx. \tag{5.45}$$

Frames of similar type have appeared [13,14,15] in 2-dimensional dilaton gravity theories. In the commutative limit the connection and the curvature which correspond to this frame are

$$\tilde{\omega}^0{}_1 = \tilde{\omega}^1{}_0 = F\tilde{\theta}^0, \tag{5.46}$$

$$\tilde{\Omega}^0{}_1 = \tilde{\Omega}^1{}_0 = -(F' + F^2)\theta^0\theta^1 = -f^{-1}f''\tilde{\theta}^0\tilde{\theta}^1. \tag{5.47}$$

The solution is a completely regular manifold of Minkowski signature which has the Rindler metric as singular limit. In the limit $\beta \to 0$

$$i\bar{k}p_0 = -x, \qquad f = 1, \tag{5.48}$$

and one finds Minkowski space. In 'tortoise' coordinate x^*,

$$x^* = \int \frac{dx}{f(x)} \tag{5.49}$$

the frame is given by

$$\theta^0 = \frac{1}{1 - x^{*2}}\, dt, \qquad \theta^1 = \frac{1}{1 - x^{*2}}\, dx^*. \tag{5.50}$$

From (5.23) we see that $x^* = -i\bar{k}p_0$.

Under a Wick rotation

$$u = 2i\beta x, \qquad v = t \tag{5.51}$$

the frame (5.12) becomes

$$\theta^0 = \tfrac{1}{2}(1 + \cos u)dv, \qquad \theta^1 = \frac{1}{2i\beta}\, du \tag{5.52}$$

and the line element in the commutative limit has the form

$$ds^2 = \tfrac{1}{4}(1 + \cos\tilde{u})^2 d\tilde{v}^2 + \tfrac{1}{4}\beta^{-2} d\tilde{u}^2. \tag{5.53}$$

This is the surface of the torus embedded in $\mathbb{R}^3$:

$$\tilde{x} = \tfrac{1}{2}(1 + \cos\tilde{u})\cos\tilde{v}, \quad \tilde{y} = \tfrac{1}{2}(1 + \cos\tilde{u})\sin\tilde{v}, \quad \tilde{z} = \tfrac{1}{2}\beta^{-1}\sin\tilde{u}, \tag{5.54}$$

and for this reason we call this metric the 'fuzzy donut'. It is a singular axially-symmetric surface of Gaussian curvature

$$\tilde{K} = 2\beta^2(1 - \tan^2 \tfrac{1}{2}\tilde{u}). \tag{5.55}$$

The donut is defined by the coordinate range $0 \leq u \leq 2\pi$, $0 \leq v \leq 2\pi$, with a singularity at the point $u = \pi$. In spite of the singularity, the Euler characteristic is given by

$$e[\mathcal{A}] = \frac{1}{4\pi}\epsilon_{ab} \int \tilde{\Omega}^{ab} = -\frac{1}{2\pi} \int \tilde{\Omega}^0{}_1 = -\frac{1}{2\pi} \int d\tilde{\omega}^0{}_1 = 0 \tag{5.56}$$

as it should be. If we suppose the same domain in the Wick rotated real-t region, then

$$0 \leq x \leq \beta^{-1}\pi, \qquad 0 \leq t \leq 2\pi. \tag{5.57}$$

As $\beta \to \infty$ the donut becomes more and more squashed, and this domain becomes an elementary domain in the limiting Minkowski space.

6. Noncommutative differential geometry

We have presented several noncommutative 'blurings' of classical geometries, all of which are of dimension two. We have concentrated our attention on the new aspects of the noncommutative theory, especially the plethora of differential calculi and the relation of the geometry to the symplectic structures. We have not, in fact, introduced the metric, the connection or the curvature on the noncommutative space. This can be done by taking the commutative limit and using the definition of a metric in terms of the frame. It can also be done [3] before the limit is taken. To complete the analysis of the family of examples discussed in Section 5, we mention the linear connections, the metric and the curvature without defining them in the full rigor; for details we refer to [8]. Note that when the momenta exist the metric is given; otherwise there is a certain ambiguity which must be determined by field equations.

To define a linear connection one needs a 'flip' [16,17],

$$\sigma(\theta^a \otimes \theta^b) = S^{ab}{}_{cd}\theta^c \otimes \theta^d, \tag{6.1}$$

which in the present notation is equivalent to a 4-index set of complex numbers $S^{ab}{}_{cd}$ which we can write as

$$S^{ab}{}_{cd} = \delta^b_c\delta^a_d + i\epsilon T^{ab}{}_{cd}. \tag{6.2}$$

The covariant derivative is given by

$$D\xi = \sigma(\xi \otimes \theta) - \theta \otimes \xi. \tag{6.3}$$

In particular

$$D\theta^a = -\omega^a{}_c \otimes \theta^c = -(S^{ab}{}_{cd} - \delta^b_c\delta^a_d)p_b\theta^c \otimes \theta^d = -i\epsilon T^{ab}{}_{cd}p_b\theta^c \otimes \theta^d, \tag{6.4}$$

so the connection-form coefficients are linear in the momenta

$$\omega^a{}_c = \omega^a{}_{bc}\theta^b = i\epsilon p_d T^{ad}{}_{bc}\theta^b. \tag{6.5}$$

On the left-hand side of the last equation is a quantity $\omega^a{}_c$ which measures the variation of the metric; on the right-hand side is the array $T^{ad}{}_{bc}$ which is directly related to the anti-commutation rules for the 1-forms, and more important the momenta p_d which define the frame. As $\hbar \to 0$ the right-hand side remains finite and

$$\omega^a{}_c \to \tilde{\omega}^a{}_c. \tag{6.6}$$

The identification is only valid in the weak-field approximation. The connection is torsion-free if the components satisfy the constraint

$$\omega^a{}_{ef} P^{ef}{}_{bc} = \tfrac{1}{2} C^a{}_{bc}. \tag{6.7}$$

The metric is a map

$$g : \Omega^1(\mathcal{A}) \otimes \Omega^1(\mathcal{A}) \to \mathcal{A}. \tag{6.8}$$

Using the frame it is defined by

$$g(\theta^a \otimes \theta^b) = g^{ab}, \tag{6.9}$$

and bilinearity of the metric implies that g^{ab} are complex numbers. In the present formalism [3] the metric is 'real' if it satisfies the condition

$$\bar{g}^{ba} = S^{ab}{}_{cd} g^{cd}. \tag{6.10}$$

'Symmetry' of the metric can be defined either using the projection

$$P^{ab}{}_{cd} g^{cd} = 0, \tag{6.11}$$

or the flip

$$S^{ab}{}_{cd} g^{cd} = c g^{ab}. \tag{6.12}$$

We usually take the frame to be orthonormal in the commutative limit, therefore one can write the metric as

$$g^{ab} = \eta^{ab} + i\epsilon h^{ab}. \tag{6.13}$$

In the linear approximation, the condition of the reality of the metric becomes

$$h^{ab} + \bar{h}^{ab} = -T^{ba}{}_{cd} \eta^{cd}. \tag{6.14}$$

The connection is metric if

$$\omega^a{}_{bc} g^{cd} + \omega^d{}_{ce} S^{ac}{}_{bf} g^{fe} = 0, \tag{6.15}$$

or linearized,

$$T^{(ac}{}_d{}^{b)} = 0. \tag{6.16}$$

In our 2-dim model the frame is of the form

$$\theta^0 = f(x) dt, \qquad \theta^1 = dx. \tag{6.17}$$

The torsion-free metric-compatible connection and the curvature are classically given by the expressions (5.47). From these expressions we see that the geometry is flat only if $f(x)$ is linear in x. We recall that $\epsilon = \hbar\mu^2$. To first order the fuzzy calculus differs from the commutative limit in the two relations

$$\begin{aligned}\theta^0\theta^1 &= P^{01}{}_{ab}\theta^a\theta^b = \tfrac{1}{2}\theta^{[0}\theta^{1]} + i\epsilon Q^{01}{}_{00}(\theta^0)^2 \\ &= \tfrac{1}{2}\theta^{[0}\theta^{1]} - i\epsilon q(\theta^0)^2 \end{aligned} \qquad (6.18)$$

$$(\theta^0)^2 = P^{00}{}_{ab}\theta^a\theta^b = i\epsilon Q^{00}{}_{01}\theta^{[0}\theta^{1]} = i\epsilon q\theta^{[0}\theta^{1]}. \qquad (6.19)$$

These can be better written as

$$\theta^{(0}\theta^{1)} = -2i\epsilon q(\theta^0)^2, \quad (\theta^0)^2 = i\epsilon q\theta^{[0}\theta^{1]}, \qquad (6.20)$$

and to first order reduce to

$$\theta^{(0}\theta^{1)} = 0, \quad (\theta^0)^2 = 2i\epsilon q\theta^0\theta^1. \qquad (6.21)$$

The quantity q which we have introduced in (6.18-6.21) is a constant, $q = 0$ in the cases of flat and de Sitter noncommutative spaces and $q = 1$ in the fuzzy donut case. We will restrict our considerations to the latter.

The differentials of the frame are given by

$$d\theta^0 = -C^0{}_{01}\theta^0\theta^1, \qquad d\theta^1 = 0, \qquad (6.22)$$

with

$$C^0{}_{01} = -4i\epsilon p_0 Q^{00}{}_{01} = -4i\epsilon p_0. \qquad (6.23)$$

The only non-vanishing components of the connection are

$$\omega^0{}_1 = \omega^1{}_0 = -4i\epsilon p_0\theta^0 = F\theta^0, \qquad (6.24)$$

and from (6.5) we find

$$T^{00}{}_{01} = T^{10}{}_{00} = -4. \qquad (6.25)$$

To first order the condition that the torsion vanish is the equation (6.7); it is satisfied by the values we obtain. The curvature 2-form has components

$$\Omega^0{}_1 = -(F' + F^2)\,\theta^0\theta^1 \qquad (6.26)$$

$$\Omega^0{}_0 = \Omega^0{}_0 = 2i\epsilon F^2\theta^0\theta^1. \qquad (6.27)$$

Therefore to lowest order from (5.44) we find the Gaussian curvature

$$\Omega^0{}_1 = -2\beta^2(1 + \tanh^2(\beta x))\theta^0\theta^1. \qquad (6.28)$$

We must define a 'real', 'symmetric' metric. There are in principle four possible ways to define it depending on which of two possible ways one chooses to define symmetry, and whether or not one includes a twist in the extension of the metric to the tensor product. In all cases the torsion-free condition yields the relation

$$T^{abcd} = 2(Q_-^{bcda} + Q_-^{bdca} + Q_-^{abdc}), \tag{6.29}$$

and the reality of the metric

$$h^{ab} + \bar{h}^{ab} = -T^{ba}{}_{cd}\eta^{cd}, \tag{6.30}$$

both in the linear approximation. Here we denote $Q_-^{ab}{}_{cd} = \frac{1}{2}Q^{ab}{}_{[cd]}$, $Q_+^{ab}{}_{cd} = \frac{1}{2}Q^{ab}{}_{(cd)}$. The projector $P^{ab}{}_{cd}$ is hermitean if

$$Q_+^{abcd} = \pm Q_-^{cdab}, \tag{6.31}$$

with plus in the case of no twist and minus with twist. If one use the flip to define symmetry, then for some γ the linearized perturbation must satisfy

$$h^{[ab]} = T^{ab}{}_{cd}\eta^{cd} - \gamma\eta^{ab} \tag{6.32}$$

if the metric is to be symmetric. If one use the product to define symmetry then

$$h^{[ab]} = -2Q_+^{abcd}\eta_{cd}. \tag{6.33}$$

In the present example the only consistent choice is the following

$$h^{[ab]} = -2Q_+^{abcd}\eta_{cd} = -2Q_-^{cdab}\eta_{cd}. \tag{6.34}$$

Thus for the symmetric and real metric we obtain

$$g^{ab} = \eta^{ab} + i\epsilon h^{ab}, \qquad h^{ab} = 2\begin{pmatrix} 0 & 1 \\ 0 & 0 \end{pmatrix}. \tag{6.35}$$

The η^{ab} here is the matrix of components of the canonical Minkowski metric; to it can be added an antisymmetric real matrix which is not fixed:

$$\eta^{ab} \mapsto \eta^{ab} + \epsilon\begin{pmatrix} 0 & a \\ -a & 0 \end{pmatrix}. \tag{6.36}$$

This ambiguity exists already at the classical level.

7. Higher-order effects

To find the second order corrections to our system, we write the 4-index tensors as matrices ordering the indices $(01, 10, 11, 00)$. Let P_0 and S_0 be respectively the canonical projector and the flip

$$P_0 = \begin{pmatrix} 1/2 & -1/2 & 0 & 0 \\ -1/2 & 1/2 & 0 & 0 \\ 0 & 0 & 0 & 0 \\ 0 & 0 & 0 & 0 \end{pmatrix}, \qquad S_0 = \begin{pmatrix} 0 & 1 & 0 & 0 \\ 1 & 0 & 0 & 0 \\ 0 & 0 & 1 & 0 \\ 0 & 0 & 0 & 1 \end{pmatrix}. \tag{7.1}$$

The projector constraints are, in matrix notation,

$$P^2 = P, \qquad \bar{P}\hat{P} = \hat{P} \tag{7.2}$$

where $\hat{A}^{ab}{}_{cd} = A^{ba}{}_{cd} = (S_0 A)^{ab}{}_{cd}$. To lowest order these conditions become

$$\bar{Q} = Q, \qquad \hat{Q}_- = Q_-. \tag{7.3}$$

The twist constraints are

$$\hat{\bar{S}}\hat{S} = 1, \tag{7.4}$$

$$\hat{S}P + \bar{P}\hat{P} = 0, \tag{7.5}$$

$$SP + P = 0. \tag{7.6}$$

The last two identities are equivalent if

$$\bar{P}\hat{P} = \hat{P}. \tag{7.7}$$

This condition was already imposed in (7.2).

One can easily check that the first order solution of the previous section which we write as $P = P_0 + i\epsilon Q$, $S = S_0 + i\epsilon T$, is given by

$$Q = Q_- + Q_+, \qquad Q_- = \begin{pmatrix} 0 & 0 \\ \tau & 0 \end{pmatrix}, \qquad Q_+ = \begin{pmatrix} 0 & -\tau^* \\ 0 & 0 \end{pmatrix}, \tag{7.8}$$

and

$$T = -2 \begin{pmatrix} 0 & \tau\tau^* \\ \tau\tau^*\sigma_1 & 0 \end{pmatrix}. \tag{7.9}$$

We introduced the matrix τ and its transpose τ^*:

$$\tau = \begin{pmatrix} 0 & 0 \\ 1 & -1 \end{pmatrix}, \qquad \tau\tau^* = 1 - \sigma_3, \quad \tau^*\tau = 1 - \sigma_1. \tag{7.10}$$

The constraints (7.2), (7.5-7.6) can be solved to second order using inner automorphisms of the matrix algebra. Denote

$$P = P_0 + i\epsilon Q + (i\epsilon)^2 Q_2, \tag{7.11}$$

$$S = S_0 + i\epsilon T + (i\epsilon)^2 T_2 \tag{7.12}$$

and introduce the automorphism $P = W^{-1} P_0 W$, where W is an arbitrary nonsingular 4×4 matrix with inverse W^{-1}. We see immediately that $P^2 = P$. To satisfy the second condition of (7.2) on P it is sufficient to require that

$$\bar{W} S_0 = S_0 W, \tag{7.13}$$

and to recall $\hat{S} = S_0 S$. Let $W = \exp(i\epsilon B)$. To second order

$$P = P_0 + i\epsilon [P_0, B] + \tfrac{1}{2}(i\epsilon)^2 [[P_0, B], B], \tag{7.14}$$

and the two expansions coincide if

$$Q = [P_0, B], \qquad Q_2 = \tfrac{1}{2}[[P_0, B], B] = \tfrac{1}{2}[Q, B]. \tag{7.15}$$

It is easy to see that an appropriate solution is

$$B = \begin{pmatrix} 0 & -\tau^* \\ -\tau & 0 \end{pmatrix}. \tag{7.16}$$

One can also check

$$P_0 B = Q_+, \quad B P_0 = -Q_-, \quad S_0 B = -B S_0 = -Q. \tag{7.17}$$

The (7.13) becomes the condition $\bar{B} S_0 = -S_0 B$, which in turn, since B is real, is the condition that B and S_0 anticommute.

The solution for T_2 is

$$T_2 = \tfrac{1}{2} T S_0 T. \tag{7.18}$$

To check whether the twist constraints hold, introduce

$$A = \begin{pmatrix} 0 & \tau\tau^* \\ \tau\tau^* & 0 \end{pmatrix}, \qquad T = -2AS_0, \quad (A - B)P_0 = 0. \tag{7.19}$$

At least to second order we have

$$S = S_0 X, \quad S = Y S_0, \qquad X = \exp(i\epsilon S_0 T), \quad Y = \exp(i\epsilon T S_0). \tag{7.20}$$

The twist constraint

$$\bar{\hat{S}}\hat{S} = S_0(S_0\bar{X})S_0(S_0 X) = S_0^2 \bar{X} X = 1 \tag{7.21}$$

follows. Further, consider the identity

$$(1 - WYW)P_0 = (1 - \exp(i\epsilon B)\exp(i\epsilon TS_0)\exp(i\epsilon B)P_0$$
$$= -i\epsilon(TS_0 + 2B)P_0 + \cdots = i\epsilon(T - 2B)P_0 + \cdots$$
$$= i\epsilon(T + 2Q)P_0 + \cdots = (i\epsilon)^2 H + o((i\epsilon)^3) \tag{7.22}$$

with

$$H = \begin{pmatrix} 1 & -1 & 0 & 0 \\ -5 & 5 & 0 & 0 \\ 0 & 0 & 0 & 0 \\ 0 & 0 & 0 & 0 \end{pmatrix}. \tag{7.23}$$

To lowest order, therefore,

$$SP + P = YS_0W^{-1}P_0W + P = YWS_0P_0W + P$$
$$= W^{-1}(1 - WYW)P_0W$$
$$= (i\epsilon)^2[H, B] + \cdots . \tag{7.24}$$

So all constraints are satisfied at least to second order.

The second-order metric is

$$g = \sqrt{\bar{X}} g_0. \tag{7.25}$$

To analyse the metric we write it as a 4-vector. We see then that if $\bar{g}_0 = g_0$,

$$\hat{\bar{g}} = S_0\sqrt{X}g_0 = S_0X\sqrt{\bar{X}}g_0 = Sg. \tag{7.26}$$

The metric is real. Since also

$$Pg = W^{-1}P_0W\sqrt{\bar{X}}g_0 = W^{-1}[P_0, W\sqrt{\bar{X}}]g_0 \tag{7.27}$$

the metric is symmetric to the extent that

$$[P_0, W\sqrt{\bar{X}}]g_0 = 0. \tag{7.28}$$

To first order this condition becomes

$$g = g_0 - \tfrac{1}{2}i\epsilon S_0 T g_0 == \begin{pmatrix} -1 & 0 \\ 0 & 1 \end{pmatrix} + 2i\epsilon \begin{pmatrix} 0 & 1 \\ 0 & 0 \end{pmatrix}. \tag{7.29}$$

We saw in Section 6 that this metric is compatible with the connection

$$\omega^a{}_b = S^{ac}{}_{db}\theta^d p_c + \theta\delta^a_b, \qquad \theta = -p_a\theta^a \tag{7.30}$$

to first order in the expansion parameter.

In general a connection is metric-compatible if the condition

$$\omega^i{}_{kl} g^{lj} + \omega^j{}_{ln} S^{il}{}_{km} g^{mn} = 0 \tag{7.31}$$

is satisfied. This can be written in a more familiar form if one introduce the 'covariant derivative'

$$D_i X^j = \omega^j{}_{ik} X^k \tag{7.32}$$

which is twisted:

$$D_i(X^j Y^k) = D_i X^j Y^k + S^{jl}{}_{im} X^m D_l Y^k. \tag{7.33}$$

Condition (7.31) becomes then

$$D_i g^{jk} = 0. \tag{7.34}$$

If $F^i{}_{jk} = 0$ then one can also express the condition as

$$S^{im}{}_{ln} g^{np} S^{jk}{}_{mp} = g^{ij} \delta^k_l. \tag{7.35}$$

We have not succeeded in finding a connection which is metric-compatible and torsion-free to second order; there are, however, solutions with torsion which are metric-compatible.

8. Conclusions

Several models have been found which illustrate a close relation between noncommutative geometry in its 'frame-formalism' version and classical gravity. Heuristically, but incorrectly, one can formulate the relation by stating that gravity is the field which appears when one quantizes the coordinates much as the Schrödinger wave function encodes the uncertainty resulting from the quantization of phase space.

The first and simplest of these is the fuzzy-sphere which is a noncommutative geometry which can be identified with the 2-dimensional (euclidean) 'gravity' of the 2-sphere. The algebra in this case is an $n \times n$ matrix algebra; if the sphere has radius r then the parameter r/n can be interpreted as a lattice length. With the identification this model illustrates how gravity can act as an ultraviolet cutoff, a regularization which is very similar to the 'point splitting' technique which has been used when quantizing a field in classical curved backgrounds. It can also be compared with the screening of electrons in plasma physics, which gives rise to a Debye length proportional to the inverse of the electron-number density n. The analogous 'screening' of an electron by virtual electron-positron pairs is responsible for the reduction of the electron self-energy from a linear to logarithmic dependence on the classical electron radius. Other models have been found which illustrate the identification including an infinite series in all dimensions.

In the present paper yet another model is given, one which although representing a classical manifold of dimension 2 is of interest because the classical 'gravity' which arises has a varying Gaussian curvature. The authors will leave to a subsequent article the delicate task of explaining exactly which property of the metric makes it 'quantizable'. This geometry could furnish a convenient model to study noncommutative effects, for example in the colliding-D-brane description of the Big-Bang proposed by Turok & Steinhardt [18]. The 2-space describing the time evolution of the separation of the branes has been shown to be conveniently described using Rindler coordinates. One can blur this geometry by using the metric and connection described here. The flat geometry would have to be replaced by the one given in this section; in the limit $q \to 0$ it would become flat.

The donut example is of importance in that is is the first explicit construction of an algebra and differential calculus which is singularity-free in the Minkowksi-signature domain and which has a non-constant curvature. There are two aspects of this problem. To construct a classical manifold from a differential calculus is relatively simple once one has constructed the frame. One takes formally the limit and uses the so constructed moving frame to define the metric. This is contained in the upper right of the following little diagram

$$
\begin{array}{ccc}
\begin{matrix}\text{Fuzzy}\\ \text{Frame}\end{matrix} & \longrightarrow & \begin{matrix}\text{Classical}\\ \text{Frame}\end{matrix} \\
\Big\downarrow & & \Big\downarrow \\
\begin{matrix}\textit{Fuzzy}\\ \textit{Geometry}\end{matrix} & \longrightarrow & \begin{matrix}\text{Classical}\\ \text{Geometry}\end{matrix}
\end{array}
\tag{8.1}
$$

More difficult is the construction of a 'fuzzy geometry' which would fill in the lower left of the diagram and would be such that the classical geometry is a limit thereof. But this step is very important since it gives an extension of the right-hand side into what could eventually be a domain of quantum geometry. It is the box in the to-be-constructed lower left corner where possibly one can find an interesting extension of the metric containing correction terms which describe the noncommutative structure.

We have not succeeded however to completely extend this geometry to all orders in the noncommutativity parameter $i\epsilon$. This will be considered in a subsequent article. There is evidence that the extension will involve a non-vanishing value of the torsion 2-form. The metric is extended into the noncommutative domain so as to maintain such formal properties as reality and symmetry. The interpretation however as a length requires more attention when the 'coordinates' do not commute.

Acknowledgment

Part of this work was done while the authors were visiting ESI in Vienna. They would like to thank the director for his hospitality as well as T. Grammatikopoulos J. Mourad, T. Schücker and G. Zoupanos for enlightening conversations.

References

1. S. L. Woronowicz, "Differential calculus on compact matrix pseudogroups," *Commun. Math. Phys.* **122** (1989) 125.
2. A. Connes, *Noncommutative Geometry.* Academic Press, 1994.
3. J. Madore, *An Introduction to Noncommutative Differential Geometry and its Physical Applications.* No. 257 in London Mathematical Society Lecture Note Series. Cambridge University Press, second ed., 2000. 2nd revised printing.
4. G. Landi, *An Introduction to Noncommutative Spaces and their Geometries*, vol. 51 of *Lecture Notes in Physics. New Series M, Monographs.* Springer-Verlag, 1997.
5. H. Figueroa, J. M. Gracia-Bondía, and J. C. Várilly, *Elements of Noncommutative Geometry.* Birkhauser Advanced Texts. Birkhäuser Verlag, Basel, 2000.
6. B. L. Cerchiai, G. Fiore, and J. J. Madore, "Geometrical tools for quantum euclidean spaces," *Commun. Math. Phys.* **217** (2001), no. 3, 521–554, `math.QA/0002007`.
7. C. Jambor and A. Sykora, "Realization of algebras with the help of ∗-products," `hep-th/0405268`.
8. M. Burić, M. Maceda, and J. Madore, "On the resolution of space-time singularities III," in *Geometric Methods In Physics*, A. Odzijewicz, A. Strasburger, S. T. Ali, J.-P. Antoine, T. Friedrich, J.-P. Gazeau, Z. Hasiewicz, and M. Schlichenmaier, eds., pp. –. 2004. Bialowieza, Poland, July 2004.
9. M. Dimitrijevic, L. Jonke, L. Moller, E. Tsouchnika, J. Wess, and M. Wohlgenannt, "Deformed field theory on κ-spacetime," *Euro. Phys. Jour. C* **C31** (2003) 129–138, `hep-th/0307149`.
10. A. Dimakis and F. Mueller-Hoissen, "Automorphisms of associative algebras and noncommutative geometry," *J. Phys. A: Math. Gen.* **37** (2004) 2307–2330, `math-ph/0306058`.
11. M. Dubois-Violette, R. Kerner, and J. J. Madore, "Classical bosons in a

noncommutative geometry," *Class. and Quant. Grav.* **6** (1989), no. 11, 1709–1724.
12. A. Aghamohammadi, "The two-parametric extension of h-deformation of $GL(2)$ and the differential calculus on its quantum plane," *Mod. Phys. Lett. A* **8** (1993) 2607.
13. J. P. S. Lemos and P. M. Sa, "The black holes of a general two-dimensional dilaton gravity theory," *Phys. Rev.* **D49** (1994) 2897–2908, `gr-qc/9311008`.
14. J. Gegenberg and G. Kunstatter, "Solitons and black holes," *Phys. Lett.* **B413** (1997) 274–280, `hep-th/9707181`.
15. D. Grumiller, W. Kummer, and D. V. Vassilevich, "Dilaton gravity in two dimensions," *Phys. Rep.* **369** (2002) 327–430, `hep-th/0204253`.
16. J. Mourad, "Linear connections in non-commutative geometry," *Class. and Quant. Grav.* **12** (1995) 965.
17. M. Dubois-Violette, J. Madore, T. Masson, and J. Mourad, "On curvature in noncommutative geometry," *J. Math. Phys.* **37** (1996), no. 8, 4089–4102, `q-alg/9512004`.
18. N. Turok and P. J. Steinhardt, "Beyond inflation: A cyclic universe scenario," *Physica Scripta* (2004) `hep-th/0403020`.

INDUCTION OF REPRESENTATIONS IN DEFORMATION QUANTIZATION

HENRIQUE BURSZTYN
Department of Mathematics
University of Toronto
Toronto, ON, M5S3G3
Canada
henrique@math.toronto.edu

STEFAN WALDMANN
Fakultät für Mathematik und Physik
Albert-Ludwigs-Universität Freiburg
Physikalisches Institut
Hermann-Herder-Straße 3
D 79104 Freiburg
Germany
Stefan.Waldmann@physik.uni-freiburg.de

We discuss the procedure of Rieffel induction of representations in the framework of formal deformation quantization of Poisson manifolds. We focus on the central role played by algebraic notions of complete positivity.

1. Introduction

In this note we describe how various concepts and constructions in the theory of C^*-algebras carry over to the purely algebraic setting of formal deformation quantization of Poisson manifolds. Our discussion centers around the construction of induced representations, due to Rieffel in the framework of C^*-algebras [15], and its interplay with notions of complete positivity. Although this note is mostly expository, we highlight some aspects of the theory that we have not made explicit before.

Deformation quantization [1] is a procedure to construct algebras of quantum observables associated with classical systems. More precisely, a classical phase space is a Poisson manifold $(M, \{\cdot,\cdot\})$ and its quantization is a formal associative deformation $\star$, also called a *star product*, of the classical observable algebra $C^\infty(M)$ in the direction of the Poisson bracket. Here $C^\infty(M)$ denotes the algebra of complex-valued smooth functions on M and $\star$ is a $\mathbb{C}[[\lambda]]$-bilinear associative multiplication on $C^\infty(M)[[\lambda]]$ given by

$$f \star g = \sum_{r=0}^{\infty} \lambda^r C_r(f, g), \tag{1}$$

where $C_0(f,g) = fg$, $C_1(f,g) - C_1(g,f) = \mathrm{i}\{f,g\}$, $1 \star f = f = f \star 1$ and all C_r are bidifferential operators. The formal parameter λ satisfies $\overline{\lambda} = \lambda$ and plays the role of Planck's constant $\hbar$. We require $\star$ to be a *Hermitian star product*, in the sense that $\overline{f \star g} = \overline{g} \star \overline{f}$, so that the $\mathbb{C}[[\lambda]]$-algebra $(C^\infty(M)[[\lambda]], \star)$ acquires a *-involution given by pointwise complex conjugation.

Other quantum mechanical concepts can be defined in deformation quantization analogously to the usual C^*-algebraic approach to quantum theory. The starting point is to regard $\mathbb{R}[[\lambda]]$ as an ordered ring by considering $\sum_{r=r_0}^{\infty} \lambda^r a_r$ to be *positive* if $a_{r_0} > 0$, where a_{r_0} is the first nonzero coefficient. Then a $\mathbb{C}[[\lambda]]$-linear functional

$$\omega : C^\infty(M)[[\lambda]] \longrightarrow \mathbb{C}[[\lambda]] \tag{2}$$

is called *positive* if $\omega(\overline{f} \star f) \geq 0$ for all $f \in C^\infty(M)[[\lambda]]$; a *state* is a positive linear functional such that $\omega(1) = 1$, and the value $\omega(f)$ is interpreted as the *expectation value* of the observable f in the state ω.

To implement the idea of superposition of states, one needs a notion of representation in deformation quantization. Given a Hermitian star product, a *representation* consists of a pre-Hilbert space $\mathcal{H}$ over $\mathbb{C}[[\lambda]]$ (here one uses the order structure of $\mathbb{R}[[\lambda]]$ for the definition of positive definite $\mathbb{C}[[\lambda]]$-valued inner products) on which $(C^\infty(M)[[\lambda]], \star)$ acts by adjointable operators. Many physically interesting examples can be found in [4], see [18] for a recent review.

As a next step, following the theory of C^*-algebras, one is led to the construction of induced representations. Recall that if $\mathcal{A}$ and $\mathcal{B}$ are C^*-algebras, the procedure of Rieffel induction consists of constructing representations of $\mathcal{B}$ from representations of $\mathcal{A}$ with the aid of a suitable $(\mathcal{B}, \mathcal{A})$-bimodule ${}_{\mathcal{B}}\mathcal{E}_{\mathcal{A}}$ possessing an $\mathcal{A}$-valued inner product $\langle \cdot, \cdot \rangle_{\mathcal{A}}$. For each *-representation of $\mathcal{A}$ on a Hilbert space $(\mathcal{H}, \langle \cdot, \cdot \rangle)$, one considers the tensor product $\mathcal{E} \otimes_{\mathcal{A}} \mathcal{H}$ over $\mathcal{A}$ and the natural left action of $\mathcal{B}$ on it. In order to turn this tensor product into a Hilbert space carrying a representation of $\mathcal{B}$, the key point is that one can combine $\langle \cdot, \cdot \rangle_{\mathcal{A}}$ and $\langle \cdot, \cdot \rangle$ to produce an inner product on $\mathcal{E} \otimes_{\mathcal{A}} \mathcal{H}$ uniquely defined by

$$(x \otimes \phi, y \otimes \psi) \mapsto \langle \phi, \langle x, y \rangle_{\mathcal{A}} \cdot \psi \rangle, \tag{3}$$

where $x, y \in {}_{\mathcal{B}}\mathcal{E}_{\mathcal{A}}$, $\phi, \psi \in \mathcal{H}$. An important point of this construction where specific properties of C^*-algebras must come into play is showing that the inner product defined by (3) is positive, see e.g. [14] for a detailed discussion.

Understanding the positivity of inner products of the form (3) in purely algebraic versions of Rieffel induction is the heart of this note, see also [8], [10]. We will discuss Rieffel induction in the framework of *-algebras over ordered rings in Section 3. Applications to deformation quantization are presented in Section 4. The last section contains a brief discussion on strong Morita equivalence, a notion closely related to algebraic Rieffel induction.

2. The general framework of *-algebras over ordered rings

In order to give a unified treatment of C^*-algebras and the *-algebras over $\mathbb{C}[[\lambda]]$ defined by Hermitian star products, we work in the following general algebraic setting, see [10] for details: we consider *-algebras $\mathcal{A}$ over a ring of the form $\mathsf{C} = \mathsf{R}(\mathrm{i})$; here R is an ordered ring, like e.g. $\mathbb{R}$ or $\mathbb{R}[[\lambda]]$, so C is a ring extension of R by a square root of -1.

Along the same lines of the discussion in the introduction, we define a C-linear functional $\omega : \mathcal{A} \longrightarrow \mathsf{C}$ to be *positive* if $\omega(a^*a) \geq 0$ for all $a \in \mathcal{A}$, which makes sense since $\mathsf{R} \subseteq \mathsf{C}$ is ordered. An algebra element $a \in \mathcal{A}$ is called *positive* if its expectation values are all non-negative, i.e. $\omega(a) \geq 0$ for all positive linear functionals ω. These notions agree with the usual ones, e.g., for C^*-algebras, and also make sense for Hermitian star products. We denote the set of positive elements by $\mathcal{A}^+$. See [16] for more general concepts of positivity in O^*-algebras and [17] for a comparison between them.

We now pass to representations. A *pre-Hilbert space* $\mathcal{H}$ over C is a C-module with a positive definite inner product $\langle \cdot, \cdot \rangle : \mathcal{H} \times \mathcal{H} \longrightarrow \mathsf{C}$, i.e. $\langle \phi, \psi \rangle = \overline{\langle \psi, \phi \rangle}$, $\langle \phi, \phi \rangle > 0$ for $\phi \neq 0$ and $\langle \cdot, \cdot \rangle$ is linear in the second argument. The *adjointable operators* from $\mathcal{H}_1$ to $\mathcal{H}_2$ are defined as C-linear maps for which adjoints exist in the usual sense. It is easy to check that when an adjoint exists, it is is unique. Note that, in the case of complex Hilbert spaces, the Hellinger-Toeplitz theorem ensures that the adjointable operators coincide with bounded operators. The adjointable operators $\mathfrak{B}(\mathcal{H})$ on a pre-Hilbert space $\mathcal{H}$ form a *-algebra in the natural way, and a **-representation* of a *-algebra $\mathcal{A}$ over C on $\mathcal{H}$ is a *-homomorphism $\pi : \mathcal{A} \longrightarrow \mathfrak{B}(\mathcal{H})$. When $\mathcal{A}$ is unital, we assume that $\pi(1_{\mathcal{A}}) = \mathsf{id}_{\mathcal{H}}$.

Example 2.1. If $\mathcal{A}$ is a *-algebra over C, an important class of examples of representations is given by an algebraic version of the GNS construction for C^*-algebras. Following [4], for each positive linear functional $\omega : \mathcal{A} \to \mathsf{C}$, one forms the space $\mathcal{H}_\omega := \mathcal{A}/\mathcal{J}_\omega$, where $\mathcal{J}_\omega$ consists of elements $a \in \mathcal{A}$ with $\omega(a^*a) = 0$. The space $\mathcal{H}_\omega$ is a pre-Hilbert space with inner product $\langle \psi_a, \psi_b \rangle := \omega(a^*b)$, where ψ_a denotes the class of $a \in \mathcal{A}$ in $\mathcal{H}_\omega$; the *GNS *-representation* of $\mathcal{A}$ on $\mathcal{H}_\omega$ is defined by $\pi(a)\psi_b := \psi_{ab}$.

This construction in deformation quantization gives rise to important formal representations of Hermitian star products, such as the Bargmann-Fock representation of Wick star products, or the Schrödinger representation of Weyl star products on cotangent bundles, see [4], [3].

For a *-algebra $\mathcal{A}$ over C, we define $^*\text{-}\mathsf{rep}(\mathcal{A})$ to be the category whose objects are *-representations of $\mathcal{A}$ on pre-Hilbert spaces over C and with adjointable intertwiners as morphisms. We refer to this category as the *representation category* (or *representation theory*) of $\mathcal{A}$. In these terms, the procedure of Rieffel induction, to be discussed in the next section, can be seen as an explicit construction of functors between representation categories. Functors which establish equivalence of categories of representations will be briefly discussed in the last section.

3. Complete positivity and algebraic Rieffel induction

In order to describe Rieffel induction in the algebraic framework of Section 2, we need to consider algebraic analogs of Hilbert C^*-modules, see e.g. [13]. The reader may consult [10] for details.

Let $\mathcal{A}$ be a *-algebra over C, and let $\mathcal{E}$ be a (right) $\mathcal{A}$-module (we may write $\mathcal{E}_{\mathcal{A}}$ to stress the $\mathcal{A}$-action). An *$\mathcal{A}$-valued inner product* on $\mathcal{E}$ is a C-sesquilinear map (linear in the second argument)

$$\langle \cdot, \cdot \rangle_{\mathcal{A}} : \mathcal{E} \times \mathcal{E} \longrightarrow \mathcal{A}, \tag{4}$$

such that $\langle x, y \rangle_{\mathcal{A}} = \langle y, x \rangle_{\mathcal{A}}^*$ and $\langle x, y \cdot a \rangle_{\mathcal{A}} = \langle x, y \rangle_{\mathcal{A}}\, a$ for all $x, y \in \mathcal{E}$ and $a \in \mathcal{A}$. We call $\langle \cdot, \cdot \rangle_{\mathcal{A}}$ *non-degenerate* if $\langle x, y \rangle_{\mathcal{A}} = 0$ for all x implies $y = 0$, in which case the pair $(\mathcal{E}, \langle \cdot, \cdot \rangle_{\mathcal{A}})$ is called an *inner-product $\mathcal{A}$-module.* The inner product $\langle \cdot, \cdot \rangle_{\mathcal{A}}$ is called *positive* if $\langle x, x \rangle_{\mathcal{A}} \in \mathcal{A}^+$. Finally, $\langle \cdot, \cdot \rangle_{\mathcal{A}}$ is called *strongly non-degenerate* if the map $\mathcal{E} \ni x \mapsto \langle x, \cdot \rangle_{\mathcal{A}} \in \mathrm{Hom}_{\mathcal{A}}(\mathcal{E}, \mathcal{A})$ is a bijection. Similar definitions hold for left modules (the only difference is that we have C- and $\mathcal{A}$-linearity in the *first* argument).

If $\mathcal{B}$ is another *-algebra over C, then a *$(\mathcal{B}, \mathcal{A})$-inner-product bimodule* is an inner-product $\mathcal{A}$-module $(\mathcal{E}, \langle \cdot, \cdot \rangle_{\mathcal{A}})$ together with a *-homomorphism

$\mathcal{B} \to \mathfrak{B}(\mathcal{E})$, where $\mathfrak{B}(\mathcal{E})$ is the *-algebra of adjointable operators with respect to $\langle \cdot, \cdot \rangle_{\mathcal{A}}$. Consider an object in *-$\mathsf{rep}(\mathcal{A})$, i.e., a pre-Hilbert space $(\mathcal{H}, \langle \cdot, \cdot \rangle)$ carrying a *-representation of $\mathcal{A}$. In order to obtain an object in *-$\mathsf{rep}(\mathcal{B})$ from ${}_{\mathcal{B}}\mathcal{E}_{\mathcal{A}}$ and $\mathcal{H}$, we follow [15] and consider the algebraic tensor product $\mathcal{E} \otimes_{\mathcal{A}} \mathcal{H}$, which carries a left $\mathcal{B}$-action, equipped with the inner product determined by

$$(x \otimes \phi, y \otimes \psi) \mapsto \langle \phi, \langle x, y \rangle_{\mathcal{A}} \cdot \psi \rangle, \tag{5}$$

for $x, y \in \mathcal{E}$ and $\phi, \psi \in \mathcal{H}$. In the framework of C^*-algebras, one can prove that if $\langle \cdot, \cdot \rangle_{\mathcal{A}}$ is positive, then so is the induced inner product (5) (see e.g [13], [14]). The following proposition indicates what is algebraically needed in general.

Proposition 3.1. *Let us assume, for simplicity, that $\mathcal{A}$ and $\mathcal{B}$ are unital, and let $({}_{\mathcal{B}}\mathcal{E}_{\mathcal{A}}, \langle \cdot, \cdot \rangle_{\mathcal{A}})$ be a $(\mathcal{B}, \mathcal{A})$-inner-product bimodule. Then the following are equivalent:*

(1) The inner product (5) *is positive for any *-representation of $\mathcal{A}$.*
(2) For all n and all $x_1, \ldots, x_n \in \mathcal{E}$, the matrix $(\langle x_i, x_j \rangle_{\mathcal{A}})$ is a positive element in $M_n(\mathcal{A})$ (viewing $M_n(\mathcal{A})$ as a *-algebra over C in the natural way).

For the proof, we need the following simple lemma:

Lemma 3.1. *Let $\mathcal{A}$ be unital. If $\Omega : M_n(\mathcal{A}) \longrightarrow \mathsf{C}$ is a positive linear functional then there exists a *-representation $(\mathcal{H}, \pi)$ of $\mathcal{A}$ and vectors $\phi_1, \ldots, \phi_n \in \mathcal{H}$ such that*

$$n\Omega(A) = \sum_{i,j} \langle \phi_i, \pi(a_{ij}) \phi_j \rangle \tag{6}$$

where $A = (a_{ij}) \in M_n(\mathcal{A})$. Conversely, for any *-representation $(\mathcal{H}, \pi)$ of $\mathcal{A}$ and any choice of vectors $\phi_1, \ldots, \phi_n \in \mathcal{H}$, the right hand side of (6) *defines a positive linear functional of $M_n(\mathcal{A})$ (and this defines a positive Ω if $1/n \in \mathsf{C}$).*

Proof. This is a simple application of the GNS construction and should be well-known. For the reader's convenience we outline the proof. Let $E_{ij} \in M_n(\mathcal{A})$ be the elementary matrices with 1 at the (i, j)-position and 0 elsewhere. Then $nA = \sum_{i,j,k,l} E_{ji}^* a_{il} E_{kl}$. Now let $(\mathcal{H}_\Omega, \Pi_\Omega)$ be the GNS representation of $M_n(\mathcal{A})$ with respect to Ω. Then define $\phi_i = \sum_j \psi_{E_{ji}} \in$

$\mathcal{H}_\Omega$. Clearly $\pi(a) := \Pi_\Omega(aE_{11})$ is a *-representation of $\mathcal{A}$ on $\mathcal{H}_\Omega$ and we now have

$$n\Omega(A) = n\left\langle \psi_{1_{n\times n}}, \Pi_\Omega(A)\psi_{1_{n\times n}} \right\rangle = \sum_{i,j} \left\langle \phi_i, \pi(a_{ij})\phi_j \right\rangle .$$

The converse statement can be easily checked. □

Proof. We can now complete the proof of the proposition. Let $A = (\langle x_i, x_j \rangle) \in M_n(\mathcal{A})$. Using the assumption of (1) and the lemma, we have $\Omega(A) \geq 0$ for all positive linear functionals $\Omega : M_n(\mathcal{A}) \longrightarrow \mathsf{C}$. The converse implication will follow in much more generality in Theorem 3.1. □

An $\mathcal{A}$-valued inner product on $\mathcal{E}$ satisfying the condition in (2) is called *completely positive* [10]. If $\langle \cdot, \cdot \rangle_{\mathcal{A}}$ is completely positive, we call $(\mathcal{E}, \langle \cdot, \cdot \rangle_{\mathcal{A}})$ a *pre-Hilbert $\mathcal{A}$-module*; a $(\mathcal{B}, \mathcal{A})$-inner product bimodule for which the $\mathcal{A}$-valued inner product is completely positive is called a *pre-Hilbert bimodule.*

Example 3.1.

(1) If $(\mathcal{H}, \langle \cdot, \cdot \rangle)$ is a pre-Hilbert space over C, then $\langle \cdot, \cdot \rangle$ is automatically completely positive;

(2) If $\mathcal{A}$ is a C^*-algebra, then any positive $\mathcal{A}$-valued inner product is completely positive;

(3) If $\mathcal{A}$ is a *-algebra over C and $\mathcal{E}$ is the right projective $\mathcal{A}$-module $P\mathcal{A}^n$, where $P \in M_n(\mathcal{A})$ is a projection, then the restriction of the natural $\mathcal{A}$-valued inner product on $\mathcal{A}^n$ to $\mathcal{E}$ is completely positive.

(4) If $\mathcal{A} = C^\infty(M)$, then any positive strongly nondegenerate $\mathcal{A}$-valued inner product on a finitely generated projective (f.g.p.) $\mathcal{A}$-module is completely positive.

To see why (4) holds, note that it follows from (3) that any f.g.p. module over $\mathcal{A}$ can be equipped with a completely positive $\mathcal{A}$-valued inner product, and in the case where $\mathcal{A} = C^\infty(M)$, any two $\mathcal{A}$-valued inner products on the same f.g.p. module are equivalent. This is because, by Serre-Swan's theorem, each f.g.p. module $\mathcal{E}$ is given by the space of sections of a complex vector bundle $E \to M$, and strongly non-degenerate $\mathcal{A}$-valued inner products on $\mathcal{E}$ correspond to hermitian fibre metrics on E. But any two such metrics on E are isometric.

With the assumption of complete positivity on inner products, it turns out that Rieffel induction can be carried out in an even broader setting, as we now recall.

Let $\mathcal{A}$, $\mathcal{B}$ and $\mathcal{D}$ be *-algebras over C (not necessarily unital), and let $({}_{\mathcal{B}}\mathcal{E}_{\mathcal{A}}, \langle\cdot,\cdot\rangle^{\mathcal{E}}_{\mathcal{A}})$ and $({}_{\mathcal{A}}\mathcal{H}_{\mathcal{D}}, \langle\cdot,\cdot\rangle_{\mathcal{D}})$ be right inner-product bimodules. Let ${}_{\mathcal{B}}\mathcal{E}_{\mathcal{A}} \otimes_{\mathcal{A}} {}_{\mathcal{A}}\mathcal{H}_{\mathcal{D}}$ be the algebraic tensor product over $\mathcal{A}$, seen as a $(\mathcal{B},\mathcal{D})$-bimodule in the usual way. It carries a $\mathcal{D}$-valued inner product, generalizing (5), determined by

$$\langle x \otimes \phi, y \otimes \psi\rangle^{\mathcal{E}\otimes\mathcal{H}}_{\mathcal{D}} := \left\langle \phi, \langle x, y\rangle^{\mathcal{E}}_{\mathcal{A}} \cdot \psi\right\rangle_{\mathcal{D}}, \tag{7}$$

for $x, y \in {}_{\mathcal{B}}\mathcal{E}_{\mathcal{A}}$ and $\phi, \psi \in {}_{\mathcal{A}}\mathcal{H}_{\mathcal{D}}$. The main observation is [10] :

Theorem 3.1. *If $\langle\cdot,\cdot\rangle^{\mathcal{E}}_{\mathcal{A}}$ and $\langle\cdot,\cdot\rangle_{\mathcal{D}}$ are completely positive, then $\langle\cdot,\cdot\rangle^{\mathcal{E}\otimes\mathcal{H}}_{\mathcal{D}}$ is completely positive.*

To obtain a pre-Hilbert module, we consider the quotient

$$\mathcal{E} \mathbin{\widehat{\otimes}_{\mathcal{A}}} \mathcal{H} = \mathcal{E} \otimes_{\mathcal{A}} \mathcal{H}/(\mathcal{E} \otimes_{\mathcal{A}} \mathcal{H})^{\perp}, \tag{8}$$

which now carries a nondegenerate, completely positive inner product induced by (7). So $\mathcal{E} \mathbin{\widehat{\otimes}_{\mathcal{A}}} \mathcal{H}$ is a $(\mathcal{B},\mathcal{D})$-pre-Hilbert bimodule. In fact, the tensor product $\widehat{\otimes}_{\mathcal{A}}$ defines a functor

$$\widehat{\otimes}_{\mathcal{A}} : {}^*\text{-}\mathsf{rep}_{\mathcal{A}}(\mathcal{B}) \times {}^*\text{-}\mathsf{rep}_{\mathcal{D}}(\mathcal{A}) \longrightarrow {}^*\text{-}\mathsf{rep}_{\mathcal{D}}(\mathcal{B}), \tag{9}$$

where ${}^*\text{-}\mathsf{rep}_{\mathcal{D}}(\mathcal{A})$ denotes the category of *-representations of $\mathcal{A}$ on (right) pre-Hilbert $\mathcal{D}$-modules (in other words, pre-Hilbert $(\mathcal{A},\mathcal{D})$-bimodules). Note that, by Example 3.1, part (1), if $\mathcal{D} = \mathsf{C}$, then ${}^*\text{-}\mathsf{rep}_{\mathcal{D}}(\mathcal{A})$ agrees with ${}^*\text{-}\mathsf{rep}(\mathcal{A})$, the representation category of $\mathcal{A}$ defined in Section 2.

By fixing the bimodule ${}_{\mathcal{B}}\mathcal{E}_{\mathcal{A}}$, we obtain the *Rieffel induction* functor

$$\mathsf{R}_{\mathcal{E}} = {}_{\mathcal{B}}\mathcal{E}_{\mathcal{A}} \mathbin{\widehat{\otimes}_{\mathcal{A}}} : {}^*\text{-}\mathsf{rep}_{\mathcal{D}}(\mathcal{A}) \longrightarrow {}^*\text{-}\mathsf{rep}_{\mathcal{D}}(\mathcal{B}), \tag{10}$$

which allows to compare the representation theories of $\mathcal{A}$ and $\mathcal{B}$ for any auxiliary *-algebra $\mathcal{D}$. When $\mathcal{A}$, $\mathcal{B}$ and $\mathcal{D}$ are C^*-algebras, one recovers the original construction of Rieffel after suitable topological completions.

Remark 3.1. Note that condition (1) in Proposition 3.1 coincides with property **P** used in [8] for the description of Rieffel induction; hence Proposition 3.1 relates the approaches of [10] and [8].

Example 3.2. Let $\mathcal{A}$ be a *-algebra over C, and let $\omega : \mathcal{A} \to \mathsf{C}$ be a positive linear functional. We consider $\mathcal{A}$ as an $(\mathcal{A},\mathsf{C})$-bimodule, with C-valued inner product $\langle a, b\rangle_{\omega} := \omega(a^*b)$. Although this is not strictly a pre-Hilbert bimodule according to our definition, since $\langle\cdot,\cdot\rangle_{\omega}$ may be degenerate, Rieffel induction goes through just as well. The representation of $\mathcal{A}$ induced by the canonical representation of C on itself by left multiplication is the GNS representation of Example 2.1.

4. Rieffel induction in deformation quantization

In this section we discuss examples of modules over Hermitian star products which carry completely positive inner products, and hence can be used to implement Rieffel induction in the context of deformation quantization. We start by recalling how classical and quantum positive linear functionals are related in this context.

Theorem 4.1. *Let* $\mathcal{A} := (C^\infty(M)[[\lambda]], \star)$ *be a Hermitian deformation quantization, and let* ω_0 *be a positive linear functional on* $C^\infty(M)$. *Then one can find* $\mathbb{C}$*-linear functionals* $\omega_r : \mathcal{A} \to \mathbb{C}$, $r = 1, 2, \ldots$, *so that* $\omega_0 + \sum_{r=1}^{\infty} \lambda^r \omega_r$ *is a positive linear functional of* $\mathcal{A}$.

In other words, any Hermitian star product is a *positive deformation* in the sense of [6]. A proof of this theorem can be found in [11].

We now turn our attention to examples of pre-Hilbert modules over Hermitian star products. Let $\mathcal{E}$ be a f.g.p. module over a Hermitian deformation quantization $\mathcal{A} = (C^\infty(M)[[\lambda]], \star)$, and let

$$h : \mathcal{E} \times \mathcal{E} \to \mathcal{A}, \quad (x, y) \mapsto h(x, y)$$

be an $\mathcal{A}$-valued inner product. Then $\mathcal{E}_0 := \mathcal{E}/(\lambda\mathcal{E})$ is a (f.g.p.) module over $C^\infty(M)$, and h naturally induces an inner product

$$h_0 : \mathcal{E}_0 \times \mathcal{E}_0 \to C^\infty(M)$$

by $h_0([x], [y]) := h(x, y) \mod \lambda$. We refer to h_0 as the *classical limit* of the inner product h. The next result is an analogue in deformation quantization of Example 3.1, part (4).

Theorem 4.2. *Let* $\mathcal{E}$ *be a f.g.p. module over a Hermitian deformation quantization* $\mathcal{A} = (C^\infty(M)[[\lambda]], \star)$, *let* h *be a positive, strongly nondegenerate* $\mathcal{A}$*-valued inner product on* $\mathcal{E}$. *Then* h *is completely positive, and its classical limit* h_0 *is a Hermitian fibre metric on the vector bundle* E *corresponding to* $\mathcal{E}_0$.

Proof. Let h_0 be the classical limit of h. We first observe that h_0 is a positive inner product on $\mathcal{E}_0$.

Given $x \in \mathcal{E}$, consider $h_0([x], [x]) = h(x, x) \mod \lambda \in C^\infty(M)$, and let ω_0 be a positive linear functional on $C^\infty(M)$. By Theorem 4.1, we can find a positive linear functional on $\mathcal{A}$ of the form $\omega = \omega_0 + \sum_{r \geq 1} \lambda^r \omega_r$. Since $\omega(h(x, x)) \geq 0$ in $\mathbb{C}[[\lambda]]$, we have that $\omega_0(h(x, x) \mod \lambda) \geq 0$ in $\mathbb{C}$. So $h_0([x], [x]) \geq 0$.

A direct computation shows that h_0 is strongly non-degenerate. Thus $(\mathcal{E}_0, h_0)$ comes from a vector bundle E over M carrying a Hermitian fibre metric h_0, and $(\mathcal{E}, h)$ is an example of a deformation quantization of a Hermitian vector bundle in the sense of [5] . By [10] , it follows that $(\mathcal{E}, h)$ is isometric to an $\mathcal{A}$-module as the one in Example 3.1, part (3). Hence h is completely positive. □

We note that checking that the classical limit h_0 is strongly nondegenerate is sufficient to guarantee the strong nondegeneracy of h. In particular, according to [5], any Hermitian vector bundle over M can be deformed into a pre-Hilbert module over $\mathcal{A}$ which can be used for the construction of induced representations. In this context, line bundles over M play a special role. This is because a deformation of a line bundle $L \to M$ with respect to a star product $\star$ defines a pre-Hilbert *bimodule* for $\star$ and another deformation quantization $\star'$ of M. If M is symplectic, the relationship between $\star$ and $\star'$ is that the difference of their characteristic classes (in the sense of e.g. [12]) is $2\pi i c_1(L)$ [9], where $c_1(L)$ denotes the first Chern class of L. One can then use Rieffel induction to transfer representations from one quantization to the other.

An interesting physical example is discussed in [9], where it is shown that the formal representations of star products on cotangent bundles with a "magnetic term" studied in [2] can be obtained by Rieffel induction of the formal Schrödinger representation of the standard Weyl star product. Here, the pre-Hilbert bimodule used to implement the induction is a deformation of the line bundle associated with a magnetic charge satisfying Dirac's quantization condition; see [18] for a detailed physical discussion of this example.

5. A unified view of strong Morita equivalence

We now briefly recall how to obtain an equivalence of categories of representations using the functor (10). This leads to a generalization of the notion of strong Morita equivalence in C^*-algebras to the algebraic framework of Section 2; details can be found in [10].

Definition 5.1. Let $\mathcal{A}$, $\mathcal{B}$ be *-algebras over C and ${}_{\mathcal{B}}\mathcal{E}_{\mathcal{A}}$ a $(\mathcal{B}, \mathcal{A})$-bimodule so that $\mathcal{B} \cdot \mathcal{E} = \mathcal{E}$ and $\mathcal{E} \cdot \mathcal{A} = \mathcal{E}$. Suppose that $\mathcal{E}$ is equipped with completely positive and non-degenerate inner products $\langle \cdot, \cdot \rangle_{\mathcal{A}}$ and ${}_{\mathcal{B}}\langle \cdot, \cdot \rangle$ such that

(1) $\langle b \cdot x, y \rangle_{\mathcal{A}} = \langle x, b^* \cdot y \rangle_{\mathcal{A}}$,

(2) ${}_{\mathcal{B}}\langle x \cdot a, y\rangle = {}_{\mathcal{B}}\langle x, y \cdot a^*\rangle$,
(3) ${}_{\mathcal{B}}\langle x, y\rangle \cdot z = x \cdot \langle y, z\rangle_{\mathcal{A}}$,
(4) $\mathsf{C}\text{-span}\,\{\, \langle x, y\rangle_{\mathcal{A}} \mid x, y \in \mathcal{E}\} = \mathcal{A}$,
(5) $\mathsf{C}\text{-span}\,\{\, {}_{\mathcal{B}}\langle x, y\rangle \mid x, y \in \mathcal{E}\} = \mathcal{B}$.

Then ${}_{\mathcal{B}}\mathcal{E}_{\mathcal{A}}$ is called a *strong Morita equivalence bimodule.* If there exists such a bimodule then $\mathcal{A}$ and $\mathcal{B}$ are called *strongly Morita equivalent.*

As discussed in [10], [7], one recovers Rieffel's notion of strong Morita equivalence of C^*-algebras from this purely algebraic definition by passing to minimal dense ideals.

The following theorem summarizes some of the properties of strong Morita equivalence that have well-known counterparts in ring theory and C^*-algebra theory.

Theorem 5.1.

*(1) Strong Morita equivalence is an equivalence relation among nondegenerate and idempotent *-algebras over* C.

(2) If ${}_{\mathcal{B}}\mathcal{E}_{\mathcal{A}}$ *is a strong equivalence bimodule then the Rieffel induction functor*

$$ {}_{\mathcal{B}}\mathcal{E}_{\mathcal{A}} \mathbin{\widehat{\otimes}_{\mathcal{A}}} : {}^*\text{-}\mathsf{Rep}_{\mathcal{D}}(\mathcal{A}) \longrightarrow {}^*\text{-}\mathsf{Rep}_{\mathcal{D}}(\mathcal{B}) \tag{11} $$

*establishes an equivalence of categories for any fixed *-algebra* $\mathcal{D}$.

(3) If ${}_{\mathcal{B}}\mathcal{E}_{\mathcal{A}}$ *is a strong Morita equivalence bimodule for unital *-algebras* $\mathcal{A}$ *and* $\mathcal{B}$*, then there exist Hermitian dual bases* (ξ_i, η_i) *and* (x_j, y_j)*, respectively, such that*

$$ x = \sum_{i=1}^{n} \xi_i \cdot \langle \eta_i, x\rangle_{\mathcal{A}} = \sum_{j=1}^{m} {}_{\mathcal{B}}\langle x, y_j\rangle \cdot x_j \tag{12} $$

for all $x \in {}_{\mathcal{B}}\mathcal{E}_{\mathcal{A}}$*. In particular,* ${}_{\mathcal{B}}\mathcal{E}_{\mathcal{A}}$ *is finitely generated and projective as right* $\mathcal{A}$*-module and also as left* $\mathcal{B}$*-module.*

(4) If $\mathcal{A}$ *and* $\mathcal{B}$ *are unital, then strong Morita equivalence implies ring-theoretic Morita equivalence.*

Some comments are in order. The crucial point in (1) is to show transitivity, which relies on the fact that completely positive inner products behave well under tensor products, see Theorem 3.1; in (2), ${}^*\text{-}\mathsf{Rep}_{\mathcal{D}}(\mathcal{A})$ denotes the subcategory of ${}^*\text{-}\mathsf{rep}_{\mathcal{D}}(\mathcal{A})$ consisting of pre-Hilbert bimodules ${}_{\mathcal{A}}\mathcal{H}_{\mathcal{D}}$ satisfying the extra nondegeneracy condition $\mathcal{A}\mathcal{H} = \mathcal{H}$; property (3) essentially implies (4) and is also used to show that the inner products on equivalence bimodules of unital *-algebras are strongly nondegenerate.

In [10], we describe a class of unital *-algebras, including both unital C^*-algebras and Hermitan deformation quantizations, for which the converse of part (3) holds, i.e., strong and ring-theoretic Morita equivalences define the same equivalence relation. The comparison between these two types of Morita equivalence becomes more interesting at the level of Picard group(oid)s, see [10] for a discussion.

The fact that, for star products, strong Morita equivalence coincides with Morita equivalence in the classical sense of ring theory is used in [9] to classify strong Morita equivalent Hermitian deformation quantizations on symplectic manifolds. An interesting problem is to investigate the precise connection between Morita equivalence for star products and their counterparts in C^*-algebraic versions of deformation quantization.

Acknowledgments

It is a pleasure to thank the organizers of the Keio workshop for the invitation and the wonderful working atmosphere. We also thank the participants for valuable discussions and remarks.

References

1. F. Bayen, M. Flato, C. Frønsdal, A. Lichnerowicz and D. Sternheimer, *Deformation Theory and Quantization.* Ann. Phys. **111** (1978), 61–151.
2. M. Bordemann, N. Neumaier, M. J. Pflaum and S. Waldmann, *On representations of star product algebras over cotangent spaces on Hermitian line bundles.* J. Funct. Anal. **199** (2003), 1–47.
3. M. Bordemann, N. Neumaier and S. Waldmann, *Homogeneous Fedosov star products on cotangent bundles II: GNS representations, the WKB expansion, traces, and applications.* J. Geom. Phys. **29** (1999), 199–234.
4. M. Bordemann and S. Waldmann, *Formal GNS Construction and States in Deformation Quantization.* Commun. Math. Phys. **195** (1998), 549–583.
5. H. Bursztyn and S. Waldmann, *Deformation Quantization of Hermitian Vector Bundles.* Lett. Math. Phys. **53** (2000), 349–365.
6. H. Bursztyn and S. Waldmann, *On Positive Deformations of *-Algebras.* In: Dito, G., Sternheimer, D. (eds.): *Conférence Moshé Flato 1999. Quantization, Deformations, and Symmetries, Mathematical Physics Studies* no. **22**, 69–80. Kluwer Academic Publishers, Dordrecht, Boston, London, 2000.
7. H. Bursztyn and S. Waldmann, **-Ideals and Formal Morita Equivalence of *-Algebras.* Int. J. Math. **12**.5 (2001), 555–577.

8. H. Bursztyn and S. Waldmann, *Algebraic Rieffel Induction, Formal Morita Equivalence and Applications to Deformation Quantization.* J. Geom. Phys. **37** (2001), 307–364.
9. H. Bursztyn and S. Waldmann, *The characteristic classes of Morita equivalent star products on symplectic manifolds.* Commun. Math. Phys. **228** (2002), 103–121.
10. H. Bursztyn and S. Waldmann, *Completely positive inner products and strong Morita equivalence.* Preprint (FR-THEP 2003/12) **math.QA/0309402** (September 2003), 36 pages. To appear in Pacific J. Math.
11. H. Bursztyn and S. Waldmann, *Hermitian star products are completely positive deformations.* Preprint (FR-THEP 2004/18) **math.QA/0410350** (October 2004), 8 pages. To appear in Letters in Mathematical Physics.
12. S. Gutt and J. Rawnsley, *Equivalence of star products on a symplectic manifold; an introduction to Deligne's Čech cohomology classes.* J. Geom. Phys. **29** (1999), 347–392.
13. E. C. Lance, *Hilbert C^*-modules. A toolkit for operator algebraists*, vol. 210 in *London Mathematical Society Lecture Note Series.* Cambridge University Press, Cambridge, 1995.
14. I. Raeburn and D. P. Williams, *Morita equivalence and continuous-trace C^*-algebras*, vol. 60 in *Mathematical Surveys and Monographs.* American Mathematical Society, Providence, RI, 1998.
15. M. A. Rieffel, *Induced representations of C^*-algebras.* Adv. Math. **13** (1974), 176–257.
16. K. Schmüdgen, *Unbounded Operator Algebras and Representation Theory*, vol. 37 in *Operator Theory: Advances and Applications.* Birkhäuser Verlag, Basel, Boston, Berlin, 1990.
17. S. Waldmann, *The Picard Groupoid in Deformation Quantization.* Lett. Math. Phys. **69** (2004), 223–235.
18. S. Waldmann, *States and Representation Theory in Deformation Quantization.* Rev. Math. Phys. **17** (2005), 15–75.

ζ-FUNCTION REGULARIZATION AND INDEX THEORY IN NONCOMMUTATIVE GEOMETRY

ALEXANDER CARDONA *

Mathematics Department
Faculty of Science and Technology
Keio University
3-14-1, Hiyoshi, Kohoku-ku
Yokohama, 223-8522, Japan.
acardona@uniandes.edu.co.

In this paper we review the ζ-function regularization approach to noncommutative index theory. In particular, we show how, through the use of a suitable generalization of ζ-function regularized quantities (as the weighted traces used in[4,5,19]) it is possible to build the basic blocks used to compute the local index formula due to Connes and Moscovici[8] (and revisited by Higson[12]) in noncommutative geometry.

Introduction

The index map associated to a spectral triple $(\mathcal{A}, \mathcal{H}, D)$ in Noncommutative Geometry[6,8,12] is the additive map $\text{ind}_D : K_*(\mathcal{A}) \to \mathbb{Z}$ defined as follows. Let us assume that the spectral triple is *even*, which means that there exists a chirality operator $\gamma \in \mathcal{L}(\mathcal{H})$ inducing a $\mathbb{Z}_2$-grading on $\mathcal{H}$ and anticommuting with D, i.e. $\gamma^2 = 1$, $\gamma = \gamma^*$, $\gamma D = -D\gamma$ and $\gamma a = a\gamma$, for all $a \in \mathcal{A}$. With respect to the decomposition $\mathcal{H} = \mathcal{H}^+ \oplus \mathcal{H}^-$, D takes the antidiagonal form

$$D = \begin{pmatrix} 0 & D^- \\ D^+ & 0 \end{pmatrix},$$

with $D^\pm : \mathcal{H}^\mp \to \mathcal{H}^\pm$. For any selfadjoint idempotent $e \in M_q(\mathcal{A})$ (the algebra of $q \times q$ matrices with entries in $\mathcal{A}$), the operator

$$e(D^+ \otimes 1)e : e(\mathcal{H}^+ \otimes \mathbb{C}^q) \to e(\mathcal{H}^- \otimes \mathbb{C}^q)$$

*Present address: Mathematics Department, Universidad de Los Andes, A.A. 4976, Bogotá, Colombia.

is Fredholm, and its index depends only on the homotopy class of e. The index is defined by

$$\text{ind}_D[e] = \text{ind}\, e(D^+ \otimes 1)e. \qquad (1)$$

It is well known that, e.g. in the case of a compact Riemannian spin manifold (M,g) of even dimension, with D_M the Dirac operator acting on its spin bundle $S = S^+ \oplus S^-$, the index corresponding to the even spectral triple $(C^\infty(M), L^2(M,S), D_M)$ coincides with the classical index. In this case, as shown in the sixties by Atiyah and Singer (see the volumes[1]), the index can be computed through a local formula.

This *locality* property of the index generalizes to the noncommutative case, and the corresponding local index formula was shown in [8] by Connes and Moscovici. Indeed, (under some –spectral– conditions on the spectral triple) the index map (1) can be computed through the pairing of the K-theory $K(\mathcal{A})$ of the algebra with a cyclic cohomology class defined by Connes.[6] The local index theorem states that there exists a cyclic cohomology class $[\varphi_{CM}]$ such that

$$\text{ind}_D(E) = \langle [\varphi_{CM}] | E \rangle, \qquad (2)$$

for any $E \in K_0(\mathcal{A})$, and the components of such an even cocycle $\varphi_{CM} = (\varphi_{2p})$ —in the (b,B)-complex of $\mathcal{A}$— are given by local formulas, i.e. by (noncommutative) integrals which generalize the ones found in classical (e.g. Atiyah-Singer) index theorems.

Another feature of classical indices is that it is possible to compute them using ζ-function regularized quantities[2], a method extensively used to define other important invariants of differential manifolds. On the other hand, it is possible to study both algebraic and geometric anomalies appearing in infinite-dimensional geometry through the use of, e.g. trace extensions defined using this kind of regularized quantities (see [17],[4]). This approach brings a direct proof of the relation between such anomalies and index-like quantities, as well as a bridge to understand how these quantities appear in the study of related phenomena such as Quantum Field Theory anomalies[5]. In noncommutative geometry [6],[10] ζ-type functions are also used to state what would correspond to some classical definitions in the noncommutative case, e.g. what would be the *dimension* of a spectral triple.

Our goal in the following pages is to show how, through the use of a suitable generalization of ζ-function regularized quantities (as the used in[4,5,17,19]) it is possible to build the basic blocks used to compute the local index formula in noncommutative geometry. We follow similar lines to those of Higson[12] and Paycha[18], reviewing some of the results stated in these works. Our approach contains also some common ideas with other works employing regularization by heat-kernel methods, as the work of Ponge[20] on local index formulas. Finally, throughout this paper we consider only *even* spectral triples, although everything applies to the odd case by the usual modifications (see e.g. [6], [8]).

1. Spectral Triples and ζ-functions

1.1. *The weighted algebra of spectral triples*

Let $(\mathcal{A}, \mathcal{H}, D)$ be a spectral triple, i.e. an involutive algebra $\mathcal{A}$ represented in a Hilbert space $\mathcal{H}$, together with a self-adjoint operator D with compact resolvent in $\mathcal{H}$ such that $[D, a]$ is bounded for any $a \in \mathcal{A}$. Given a fixed positive unbounded self-adjoint operator Q acting on $\mathcal{H}$ —which we refer to as the *weight*— such that

$$\mathcal{H}^\infty = \bigcap_{k=1}^{\infty} \mathsf{Dom}\,(Q^k) \subseteq \mathcal{H},$$

and any $a \in \mathcal{A}$ maps $\mathcal{H}^\infty$ into itself, we want to associate to Q and $(\mathcal{A}, \mathcal{H}, D)$ an algebra of operators on $\mathcal{H}^\infty$.

We say that $\mathcal{D}_Q$ is a *q-order weighted algebra of abstract differential operators*, and Q its associated *weight* of order q (see [12] and [18]), if it is an associative algebra of linear operators on $\mathcal{H}^\infty$ which is

(1) *Filtered*, i.e.

$$\mathcal{D}_Q = \bigcup_{p=0}^{\infty} \mathcal{D}_Q^p,$$

and we say $\mathsf{ord}(P) \leq p$ if $P \in \mathcal{D}_Q^p$.

(2) *Closed under commutation with the weight*, i.e. for any $P \in \mathcal{D}_Q^p$,

$$[P, Q] \in \mathcal{D}_Q,$$

and $\mathsf{ord}([P, Q]) \leq p + q - 1$.

(3) *Regular*,[11] i.e. it satisfies the following 'elliptic estimate': For any $P \in \mathcal{D}_Q^p$ there exists a positive constant ϵ such that

$$\| Q^{\frac{p}{q}} X \| + \| X \| \geq \epsilon \| PX \|, \qquad \forall X \in \mathcal{H}^\infty.$$

Notice that, as follows from the definition, if $P \in \mathcal{D}_Q^p$, for any $s \geq 0$ then P extends to a bounded operator from $\mathcal{H}^{s+p}$ to $\mathcal{H}^s$.

As shown by Higson (Section 4 in [12], see also [13]), for any regular spectral triple $(\mathcal{A}, \mathcal{H}, D)$ as above, taking as weight $Q = \Delta = D^2$, it is possible to associate a $2\,\mathsf{ord}(D)$-weighted algebra $\mathcal{D}_\Delta$: the smallest algebra of operators on $\mathcal{H}^\infty$ containing $\mathcal{A}$, the commutator $[\mathcal{A}, D]$ which is closed under commutation with the weight $P \mapsto [Q, P]$. The filtration is, in this case, the one in which elements of $\mathcal{A}$ and $[\mathcal{A}, D]$ have order zero, and the order is raised by (at most) one by commutation with the weight D^2 —it is assumed that the degree of D is 1. An immediate example is the one of the spectral triple $(C^\infty(M), L^2(M, S), D_M)$ associated to a compact Riemannian manifold (M, g), taking $\mathcal{D}_\Delta$ to be the algebra of differential operators acting on spinors, weighted by the Laplacian $\Delta = D_M^2$.

The role of the weight in the context of noncommutative spectral triples will be, as we will see in the next section, the one of parameter for the regularization of the quantities in terms of which the local formula for the index can be written. Indeed, the preliminary definitions introduced in this section are there to ensure that we can define the ζ-functions we will work with in the following.

1.2. *ζ-function regularization, traces and dimension spectrum*

In classical (commutative) geometry the ζ-function regularization method gives rise to important invariants of differential manifolds. It is also used, among other things, for the construction of a —unique in many cases— trace on the algebra of pseudo-differential operators acting on sections of vector fibrations. In this section we recall some of these classical results, as well as the definition of the dimension spectrum of a spectral triple given in [8] using these methods.

1.2.1. *ζ-function regularization and the noncommutative residue*

Let E be a vector bundle over a smooth n-dimensional closed Riemannian manifold M, and let $Cl(E)$ denote the algebra of classical pseudo-differential operators acting on smooth sections of E. Let $Q \in \mathcal{A}d(E)$ be an *admissible operator* —in the sense of references [16] and [5], i.e. a pseudo-differential operator for which arbitrary complex powers can be defined as it was by Seeley[21],

$$Q^{-z} = \frac{1}{2\pi i}\int_\Gamma \lambda^{-z}(Q-\lambda I)^{-1}d\lambda, \tag{3}$$

where Γ denotes the contour of integration, coming from infinity and separating zero from the spectrum of Q. Since, for any $A \in Cl(E)$, $Q \in \mathcal{A}d(E)$, the map $z \mapsto \mathrm{tr}(AQ^{-z})$ extends meromorphically to a map with a simple pole at zero,[16]

$$\mathsf{res}(A) = q\,\mathrm{Res}_{z=0}\left(\mathrm{tr}(AQ^{-z}))\right), \tag{4}$$

where $q = \mathsf{ord}(Q)$, defines a quantity which is independent of the reference operator Q. More importantly, this quantity indeed defines a trace (i.e. $\mathsf{res}([A, B])=0$ for any $A, B \in Cl(E)$), the so-called *non-commutative* or *Wodzicki residue*[22] (see also [15] for a review). It also has the remarkable locality property

$$\mathsf{res}(A) = \frac{1}{(2\pi)^n}\int_M\int_{|\xi|=1} \mathrm{tr}_x\left(\sigma_{-n}(x,\xi)\right)d\xi d\mu_M(x), \tag{5}$$

where n is the dimension of M, μ_M the volume measure on M, tr_x the trace on the fibre above x and σ_{-n} the homogeneous component of order $-n$ of the symbol of A. Notice that the Wodzicki residue is *not* an extension of the trace in finite dimensions, as follows from the fact that it vanishes on any finite-rank operator.

Remark 1.1. The meromorphic extension of the map $z \mapsto \mathrm{tr}(AQ^{-z})$ has simple poles located in the set[16] $\left\{\frac{a+m-k}{q}; k \in \mathbb{Z}^+\right\}$, where $a = \mathsf{ord}(A)$, $q = \mathsf{ord}(Q)$ and $m = \dim M$. Notice then that, if $a < -m$ there is no pole at zero, so that $\mathsf{res}(A) = 0$.

1.2.2. *Weighted traces*

The same ζ-function regularization employed to define the Wodzicki residue has been used to define functionals which, although non-tracial, extend the usual trace in finite dimensions[a] (see [4,5,17]). Indeed, starting from the same map $z \mapsto \mathrm{tr}(AQ^{-z})$, for A classical pseudo-differential operator and Q admissible, but considering its finite part instead of its residue at the origin, gives rise to *weighted trace functionals* [4] which, although non-tracial and dependent of the reference operator Q, are very useful tools to understand the origin of index-like terms and their appearance in Quantum Field Theory anomalies, which have well known locality features in relation with classical (commutative) index theorems.[5] For $A \in Cl(E)$ and $Q \in \mathcal{A}d(E)$, the Q-weighted trace of A is defined by the expression

$$\mathrm{tr}^Q(A) = \mathrm{f.p.}|_{z=0}\left(\mathrm{tr}(AQ^{-z})\right). \tag{6}$$

It follows from the definition that weighted traces extend usual finite-dimensional traces, i.e. $\mathrm{tr}^Q(A) = \mathrm{tr}(A)$ whenever A is a finite rank operator. Formulae for anomalies associated with weighted trace functionals can be found in references [4] and [5], where their relationship with index theory and field theory anomalies are also discussed.

Remark 1.2. Another very important regularization method used in infinite-dimensional geometry is the *heat kernel method*, in which the map $\mathrm{tr}(e^{-tQ})$, for t real and positive, is used instead of the complex map $\mathrm{tr}(Q^{-z})$ considered above, where Q denotes any admissible reference operator with positive leading symbol. Both methods coincide modulo noncommutative residues. Indeed, notice that, for $A \in Cl(E)$ and an admissible operator Q with positive leading symbol, we can recover the ζ-regularized trace (6) using a heat-kernel expansion, through a *Mellin Transform*. The zeta regularized trace $\mathrm{tr}(AQ^{-z})$ is the Mellin transform $\mathrm{M}[f](z) = \frac{1}{\Gamma(z)}\int_0^\infty f(t)t^{z-1}\,dt$ of the map $f(t) = \mathrm{tr}(Ae^{-tQ}) \in C^\infty(\mathbb{R}^+)$. Since $f(t)$ has an asymptotic expansion for small t of the form $f(t) \sim \sum_{k\geq -n} f_k t^{\frac{k}{q}} + c\log t$, where $q = \mathsf{ord}(Q)$, properties of the Mellin transform [10] imply that

$$\mathrm{f.p.}|_{t=0}\left(\mathrm{tr}(Ae^{-tQ})\right) = \mathrm{f.p.}|_{z=0}\left(\mathrm{tr}(AQ^{-z})\right) + \gamma_E \cdot \mathsf{res}(A) \tag{7}$$

where γ_E is the Euler constant. Thus, if $\mathsf{res}(A) = 0$, the two regularization methods coincide, i.e. $\mathrm{tr}^Q(A) = \mathrm{f.p.}|_{t=0}\left(\mathrm{tr}(Ae^{-tQ})\right)$.

[a]Traces and trace extensions are not the only objects it is possible to build by this method. There are also, among others, the ζ-determinants giving rise to analytic torsion and the eta invariant of Atiyah, Patodi and Singer.

In the more general context of noncommutative spectral triples there is a trace (defined by Dixmier[9] in the 60's on certain ideals of linear operators acting on a Hilbert space) which has been extensively used by Alain Connes as noncommutative integral. [6] Indeed, it coincides with the Wodzicki residue on the algebra of classical pseudodifferential operators acting on smooth sections of a vector bundle over a Riemannian manifold, so that in that case it is given by an integral. The index formulas arising in Connes-Moscovici theorem are local in the sense that they can be written using noncommutative integrals (Dixmier traces), we want to illustrate in the following how they can also be build from ζ-function regularized quantities.

1.2.3. *ζ-functions, analytic dimension and the dimension spectrum*

Let $(\mathcal{A}, \mathcal{H}, D)$ be a spectral triple such that $\mathcal{H}^\infty = \bigcap_{k=1}^\infty \mathsf{Dom}\,(\Delta^k) \subseteq \mathcal{H}$, and any $a \in \mathcal{A}$ maps $\mathcal{H}^\infty$ into itself. Let $\mathcal{D}_\Delta$ be the order 2 weighted algebra of abstract differential operators considered in Section 1.1, where $\Delta = D^2$ denotes its associated weight[b] of order 2. An abstraction of the usual properties of the algebra of classical pseudodifferential operators (e.g. the relative to ζ-functions given in Remarks 1.1 and 1.2) suggests what should be the idea of (algebraic and geometric) dimension of a given spectral triple. Let us begin by recalling Higson's definition of analytic dimension for a weighted algebra of abstract differential operators.

Definition 1.1. Let $\mathcal{D}_Q$ be a order q weighted algebra of abstract differential operators. The *analytic dimension* d of $\mathcal{D}_Q$ is the smallest value —provided it exists— $d \geq 0$ for which PQ^{-z} extends to a trace class operator on $\mathcal{H}$, for $P \in \mathcal{D}_Q^p$ and $z \in \mathbb{C}$ such that $Re(z) > \frac{p+d}{q}$. If $\mathcal{D}_Q$ has finite analytic dimension and, for any $P \in \mathcal{D}_Q$, the function $\mathrm{tr}(PQ^{-z})$ extends to a meromorphic function on $\mathbb{C}$, the algebra $\mathcal{D}_Q$ is said to have the *analytic continuation property.*

Now, a spectral triple $(\mathcal{A}, \mathcal{H}, D)$ such that any $a \in \mathcal{A}$ maps $\mathcal{H}^\infty$ into itself is said *p-summable* if [10] for some finite $p \geq 1$

$$a \cdot (1 + D^2)^{-1/2} \in \mathcal{L}^p(\mathcal{H}) \qquad \text{for any } a \in \mathcal{L}^p(\mathcal{H}). \tag{8}$$

Consequently, for any element $P \in \mathcal{D}_\Delta$ there exists some half-plane on which the ζ-function $z \mapsto \mathrm{tr}(P\Delta^{-z})$ is holomorphic. In [8] the following

[b]Here, and in what follows, we assume Δ to be invertible, if it is not the case it should be replaced by $\Delta' = \Delta + \Pi_{\ker \Delta}$.

definition is given for the "dimension" of a spectral triple satisfying the above conditions:

Definition 1.2. The *dimension spectrum* Sd_Δ of a finite summable and regular spectral triple $(\mathcal{A}, \mathcal{H}, D)$, is the subset of $\mathbb{C}$ such that, for any $P \in \mathcal{D}^p_\Delta$, the zeta function

$$\zeta_P(z) = \operatorname{tr}(P\Delta^{-\frac{z}{2}})$$

extends holomorphically to $\mathbb{C} \setminus \mathsf{Sd}_\Delta$.

As follows from Remark 1.1, in the case of a compact Riemannian spin manifold (M, g) of even dimension with its associated even spectral triple $(C^\infty(M), L^2(M, S), D_M)$, the corresponding dimension spectrum is discrete and, moreover, the poles of the corresponding ζ-functions are simple.

Whenever the functions $\zeta_P(z)$, for $P \in \mathcal{D}^p_\Delta$, have a discrete set of poles and such poles are at most simple poles, we call the dimension spectrum *discrete* and *simple*, respectively. Examples of spectral triples with these properties have been discussed in, e.g. Ref. [7].

Under the assumption of simplicity and discreteness for the dimension spectrum associated to a finite summable and regular spectral triple, it is possible to define a pseudodifferential calculus associated to the generalized algebra $\mathcal{D}_\Delta$,[12] and generalize on it the ζ-function machinery used in the analysis of classical pseudodifferential operators. In Section 2.2 we study such objects and we recall how to write local formulas, for the index associated to a noncommutative spectral triple, in terms of such ζ-function quantities.

2. ζ-function regularization and the Connes-Moscovici cocycle

2.1. *The local index formula for even spectral triples*

Recall that a cochain ψ in the spaces $C^k(\mathcal{A})$ of $(k+1)$-linear forms on $\mathcal{A}$, for $k \in \mathbb{N}$, is called *cyclic* if it satisfies

$$\psi(a_1, \cdots, a_k, a_0) = (-1)^k \psi(a_0, a_1, \cdots, a_k) \qquad a_j \in \mathcal{A}, \qquad (9)$$

a *cyclic cocycle* is a cyclic cochain ψ such that $b\psi = 0$, and the *cyclic cohomology groups* $HC^*(\mathcal{A})$ of the algebra $\mathcal{A}$ are obtained from by restricting the Hochschild coboundary,

$$b\psi(a_0, \cdots, a_{k+1}) = \sum_{j=0}^{k}(-1)^j \psi(a_0, \cdots, a_j a_{j+1}, \cdots, a_{k+1}) \\ + (-1)^{k+1}\psi(a_{k+1}a_0, \cdots, a_k), \qquad a_j \in \mathcal{A}, \quad (10)$$

to those cochains. It can equivalently be described as the second filtration of the (b, B)-bicomplex of (not necessarily cyclic) cochains, where b is the same as before and $B = AB_0 : C^k(\mathcal{A}) \to C^{k-1}(\mathcal{A})$, with

$$(A\psi)(a_0, \cdots, a_{k-1}) = \sum_{j=0}^{k}(-1)^{(k-1)j}\psi(a_j, \cdots, a_{j-1}),$$

and $B_0\psi(a_0, \cdots, a_{k-1}) = \psi(1, a_0, \cdots, a_{k-1})$, for $a_j \in \mathcal{A}$. The *periodic cyclic cohomology* is the cohomology of the short complex

$$C^{even}(\mathcal{A}) \overset{b+B}{\leftrightarrows} C^{odd}(\mathcal{A}), \qquad C^{even/odd}(\mathcal{A}) = \bigoplus_{k \text{ even/odd}} C^k(\mathcal{A}),$$

whose cohomology groups are denoted $HC^+(\mathcal{A})$ and $HC^-(\mathcal{A})$, respectively.

The pairing between $HC^+(\mathcal{A})$ and $K_0(\mathcal{A})$ used to compute the local formula for the index (1) is given by [6, 10]

$$\langle[\varphi]|[e]\rangle = \sum_{k\geq 0}(-1)^k \frac{(2k)!}{k!}(\varphi_{2k}\#\mathrm{tr})(e, \cdots, e), \qquad (11)$$

for any cocycle $\varphi = (\varphi_{2k})$ in $C^{even}(\mathcal{A})$ and any self-adjoint idempotent e in $M_q(\mathcal{A})$. Here $\varphi_{2k}\#\mathrm{tr}$ denotes the $(2k+1)$-linear map on $M_q(\mathcal{A}) = M_q(\mathbb{C}) \otimes \mathcal{A}$ given by

$$(\varphi_{2k}\#\mathrm{tr})(\mu^0 \otimes a_0, \cdots, \mu^{2k} \otimes a_{2k}) = \mathrm{tr}(\mu^0 \ldots \mu^{2k})\varphi_{2k}(a_0, \cdots, a_{2k}),$$

for $\mu^j \in M_q(\mathbb{C})$ and $a_j \in \mathcal{A}$.

The following theorem provides a local formula to compute the index in even noncommutative spectral triples:

Theorem 2.1. [8] *Suppose that $(\mathcal{A}, \mathcal{H}, D)$ is even, finitely summable and has a discrete and simple dimension spectrum. Then, for any $E \in K_0(\mathcal{A})$,*

$$\mathrm{ind}\,_D(E) = \langle[\varphi_{CM}^+]|E\rangle, \qquad (12)$$

where the even cocycle $\varphi_{CM}^{+} = (\varphi_{2k})$ *in the* (b, B)*-complex of* $\mathcal{A}$ *is given, for* $k = 0$*, by*

$$\varphi_0(a_0) = \text{f.p.}|_{t=0}\text{str}(a_0 e^{-t\Delta}), \tag{13}$$

and for $k > 0$ *by*

$$\varphi_{2k}\ (a_0, \ldots, a_{2k}) = \sum_{\alpha\geq 0} c_{k,\alpha}\text{Res}_{z=0}Tr\left(\gamma a_0[D, a_1]^{[\alpha_1]}\ldots[D, a_{2k}]^{[\alpha_{2k}]}\Delta^{-|\alpha|-k-z}\right), \tag{14}$$

where the sum is over the multi-index $\alpha = (\alpha_1, \ldots, \alpha_k)$*, with* $\alpha_j \geq 0$*, and*

$$c_{k,\alpha} = \frac{(-1)^{|\alpha|}\Gamma(|\alpha| + k)}{\alpha!(\alpha_1 + 1)\cdots(\alpha_1 + \cdots + \alpha_{2k} + 2k)}, \tag{15}$$

and the symbol $X^{[j]}$ *denotes the* j*-th iterated commutator of* X *with* $\Delta = D^2$.

2.2. *Weighted cochains and* ζ*-function cyclic cocicles*

2.2.1. *JLO Functionals*

Let $(\mathcal{A}, \mathcal{H}, D)$ be an even spectral triple, with chirality operator γ (with respect to the decomposition $\mathcal{H} = \mathcal{H}^+ \oplus \mathcal{H}^-$) and Q a weight in the sense of definition given above. Let us assume that the spectral triple is Q-summable, i.e. that for any $t > 0$ the operator e^{-tQ} is trace class [c]. A natural extension of the functional (6) to the context of noncommutative spectral triples can be done by considering the Mellin transform of the following JLO-type multifunctional [14]

$$\phi_p^Q(t)(A_0, \ldots, A_p) = \int_{\Delta_p} \text{tr}(\gamma A_0 e^{-u_0 tQ} A_1 e^{-u_1 tQ} \cdots A_p e^{-u_p tQ})du, \tag{16}$$

where $t \in \mathbf{R}^+$, $p \in \mathbb{Z}^+$, $A_0, \ldots, A_p \in \mathcal{A}$ and the integration is over the k-simplex $\Delta_p = \{(u_0, \ldots, u_p) : u_0 + \cdots + u_p = 1, u_i \geq 0\}$. This functional, defined using heat kernel regularization methods, was originally used in [14] in order to obtain an infinite-dimensional Chern character for (θ-summable) Fredholm modules. It is also at the origin of the local formula for the index given by Connes and Moscovici.[8]

---

[c]For $Q = D^2$ the spectral triple is called θ-summable.

Looking at (16) as a function in $C^\infty(\mathbf{R}^+)$ associated to any set of p elements of $\mathcal{A}$, it is natural to consider its Mellin transform

$$\Phi_p^Q(A_0,\dots,A_p)(z) = \frac{1}{\Gamma(z)}\int_0^\infty \phi_p^Q(t)(A_0,\dots,A_p)t^{z-1}\,dt, \tag{17}$$

i.e. a complex function associated to any set of p elements of $\mathcal{A}$, defined in a region of the complex plane. As a matter of fact, in [12] it has been shown that this object is well defined whenever $A_0,\dots,A_p \in \mathcal{D}_Q$, the associated weighted algebra, if it has finite analytic dimension and satisfies the analytic continuation property defined above (see Section 4 in [12] for a careful exposition), properties which we assume in what follows. As prompted by Remark 1.2 —which is only valid in the classical pseudodifferential algebra setting, it is possible to write down a complex power expression for $\Phi_\bullet^Q(z)$.

Lemma 2.1. [12] *Let $(\mathcal{A},\mathcal{H},D)$ be a spectral triple with associated algebra of generalized differential operators $\mathcal{D}_Q$, satisfying the analytic continuation property. Then, for $a_0,\dots,a_p \in \mathcal{A}$, $\Phi_p^Q(a_0,\dots,a_p)(z)$ defines a meromorphic function on $\mathbb{C}$, and*

$$\Phi_p^Q(a_0,\dots,a_p)(z) = \frac{(-1)^p}{2\pi i}\int \lambda^{-z+p}\,\mathrm{tr}(\gamma a_0(\lambda-Q)^{-1}\cdots a_p(\lambda-Q)^{-1})d\lambda, \tag{18}$$

where the integral is performed following a line in the complex plane which separates the spectrum of Q from 0. Moreover, if p is bigger than the analytic dimension of $\mathcal{D}_Q$, and $Q=\Delta=D^2$, $\Phi_p^Q(a_0,\dots,a_p)(z+\frac{p}{2})$ is holomorphic at zero.

This crucial result follows from some asymptotic expansion properties of commutators between elements of $\mathcal{D}_Q$ and resolvents of the weight (see Lemma 4.20 in [12]), and gives us some clues about the residue generalizing the noncommutative one in the context of classical pseudodifferential algebras. First, notice that in the context of classical pseudodifferential operator algebras, when $p=0$, this map (17) is the one taking $z\in\mathbb{C}$ to

$$\Phi_0^Q(A)(z) = \mathrm{str}(AQ^{-z}), \tag{19}$$

where $\mathrm{str}(A)=\mathrm{tr}(\gamma A)$ denotes the usual supertrace. Thus, it seems natural to consider the different options we know to take traces or residues over these functionals —the result is that doing it, on the more general context of weighted algebras of abstract differential operators associated to a spectral triple, it will be possible to recover the cyclic cohomology cocycles used to give a local formula for the index map (1). Before doing that, let us remark that from (19) it follows that

$$\mathrm{Res}_{z=0}\Phi_0^Q(A)(z) = \frac{1}{q}\mathsf{res}(\gamma A),$$

for any classical pseudodifferential operator acting on sections of a $\mathbb{Z}_2$-graded bundle, where res denotes the Wodzicki residue. In general, [18] for $A_0, \ldots, A_p \in Cl(E)$,

$$\mathrm{Res}_{z=0}\Phi_p^Q(A_0, \ldots, A_p)(z) = \frac{1}{q}\mathsf{res}(\gamma A_0 \cdots A_p), \tag{20}$$

so that regularized traces built from the multifunctionals $\Phi_\bullet^Q$ are in this context, as expected, proportional to res. Moreover, it follows from Lemma 2.1 that, for $p > d =$ analytic dimension of $\mathcal{D}_Q$, $\Phi_p^Q(a_0, \ldots, a_p)(z + \frac{p}{2})$ is holomorphic at zero, therefore

$$\mathrm{Res}_{z=0}\Phi_p^Q(a_0, \ldots, a_p)(z + \frac{p}{2}) = 0.$$

Comparing this with Remark 1.1, i.e. $\mathrm{Res}_{z=0}\,\mathrm{tr}(AQ^{-z}) = 0$ if $a < -m$, where $a = \mathsf{ord}(A)$ and $m = \dim M$, we see that the quantity $q \cdot \mathrm{Res}_{z=0}\Phi_p^Q(a_0, \ldots, a_p)(z + \frac{p}{2})$ generalizes the Wodzicki residue in the more general context of algebras of abstract differential operators associated to spectral triples.

2.2.2. *Weighted cochains and holomorphic cocycles*

The JLO functional $\phi_\bullet^Q(t)$ defined in (16) gives rise to an improper cocycle (with coefficients in $C^\infty(\mathbb{R}^+)$) in (b, B)-cohomology,

$$(a_0, \ldots, a_p) \mapsto \langle a_0, [D, a_1], \ldots, [D, a_p]\rangle_t^{JLO}, \tag{21}$$

where

$$\langle a_0, [D, a_1], \ldots, [D, a_p]\rangle_t^{JLO} := t^{\frac{p}{2}}\phi_p^Q(t)(a_0, [D, a_1], \ldots, [D, a_p]),$$

called *JLO cocycle.*[14] Improper means that it does not vanish in general in any component (i.e. it has an infinite number of components). The same holds true for the functional $\Phi_\bullet^Q(z)$ defined in (17), it gives rise to an improper cocycle (with coefficients in the space of holomorphic functions on a half-plane in $\mathbb{C}$) in (b, B)-cohomology,

$$(a_0, \ldots, a_p) \mapsto \langle a_0, [D, a_1], \ldots, [D, a_p]\rangle_{z-\frac{p}{2}}^H, \tag{22}$$

where

$$\langle a_0, [D, a_1], \ldots, [D, a_p]\rangle_z^H := \Gamma\,(z+p)\,\Phi_p^Q(a_0, [D, a_1], \ldots, [D, a_p])(z+p).$$

This improper cocycle was considered in detail by N. Higson in [12], we will call it *holomorphic cocycle.*

The usual way to find proper cocycles from the improper ones is the computation of traces (residues) which cancel an infinite number of components of these improper cocycles . Indeed, in the classical pseudodifferential setting, when the order of a pseudodifferential operator A is less than $-\dim M$, $\mathrm{res}(A) = 0$. Since this feature appears also for the multifunctional Φ_p^Q in the context of weighted algebras of abstract differential operators associated to noncommutative spectral triples, as follows from Lemma 2.1, a simple degree counting argument shows that —provided that the poles are simple at zero, which is true in the cases we consider— for p big enough, the residues of the components of the holomorphic functional (22) will vanish. This is the way the cyclic cocycles associated to the local index formula can be built in (see [12] and [6]).

Following the lines of the construction of trace extensions for pseudodifferential operator algebras in [4],[17], it is natural to consider the cochains obtained from the holomorphic mappings $\Phi_p^Q(a_0, a_1, \ldots, a_p)(z)$ when, instead of the residues at the origin, the finite part is taken. This was done in [18] by S. Paycha, in the context of the algebra of classical pseudodifferential operators acting on sections of finite rank vector bundles, but can be easily adapted to the case of the weighted algebra of generalized abstract differential operators associated to a regular and finite-summable spectral triple $(\mathcal{A}, \mathcal{H}, D)$.

Definition 2.1. Let us define *weighted cochains* $\chi_\bullet^Q$ as finite parts of the meromorphic functions (with simple poles) $\Phi_\bullet^Q$,

$$\chi_p^Q(a_0, \ldots, a_p) = \mathrm{f.p.}|_{z=0}(\Phi_p^Q(a_0, \ldots, a_p)(z)), \tag{23}$$

where $a_0, \ldots, a_p \in \mathcal{A}$ and Q denotes a reference weight operator.

These weighted cochains have been used to study several types of (algebraic and geometric) anomalies in [18]. The same asymptotic expansions used to show Lemma 2.1 (see Lemma 4.20 and Lemma 4.30 in [12]) can be used to show the following

Proposition 2.1. *Let $a_0, \ldots, a_p \in \mathcal{A}$ for a regular and finite-summable spectral triple $(\mathcal{A}, \mathcal{H}, D)$ with associated weighted algebra of generalized differential operators $\mathcal{D}_Q$ with the analytic continuation property. Then*

$$\chi_p^Q(a_0, \ldots, a_p) - \text{f.p.}|_{z=0}\Phi_0^Q(a_0 \cdots a_p) = \sum_{|k|\geq 1} c_{|k|}\text{Res}_{z=0}Tr\left(\gamma a_0 a_1^{[k_1]} \cdots a_p^{[k_p]} Q^{-|k|-z}\right), \tag{24}$$

where $c_{|k|} = \frac{(-1)^{|k|}(|k|-1)!}{(|k|+1)!}$ *for any multiindex k, $|k| = k_1 + \cdots k_p$ and $q = ord(Q)$.*

Let us come back to the case of the classical pseudodifferential operator algebra. Consider, for example, $A \in Cl(E)$. Then —as follows from (19)—

$$\chi_0^Q(A) = \text{f.p.}|_{z=0}\left(\Phi_0^Q(A)(z)\right) = \text{str}^Q(A). \tag{25}$$

Thus, at the level $p = 0$, weighted cochains and weighted traces are the same (thus, since the weighted traces are non tracial in general, this shows that weighted cochains are not cocycles). However, at higher levels this is no longer true, although the difference between $\chi_p^Q(A_0, \ldots, A_p)$ and the Q-weighted (super)trace of the operator $A_0 \cdots A_p \in Cl(E)$ is given by a finite linear combination of noncommutative residues (see [18], Propostion 2): If $A_0, \ldots, A_p \in Cl(E)$, whenever Q has scalar leading symbol, then

$$\chi_p^Q(A_0, \ldots, A_p) - \text{str}^Q(A_0 \cdots A_p) = \frac{1}{q}\sum_{|k|\geq 1}\frac{(-1)^{|k|}(|k|-1)!}{(|k|+1)!}\text{res}\left(\gamma A_0 A_1^{[k_1]} \cdots A_p^{[k_p]} Q^{-|k|}\right). \tag{26}$$

Notice that, since the order of the operator within the res term in (26) decreases as $|k|$ increases, the residues will cancel from the moment at which the map $z \mapsto \text{tr}(\gamma A_0 A_1^{[k_1]} \cdots A_p^{[k_p]} Q^{-|k|-z})$ has no pole at the origin, so that in the sum there are only a finite number of terms. The same holds true, by Lemma 2.1, for formula (24).

Since weighted cochains are not cocycles, it is clear that it is not possible to build from them the cocycles necessary to write the local index theorem terms. Indeed, formula (26) shows that in the context of pseudodifferential operator algebras weighted cochains are no local in general, but that differences between them and appropriate weighted (super)traces can be local. But actually we have no need of weighted cochains to write down a local expression for the index (1), taking residues of the holomorphic cocycles will be enough to recover the Connes-Moscovici cocycle φ^{+}_{CM}. After computation of the corresponding residues, the asymptotic expansion giving rise to the results stated in Lemma 2.1 imply the following

Lemma 2.2. *[Higson[12]] Let $(\mathcal{A}, \mathcal{H}, D)$ be a regular and finite-summeble spectral triple, $\Delta = D^2$ and $a_0, \ldots, a_p \in \mathcal{A}$, for $p \geq 0$. Then,*

$$\begin{aligned} & \mathrm{Res}_{z=0}\langle a_0, [D, a_1], \ldots, [D, a_p]\rangle^{H}_{z-\frac{p}{2}} \\ = \; & \mathrm{Res}_{z=0}\left[\Gamma(z+\frac{p}{2})\Phi^{\Delta}_{p}(a_0, [D, a_1], \ldots, [D, a_p])(z+\frac{p}{2})\right] \\ = \; & \sum_{|k|\geq 0} c_{p,k}\mathrm{Res}_{z=0}Tr\left(\gamma a_0[D, a_1]^{(k_1)}\cdots[D, a_p]^{(k_p)}\Delta^{-(z+\frac{p}{2})-|k|}\right), \end{aligned} \tag{27}$$

where $|k| = k_1 + \cdots k_p$, for the multiindex $k = (k_1, \ldots, k_p)$, and the constant $c_{p,k}$ is given by the formula (15) above.

Once again we observe the (residues of the) multifunctional $\Phi^{\Delta}_{p}(z + \frac{p}{2})(a_0, [D, a_1], \ldots, [D, a_p])$ appearing in this expression, for $p > 0$, which –in some sense– generalizes the classical Wodzicki residue.[d] This *a priori* locality is indeed confirmed as taking residues of Higson's holomorphic cocycle (22) gives us back the local index formulas given by Connes-Moscovici in Theorem 2.1.

Theorem 2.2. *[Connes-Moscovici[8], Higson[12]] Let $(\mathcal{A}, \mathcal{H}, D)$ be a regular and finite-summeble spectral triple, with $\Delta = D^2$. If the Δ-weighted algebra of generalized differential operators $\mathcal{D}_{\Delta}$ has finite analytic dimension and the analytic continuation property then, for any idempotent e,*

$$\langle \mathrm{Res}_{z=0}\Psi^{H}|e\rangle = \mathrm{ind}\,_{D}(e),$$

where $\mathrm{Res}_{z=0}\Psi^{H}$ *denotes the residue cocycle* $\mathrm{Res}_{z=0}\Psi^{H}_{p}(a_0, a_1, \ldots, a_p) = \mathrm{Res}_{z=0}\langle a_0, [D, a_1], \ldots, [D, a_p]\rangle^{H}_{z-\frac{p}{2}}$.

Thus, the ζ-function regularized objects giving rise to local formulae in the framework of classical pseudodifferential operator theory, such as the formulae for indexes, anomalies, etc., can be defined in the more general setting of noncommutative spectral triples, giving rise to results that could

[d] Notice that, for $p = 0$, since $\mathrm{Res}_{z=0}\left[\Gamma(z)\Phi^{\Delta}_{0}(a_0)(z)\right] = \mathrm{f.p.}|_{z=0}\Phi^{\Delta}_{0}(a_0)(z)$, the complex residue gives rise to a finete part which generalizes a weighted trace rather than a Wodzicki residue (see equation (25)).

be interpreted as examples of the corresponding locality phenomena for noncommutative spaces. This locality has begun to be discussed on concrete examples recently (see e.g. [7], also [3]), but its possible application to physical models —where the locality features could be fundamental— remains to be done.

Acknowledgments

Part of this work was done while the author was visiting the *Institut des Hautes Etudes Scientifiques*, Bures-sur-Yvette, France, and the author wishes to tank warmly professor Jean-Pierre Bourgignon for his kind invitation. Thanks also goes to all the colleagues with whom I discussed about it, specially to Joseph Varilly, Yoshiaki Maeda and Sylvie Paycha, and to the referee for the careful reading of the text and the remarks which improve its final presentation. Finally, many thanks to the members of the Mathematics Department of Keio University and the organizers of the conference on *Noncommutative Geometry and Physics*, 2004.

References

1. Atiyah, M. *Collected Works. Index Theory (Vols. 3 and 4).* Oxford Science Publications. The Clarendon Press, Oxford University Press, New York, 1988.
2. Atiyah, M., Bott, R., Patodi, V.: *On the heat equation and the index theorem.* Invent. Math. **19**, 279–330, 1973.
3. Benameur, Moulay-Tahar *Noncommutative geometry and abstract integration theory.* Geometric and topological methods for quantum field theory (Villa de Leyva, 2001), 157–227, World Sci. Publishing, River Edge, NJ, 2003.
4. Cardona, A., Ducourtioux, C. Magnot, J-P. and Paycha, S. *Weighted traces on algebras of pseudo-differential operators and geometry on loop groups.* Inf.Dim.Anal. Quant.Prob. and Rel.Top. **5**, 503–540, 2002.
5. Cardona, A., Ducourtioux, C. and Paycha, S. *From tracial anomalies to anomalies in Quantum Field Theory.* Comm.Math.Phys. **242**, 31-65, 2003.
6. Connes, A. *Noncommutative Geometry.* Academic Press, San Diego, 1994.
7. Connes, A. *Cyclic cohomology, quantum group symmetries and the local index formula for* $\mathrm{SU}_q(2)$. J. Inst. Math. Jussieu **3**, no. 1, 17–68, 2004.
8. Connes, A., Moscovici, H.: *The local index formula in noncommutative geometry.* GAFA **5**, 174–243, 1995.
9. Dixmier, J. *Existence de traces non-normales.* C.R. Acad. Sci. Paris A-B **262**, pp. A1107 洵 1108, 1966.
10. Gracia-Bonda, J.; Várilly, J. and Figueroa, H. *Elements of noncommutative geometry.* Birkhäuser, Boston, MA, 2001.
11. Higson, N. *A note on regular spectral triples.* Preprint, 2003.

12. Higson, N. *The residue index theorem of Connes and Moscovici.* Preprint, 2004.
13. Higson, N. *Meromorphic continuation of zeta functions associated to elliptic operators.* Operator algebras, quantization, and noncommutative geometry, *Contemp. Math.*, **365**, Amer. Math. Soc., 129–142, 2004.
14. Jaffe, A., Lesniewski, A. and Osterwalder, K. *Quantum K-theory. I. The Chern character.* Comm. Math. Phys. **118**, no. 1, 1–14, 1988.
15. Kassel, C. *Le residu non commutatif (d'aprs M. Wodzicki).* Seminaire Bourbaki, Vol. 1988/89. Asterisque No. 177-178, Exp. No. 708, 199–229, 1989.
16. Kontsevich, M. and Vishik, S. *Determinants of elliptic pseudo-differential operators*, Max Planck Institut Preprint, 1994.
17. Paycha, S. *Renormalized traces as a looking glass into infinite-dimensional geometry*, Inf.Dim.Anal.Quant.Prob. and Rel.Top., Vol **4**, N.2 p.221-266, 2001.
18. Paycha, S. *Weighted Trace Cochains, the geometric setup for anomalies*, U.B.P. Preprint, 2004.
19. Paycha, S. and Rosenberg, S. *Curvature on determinant bundles and first Chern forms*, J. Geom. Phys., Vol **45**, no. 3-4, 393–429, 2003.
20. Ponge, R. *A new short proof of the local index formula and some of its applications.* Comm. Math. Phys. **241** , no. 2-3, 215–234, 2003. Erratum, Comm. Math. Phys. **248** , 639, 2004.
21. Seeley, R. T. *Complex powers of an elliptic operator*, in Proc. Sympos. Pure Math., Vol. 10, pp. 288–307, Amer. Math. Soc., Providence, 1967.
22. Wodzicki, M. *Non commutative residue*, in Lecture Notes in Mathematics **1289**, Springer Verlag, 1987.

LINE BUNDLES ON FUZZY $\mathbb{C}P^N$

URSULA CAROW-WATAMURA AND SATOSHI WATAMURA

Department of Physics
Graduate School of Science
Tohoku University
Sendai, Japan
watamura@tuhep.phys.tohoku.ac.jp
ursula@tuhep.phys.tohoku.ac.jp

We construct line bundles over fuzzy complex projective spaces as projective modules. Their corresponding Chern classes are calculated. They reduce to the monopole charges in the $N \to \infty$ limit, where N labels the representation of the fuzzy algebra. This talk is based on a paper in collabolation with H. Steinacker[23].

1. Introduction

Since the formulation of the Gel'fand-Naimark theorem, there has been an intensive research on the various possibilities to generalize geometry. The theorem states that one may construct geometry from a function algebra. Thus, expressing geometry in algebraic terms opens the possibility to replace such a function algebra by a more general one, includig the noncommutative algebra.

Recently, topologically nontrivial solutions, e.g. the instantons on noncommutative space, are attracting a great deal of attention, as the topological invariants carry rich information about geometrical data. Topological nontrivial solutions are conveniently described in terms of projective modules over the algebra $\mathcal{A}$ of functions on the noncommutative space.

On the other hand, much effort is undertaken to understand the gauge theory on curved noncommutative spaces. Topological aspects of field theory on the so-called fuzzy sphere [14] have been first discussed in [7]. Their formulation as projective modules has been elaborated in [3,19,8] and by matrix model approach in [18]. Dirac operators on the fuzzy sphere have been proposed in [7] and [20,21].

The understanding of field theory and in particular gauge theory on fuzzy curved spaces is important since fuzzy spaces also arize in string theory. A very well known example is the fuzzy sphere and its q-deformed version which appear as a D-brane in the $SU(2)$ WZW model, as discussed by several authors [2,5]; see also [10] and references therein. Gauge theory on the fuzzy sphere appears as an effective theory of the D-branes in S^3, in the $k \to \infty$ limit of the $SU(2)_k$ WZW model at level k.

Therefore, a proper geometrical description and interpretation of such a system, in particular of the topologically nontrivial configurations is clearly desirable. For physical applications, we would also like to consider spaces of dimension 4 and higher. See for example [9,1,4,11].

Here, we want to concentrate on the construction of projective modules for fuzzy $\mathbb{C}P^2$ and $\mathbb{C}P^n$. The result of our investigation is a simple formulation of monopole bundles on fuzzy $\mathbb{C}P^n$ using projective modules. Furthermore, with a suitable differential calculus we compute the canonical connection and field strength explicitly, and we will compute the first Chern classes and discuss its properties.

In the next section we formulate the geometry of classical $\mathbb{C}P^n$ by using two different approaches: (a) adjoint orbits of $su(n+1)$, and (b) a generalized Hopf fibration. Both lead to a characterization in terms of $(n+1) \times (n+1)$ matrices satisfying a quadratic characteristic equation.

In section 3, we study the fuzzy spaces $\mathbb{C}P^2_N$ and $\mathbb{C}P^n_N$ from these two points of view. Similarly to the commutative case we obtain quadratic characteristic equations for algebra-valued $(n+1) \times (n+1)$ matrices. These contain the commutation relations of the coordinate algebra in a compact way. We will see that the Hopf fibration is quantized in terms of a Fock space representation.

In section 4, the construction of projective modules corresponding to the monopole bundles for fuzzy $\mathbb{C}P^n_N$ is performed. We do this by using an explicit form of the projection operators, thus we will see how a section of the constructed line bundles corresponds to a complex scalar field.

Finally, a differential calculus is introduced. In the fuzzy case it involves more degrees of freedom than the classical case. This feature is typical for fuzzy spaces, and we explain in what sense the classical calculus is recovered in the commutative limit. Using this calculus we compute the field strength and Chern class for the monopole bundles. We show that the usual (integer) Chern numbers are recovered in the limit of large N.

2. The geometry of $\mathbb{C}P^n$

As mentioned above, we discuss here two descriptions of $\mathbb{C}P^n$. One is in terms of adjoint orbits of $su(n+1)$, and the other one is based on a generalized Hopf fibration $U(1) \to S^{2n+1} \to \mathbb{C}P^n$. Both descriptions are manifestly covariant under $SU(n+1)$, and this symmetry is maintained in their quantization as fuzzy $\mathbb{C}P^n$.

2.1. *Adjoint orbits*

Consider a finite-dimensional matrix Lie group G with Lie algebra $\mathfrak{g}$ and some $t \in \mathfrak{g}$. Then an adjoint orbit is given by

$$\mathcal{O}(t) = \{gtg^{-1};\ \ g \in G\} \subset \mathfrak{g}\ . \tag{1}$$

$\mathcal{O}(t)$ can be viewed as a homogeneous space:

$$\mathcal{O}(t) \cong G/K_t\ . \tag{2}$$

and $K_t = \{g \in G :\ \ [g,t] = 0\}$ is the stabilizer of t. Any such conjugacy class is invariant under the adjoint action of G. "Regular" conjugacy classes are those with K_t being the maximal torus, and have maximal dimension $dim(\mathcal{O}(t)) = dim(G) - rank(G)$. The degenerate orbits such as $\mathbb{C}P^2$ or $\mathbb{C}P^n$. correspond to degenerate t, and have dimension $dim(\mathcal{O}(t)) = dim(G) - dim(K_t)$.

To characterize the type of the orbit (for matrix Lie groups) we can use its characteristic equation $\chi(Y) = 0$ for $Y \in \mathcal{O}(t)$, which is invariant under conjugation, for it depends only on the eigenvalues of t. The scale of the resulting $\mathbb{C}P^n$ is given by the normalization of the matrix t.

Take for example $\mathbb{C}P^2$: Its characteristic equation for $\mathbb{C}P^2$ obtained in this way has the form

$$\chi(Y) = (Y+1)(Y-2) = 0\ . \tag{3}$$

It characterizes $\mathbb{C}P^2$ as a submanifold in the embedding space $\mathbb{R}^8$.

For $\mathbb{C}P^n \cong SU(n{+}1)/(SU(n){\times}U(1))$, we consider a matrix $t \in su(n{+}1)$ with 2 distinct eigenvalues and multiplicities $(n,1)$. A natural choice is $t = diag(-1,-1,...,-1,n)$ up to normalization. It satisfies the characteristic equation

$$\chi(Y) = (Y+1)(Y-n) = 0\ . \tag{4}$$

This can be understood by considering the $(n+1)\times(n+1)$ matrix

$$P = \frac{1}{n+1}(Y+1) \qquad \in Mat(n+1,\mathbb{C}) \tag{5}$$

which is a projector of rank 1. Hence $P \in Mat(n+1, \mathbb{C})$ can be written as

$$P = |z^i\rangle\langle z^i|$$

where $\langle z^i|z_i\rangle = 1$. This is the relation to the second description of $\mathbb{C}P^n$ as $S^{2n+1}/U(1)$.

The next step is to introduce global coordinates for $\mathbb{C}P^2$ and $\mathbb{C}P^n$. Let us first consider $\mathbb{C}P^2$, described as above by the matrix $Y = g^{-1}tg \in Mat(3, \mathbb{C})$. It is convenient to use an overcomplete set of global coordinates, which is easily generalized to the fuzzy case. We express the matrix Y in terms of the Gell-Mann matrices λ_a of $su(3)$ as

$$Y = y_a \lambda_a \ . \tag{6}$$

The λ_a are related to the generators T_a of the Lie algebra via

$$\lambda_a = 2\pi_{\Lambda_{(1)}}(T_a) \ . \tag{7}$$

$\pi_{\Lambda_{(1)}}$ denotes the fundamental representation of $su(3)$ with highest weight $\Lambda_{(1)}$. The characteristic equation reads

$$Y^2 = Y + 2 \ . \tag{8}$$

In terms of the coordinates y_a in eq.(6) it takes the form

$$g_{ab}\ y_a y_b = 3 \ , \qquad d_{abc}\ y_a y_b = y_c \ . \tag{9}$$

From the above construction we see that this set of relations characterizes the appropriate adjoint orbit in $su(3)$.

For $\mathbb{C}P^n$, we consider the generalized Gell-Mann matrices of $su(n+1)$:

$$\lambda_a = 2\pi_{\Lambda_{(1)}}(T_a) \tag{10}$$

where T_a are the generators of $su(n+1)$, satisfying

$$\lambda_a \lambda_b = \frac{2}{n+1}\delta_{ab} + (if_{abc} + d_{abc})\lambda_c \ . \tag{11}$$

Using the expansion $Y = y_a\lambda_a = g^{-1}tg$ the characteristic equation is

$$Y^2 = (n-1)Y + n \ , \tag{12}$$

where

$$g_{ab}\ y_a y_b = \frac{n(n+1)}{2} \ , \qquad d_{abc}\ y_a y_b = (n-1)\ y_c \ . \tag{13}$$

3. From $\mathbb{C}P^n$ to Fuzzy $\mathbb{C}P^n_N$.

Let us now consider the formulation of fuzzy complex projective spaces using the approach of adjoint orbits (1) on G which can be quantized in terms of a simple matrix algebra $End_{\mathbb{C}}(V_N)$, where V_N are suitable representations of G. The appropriate representations V_N can be identified by matching the spaces of harmonics (i.e. using harmonic analysis [17]) for the general case.

3.1. *The algebra of* $\mathbb{C}P^n_N$

We first consider $\mathbb{C}P^2$ as a concrete example. To make the correspondence with classical $\mathbb{C}P^2$ explicit, we consider the $3D_N \times 3D_N$ matrix

$$X = \sum_a \xi_a \otimes \lambda_a \tag{14}$$

where λ_a are the Gell-Mann matrices as before, and

$$\xi_a = \pi_{V_N}(T_a) \qquad \in \mathbb{C}P^2_N \tag{15}$$

denotes the representation of $T_a \in su(3)$ on V_N. The coordinate functions $x_a = (x_1, ...x_8)$ on fuzzy $\mathbb{C}P^2$ are defined by

$$x_a = \Lambda_N \xi_a \qquad \in \mathbb{C}P^2_N \ . \tag{16}$$

They are operators acting on V_N. Λ_N is a scaling parameter which will be fixed below. By construction, the x_a transform in the adjoint under $su(3)$, just like the classical coordinate functions y_a introduced in eq.6.

The relations among these generators x_a can be easily found by use of the corresponding characteristic equation of X which is derived as [23]:

$$\begin{aligned} X^2 &= \xi_a \xi_b \, (\frac{2}{3}\delta_{ab} + (if_{abc} + d_{abc})\lambda_c) \\ &= \frac{2}{3}(\frac{1}{3}N^2 + N) + (\frac{N}{3} - 1)X \ . \end{aligned} \tag{17}$$

The scale parameter Λ_N is determined as

$$\Lambda_N = \frac{1}{\sqrt{\frac{1}{3}N^2 + N}} \ , \tag{18}$$

thus, we find the defining relation of the algebra $\mathbb{C}P_N^2$:

$$[x_a, x_b] = \frac{i}{\sqrt{\frac{1}{3}N^2 + N}} f_{abc}\, x_c, \tag{19}$$

$$g_{ab}\, x_a x_b = 1, \tag{20}$$

$$d_{abc}\, x_a x_b = \frac{\frac{N}{3} + \frac{1}{2}}{\sqrt{\frac{1}{3}N^2 + N}}\, x_c\ . \tag{21}$$

The "angular momentum" operators (generators of $SU(3)$) now become inner derivations:

$$J_a f(x) = [\xi_a, f]\ , \tag{22}$$

and $J_a x_b = [\xi_a, x_b] = i f_{abc} x_c$, as classically.

The integral on $\mathbb{C}P_N^2$ is given by the suitably normalized trace,

$$\int f(x) = \frac{1}{D_N}\, Tr(f) = \frac{2}{(N+1)(N+2)}\, Tr(f) \tag{23}$$

which is clearly invariant under $SU(3)$.

Fuzzy $\mathbb{C}P^n$ is obtained entirely analogous to fuzzy $\mathbb{C}P^2$. To identify the correct representations V_N of $su(n+1)$, we must match the space of harmonics with the decomposition of $End(V_N) = V_N \otimes V_N^* = \oplus_\lambda n_\lambda V_\lambda$ for certain multiplicities n_λ (up to some cutoff N).

In analogy to the case of CP_2^n we define the algebra of the fuzzy projective space by

$$\mathbb{C}P_N^n := End_{\mathbb{C}}(V_N) = Mat(D_N, \mathbb{C}) \tag{24}$$

where

$$D_N = dim(V_N) = \frac{(N+n)!}{n!\, N!} \tag{25}$$

from Weyl's dimension formula. The fuzzy coordinates and their commutation relations are again obtained by considering the $(n+1)D_N \times (n+1)D_N$ matrix

$$X = \sum_a \xi_a \otimes \lambda_a\ , \tag{26}$$

where λ_a are the Gell-Mann matrices of $su(n+1)$ and

$$\xi_a = \pi_{V_N}(T_a) \qquad \in \mathbb{C}P_N^n\ . \tag{27}$$

We define the coordinate functions x_a for $a = 1, ..., n^2 + 2n$ on fuzzy $\mathbb{C}P^n$ by

$$x_a = \Lambda_N \xi_a \qquad \in \mathbb{C}P^n_N \, , \tag{28}$$

where Λ_N is a scaling parameter which will be fixed below. By construction, the x_a transform in the adjoint under $su(n+1)$, just like the classical coordinate functions y_a introduced previously. Using the characteristic equation of X,

$$(X - \frac{nN}{n+1})(X + \frac{N}{n+1} + 1) = 0 \tag{29}$$

we obtain

$$\begin{aligned} X^2 &= \xi_a \xi_b \, (\frac{2}{n+1}\delta_{ab} + (if_{abc} + d_{abc})\lambda_c), \\ &= \frac{n}{n+1}(\frac{1}{n+1}N^2 + N) + (\frac{N(n-1)}{n+1} - 1)X \, . \end{aligned} \tag{30}$$

With $f_{abc}f_{dbc} = (n+1)\delta_{ad}$. we get

$$\Lambda_N = \frac{1}{\sqrt{\frac{n}{2(n+1)}N^2 + \frac{n}{2}N}} \tag{31}$$

and we find

$$[x_a, x_b] = i\Lambda_N f_{abc}\, x_c, \tag{32}$$

$$g_{ab}\, x_a x_b = 1, \tag{33}$$

$$d_{abc}\, x_a x_b = (n-1)(\frac{N}{n+1} + \frac{1}{2})\Lambda_N \, x_c \, . \tag{34}$$

For large N, this reduces to (13).

The symmetry group $SU(n+1)$ acts by inner derivation as for $\mathbb{C}P^2$, and the integral is given by the suitably normalized trace over V_N.

3.2. *Representation on a Fock space*

To introduce nontrivial line bundles, we directly quantize the fibration $U(1) \to S^{2n+1} \to \mathbb{C}P^n$. For this end we proceed as follows: First introduce the noncommutative $\mathbb{C}^{n+1}$ in terms of operators a^i, a^+_i $(i = 1, ..., n+1)$ which are quantizations of the coordinate functions $z^i, \overline{z}_i$ of $\mathbb{C}^{n+1} \supset S^{2n+1}$. Then, the fuzzy $\mathbb{C}P^n$ will be obtained as a subalgebra of this noncommutative $\mathbb{C}^{n+1}$. The equivalence to the definition given previously will become manifest.

The generators a_i^+, a^i of noncommutative $\mathbb{C}^{n+1}$ are creation- resp. annihilation operators which transform as $V_{(1,0...,0)}$ resp. $V^*_{(1,0,...,0)} = V_{(0,..,0,1)}$. They satisfy the canonical commutation relations

$$[a^i, a^j] = [a_i^+, a_j^+] = 0 \ , \quad [a^i, a_j^+] = \delta^i_j \ . \tag{35}$$

We denote the resulting algebra as $\mathbb{C}_\theta^{n+1}$.

As in the previous section, we consider the $U(1)$ defined by

$$\omega \circ (a^i, a_i^\dagger) = (a^i\omega, a_i^+\overline{\omega}) \tag{36}$$

where $\omega \in \mathbb{C}$ and $|\omega| = 1$. With this $U(1)$ action, the equivariant operators $C(\kappa, \mathbb{C}_\theta^{n+1})$ are defined by

$$C(\kappa, \mathbb{C}_\theta^{n+1}) = \{f | f \in Pol(a^i, a_i^+), i = 1, ..., n+1 \quad \text{and} \quad \omega \circ f = f\omega^\kappa\} \ . \tag{37}$$

This means that κ counts the difference of the number of creation and annihilation operators, and thus an element $f \in C(\kappa, \mathbb{C}_\theta^{n+1})$ satisfies

$$[\mathbf{N}, f] = \kappa f \ , \tag{38}$$

where $\mathbf{N}$ is the number operator: $\mathbf{N} = \sum_{i=1}^{n+1} a_i^+ a^i$.

Also in the noncommutative case, there is a natural multiplication of the equivariant operators $C(\kappa, \mathbb{C}_\theta^{n+1})$ and $C(\kappa', \mathbb{C}_\theta^{n+1})$ such that

$$C(\kappa, \mathbb{C}_\theta^{n+1}) \times C(\kappa', \mathbb{C}_\theta^{n+1}) \longrightarrow C(\kappa + \kappa', \mathbb{C}_\theta^{n+1}) \ . \tag{39}$$

Therefore, the equivariant operators $C(\kappa, \mathbb{C}_\theta^{n+1})$ can be interpreted as $C(0, \mathbb{C}_\theta^{n+1})$-module. To see this, consider the following generators of $C(0, \mathbb{C}_\theta^{n+1})$:

$$\tilde{\xi}_a = a_i^+ (T_a)^i{}_j a^j, \qquad T_a = \frac{1}{2}\lambda_a \ . \tag{40}$$

By construction, they transform in the adjoint of $su(n+1)$, and satisfy the relations

$$[\tilde{\xi}_a, \tilde{\xi}_b] = i f_{abc} \tilde{\xi}_c \tag{41}$$

and

$$\sum_{a=1,...,n^2+2n} \tilde{\xi}_a \tilde{\xi}_a = \frac{n}{2(n+1)} \mathbf{N}(\mathbf{N} + n + 1) \ . \tag{42}$$

To obtain fuzzy $\mathbb{C}P_N^n$, we fix the "radius" in $\mathbb{C}_\theta^{n+1}$, i.e. choose a Fock space representation with fixed particle number N. As usual, the states of the Fock space are defined by acting with the creation operators a_i^+ on the vacuum state $|0\rangle$, with $a_i|0\rangle = 0$. The N-particle subspace $\mathcal{F}_N$ is obtained by

$$|\vec{m}\rangle = \sqrt{\frac{m_1!m_2!\cdots m_{n+1}!}{N!}}(a_1^+)^{m_1}(a_2^+)^{m_2}\cdots(a_{n+1}^+)^{m_{n+1}}|0\rangle \tag{43}$$

where the label $\vec{m}$ is a set of positive integers m_i, $\vec{m} = (m_1, ..., m_{n+1})$ satisfying $\sum_i m_i = N$.

We want to show that on the N-particle subspace $\mathcal{F}_N$, one can recover the characteristic equation (29). To verify this, using the Fierz identity for the generators T^a and the definition of the generators $\tilde{\xi}$ we obtain

$$a_j^+ a^i = 2\sum T^{a\,i}_{\;j}\tilde{\xi}^a + \frac{1}{n+1}\delta^i_j\mathbf{N} = a^i a_j^+ - \delta^i_j \ . \tag{44}$$

On the other hand we know that

$$P^i{}_j = \frac{1}{\mathbf{N}+1}a^i a_j^+ \tag{45}$$

is indeed a projection operator, i.e. $P^2 = P$. Then, defining the matrix $\tilde{X}^i_j = 2\sum T^{a\,i}_{\;j}\tilde{\xi}^a$ and using (44), the operator P can be written as

$$P = \frac{\tilde{X} + 1 + \frac{\mathbf{N}}{n+1}}{\mathbf{N}+1} \ . \tag{46}$$

In terms of $\tilde{X}$, the relation $P^2 = P$ gives

$$0 = P(P-1) = \frac{1}{(\mathbf{N}+1)^2}(\tilde{X} + 1 + \frac{\mathbf{N}}{n+1})(\tilde{X} - \mathbf{N} + \frac{\mathbf{N}}{n+1}) \ . \tag{47}$$

This is precisely the characteristic equation (29) for X, if the number operator $\mathbf{N}$ is replaced by the number N. Hence, $\tilde{\xi}_a$ and ξ_a are equivalent, i.e. $x_a = \Lambda_N\tilde{\xi}_a$ and the relations of the fuzzy $\mathbb{C}P^n$ hold on the Fock space representation $\mathcal{F}_N$ of $C(0, \mathbb{C}^{n+1}_\theta)$.

Therefore any $f \in C(0, \mathbb{C}^{n+1}_\theta)$ defines a map

$$f : \mathcal{F}_N \longrightarrow \mathcal{F}_N \ . \tag{48}$$

Now we observe that the Fock space $\mathcal{F}_N$ is in fact the irreducible representation $V_N = V_{(N,0..,0)}$ of $su(n+1)$. Since the generators $\tilde{\xi}_a$ act on $\mathcal{F}_N$ and generate the $su(n+1)$ algebra, they generate $End_{\mathbb{C}}(\mathcal{F}_N)$, and the equivalence of this definition of $\mathbb{C}P^n_N$ with the one given in the previous section in terms of $Mat(D_N, \mathbb{C})$ (or more precisely $End_{\mathbb{C}}(V_N)$) is manifest[a].

[a]The conjugated version $\mathbb{C}P^{2*}$ would be obtained using the conjugated λ_a matrices $\tilde{\lambda}_a$, and b^{+i}, b_j transforming in the dual representations.

4. Projective $\mathbb{C}P^n_N$ modules

Due to the Serre-Swan theorem it has been known already for a longer time that the algebraic object corresponding to a vector bundle is the projective module of finite type. Thus, from the noncommutative point of view, the projective modules are the relevant objects to deal with.

A projective $\mathcal{A}$-module can be constructed from the free module $\mathcal{A}^p$ together with a projection operator $\mathcal{P}$, which is an element of $Mat(p, \mathcal{A})$, the space of $p \times p$ matrices with elements in the base algebra $\mathcal{A}$.

Since we are going to consider the noncommutative analogue of the monopole bundles, i.e. the $U(1)$ bundles over the fuzzy $\mathbb{C}P^n$, we construct a rank 1 projection operator which determines the module associated with the complex rank 1 vector bundle. For this we apply the approach taken in ref. [12,13] where the construction of the projective module over S^2 is formulated. The advantage of this formulation compared to [9] is that it also provides a canonical connection, which can be used to calculate Chern numbers.

Consider an p-component vector defined as

$$\boldsymbol{v} = (v_\mu) \ , \quad \mu = 1, ..., p \tag{49}$$

where v_μ is an element of a $\mathcal{B}$-$\mathcal{A}$ bimodule $\mathcal{M}$. Here $\mathcal{B}$ is also an algebra, but not necessarily equivalent to $\mathcal{A}$. The only condition needed is the normalization condition

$$\boldsymbol{v}^\dagger \boldsymbol{v} = \sum_\mu v_\mu^\dagger v_\mu = \mathbf{1}_\mathcal{B} \tag{50}$$

where $\mathbf{1}_\mathcal{B}$ denotes the identity of the algebra $\mathcal{B}$. Using this vector $\boldsymbol{v}$, we define the projection operator as

$$\mathcal{P} = \boldsymbol{v}\boldsymbol{v}^\dagger \ . \tag{51}$$

By the normalization condition it is apparent that

$$\mathcal{P}\mathcal{P} = \boldsymbol{v}(\boldsymbol{v}^\dagger \boldsymbol{v})\boldsymbol{v}^\dagger = \boldsymbol{v}\mathbf{1}_\mathcal{B}\boldsymbol{v}^\dagger = \mathcal{P} \ . \tag{52}$$

To define the projective module, the matrix elements of the projection operator $\mathcal{P}$ must be elements of the algebra $\mathcal{A}$. However, this does not mean that each element of the vector $\boldsymbol{v}$ is also an element of $\mathcal{A}$. This is similar to the situation in the ADHM construction of the so-called localized instanton in $\mathbb{R}^4_\theta$ [15,6,16].

Therefore, we take the elements of $\boldsymbol{v}$ to be in $C(\kappa, \mathbb{C}^{n+1}_\theta)$. Then the matrix elements of the projection operator $\mathcal{P}_\kappa$ defined as in (51) are indeed elements of $C(0, \mathbb{C}^{n+1}_\theta)$.

To define the vector $\boldsymbol{v}$, we have to distinguish two cases depending on the sign of the integer κ:

(1) For $0 < \kappa$:

$$v(\vec{j}) = (a^1)^{j_1}(a^2)^{j_2}\cdots(a^{n+1})^{j_{n+1}}c_+(\vec{j}) \tag{53}$$

where the dimension of the vector $\boldsymbol{v}$ is $D_\kappa = \frac{(n+\kappa)!}{n!\kappa!}$, $\vec{j} = (j_1, ..., j_n)$ with $j_i \geq 0$ integers satisfying $\sum j_i = \kappa$. The normalization factor is

$$(c_+(\vec{j}))^2 = \frac{\kappa!}{j_1!j_2!\cdots j_{n+1}!\mathbf{N}(\mathbf{N}-1)\cdots(\mathbf{N}-\kappa+1)} \tag{54}$$

(2) For $-N < \kappa < 0$:

$$v_\mu = (a_1^+)^{j_1}(a_2^+)^{j_2}\cdots(a_{n+1}^+)^{j_{n+1}}c_-(\vec{j}) \tag{55}$$

with dimension $D_\kappa = \frac{(n+|\kappa|)!}{n!\,|\kappa|!}$, and the normalization is

$$(c_-(\vec{j}))^2 = \frac{|\kappa|!}{j_1!j_2!\cdots j_{n+1}!(\mathbf{N}+n+1)(\mathbf{N}+n+2)\cdots(\mathbf{N}+n+|\kappa|)} \tag{56}$$

It is easy to verify in both cases that $\boldsymbol{v}^\dagger\boldsymbol{v} = \mathbf{1}_\mathcal{B}$.

One might worry that the denominator of the normalization factor $c_+(\vec{j})$ can become 0 for large $\kappa > 0$. However, when constructing the projection operator $\mathcal{P}_\kappa = \boldsymbol{v}\boldsymbol{v}^\dagger$ and specifying the representation space to be $\mathcal{F}_N$, the number operator $\mathbf{N}$ in c_+ is replaced by the value $N+\kappa$. Therefore for $\kappa > 0$ all expressions are well-defined.

On the other hand, for $\kappa < 0$, there is a limit for the admissible values of κ. The reason is that $\boldsymbol{v}^\dagger$ contains $|\kappa|$ annihilation operators, hence

$$\mathcal{P}_\kappa = \boldsymbol{v}\boldsymbol{v}^\dagger$$

acting on $\mathcal{F}_N$ is ill-defined if $|\kappa| > N$. Therefore we impose the bound $\kappa + N \geq 0$ in the case $\kappa < 0$.

4.1. *Sections in the Monopole Bundle*

Now we can consider the projective module $\Gamma_\kappa(\mathbb{C}P_N^n) = \mathcal{P}_\kappa(\mathbb{C}P_N^n)^{D_{|\kappa|}}$. An element ξ of the module $\Gamma_\kappa(\mathbb{C}P_N^n)$ is a $D_{|\kappa|}$ dimensional vector $\xi = \{\xi_\mu\}$, the components of which are $\xi_\mu \in \mathbb{C}P_N^n$. This is a section of the line bundle, corresponding to a complex scalar field in $U(1)$ gauge theory. However, the complex scalar field in $U(1)$ gauge theory has a single component in the conventional field theory formulation. We will explain the relation between these formulations in the following.

Assume that $\kappa + N \geq 0$. The single-component scalar field, i.e. section of the monopole bundle, is given by

$$\hat{\xi} = \boldsymbol{v}^\dagger \xi = \sum_\mu v_\mu^\dagger \xi_\mu \; . \tag{57}$$

On the other hand, we can act with $\hat{\xi}$ on an element $\psi = \sum_m f_m|m\rangle \in \mathcal{F}_N$, with the result

$$\begin{aligned} \hat{\xi}\psi &= (\sum_\mu v_\mu^\dagger \xi_\mu)(\sum_m f_m|m\rangle) \\ &= \sum_{\substack{p_i \in \mathbf{Z}_+, \\ p_1+\ldots+p_{n+1}=N+\kappa}} f'_{p_1\cdots p_{n+1}}(a_1^+)^{p_1}\cdots(a_{n+1}^+)^{p_{n+1}}|0\rangle \; , \end{aligned} \tag{58}$$

i.e.

$$\hat{\xi}\psi \in \mathcal{F}_{N+\kappa} \; . \tag{59}$$

Thus we can identify an element $\hat{\xi} \in C(\kappa, \mathbb{C}P_N^n)$ with a map

$$\hat{\xi} \in Hom_{\mathbb{C}}(\mathcal{F}_N, \mathcal{F}_{N+\kappa}) : \mathcal{F}_N \longrightarrow \mathcal{F}_{N+\kappa} \; . \tag{60}$$

In other words, we can identify the scalar field $\hat{\xi}$ on $\mathbb{C}P_N^n$ with monopole charge κ with a $D_N \times D_{(N+\kappa)}$ rectangular matrix. Equivalently, we can identify the section of a line bundle over $\mathbb{C}P_N^n$ given by $\hat{\xi} \in C(\kappa, \mathbb{C}_\theta^{n+1})_N$ as a $\mathbb{C}P_{N+\kappa}^n$ - $\mathbb{C}P_N^n$ bimodule. From this construction it is apparent that we must impose the bound $\kappa + N \geq 0$.

Now, we see that there are two pictures of the monopole bundle $\mathbb{C}P_{N+\kappa}^n$ - $\mathbb{C}P_N^n$ bimodule. Namely, the same bimodule can be obtained from fuzzy $\mathbb{C}P_{N+\kappa}^n$ with monopole charge $-\kappa$ (assuming $N+\kappa > 0$). Since in noncommutative algebras we have to make a choice of left and right multiplication on the module, we can find two equivalent bimodules as

(1) monopole with charge $|\kappa|$ on $\mathbb{C}P_N^n$
(2) monopole with charge $-|\kappa|$ on $\mathbb{C}P_{N+|\kappa|}^n$.

Hence, there is a duality between $\mathbb{C}P_N^n$ with monopole charge κ and $\mathbb{C}P_{N+\kappa}^n$ with monopole charge $-\kappa$. We see that $\mathbb{C}P_N^n$ and $\mathbb{C}P_{N+|\kappa|}^n$ are Morita equivalent, and the scalar field is the equivalence bimodule (the inverse of $\hat{\xi}$ is given by its conjugate). This is an example showing the relation of Morita equivalence and duality of the noncommutative space.

5. Chern Number on Fuzzy $\mathbb{C}P^n_N$

5.1. *Canonical Connection and Field Strength*

A few methods have been proposed to introduce the differential calculus on Fuzzy spaces. We take here the calculus given in [14], i.e. we consider a basis of one-forms θ_a, $a = 1, 2, ..., n(n+2)$ such that

$$[\theta_a, f] = 0, \qquad \theta_a \theta_b = -\theta_b \theta_a \ . \tag{61}$$

This defines a space of exterior forms on fuzzy $\mathbb{C}P^n_N$, denoted by $\Omega^*_N := \Omega^*(\mathbb{C}P^n_N)$. The anticommuting generators θ_a define a gradation. The highest non-vanishing form is the (n^2+2n)-form corresponding to the volume form of $su(n+1)$.

The exterior derivative $d : \Omega^k_N \to \Omega^{k+1}_N$ is defined such that $d^2 = 0$ and the graded Leibniz rule hold. The action of d on the algebra elements $f \in \Omega^0_N$ is given by the commutator with a special one-form:

$$df := [\Theta, f] = [\xi_a, f]\theta_a \ . \tag{62}$$

where $\Theta = \xi_a \theta_a$.

The 2-form can be defined by imposing the consistency of the relations among the algebra elements and 1-forms with the exterior derivative d. For example

$$d(\xi_a \theta_b - \theta_b \xi_a) = 0 \ . \tag{63}$$

and $dd\xi_a = 0$. Then we obtain

$$d\theta_a = -\frac{i}{2} f^a_{bc} \theta_b \theta_c \ . \tag{64}$$

We also see here that the anticommutativity of the θ_a is consistent with this differential calculus.

For later purpose, let us note here that one can show also:

$$d\Theta = \Theta^2 \tag{65}$$

where

$$\Theta^2 = \frac{i}{\Lambda_N} \eta \ . \tag{66}$$

On the other hand, η is proportional to

$$\eta = C \frac{1}{2} f_{abc} dx_a dx_b dx_c \ , \tag{67}$$

where C is a constant which will be determined later.

Using this differential calculus, we can define a canonical connection ∇ over the projective module $\Gamma_\kappa(\mathbb{C}P^n_N)$ defined by the projection $\mathcal{P}_\kappa$ as

$$\mathcal{P}_\kappa d\xi \tag{68}$$

where $\xi \in \Gamma_\kappa(\mathbb{C}P^n_N)$. The curvature 2-form of this canonical connection is given by $\mathcal{P}_\kappa d\mathcal{P}_\kappa d\mathcal{P}_\kappa$.

We represent the connection by the covariant derivative on the scalar field $\hat{\xi}$ as

$$\nabla\hat{\xi} \equiv \boldsymbol{v}^\dagger(\mathcal{P}_\kappa d\xi) = (d + \boldsymbol{v}^\dagger d\boldsymbol{v})\hat{\xi}\ . \tag{69}$$

Then the gauge field and field strength of this connection ∇ are given by

$$A = \boldsymbol{v}^\dagger d\boldsymbol{v} = \boldsymbol{v}^\dagger \Theta \boldsymbol{v} - \Theta\ , \tag{70}$$

$$F = \boldsymbol{v}^\dagger \mathcal{P}_\kappa d\mathcal{P}_\kappa d\mathcal{P}_\kappa \boldsymbol{v} = \boldsymbol{v}^\dagger \Theta \boldsymbol{v}\boldsymbol{v}^\dagger \Theta \boldsymbol{v} - \boldsymbol{v}^\dagger \Theta^2 \boldsymbol{v}\ . \tag{71}$$

To evaluate them we extend the differential calculus from $\mathbb{C}P^n_N$ to $\mathbb{C}^{n+1}_\theta$ as discussed above, postulating that the θ_a commute with all a^i, a_j (this can also be interpreted as a calculus on the $U(1)$ principal bundle over $\mathbb{C}P^n_N$).

First we consider $\kappa > 0$. Using

$$a^j\mathbf{N} = (\mathbf{N}+1)a^j\ , \qquad \mathbf{N}a_i^+ = a_i^+(\mathbf{N}+1) \tag{72}$$

we have

$$\begin{aligned} \boldsymbol{v}^\dagger \Theta \boldsymbol{v} &= \boldsymbol{v}^\dagger a_i^+ a^j \boldsymbol{v}\, T^i_{a\,j}\theta_a \\ &= \frac{N-\kappa}{N}\Theta \end{aligned} \tag{73}$$

where we have used

$$\boldsymbol{v}^\dagger(\mathbf{N}+1)\boldsymbol{v}(\mathbf{N}+1) = \frac{\mathbf{N}+1-\kappa}{\mathbf{N}+1}\ . \tag{74}$$

Similarly,

$$\begin{aligned} \boldsymbol{v}^\dagger \Theta^2 \boldsymbol{v} &= \boldsymbol{v}^\dagger(a_i^+ T^i_{a\,j} a^j\, \theta_a)(a_l^+ T^l_{b\,k} a^k\, \theta_b)\boldsymbol{v} \\ &= \frac{N-\kappa}{N}\Theta^2\ . \end{aligned} \tag{75}$$

Therefore

$$\begin{aligned} F &= \left((\frac{\mathbf{N}-\kappa}{\mathbf{N}})^2 - (\frac{\mathbf{N}-\kappa}{\mathbf{N}}) \right)\Theta^2 = \frac{-\kappa(\mathbf{N}-\kappa)}{\mathbf{N}^2\Lambda_N} i\eta \\ &= \frac{-\kappa N}{(N+\kappa)^2\Lambda_N} i\eta\ . \end{aligned} \tag{76}$$

Here we have used that when the field strength F is evaluated over the scalar field $\hat{\xi}$ and the number operator takes the value $\mathbf{N} = N + \kappa$.

For $\kappa < 0$, we have

$$F = \frac{|\kappa|(\mathbf{N} + n + |\kappa| + 1)}{(\mathbf{N} + n + 1)^2 \Lambda_N} i\eta = \frac{|\kappa|(N + n + 1)}{(N - |\kappa| + n + 1)^2 \Lambda_N} i\eta \ . \tag{77}$$

Thus in the large N limit, we obtain for all κ

$$F = -\kappa \sqrt{\frac{n}{2(n+1)}}\, i\eta \ . \tag{78}$$

5.2. *Computation of the Chern Number*

The case $N \to \infty$

In the classical limit, the first Chern number c_1 is given by $-\kappa$. In this case, we can integrate the symplectic form η' over the cycle $y_1^2 + y_2^2 + y_3^2 = \frac{n+1}{2n}$ in $\mathbb{C}P^n$ [23]. Using $f_{abc} = \varepsilon_{abc}$ for $a, b, c \in \{1, 2, 3\}$, we get

$$\int_{S_R^2} \eta' = \int_{S_R^2} \frac{1}{2} \varepsilon_{abc} y_a dy_b dy_c = 4\pi R^3 \tag{79}$$

where the sphere has radius $R^2 = \frac{n+1}{2n}$. Therefore using (78), the first Chern number is

$$c_1 = \frac{i}{2\pi} \int_{S_R^2} F = -\kappa \tag{80}$$

in the commutative limit $N \to \infty$. This means that the bundles constructed above should be interpreted as noncommutative versions of the classical monopole bundles with Chern number $c_1 = -\kappa$.

The case for finite N

In the fuzzy case (i.e. for finite N), it is difficult to give a satisfactory definition of Chern numbers. One reason for this is the lack of a differential calculus with appropriate dimensions for finite N. However, it is known e.g. from recent investigations of fuzzy spheres [8] that it is still possible to write down suitable integrals in the fuzzy case, which reproduce the usual Chern numbers in the limit of large N (i.e. the commutative limit). These integrals are neither topological nor integer for finite N. We will refer to them as "asymptotic" Chern numbers.

We will illustrate this for fuzzy $\mathbb{C}P_N^n$ by giving a prescription to calculate such an "asymptotic" Chern number c_1, integrating $\frac{i}{2\pi} F$ over a suitable "fuzzy sub-sphere".

5.3. *Fuzzy sub-spheres*

In the classical case, c_1 is obtained by integrating $\frac{i}{2\pi}F$ over any 2-sphere in $\mathbb{C}P^n$. Since $\mathbb{C}P^n_N$ is defined in terms of a simple matrix algebra, it does not admit non-trivial subspaces $\mathbb{C}P^n_N/\mathcal{I}$ defined by some two-sided ideal $\mathcal{I}$. To compute the "asymptotic" Chern number in the fuzzy geometry we have to relax the concept of a subspace. A natural way to do this is the following:

Given a root α of $su(n+1)$, then $V_N \cong \mathcal{F}_N$ decomposes into a direct sum of irreps of $su(2)_\alpha \subset su(n+1)$. Let us fix a root α. Then there is precisely one irrep denoted by $H^{\alpha,N}$ with maximal dimension $N+1$ (note that the weights of V_N form a simplex in weight lattice of size $N+1$). The other irreps have smaller dimension $M+1 \leq N+1$. We denote them by $H^{\alpha,M}$. Let $P^{\alpha,M} : V_N \to H^{\alpha,M}$ be the projector on $H^{\alpha,M}$. Then we can define maps

$$\begin{aligned} \mathbb{C}P^n_N \cong End(V_N) &\to End(H^{\alpha,M})) \\ f &\mapsto \hat{f} := P^{\alpha,M} f P^{\alpha,M} \; . \end{aligned} \tag{81}$$

In principle, each

$$S^2_{\alpha,M} := End(H^{\alpha,M}) \tag{82}$$

could be considered as a fuzzy sphere, but not necessarily as a sub-sphere of $\mathbb{C}P^n_N$. The reason is that the projected coordinate generators do not satisfy the constraint $g_{ab}x_a x_b = 1$ of $\mathbb{C}P^n_N$. However by relaxing this constraint and allowing for 'quantum' corrections of order $1/N$,

$$g_{ab}\hat{x}_a\hat{x}_b = 1 - O(1/N) \; . \tag{83}$$

we can show that eq.83 is satisfied by $H^{\alpha,N}$ only, i.e. the shpere with maximal dimension $N+1$ [23].

Therefore, we can consider the maximal $S^2_{\alpha,N}$ as an "asymptotic sub-sphere" of $\mathbb{C}P^n_N$, in the sense that it becomes the algebra of functions on a sub-sphere of $\mathbb{C}P^n$ in the large N limit. For $N \to \infty$ it coincides with the non-trivial 2-cycles found before, for a suitable choice of the root α.

We shall use this property in the calculation of asymptotic first Chern numbers below.

5.4. *Asymptotic first Chern number*

We define the first Chern number c_1 as the integral of $\frac{i}{2\pi}F$ over a specific fuzzy $S^2_{\alpha,N} \cong \mathbb{C}P^1_N$ which is asymptotic subspace of $\mathbb{C}P^n_N$ in the sense discussed above.

To specify this $S^2_{\alpha,N}$, we first choose a Hilbert space $H^{\alpha,N}$ with maximal dimension, denoted by $\mathcal{F}_S$ as

$$\mathcal{F}_S = \{|\psi\rangle \in \mathcal{F}_N;\ \ a_i|\psi\rangle = 0 \text{ for } \ i = 3, ...n+1 \text{ and } \mathbf{N}|\psi\rangle = N|\psi\rangle\}\ . \quad (84)$$

The space $\mathcal{F}_S$ has the same dimension as the Hilbert space of fuzzy $\mathbb{C}P^1_N$. The generators associated with the $su(2)$ rotations of this fuzzy $\mathbb{C}P^1$ are

$$T_m = \frac{1}{2}\begin{pmatrix} \sigma^m & \cdots & 0 \\ \vdots & \ddots & \vdots \\ 0 & \cdots & 0 \end{pmatrix}\ . \quad (85)$$

The corresponding coordinates are linear combinations of coodinate operators of $\mathbb{C}P^n_N$, and we denote them by $\hat{x}_m = \Lambda_N a_i^+ T^m_{ij} a_j$. The radius of fuzzy $\mathbb{C}P^1$ is defined by these coordinates $\hat{x}_m$ as

$$\sum_{m=1}^{3} (\hat{x}_m)^2 = \frac{1}{4}\Lambda_N^2 N(N+2) = R_N^2\ . \quad (86)$$

Representing the algebra generated by $\hat{x}_m$ on $\mathcal{F}_S$, we obtain precisely the matrix algebra of fuzzy $\mathbb{C}P^1_N$. Therefore we can use the standard results of the integration over $\mathbb{C}P^1_N$.

We introduce the volume element of this fuzzy $\mathbb{C}P^1_N$,

$$\omega = \frac{1}{2R_N}\epsilon_{mnp}\hat{x}_m d\hat{x}_n d\hat{x}_p \quad \in \Omega^2\ .$$

This is an invariant 2-form and it agrees with the volume element of the sphere with radius $R = \sqrt{\frac{n+1}{2n}}$ in the commutative limit $N \to \infty$. Then the integration $\int : \Omega^2 \to \mathbb{C}$ over this fuzzy $\mathbb{C}P^1_N$ is defined by [8]

$$\frac{1}{2}\int_{\mathbb{C}P^1_N} f\epsilon_{mnp}\hat{x}_m d\hat{x}_n d\hat{x}_p \equiv 4\pi R_N^3 Tr_{\mathcal{F}_S}\{f\}\ , \quad (87)$$

where $Tr_{\mathcal{F}_S}$ denotes the trace over $\mathcal{F}_S$ normalized such that $Tr_{\mathcal{F}_S}\{1\} = 1$. With the commutation relations of $\hat{x}_m$ and the definition of the derivatives defined previously we obtain

$$\omega = R_N\left(\frac{1}{\mathbf{N(N+2)}} - \frac{1}{2}\right)\epsilon_{mnp}\hat{x}_m\hat{\theta}_n\hat{\theta}_p. \quad (88)$$

where the $\hat{\theta}_m$, m=1,2,3 is the one form over $\mathbb{C}P^1_N$.

The Chern character c_1 is defined by using the projection operators $\mathcal{P}_\kappa$. To evaluate it we first take the $\mathbb{C}P^n_N$ -valued trace $Tr_\kappa\{\cdot\}$ over $End(\Gamma_\kappa(\mathbb{C}P^n_N))$, and then integrate over $\mathbb{C}P^1_N$:

$$c_1 = \frac{i}{2\pi}\int_{\mathbb{C}P^1_N} Tr_\kappa\{\mathcal{P}_\kappa d\mathcal{P}_\kappa d\mathcal{P}_\kappa\} = \frac{i}{2\pi}\int_{\mathbb{C}P^1_N}\sum_\mu v_\mu F_\kappa v_\mu^\dagger \tag{89}$$

where F_κ is the field strength F given either in eq.(76) or eq.(77), depending on the value of κ.

Using the definitions (53), (55) of v_μ we can perform the summation over μ and find the following expression for the integrand:

For $\kappa > 0$:

$$\sum_\mu v_\mu F_\kappa v_\mu^\dagger = \frac{(\mathbf{N}+n+\kappa+1)\cdots(\mathbf{N}+n+2)}{(\mathbf{N}+1)\cdots(\mathbf{N}+\kappa)}\frac{-i\kappa\mathbf{N}}{2(\mathbf{N}+\kappa)^2\Lambda_N} f_{abc}\, x_a\theta_b\theta_c\,. \tag{90}$$

For $\kappa < 0$:

$$\sum_\mu v_\mu F_\kappa v_\mu^\dagger$$
$$= \frac{-i\kappa(\mathbf{N}-|\kappa|)\cdots(\mathbf{N}-1)}{(\mathbf{N}+n-|\kappa|+1)\cdots(\mathbf{N}+n)}\frac{(\mathbf{N}+n+1)}{2(\mathbf{N}-|\kappa|+n+1)^2\Lambda_N} f_{abc}\, x_a\theta_b\theta_c\,. \tag{91}$$

Theses are 2-forms over $\mathbb{C}P^n_N$. In order to integrate them over the fuzzy sub-space $\mathbb{C}P^1_N$, we should also pull-back these 2-forms to $\mathbb{C}P^1_N$. To do this we split the coordinates into the two orthogonal sets, $(x_m, x_{m^\perp})$, where x_m corresponds to the $SU(2)_\alpha$. Correspondingly, we split the one-forms into $(\theta_m, \theta_{m^\perp})$. This means that θ_m is the dual of $\partial_m = \frac{1}{\Lambda_N} ad_{x_m}$, analogously to the commutative case. Now we define the pull-back by projecting out $\theta_{m^\perp}$ and identify $\hat\theta_m$ with θ_m.

Since $\mathbb{C}P^1_N$ corresponds to a $su(2)$ subalgebra of $su(n+1)$, this implies that the pull-back of $f_{abc}\, x_a\theta_b\theta_c$ is $\epsilon_{mnp}\hat{x}_m\hat\theta_n\hat\theta_p$. This can now be integrated over fuzzy $\mathbb{C}P^1_N$ with radius R_N according to (87), and we get from (89) for $\kappa > 0$

$$\begin{aligned} c_1 &= \frac{i}{2\pi}\frac{(N+n+\kappa+1)\cdots(N+n+2)}{(N+1)\cdots(N+\kappa)}\frac{-i\kappa N}{2(N+\kappa)^2\Lambda_N}\int_{\mathbb{C}P^1_N}\epsilon_{mnp}(\hat{x}_m\hat\theta_n\hat\theta_p) \\ &= -\kappa\frac{(N+n+\kappa+1)\cdots(N+n+2)}{(N+1)\cdots(N+\kappa)}\frac{N}{(N+\kappa)^2}\frac{\sqrt{N(N+2)}}{(1-\frac{2}{N(N+2)})}\,. \end{aligned} \tag{92}$$

For large N this yields

$$c_1 = -\kappa + \frac{1}{N}(\kappa(n-1)+1) + \cdots\,. \tag{93}$$

In the same way we get for $\kappa < 0$:

$$c_1 = -\kappa \frac{(N-|\kappa|)\cdots(N-1)}{(N+n-|\kappa|+1)\cdots(N+n)} \frac{(N+n+1)}{(N-|\kappa|+n+1)^2} \frac{\sqrt{N(N+2)}}{(1-\frac{2}{N(N+2)})} . \tag{94}$$

The expansion with respect to $\frac{1}{N}$ is

$$c_1 = -\kappa(1 + \frac{1}{N}(-\kappa(n-1) - n) + \cdots) . \tag{95}$$

6. Discussion

To summarize, we investigated the definition of fuzzy complex projective space $\mathbb{C}P^n_N$ from two different points of view, and constructed the nontrivial $U(1)$ bundles over those spaces. The corresponding Chern classes were calculated.

One approach taken here is to consider $\mathbb{C}P^n$ as coadjoint orbit, given by $(n+1)\times(n+1)$ matrices Y which satisfy a certain characteristic equation. The quantized function algebra is given by a simple matrix algebra $Mat(D_N, \mathbb{C})$, i.e. $End_{\mathbb{C}}(V_N)$ for certain irreducible representations V_N of $su(n+1)$. This leads to an algebra-valued $(n+1)\times(n+1)$ matrix X, whose characteristic equation contains the explicit relations satisfied by the fuzzy coordinate functions.

The second approach uses the generalized Hopf fibration $U(1) \to S^{2n+1} \to \mathbb{C}P^n$. This also leads to a characteristic equation for a certain operator-valued matrix, which coincides with the one of the first approach when we specify the Fock space representation $\mathcal{F}_N \cong V_N$. The second construction is very useful to define projective modules.

Then the projective modules are defined by specifying the projection operator in terms of a normalized vector, following the approach for monopoles on S^2. We find nontrivial projective modules of $\mathbb{C}P^n_N$ labeled by an integer κ, which are interpreted as fuzzy version of the monopole bundles on $\mathbb{C}P^n$ with monopole number κ. Using a suitable differential calculus, we calculated the field strength over the monopole bundle, i.e. the first Chern class. We verified that the usual Chern number c_1 is recovered in the commutative limit.

Since fuzzy spaces arise naturally in string theory, for example as D-branes on group manifolds or as solutions of the IKKT matrix model, one can expect that the low-energy effective action should be given by an induced gauge theory.

Since noncommutativity changes the geometrical structure, a consequence of this is that noncommutative gauge theories have degrees of freedom which are not tangential, unlike as in conventional field theories [22]. Such degrees of freedom also appear in the differential calculus on fuzzy spaces. Another feature is the fact that we do not have sub-spaces in the strict sense, which causes complications when we want to compute topological numbers. Especially this latter property requires further investigation if we want to generalize this scheme to instantons. Also the formulation of fermions, so far only discussed on the fuzzy sphere, still lacks of a fully satisfactory formulation for $\mathbb{C}P^n$.

During the preparation of this proceedings the article [24] appeared in which similar problems are discussed.

References

1. G.Alexanian, A.P.Balachandran, G.Immirzi and B.Ydri, *Fuzzy $\mathbb{C}P^2$.* J.Geom.Phys. **42** (2002), 28-53.
2. A. Yu. Alekseev, A. Recknagel and V. Schomerus, *Non-commutative world-volume geometries: Branes on SU(2) and fuzzy spheres.* JHEP **9909** (1999), 023.
3. S. Baez, A. P. Balachandran, S. Vaidya, B. Ydri, *Monopoles and Solitons in Fuzzy Physics.* Commun.Math.Phys. **208** (2000), 787-798.
4. A.P. Balachandran, B. P. Dolan, J. Lee, X. Martin, D. O'Connor, *Fuzzy Complex Projective Spaces and their Star-products.* J.Geom.Phys. **43** (2002), 184-204
5. G. Felder, J. Fröhlich, J. Fuchs, C. Schweigert, *The geometry of WZW branes,* J.Geom.Phys. **34** (2000), 162-190.
6. K. Furuuchi, *Instantons on Noncommutative $\mathbb{R}^4$ and Projection Operators* Prof. Theor. Phys. **103** (2000), 1043-1068.
7. H. Grosse, C. Klimcik and P. Presnajder, *Topologically nontrivial field configurations in noncommutative geometry.* Commun.Math.Phys. **178** (1996), 507-526.
8. H. Grosse, C.W. Rupp, A. Strohmaier, *Fuzzy Line Bundles, the Chern Character and Topological Charges over the Fuzzy Sphere.* J.Geom.Phys. **42** (2002), 54-63.
9. H. Grosse and A. Strohmaier, *Towards a Nonperturbative Covariant Regularization in 4D Quantum Field Theory.* Lett.Math.Phys. **48** (1999), 163-179.
10. H. Ishikawa and S. Watamura, *Free Field Realization of D-brane in Group Manifold.* JHEP **0008** (2000), 044.
11. D. Karabali and V.P. Nair, *Quantum Hall Effect in Higher Dimensions.* Nucl. Phys. **B641** (2002), 533-546.
12. G. Landi, *Deconstructing Monopoles and Instantons.* Rev.Math.Phys. **12** (2000), 1367-1390.

13. G. Landi, *Projective Modules of Finite Type and Monopoles over S^2*. J. Geom. Phys. **37** (2001), 47-62.
14. J. Madore, *Quantum mechanics on a fuzzy sphere.* Phys. Lett. **B263** (1991), 245-247.
15. N. Nekrasov and A. Schwarz, *Instantons on noncommutative $\mathbb{R}^4$ and (2,0) superconformal six dimensional theory.* Commun. Math. Phys. **198** (1998), 689-703.
16. N. Nekrasov, *Trieste lectures on solitons in noncommutative gauge theories.* **hep-th/0011095**
17. J. Pawelczyk, H. Steinacker *A quantum algebraic description of D-branes on group manifolds.* Nucl.Phys. **B638** (2002), 433-458.
18. H. Steinacker, *Quantized Gauge Theory on the Fuzzy Sphere as Random Matrix Model.* Nucl.Phys. **B679** (2004), 66-98.
19. P. Valtancoli, *Projectors for the fuzzy sphere.* Mod.Phys.Lett. **A16** (2001), 639-646; P. Valtancoli, *Projectors, matrix models and noncommutative monopoles. hep-th/0404045.*
20. U. Carow-Watamura and S. Watamura, *Chirality and Dirac Operator on Noncommutative Sphere.* Commun. Math. Phys. **183** (1997), 365-382; *Noncommutative Geometry and Gauge Theory on Fuzzy Sphere Commun. Math. Phys.* **212** (2000), 395-413.
21. U. Carow-Watamura and S. Watamura, *Differential Calculus on Fuzzy Sphere and Scalar Field.* Int. J. Mod. Phys. **A13** (1998), 3235-3244.
22. U. Carow-Watamura and S. Watamura, *Gauge Theory on the Quantum Sphere.* talk given at the International Workshop on Noncommutative Geometry and String Theory, March 16-22, 2001, Keio U., Yokohama.
23. U. Carow-Watamura, H. Steinacker and S. Watamura, *Monopole Bundles over Fuzzy Complex Projective Spaces.* J. Geom. Phys. (in press), **hep-th/0404130**.
24. P. Valtancoli, *Projectors, matrix models and noncommutative instantons.* **hep-th/0404046**

CONSTRUCTION OF LAGRANGIAN EMBEDDINGS USING HAMILTONIAN ACTIONS

R. CHIANG

Department of Mathematics,
National Cheng Kung University,
Tainan 701, Taiwan
riverch@mail.ncku.edu.tw

This note describes certain ways to construct Lagrangian embeddings in closed manifolds. In the presence of Hamiltonian G-actions, we use symplectic reductions and symplectic cuts to produce examples. The Lagrangian submanifolds we obtain in this fashion are G-invariant.

1. Hamiltonian Actions

A symplectic manifold is a pair (M, ω) where M is a manifold and ω is a closed, nondegenerate 2-form. That is to say, the form ω satisfies $d\omega = 0$, and $\wedge^n \omega_x \neq 0$ for $n = \frac{1}{2}\text{dim}M$. One says that a G-action on the symplectic manifold is *symplectic* if any element g of G defines a diffeomorphism which preserves the symplectic form ω, namely, $g^*\omega = \omega$. The infinitesimal version of this can be written as $\mathcal{L}_{\xi_M}\omega = 0$ for all $\xi \in \mathfrak{g}$. Here ξ_M denotes the induced vector field $\frac{d}{dt}\exp t\xi \cdot x$ of ξ on M. By the Cartan formula and the fact that ω is closed, we see that $d\iota(\xi_M)\omega = 0$. We say that a symplectic G-action is *Hamiltonian* if there exists a *moment map*

$$\Phi\colon M \to \mathfrak{g}^*$$

such that it is equivariant with respect to the G-action on M and the coadjoint action on $\mathfrak{g}^*$, and such that

$$\iota(\xi_M)\omega = -d\langle \Phi, \xi \rangle$$

for all ξ in $\mathfrak{g}$.

We define the *symplectic quotient* or *reduced space* at $\alpha \in \mathfrak{g}^*$ to be the topological space

$$M_\alpha = \Phi^{-1}(G \cdot \alpha)/G = \Phi^{-1}(\alpha)/G_\alpha$$

where $G \cdot \alpha$ is the coadjoint orbit through α and G_α is the stabilizer of α for the coadjoint action. If G acts on $\Phi^{-1}(G \cdot \alpha)$ freely, the Marsden-Weinstein-Marle Theorem asserts that the reduced space is a symplectic manifold. In the following diagram

$$\begin{array}{ccc} Z_\alpha & \xrightarrow{\ i\ } & (M, \omega) \\ \downarrow{\scriptstyle \pi} & & \\ (M_\alpha, \omega_\alpha) & & \end{array}$$

where Z_α denotes the preimage $\Phi^{-1}(G \cdot \alpha)$ and ω_α denotes the symplectic form on the reduced space, we have $i^*\omega = \pi^*\omega_\alpha$. In general, the reduced space is a symplectic stratified space; see for instance [9].

Example 1.1. Let S^1 act on $(\mathbb{C}, idz \wedge d\bar{z}) = (\mathbb{C}, 2dx \wedge dy)$ by

$$t \cdot z = t^k z.$$

This is a Hamiltonian action with moment map

$$\Phi(z) = -k|z|^2 + \text{const} \in \mathbb{R}.$$

Example 1.2. It is customary to play with this constant to achieve one's purposes. Consider $\mathbb{C}^{n+1}$ with the diagonal action of S^1 and moment map $\Phi(z) = 1 - |z|^2$. The zero level set $Z_0 = \Phi^{-1}(0)$ is the unit sphere S^{2n+1}. The symplectic quotient at 0 is

$$M_0 = Z_0/S^1 = \mathbb{C}\mathrm{P}^n .$$

Example 1.3. Consider a sphere S^2 with its standard area form. Let S^1 act on S^2 by rotations about a vertical axis. This action is the compact version of Example 1.1. The moment map is the height function.

There are some remarks one can make regarding to the existence of moment maps. One of them states that if G is a semisimple group, then any symplectic G-action is Hamiltonian. Therefore we compute the moment maps for all the irreducible complex representations of SU(2) in the following example.

Example 1.4. The moment maps $\Phi\colon \mathbb{C}^n \to \mathfrak{su}(2)^* = \mathbb{R}^3$ (up to constants) of the n-dimensional irreducible complex representations of SU(2) with respect to the standard symplectic forms are

$$\Phi(z) = \begin{pmatrix} \frac{n-1}{2}|z_0|^2 + \frac{n-3}{2}|z_1|^2 + \cdots + \frac{3-n}{2}|z_{n-2}|^2 + \frac{1-n}{2}|z_{n-1}|^2 \\ z_0\bar{z}_1 + z_1\bar{z}_2 + \cdots + z_{n-2}\bar{z}_{n-1} \end{pmatrix}$$

where $z = (z_0, \ldots, z_{n-1}) \in \mathbb{C}^n$.

The symplectic reduction can be carried out in stages. If two Hamiltonian group actions commute with each other, then one group action would survive the reduction of the other and descend to a Hamiltonian group action on the reduced space. Under the usual assumptions of regularity, it makes sense to take two consecutive reductions. It turns out that the order does not matter. The resulting spaces are isomorphic as symplectic manifolds.

2. Lagrangian Submanifolds

A submanifold L of a symplectic manifold (M, ω) is *Lagrangian* if at every point $q \in L$ the tangent space T_qL is a Lagrangian subspace of T_qM. This means that $(T_qL)^\perp = T_qL$, or equivalently, $\dim L = \frac{1}{2}\dim M$ and $\omega|_{TL} = 0$. An embedding (or an immersion) $f\colon N \to (M, \omega)$ is Lagrangian if $2\dim N = \dim M$ and $f^*\omega = 0$.

Symplectomorphisms $\varphi\colon M_1^n \to M_2^n$ between symplectic manifolds (M_1, ω_1) and (M_2, ω_2) with the same dimension give rise to Lagrangians in the product manifold $(M_1, \omega_1) \times (M_2, -\omega_2)$, i.e., the symplectic manifold $M_1 \times M_2$ with symplectic form $\omega = \pi_1^*\omega_1 + \pi_2^*(-\omega_2)$, where π_i is the projection. In fact, the graph $\Gamma_\varphi = \{(x, \varphi(x)) \in M_1 \times M_2\}$ of a diffeomorphism $\varphi\colon M_1^n \to M_2^n$ is Lagrangian in $(M_1, \omega_1) \times (M_2, -\omega_2)$ if and only if φ is a symplectomorphism. Just consider the embedding

$$f\colon M_1 \to M_1 \times M_2, \qquad f(x) = (x, \varphi(x))$$

and verify that

$$\begin{aligned} f^*\omega &= f^*\pi_1^*\omega_1 + f^*\pi_2^*(-\omega_2) \\ &= \omega_1 - \varphi^*\omega_2 = 0 \end{aligned}$$

There is a similar construction in the presence of a Hamiltonian action:

Lemma 2.1. *Let G be a compact Lie group and let (M, ω) be a symplectic manifold. Assume G acts on M in a Hamiltonian fashion with moment map Φ. Denote by Z the preimage of zero of the moment map and by M_{red} the reduced space Z/G. If G acts on Z freely, there exists a Lagrangian embedding of Z into $M \times M_{red}$ with the symplectic form $\omega \oplus -\omega_{red}$.*

Proof. Since G acts on $Z = \Phi^{-1}(0)$ freely, 0 is a regular value for the moment map Φ, and $Z \to M_{\text{red}}$ is a principal G-bundle. The level set $\Phi^{-1}(0)$ is a manifold of dimension $\dim M - \dim G$ and the reduced space M_{red} is a manifold of dimension $\dim M - 2\dim G$. So

$$2\dim Z = \dim(M \times M_{\text{red}}).$$

Let $i\colon Z \to M$ be the inclusion map and $\pi\colon Z \to M_{\text{red}} \cong Z/G$ be the projection. There exists a unique symplectic form ω_{red} on the reduced space M_{red} such that $\pi^*\omega_{\text{red}} = i^*\omega$. The two-form $\tilde{\omega} = \mathrm{pr}_1^*\,\omega - \mathrm{pr}_2^*\,\omega_{\text{red}}$ is then a symplectic form on $M \times M_{\text{red}}$. We have a natural embedding $f\colon Z \to M \times M_{\text{red}}$ given by

$$z \mapsto (i(z), \pi(z)).$$

Now

$$\begin{aligned} f^*\tilde{\omega} &= f^*\mathrm{pr}_1^*\,\omega - f^*\mathrm{pr}_2^*\,\omega_{\text{red}} \\ &= (\mathrm{pr}_1 \circ f)^*\omega - (\mathrm{pr}_2 \circ f)^*\omega_{\text{red}} \\ &= i^*\omega - \pi^*\omega_{\text{red}}. \end{aligned}$$

Therefore $f^*\tilde{\omega} = 0$ and f is a Lagrangian embedding. □

The following example uses the circle action we have seen before.

Example 2.1. Consider the circle action on $\mathbb{C}^{n+1}$ as in Example 1.2 and its moment map $\Phi(z) = 1 - \sum |z_j|^2$. The zero level set $Z = \{\sum |z_j|^2 = 1\} \cong S^{2n+1}$ and the reduced space M_{red} is $Z/S^1 \cong \mathbb{C}\mathrm{P}^n$. We have the following diagram

$$\begin{array}{ccc} S^{2n+1} & \longrightarrow & \mathbb{C}^{n+1} \\ \downarrow & & \\ \mathbb{C}\mathrm{P}^n & & \end{array}$$

Denote by $S^{2n+1} \subset \mathbb{C}^{n+1}$ the unit sphere and by $h\colon S^{2n+1} \to \mathbb{C}P^n$ the Hopf map. Then

$$S^{2n+1} \ni z \longmapsto (h(z), \bar{z}) \in (\mathbb{C}P^n \times \mathbb{C}^{n+1}, \sigma_{\text{std}} \oplus \omega_{\text{std}})$$

is a Lagrangian embedding. Here ω_{std} is the standard symplectic form of $\mathbb{C}^{n+1} = \mathbb{R}^{2n+2}$ and σ_{std} is the standard Kähler form on $\mathbb{C}P^n$.

Since the image is bounded, we can embed the image into $\mathbb{C}\mathrm{P}^n \times T$, where T denotes a torus in $\mathbb{C}^{n+1}$. More generally, note that the image of the above embedding lies in $\mathbb{C}P^n \times B^{2n+2}$, where B^{2n+2} is the $(2n+2)$-dimensional closed unit ball in $\mathbb{C}^{n+1}$. Given a $(2n+2)$-dimensional symplectic manifold (X^{2n+2}, ω_X), we can choose $a > 0$ large enough so that $(X, a\omega_X)$ admits a symplectic embedding of a closed ball of radius 1, say $\varphi\colon B^{2n+2} \to (X, a\omega_X)$. Then

$$S^{2n+1} = \partial B^{2n+2} \ni z \longmapsto \big(h(z), \varphi(\bar{z})\big) \in (\mathbb{C}P^n \times X, \sigma_{\text{std}} \oplus a\omega_X)$$

is a Lagrangian embedding into a closed manifold.

The assumption that G acts on $Z = \Phi^{-1}(0)$ freely guarantees that the dimension of Z is half of the dimension of the product of the manifold M and the reduced space M_{red}. For example, if Z is a fixed point set, then $\dim Z = \dim M_{\text{red}}$. This assumption also asserts that Z and M_{red} are smooth manifolds. In general, M_{red} is a stratified space and Z consists of singularities.

One way to relax the assumption is to consider the orbits in Z. We have the following lemma.

Lemma 2.2. *Let a compact Lie group G act on a symplectic manifold (M, ω) in a Hamiltonian fashion with a moment map $\Phi\colon M \to \mathfrak{g}^*$. Assume $\alpha \in \mathfrak{g}^*$ is a fixed point of the coadjoint action. Then for any $m \in \Phi^{-1}(\alpha)$, the orbit $G \cdot m$ is an isotropic submanifold of M.*

Proof. An orbit is smooth. We only need to show that it is isotropic. Let ξ and η be any vector fields in the Lie algebra $\mathfrak{g}$ of G. Let ξ_M and η_M denote their induced vector fields on M and $\eta_{\mathfrak{g}^*}$ denote the induced vector field of η on $\mathfrak{g}^*$. For $\Phi(m) = \alpha$, we have

$$\begin{aligned}\omega(m)(\xi_M(m), \eta_M(m)) &= -d\langle \Phi(m), \xi\rangle(\eta_M(m)) \\ &= -\langle \eta_{\mathfrak{g}^*}(\Phi(m)), \xi\rangle \\ &= -\langle 0, \xi\rangle\end{aligned}$$

Therefore $\omega(m) = 0$ for all $m \in \Phi^{-1}(\alpha)$. In particular, the orbit $G \cdot m$ is a smooth isotropic submanifold of M. □

A Lagrangian submanifold is an isotropic submanifold whose dimension is half of the dimension of the ambient manifold. By equivariant isotropic embedding theorem and a dimension count, we know that if there is a Lagrangian orbit in Z, it is the only orbit in Z.

Example 2.2. Consider the irreducible SU(2) representation on $\mathbb{C}^4$. By Example 1.4, the moment map for this representation as a map from $\mathbb{C}^4$ into $\mathbb{R} \times \mathbb{C} \cong \mathfrak{su}(2)^*$ is given by

$$(u_0, u_1, u_2, u_3) \mapsto \begin{pmatrix} \frac{3}{2}|u_0|^2 + \frac{1}{2}|u_1|^2 - \frac{1}{2}|u_2|^2 - \frac{3}{2}|u_3|^2 \\ u_0\bar{u}_1 + u_1\bar{u}_2 + u_2\bar{u}_3 \end{pmatrix}$$

The circle S^1 acts diagonally on $\mathbb{C}^4$ as in Example 1.2. Its reduced space is $\mathbb{CP}^3$. Note that these two actions commute, and therefore the action of SU(2) descends to an action on $\mathbb{CP}^3$. This SU(2) action is not effective on $\mathbb{CP}^3$, but it induces an effective SO(3) action.

To investigate the isotropic orbits for the SO(3) action on $\mathbb{C}P^3$, we use the procedure of reduction in stages. We note that $(1,0,0,1)$ lies in the zero level set of the SU(2)-moment map. Therefore, the point $[1,0,0,1]$ lies in the zero level set of the induced SO(3)-moment map on $\mathbb{C}P^3$. By studying the stabilizer, we derive that $SO(3)/D_3 \cong \mathbb{R}P^3/D_3$ is a Lagrangian submanifold in $\mathbb{C}P^3$, where D_3 denotes the dihedral group.

3. Symplectic Cuts

In principle, a neighborhood of an orbit is well-understood. When there is a unique Lagrangian orbit in the zero level set of the moment map, naively we may attempt to compactify its neighborhood to obtain a Lagrangian embedding of the same orbit into a closed manifold.

Let's take a second look at Hamiltonian SU(2) or SO(3) actions more carefully. As Example 2.2 suggests, since SU(2) is a double cover of SO(3), their treatments are often similar, and it is convenient to consider both groups at the same time. It is well-known that SU(2) and SO(3) are rank one nonabelian groups. Hamiltonian SU(2) and SO(3) actions naturally induce Hamiltonian circle actions. With the help of these Hamiltonian circle actions, we may employ symplectic cutting, which is a procedure originated from symplectic reduction.

We take a digression here to recall what is a symplectic cut. Consider the standard circle action on the complex plane, $(\mathbb{C}, -idz \wedge d\bar{z})$, with moment map $|z|^2$. Suppose that (M, ω) is a symplectic manifold with a Hamiltonian circle action and a moment map $\mu\colon M \to \mathbb{R}$. Then the diagonal action makes $(M \times \mathbb{C}, \omega \oplus -idz \wedge d\bar{z})$ a Hamiltonian S^1-space with moment map $\Phi(m,z) = \mu(m) + |z|^2$. Let $\epsilon \in \mathbb{R}$. The *cut space* is defined to be the reduced space

$$\begin{aligned} \overline{M}_{\mu\leq\epsilon} &= \Phi^{-1}(\epsilon)/S^1 = \{(m,z) \in M \times \mathbb{C} : \mu(m) + |z|^2 = \epsilon\}/S^1 \\ &= \mu^{-1}(\epsilon)/S^1 \cup \{\mu < \epsilon\}. \end{aligned}$$

Proposition 3.1. *The cut space is smooth if S^1 acts freely on $\mu^{-1}(\epsilon)$. The cut space is symplectic. If M is equipped with a G-action that commutes with the circle action, then the cut space inherits the G-action.*

For details of the proof, see [7]. Note the symplectic cut is a local procedure. It can be carried out as long as the circle action is well-defined near the cut.

Now let's consider symplectic cutting on a manifold with Hamiltonian SU(2) or SO(3) actions. Let G denote SU(2) or SO(3), and let $\Phi\colon M \to \mathbb{R}^3$

denote the moment map. For each $\alpha \neq 0 \in \mathfrak{g}^*$, its stabilizer G_α of the coadjoint action is S^1, and its coadjoint orbit $G \cdot \alpha$ is a sphere S^2. The preimage of Φ of an open ray $\mathbb{R}_{>0} = \mathbb{R}_+ \smallsetminus \{0\}$ is called the *symplectic cross section.* Denote $X = \Phi^{-1}(\mathbb{R}_{>0})$. By the symplectic cross section theorem, X is symplectic, connected, of codimension 2 and it inherits a Hamiltonian circle action. In particular, the preimage $\Phi^{-1}(\mathbb{R}^3 \smallsetminus 0)$ is isomorphic to the twisted product $G \times_{S^1} X$, where S^1 denotes the circle fixing the chosen open ray $\mathbb{R}_{>0}$. X is equipped with a Hamiltonian action of this S^1.

In dimension six, let the zero level set $\Phi^{-1}(0) \cong G$. By isotropic embedding theorem, its neighborhood is modeled by $G \times \mathbb{R}^3$. Since $\Phi^{-1}(\alpha) \cong G$ for any $\alpha \in \mathbb{R}_{>0}$, the symplectic cross section $X \cong G \times (0, \infty)$. If we perform a symplectic cut on X along the open ray, at $\epsilon \in \mathbb{R}_{>0}$, the cut space $\overline{X}$ contains a two-sphere S^2 fixed under the circle action at ϵ. Namely, $\overline{X} = S^2 \cup (G \times (0, \epsilon))$. The resulting space M from $G \times \mathbb{R}^3$ after the cut satisfies $M \smallsetminus G \cong G \times_{S^1} \overline{X}$.

The moment image of a circle action gives some topology information regarding the manifold. For a circle action, the moment map is a perfect Morse-Bott function whose critical points are the fixed points of the circle action. Therefore $\dim H^j(M) = \sum \dim H^{j-i(F)}(F)$, where we sum over the connected components of the fixed point set, and $i(F)$ denotes the index of each fixed point set.

On a manifold equipped with a Hamiltonian G-action, we can always restrict to a circle subgroup and consider the manifold as a Hamiltonian S^1-space. The above-mentioned symplectic cut gives rise to fixed spheres of index 0 and 4 at $-\epsilon$ and ϵ respectively. Therefore the cohomology groups of the resulting space M are

$$H^k(M) = \begin{cases} \mathbb{Z}, & k = 0, 2, 4, 6; \\ 0, & k = 1, 3, 5. \end{cases}$$

Note that Seidel [8], and Biran and Cieliebak [2] have proved separately that $\mathbb{CP}^n$ has no simply connected Lagrangian submanifolds. We recover the following two Hamiltonian actions.

Example 3.1. The real projective space $\mathbb{RP}^3$ is a Lagrangian submanifold of the complex projective space $\mathbb{CP}^3$.

Let SU(2) act on $\mathbb{C}^2 \times \mathbb{C}^2$ diagonally. Since this action commutes with the circle action described in Example 1.2, it induces an SO(3) action on

$\mathbb{CP}^3$. As in Example 1.4, the moment map is

$$\Phi((x,y),(z,w)) = \begin{pmatrix} |x|^2 - |y|^2 + |z|^2 - |w|^2 \\ x\bar{y} + z\bar{w} \end{pmatrix}$$

for $(x, y) \in \mathbb{C}^2$, and $(z, w) \in \mathbb{C}^2$. Then $\Phi^{-1}(0) = \mathbb{RP}^3 = \mathrm{SO}(3) = \mathrm{SU}(2)/\mathbb{Z}_2$ is a Lagrangian orbit in $\mathbb{CP}^3$. The principal fiber $\Phi^{-1}(\alpha)$ for a generic $\alpha \in \Phi(\mathbb{CP}^3)$ is SO(3). Another special fiber $\Phi^{-1}(\epsilon) = S^2$ occurs when $\epsilon \in \mathbb{R}_{>0} \cap \Phi(\mathbb{CP}^3)$ is of maximal length. Moreover, $\Phi^{-1}(G \cdot \epsilon) = S^2 \times S^2$.

Example 3.2. The 3-sphere S^3 is a Lagrangian submanifold of $G_2^+(\mathbb{R}^5)$, the real Grassmannian of oriented 2-planes in $\mathbb{R}^5$.

Note that $G_2^+(\mathbb{R}^5)$ is a coadjoint orbit of SO(5) and it is of real dimension six. Let SU(2) act on $G_2^+(\mathbb{R}^5)$ by the natural embedding of $\mathrm{SU}(2) \subset \mathrm{SO}(4) \subset \mathrm{SO}(5)$. The moment map Φ is the projection $\mathfrak{so}^*(5) \to \mathfrak{su}^*(2)$. The zero level set is therefore $\Phi^{-1}(0) = \mathrm{SU}(2) = S^3$, a Lagrangian orbit in $G_2^+(\mathbb{R}^5)$. Its principal fiber $\Phi^{-1}(\alpha)$ for a generic α in the moment image is S^3. For ϵ of maximal length in the moment image, $\Phi^{-1}(\epsilon) = S^2$ and $\Phi^{-1}(G \cdot \epsilon)$ is $S^2 \times S^2$.

4. Remarks and Generalizations

In the previous section, we perform a symplectic cut on the simple space $G \times \mathbb{R}^3$ where $G = \mathrm{SU}(2)$ or SO(3). If we begin with an arbitrary G-space, where we perform the symplectic cut makes a difference to the resulting space. Moreover, if the circle action is not free on $\mu^{-1}(\epsilon)$, the cut space is a symplectic stratified space. The construction of symplectic cuts generalizes to torus actions. If G is a compact Lie group of higher rank, we may generalize the construction in Section 3 using symplectic cuts with respect to its maximal torus.

The two examples in Section 3 can be extended using analogous actions in higher dimensions.

Consider the action of $\mathrm{SO}(n+1)$ on $\mathbb{CP}^n$ induced from the diagonal action on $\mathbb{R}^{n+1} \times \mathbb{R}^{n+1}$. For every $n \geq 2$, this action produces the real projective space $\mathbb{RP}^n$ as a Lagrangian submanifold in the complex projective space $\mathbb{CP}^n$.

The real Grassmannian $G_2^+(\mathbb{R}^{n+2}) \cong \frac{\mathrm{SO}(n+2)}{\mathrm{SO}(2)\times\mathrm{SO}(n)}$ of oriented 2-planes in $\mathbb{R}^{n+2}$ is a $2n$-dimensional coadjoint orbit of $\mathrm{SO}(n+2)$. If n is even, $\mathrm{SO}(n+1)$ acts through the natural embedding $\mathrm{SO}(n+1) \subset \mathrm{SO}(n+2)$. If n is odd, $\mathrm{SU}(k)$ acts through the natural embeddings $\mathrm{SU}(k) \subset \mathrm{SO}(n+1) \subset \mathrm{SO}(n+2)$ for $k = \frac{1}{2}(n+1)$. These are Hamiltonian actions. The zero level sets of the moment maps are n-spheres. Therefore we obtain Lagrangian spheres S^n in the real Grassmannian $G_2^+(\mathbb{R}^{n+2})$. Note $G_2^+(\mathbb{R}^4)$ is $S^2 \times S^2$.

These two sets of examples are related as in the following diagram.

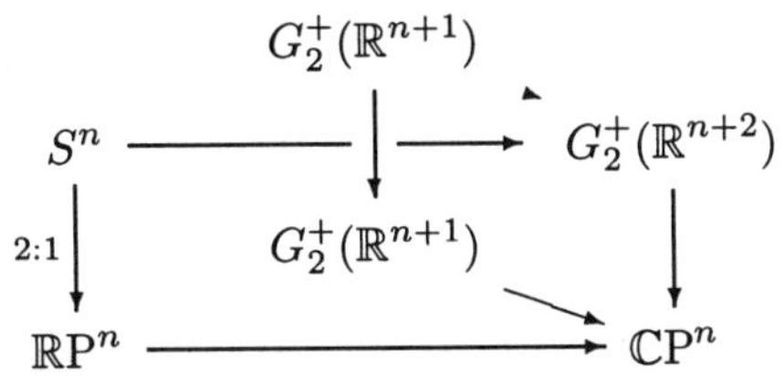

Finally, while many symplectic actions remain non-Hamiltonian, we make a remark on where to look for Lagrangian submanifolds invariant under a Hamiltonian action.

Lemma 4.1. *Let a compact Lie group G act on a symplectic manifold (M, ω) in a Hamiltonian fashion with moment map $\Phi\colon M \to \mathfrak{g}^*$. Assume L is a connected Lagrangian submanifold invariant under the action. Then L lies in a fiber $\Phi^{-1}(\alpha)$ of the moment map where α is a fixed point of the coadjoint action.*

Proof. Since L is Lagrangian, $\omega|_{TL} = 0$. Since L is G-invariant, $\xi_M|_L$ is tangent to L for all $\xi \in \mathfrak{g}$. By definition, $\iota(\xi_M)\omega = -d\langle\Phi, \xi\rangle$. We have

$$\omega(x)(\xi_M(x), X) = 0 = -d\langle\Phi(x), \xi\rangle(X)$$

for all $x \in L$, $X \in T_xL$, and $\xi \in \mathfrak{g}$. So the moment map $\Phi|_L$ is constant, and therefore $L \subset \Phi^{-1}(\alpha)$ for some $\alpha \in \mathfrak{g}^*$.

Because the moment map Φ is equivariant, if L is G-invariant and $\Phi(L) = \alpha$, then α is a fixed point of the coadjoint action. □

Acknowledgements

This note was partially written under the COE program at Hokkaido University in Sapporo, Japan. The author would like to thank K. Ono for his hospitality and delightful discussions.

References

1. M. Audin, F. Lalonde and L. Polterovich, *Symplectic rigidity: Lagrangian submanifolds.* In Holomorphic curves in symplectic geometry. Edited by M.

Audin and J. Lafontaine. Progress in Mathematics, **117**, Birkhäuser Verlag, Basel, (1994).

2. P. Biran and K. Cieliebak, *Symplectic Topology on subcritical manifolds.* Comment. Math. Helv. **76** (2001), 712–753.
3. A. Cannas da Silva, *Lectures on symplectic geometry.* Lecture Notes in Mathematics, 1764. Springer-Verlag, Berlin, (2001).
4. R. Chiang, *Complexity one Hamiltonian* SU(2) *and* SO(3) *actions.* Amer. J. Math. **127** (2005), no.1, 129–168.
5. R. Chiang, *New Lagrangian submanifolds of* $\mathbb{CP}^n$. Int. Math. Res. Not. (2004), no. 45, 2437–2441.
6. V. Ginzburg, V. Guillemin, and Y. Karshon, *Moment maps, cobordisms, and Hamiltonian group actions.* Mathematical Surveys and Monographs, 98. American Mathematical Society, Providence, RI, (2002).
7. E. Lerman, *Symplectic cuts.* Math. Res. Lett. **2** (1995), no. 3, 247–258.
8. P. Seidel, *Graded Lagrangian submanifolds.* Bull. Soc. Math. France **128** (2000), no. 1, 103–149.
9. R. Sjamaar and E. Lerman, *Stratified symplectic spaces and reduction.* Ann. of Math. (2) **134** (1991), no. 2, 375–422.
10. C. Viterbo, *Functors and Computations in Floer Homology with Applications, I.* GAFA, Geom. Funct. Anal. **9** (1999), no. 5, 985–1033.
11. A. Weinstein, *Symplectic manifolds and their Lagrangian submanifolds.* Advances in Math. **6** (1971), 329–346.

DEFORMATION QUANTIZATION OF COMPLEX INVOLUTIVE SUBMANIFOLDS

ANDREA D'AGNOLO

Università di Padova
Dipartimento di Matematica Pura ed Applicata
via G. Belzoni, 7
35131 Padova, Italy
dagnolo@math.unipd.it
Web-page: www.math.unipd.it/~dagnolo/

PIETRO POLESELLO*

Università di Padova
Dipartimento di Matematica Pura ed Applicata
via G. Belzoni, 7
35131 Padova, Italy
pietro@math.unipd.it

Introduction

Let M be a complex manifold, and T^*M its cotangent bundle endowed with the canonical symplectic structure. The sheaf of rings $\mathcal{W}_M$ of WKB microdifferential operators (WKB operators, for short) provides a deformation quantization of T^*M (we refer to [15,1,16] for the definition of $\mathcal{W}_M$ via the sheaf of microdifferential operators). On a complex symplectic manifold X there may not exist a sheaf of rings locally isomorphic to $i^{-1}\mathcal{W}_M$, for any symplectic local chart $i\colon X \supset U \to T^*M$. The idea is then to consider the whole family of locally defined sheaves of WKB operators as the deformation quantization of X. To state it precisely, one needs the notion of algebroid stack, introduced by Kontsevich [12]. In particular, the stack of WKB modules over X defined in Polesello-Schapira [16] (see also Kashiwara [9] for the contact case) is better understood as the stack of $\mathfrak{W}_X$-modules, where $\mathfrak{W}_X$ denotes the algebroid stack of deformation quantization of X.

*The second named author was partially supported by Fondazione Ing. Aldo Gini during the preparation of this paper.

Let $V \subset X$ be an involutive (*i.e.* coisotropic) submanifold. Assume for simplicity that the quotient of V by its bicharacteristic leaves is isomorphic to a complex symplectic manifold Z, and denote by $q\colon V \to Z$ the quotient map. If $\mathcal{L}$ is a simple WKB module along V, then the algebra of its endomorphisms is locally isomorphic to $q^{-1}i^{-1}\mathcal{W}_N^{\mathrm{op}}$, for $i\colon Z \supset U \to T^*N$ a symplectic local chart. Hence one may say that $\mathcal{L}$ provides a deformation quantization of V. Again, since in general there do not exist globally defined simple WKB modules, the idea is to consider the algebroid stack of locally defined simple WKB modules as the deformation quantization of V. We then establish a relation between this algebroid stack and that of deformation quantization of Z. This generalizes a result of D'Agnolo-Schapira [6] for the Lagrangian case (see also [9] for the contact case), which turns out to be an essential ingredient of our proof.

In this paper we start by defining what an algebroid stack is, and how it is locally described. We then discuss the algebroid stack of WKB operators on a complex symplectic manifold X and define the deformation quantization of an involutive submanifold $V \subset X$ by means of simple WKB modules along V. Finally, we relate this deformation quantization to that given by WKB operators on the quotient of V by its bicharacteristic leaves.

1. Algebroid stacks

We start here by recalling the categorical realization of an algebra as in [13], and we then sheafify that construction. We assume that the reader is familiar with the basic notions from the theory of stacks which are, roughly speaking, sheaves of categories. (The classical reference is [7], and a short presentation is given *e.g.* in [9,4].)

Let R be a commutative ring. An R-linear category (R-category for short) is a category whose sets of morphisms are endowed with an R-module structure, so that composition is bilinear. An R-functor is a functor between R-categories which is linear at the level of morphisms.

If A is an R-algebra, we denote by A^+ the R-category with a single object, and with A as set of morphisms. This gives a fully faithful functor from R-algebras to R-categories. If $f, g\colon A \to B$ are R-algebra morphisms, transformations $f^+ \Rightarrow g^+$ are in one-to-one correspondence with the set

$$\{b \in B\colon bf(a) = g(a)b,\ \forall a \in A\}, \tag{1}$$

with vertical composition of transformations corresponding to multiplication in B. Note that the category $\mathsf{Mod}(A)$ of left A-modules is R-equivalent to the category $\mathsf{Hom}_R(A^+, \mathsf{Mod}(R))$ of R-functors from A^+ to $\mathsf{Mod}(R)$ and that the Yoneda embedding

$$A^+ \longrightarrow \mathsf{Hom}_R((A^+)^{\mathrm{op}}, \mathsf{Mod}(R)) \approx_R \mathsf{Mod}(A^{\mathrm{op}})$$

identifies A^+ with the full subcategory of right A-modules which are free of rank one. (Here $\approx_R$ denotes R-equivalence.)

Let X be a topological space, and $\mathcal{R}$ a (sheaf of) commutative algebra(s). As for categories, there are natural notions of $\mathcal{R}$-linear stacks ($\mathcal{R}$-stacks for short), and of $\mathcal{R}$-functor between $\mathcal{R}$-stacks.

If $\mathcal{A}$ is an $\mathcal{R}$-algebra, we denote by $\mathcal{A}^+$ the $\mathcal{R}$-stack associated with the separated prestack $X \supset U \mapsto \mathcal{A}(U)^+$. This gives a functor from $\mathcal{R}$-algebras to $\mathcal{R}$-categories which is faithful and locally full. If $f, g\colon \mathcal{A} \longrightarrow \mathcal{B}$ are $\mathcal{R}$-algebra morphisms, transformations $f^+ \Rightarrow g^+$ are described as in (1). As above, the stack $\mathfrak{Mod}(\mathcal{A})$ of left $\mathcal{A}$-modules is $\mathcal{R}$-equivalent to the stack of $\mathcal{R}$-functors $\mathfrak{Hom}_{\mathcal{R}}(\mathcal{A}^+, \mathfrak{Mod}(\mathcal{R}))$, and the Yoneda embedding gives a fully faithful functor

$$\mathcal{A}^+ \longrightarrow \mathfrak{Hom}_{\mathcal{R}}((\mathcal{A}^+)^{\mathrm{op}}, \mathfrak{Mod}(\mathcal{R})) \approx_{\mathcal{R}} \mathfrak{Mod}(\mathcal{A}^{\mathrm{op}}) \tag{2}$$

into the stack of right $\mathcal{A}$-modules. This identifies $\mathcal{A}^+$ with the full substack of locally free right $\mathcal{A}$-modules of rank one.

Recall that one says a stack $\mathfrak{A}$ is non-empty if $\mathfrak{A}(X)$ has at least one object; it is locally non-empty if there exists an open covering $X = \bigcup_i U_i$ such that $\mathfrak{A}|_{U_i}$ is non-empty; and it is locally connected by isomorphisms if for any open subset $U \subset X$ and any $F, G \in \mathfrak{A}(U)$ there exists an open covering $U = \bigcup_i U_i$ such that $F|_{U_i} \simeq G|_{U_i}$ in $\mathfrak{A}(U_i)$.

Lemma 1.1. *Let $\mathfrak{A}$ be an $\mathcal{R}$-stack. The following are equivalent*

(i) $\mathfrak{A} \approx_{\mathcal{R}} \mathcal{A}^+$ *for an $\mathcal{R}$-algebra $\mathcal{A}$,*
(ii) $\mathfrak{A}$ *is non-empty and locally connected by isomorphisms.*

Proof. By (2), $\mathcal{A}^+$ is equivalent to the stack of locally free right $\mathcal{A}$-modules of rank one. Since $\mathcal{A}$ has a structure of right module over itself, (i) implies (ii). Conversely, let $\mathfrak{A}$ be an $\mathcal{R}$-stack as in (ii) and $\mathcal{P}$ an object of $\mathfrak{A}(X)$. Then $\mathcal{A} = \mathcal{E}nd_{\mathfrak{A}}(\mathcal{P})$ is an $\mathcal{R}$-algebra and the assignment $\mathcal{Q} \mapsto \mathcal{H}om_{\mathfrak{A}}(\mathcal{P}, \mathcal{Q})$ gives an $\mathcal{R}$-equivalence between $\mathfrak{A}$ and $\mathcal{A}^+$. □

We are now ready to give a definition of algebroid stack, equivalent to that in Kontsevich [12]. It is the linear analogue of the notion of gerbe (groupoid stack) from algebraic geometry [7].

Definition 1.1. An $\mathcal{R}$-algebroid stack is an $\mathcal{R}$-stack $\mathfrak{A}$ which is locally non-empty and locally connected by isomorphisms.

The $\mathcal{R}$-stack of $\mathfrak{A}$-modules is $\mathfrak{Mod}(\mathfrak{A}) = \mathfrak{Hom}_{\mathcal{R}}(\mathfrak{A}, \mathfrak{Mod}(\mathcal{R}))$.

Note that $\mathfrak{Mod}(\mathfrak{A})$ is an example of stack of twisted modules over not necessarily commutative rings (see [10,4]). As above, the Yoneda embedding identifies $\mathfrak{A}$ with the full substack of $\mathfrak{Mod}(\mathfrak{A}^{\mathrm{op}})$ consisting of locally free objects of rank one.

2. Cocycle description of algebroid stacks

We will explain here how to recover an algebroid stack from local data. The parallel discussion for the case of gerbes can be found for example in [2,3].

Let $\mathfrak{A}$ be an $\mathcal{R}$-algebroid stack. By definition, there exists an open covering $X = \bigcup_i U_i$ such that $\mathfrak{A}|_{U_i}$ is non-empty. By Lemma 1.1 there are $\mathcal{R}$-algebras $\mathcal{A}_i$ on U_i such that $\mathfrak{A}|_{U_i} \approx_{\mathcal{R}} \mathcal{A}_i^+$. Let $\Phi_i \colon \mathfrak{A}|_{U_i} \to \mathcal{A}_i^+$ and $\Psi_i \colon \mathcal{A}_i^+ \to \mathfrak{A}|_{U_i}$ be quasi-inverse to each other. On double intersections $U_{ij} = U_i \cap U_j$ there are equivalences $\Phi_{ij} = \Phi_i \Psi_j \colon \mathcal{A}_j^+|_{U_{ij}} \to \mathcal{A}_i^+|_{U_{ij}}$. On triple intersections U_{ijk} there are invertible transformations $\alpha_{ijk} \colon \Phi_{ij}\Phi_{jk} \Rightarrow \Phi_{ik}$ induced by $\Psi_j \Phi_j \Rightarrow \mathrm{id}$. On quadruple intersections U_{ijkl} the following diagram commutes

$$\begin{array}{ccc} \Phi_{ij}\Phi_{jk}\Phi_{kl} & \xRightarrow{\alpha_{ijk}\,\mathrm{id}_{\Phi_{kl}}} & \Phi_{ik}\Phi_{kl} \\ \Big\Downarrow \mathrm{id}_{\Phi_{ij}}\,\alpha_{jkl} & & \Big\Downarrow \alpha_{ikl} \\ \Phi_{ij}\Phi_{jl} & \xRightarrow{\alpha_{ijl}} & \Phi_{il}. \end{array} \tag{3}$$

These data are enough to reconstruct $\mathfrak{A}$ (up to equivalence), and we will now describe them more explicitly.

On double intersections U_{ij}, the $\mathcal{R}$-functor $\Phi_{ij} \colon \mathcal{A}_j^+ \to \mathcal{A}_i^+$ is locally induced by $\mathcal{R}$-algebra isomorphisms. There thus exist an open covering $U_{ij} = \bigcup_\alpha U_{ij}^\alpha$ and isomorphisms of $\mathcal{R}$-algebras $f_{ij}^\alpha \colon \mathcal{A}_j \to \mathcal{A}_i$ on U_{ij}^α such that $(f_{ij}^\alpha)^+ = \Phi_{ij}|_{U_{ij}^\alpha}$.

On triple intersections $U_{ijk}^{\alpha\beta\gamma} = U_{ij}^\alpha \cap U_{ik}^\beta \cap U_{jk}^\gamma$, we have invertible transformations $\alpha_{ijk}|_{U_{ijk}^{\alpha\beta\gamma}} \colon (f_{ij}^\alpha)^+ (f_{jk}^\gamma)^+ \Rightarrow (f_{ik}^\beta)^+$. There thus exist invertible sections $a_{ijk}^{\alpha\beta\gamma} \in \mathcal{A}_i^\times(U_{ijk}^{\alpha\beta\gamma})$ such that

$$f_{ij}^\alpha f_{jk}^\gamma = \mathrm{Ad}(a_{ijk}^{\alpha\beta\gamma}) f_{ik}^\beta.$$

On quadruple intersections $U_{ijkl}^{\alpha\beta\gamma\delta\epsilon\varphi} = U_{ijk}^{\alpha\beta\gamma} \cap U_{ijl}^{\alpha\delta\epsilon} \cap U_{ikl}^{\beta\delta\varphi} \cap U_{jkl}^{\gamma\epsilon\varphi}$, the diagram (3) corresponds to the equalities

$$a_{ijk}^{\alpha\beta\gamma} a_{ikl}^{\beta\delta\varphi} = f_{ij}^{\alpha}(a_{jkl}^{\gamma\epsilon\varphi}) a_{ijl}^{\alpha\delta\epsilon}.$$

Indices of hypercoverings are quite cumbersome, and we will not write them explicitly anymore[a].

Let us summarize what we just obtained.

Proposition 2.1. *Up to equivalence, an $\mathcal{R}$-algebroid stack is given by the following data:*

(1) an open covering $X = \bigcup_i U_i$,
(2) $\mathcal{R}$-algebras $\mathcal{A}_i$ on U_i,
(3) isomorphisms of $\mathcal{R}$-algebras $f_{ij}\colon \mathcal{A}_j \to \mathcal{A}_i$ on U_{ij},
(4) invertible sections $a_{ijk} \in \mathcal{A}_i^\times(U_{ijk})$,

such that

$$\begin{cases} f_{ij} f_{jk} = \mathrm{Ad}(a_{ijk}) f_{ik}, & \textit{as morphisms } \mathcal{A}_k \to \mathcal{A}_i \textit{ on } U_{ijk}, \\ a_{ijk} a_{ikl} = f_{ij}(a_{jkl}) a_{ijl} & \textit{in } \mathcal{A}_i(U_{ijkl}). \end{cases}$$

Example 2.1. If the $\mathcal{R}$-algebras $\mathcal{A}_i$ are commutative, then the 1-cocycle $f_{ij} f_{jk} = f_{ik}$ defines an $\mathcal{R}$-algebra $\mathcal{A}$ on X and $\{a_{ijk}\}$ induces a 2-cocycle with values in $\mathcal{A}^\times$. In particular, if $\mathcal{A}_i = \mathcal{R}|_{U_i}$, then $f_{ij} = \mathrm{id}$ and $\mathcal{A} = \mathcal{R}$. Hence, $\mathcal{R}$-algebroid stacks locally $\mathcal{R}$-equivalent to $\mathcal{R}^+$ are determined by the 2-cocycle $a_{ijk} \in \mathcal{R}^\times(U_{ijk})$. One checks that two such stacks are (globally) $\mathcal{R}$-equivalent if and only if the corresponding cocycles give the same cohomology class in $H^2(X; \mathcal{R}^\times)$.

Example 2.2. Let X be a complex manifold, and $\mathcal{O}_X$ its structural sheaf. A line bundle $\mathcal{L}$ on X is determined (up to isomorphism) by its transition functions $f_{ij} \in \mathcal{O}_X^\times(U_{ij})$, where $X = \bigcup_i U_i$ is an open covering such that $\mathcal{L}|_{U_i} \simeq \mathcal{O}_{U_j}$. Let $\lambda \in \mathbb{C}$, and choose determinations g_{ij} of the multivalued functions f_{ij}^λ. Since $g_{ij} g_{jk}$ and g_{ik} are both determinations of f_{ik}^λ, one has $g_{ij} g_{jk} = c_{ijk} g_{ik}$ for $c_{ijk} \in \mathbb{C}_X^\times(U_{ijk})$.

Let us denote by $\mathbb{C}_{\mathcal{L}^\lambda}$ the $\mathbb{C}$-algebroid stack associated with the cocycle $\{c_{ijk}\}$ as in the previous example. For $\lambda = m \in \mathbb{Z}$ we have $\mathbb{C}_{\mathcal{L}^m} \approx_{\mathbb{C}} \mathbb{C}_X^+$, but in general $\mathbb{C}_{\mathcal{L}^\lambda}$ is non trivial. On the other hand, $\mathcal{L}^\lambda$ defines a global

[a] Recall that, on a paracompact space, usual coverings are cofinal among hypercoverings.

object of the algebroid stack[b] $\mathcal{O}_X^+ \otimes_{\mathbb{C}} \mathbb{C}_{\mathcal{L}^\lambda}$, so that $\mathcal{O}_X^+ \otimes_{\mathbb{C}} \mathbb{C}_{\mathcal{L}^\lambda} \approx_{\mathbb{C}} \mathcal{O}_X^+$ is always trivial. Forgetting the $\mathcal{O}$-linear structure, the Yoneda embedding identifies $\mathcal{L}^\lambda$ with a twisted sheaf in (*i.e.* a global object of) $\mathfrak{Mod}(\mathbb{C}_{\mathcal{L}^{-\lambda}})$. (Here we used the equivalence $(\mathbb{C}_{\mathcal{L}^\lambda})^{\mathrm{op}} \approx_{\mathbb{C}} \mathbb{C}_{\mathcal{L}^{-\lambda}}$.)

3. Quantization of complex symplectic manifolds

The relation between microdifferential operators (see [17]) and WKB operators[c] is classical, and is discussed *e.g.* in [15,1] for the cotangent bundle of the complex line. We follow here the presentation in [16].

Let M be a complex manifold, and denote by $\rho\colon J^1M \to T^*M$ the projection from the 1-jet bundle to the cotangent bundle. Let $(t;\tau)$ be the system of homogeneous symplectic coordinates on $T^*\mathbb{C}$, and recall that J^1M is identified with the affine chart of the projective cotangent bundle $P^*(M\times\mathbb{C})$ given by $\tau \neq 0$. Denote by $\mathcal{E}_{M\times\mathbb{C}}$ the sheaf of microdifferential operators on $P^*(M\times\mathbb{C})$. Its twist by half-forms $\mathcal{E}^{\sqrt{v}}_{M\times\mathbb{C}} = \pi^{-1}\Omega^{1/2}_{M\times\mathbb{C}} \otimes_{\pi^{-1}\mathcal{O}} \mathcal{E}_{M\times\mathbb{C}} \otimes_{\pi^{-1}\mathcal{O}} \pi^{-1}\Omega^{-1/2}_{M\times\mathbb{C}}$ is endowed with a canonical anti-involution. (Here $\pi\colon P^*(M\times\mathbb{C}) \to M\times\mathbb{C}$ denotes the natural projection.)

In a local coordinate system (x,t) on $M\times\mathbb{C}$, consider the subring $\mathcal{E}^{\sqrt{v}}_{M\times\mathbb{C},\hat{t}}$ of operators commuting with ∂_t. The ring of WKB operators (twisted by half-forms) is defined by

$$\mathcal{W}^{\sqrt{v}}_M = \rho_*(\mathcal{E}^{\sqrt{v}}_{M\times\mathbb{C},\hat{t}}|_{J^1M}).$$

It is endowed with a canonical anti-involution $*$, and its center is the constant sheaf k_{T^*M} with stalk the subfield $k = \mathcal{W}_{\mathrm{pt}} \subset \mathbb{C}[\![\tau^{-1},\tau]$ of WKB operators over a point.

In a local coordinate system (x) on M, with associated symplectic local coordinates $(x;u)$ on T^*M, a WKB operator P of order m defined on a open subset U of T^*M has a total symbol

$$\sigma_{\mathrm{tot}}(P) = \sum_{j=-\infty}^{m} p_j(x;u)\tau^j,$$

where the p_j's are holomorphic functions on U subject to the estimates

$$\begin{cases} \text{for any compact subset } K \text{ of } U \text{ there exists a constant} \\ C_K > 0 \text{ such that for all } j<0, \sup_K |p_j| \le C_K^{-j}(-j)!. \end{cases} \tag{4}$$

---

[b]We denote here by $\otimes_{\mathcal{R}}$ the tensor product of $\mathcal{R}$-linear stacks. In particular, $(\mathcal{A}_1 \otimes_{\mathcal{R}} \mathcal{A}_2)^+ \approx \mathcal{A}_1^+ \otimes_{\mathcal{R}} \mathcal{A}_2^+$ for $\mathcal{R}$-algebras $\mathcal{A}_1$ and $\mathcal{A}_2$.
[c]WKB stands for Wentzel-Kramer-Brillouin.

If Q is another WKB operator defined on U, of total symbol $\sigma_{\mathrm{tot}}(Q)$, then

$$\sigma_{\mathrm{tot}}(P \circ Q) = \sum_{\alpha \in \mathbb{N}^n} \frac{\tau^{-|\alpha|}}{\alpha!} \partial_u^\alpha \sigma_{\mathrm{tot}}(P) \partial_x^\alpha \sigma_{\mathrm{tot}}(Q).$$

Remark 3.1. The ring $\mathcal{W}_M^{\sqrt{v}}$ is a deformation quantization of T^*M in the following sense. Setting $\hbar = \tau^{-1}$, the sheaf of formal WKB operators (obtained by dropping the estimates (4)) of degree less than or equal to 0 is locally isomorphic to $\mathcal{O}_{T^*M}[\![\hbar]\!]$ as $\mathbb{C}_{T^*M}$-modules (via the total symbol), and it is equipped with an unitary associative product which induces a star-product on $\mathcal{O}_{T^*M}[\![\hbar]\!]$.

Let X be a complex symplectic manifold of dimension $2n$. Then there are an open covering $X = \bigcup_i U_i$ and symplectic embeddings $\Phi_i \colon U_i \to T^*M$, for $M = \mathbb{C}^n$. Let $\mathcal{A}_i = \Phi_i^{-1} \mathcal{W}_M^{\sqrt{v}}$. Adapting Kashiwara's construction (cf [9]), Polesello-Schapira [16] proved that there exist isomorphisms of k-algebras f_{ij} and invertible sections a_{ijk} as in Proposition 2.1. Their result may thus be restated as

Theorem 3.1. *(cf [16]) On any symplectic complex manifold X there exists a canonical k-algebroid stack $\mathfrak{W}_X$ locally equivalent to $(i^{-1}\mathcal{W}_M^{\sqrt{v}})^+$ for any symplectic local chart $i \colon X \supset U \to T^*M$.*

Note that the canonical anti-involution $*$ on $\mathcal{W}_M^{\sqrt{v}}$ extends to an equivalence of k-stacks $\mathfrak{W}_X \approx_k \mathfrak{W}_X^{\mathrm{op}}$. Note also that, by Lemma 1.1, there exists a deformation quantization algebra on X if the k-algebroid stack $\mathfrak{W}_X$ has a global object, or equivalently if the stack $\mathfrak{Mod}(\mathfrak{W}_X)$ has a global object locally isomorphic to $i^{-1}\mathcal{W}_M^{\sqrt{v}}$ for any symplectic local chart $i \colon X \supset U \to T^*M$.

4. Quantization of involutive submanifolds

Let M be a complex manifold and $V \subset T^*M$ an involutive[d] submanifold. Similarly to the case of microdifferential operators (for which we refer to [11] and [8]), one introduces the sub-sheaf of rings $\mathcal{W}_V^{\sqrt{v}}$ of $\mathcal{W}_M^{\sqrt{v}}$ generated over $\mathcal{W}_M^{\sqrt{v}}(0)$ by the WKB operators $P \in \mathcal{W}_M^{\sqrt{v}}(1)$ such that $\sigma_1(P)$ vanishes on V. Here $\mathcal{W}_M^{\sqrt{v}}(m)$ denotes the sheaf of operators of order less or equal to m,

[d] Recall that V is involutive if for any pair f, g of holomorphic functions vanishing on V, their Poisson bracket $\{f, g\}$ vanishes on V.

and $\sigma_m(\cdot)\colon \mathcal{W}_M^{\sqrt{v}}(m) \to \mathcal{W}_M^{\sqrt{v}}(m)/\mathcal{W}_M^{\sqrt{v}}(m-1) \simeq \mathcal{O}_{T^*M} \cdot \tau^m$ is the symbol map of order m (which does not depend on the local coordinate system on M).

Definition 4.1. Let $\mathcal{M}$ be a coherent $\mathcal{W}_M^{\sqrt{v}}$-module.

(i) $\mathcal{M}$ is a regular WKB module along V if locally there exists a coherent sub-$\mathcal{W}_M^{\sqrt{v}}(0)$-module $\mathcal{M}_0$ of $\mathcal{M}$ which generates it over $\mathcal{W}_M^{\sqrt{v}}$, and such that $\mathcal{W}_V^{\sqrt{v}} \cdot \mathcal{M}_0 \subset \mathcal{M}_0$.

(ii) $\mathcal{M}$ is a simple WKB module along V if locally there exists a $\mathcal{W}_M^{\sqrt{v}}(0)$-module $\mathcal{M}_0$ as above such that $\mathcal{M}_0/\mathcal{W}_M^{\sqrt{v}}(-1)\cdot\mathcal{M}_0 \simeq \mathcal{O}_V$.

Example 4.1. Recall that locally, any involutive submanifold $V \subset T^*M$ of codimension d may be written as:

$$V = \{(x;u); u_1 = \cdots = u_d = 0\},$$

in a local symplectic coordinate system $(x;u) = (x_1,\ldots,x_n;u_1,\ldots,u_n)$ on T^*M. In this case, any simple WKB module along V is locally isomorphic to

$$\mathcal{W}_M^{\sqrt{v}}/\mathcal{W}_M^{\sqrt{v}} \cdot (\partial_{x_1},\ldots,\partial_{x_d}).$$

Let X be a complex symplectic manifold of dimension $2n$, and $V \subset X$ an involutive submanifold. The notions of regular and simple module along V being local, they still make sense in the stack $\mathfrak{Mod}_{coh}(\mathfrak{W}_X)$ of coherent WKB modules on X.

Definition 4.2. Denote by $\mathfrak{Mod}_{V\text{-}reg}(\mathfrak{W}_X)$ the full substack of regular objects along V in $\mathfrak{Mod}_{coh}(\mathfrak{W}_X)$, and by $\mathfrak{S}_V$ its full substack of simple objects along V.

By Example 4.1, $\mathfrak{S}_V$ is locally non-empty and locally connected by isomorphisms. Hence it is a k-algebroid stack on X. Since it is supported by V, we consider $\mathfrak{S}_V$ as a stack on V.

The first equivalence in the following theorem asserts that the deformation quantization of V by means of simple WKB modules is equivalent, up to a twist, to that given by WKB operators on the quotient of V by its bicharacteristic leaves.

Theorem 4.1. *Let X be a complex symplectic manifold, and $V \subset X$ an involutive submanifold. Assume that there exist a complex symplectic manifold Z and a map $q\colon V \to Z$ whose fibers are the bicharacteristic leaves of*

V. Then there are an equivalence of k-algebroid stacks[e] *on V*

$$\mathfrak{S}_V \approx_k q^{-1}\mathfrak{W}_Z \otimes_{\mathbb{C}} \mathbb{C}_{\Omega_V^{1/2}}, \tag{5}$$

and a k-equivalence

$$\mathfrak{Mod}_{V\text{-}reg}(\mathfrak{W}_X) \approx_k \mathfrak{Mod}_{coh}(q^{-1}\mathfrak{W}_Z^{\mathrm{op}} \otimes_{\mathbb{C}} \mathbb{C}_{\Omega_V^{-1/2}}). \tag{6}$$

Note that the statement still holds for a general involutive submanifold $V \subset X$, replacing $q^{-1}\mathfrak{W}_Z$ with the algebroid stack obtained by adapting [16] for the symplectic case.

If $V = X$, then $\mathfrak{S}_X \approx_k \mathfrak{W}_X^{\mathrm{op}}$ is the stack of locally free left WKB modules of rank one, and $\mathfrak{Mod}_{X\text{-}reg}(\mathfrak{W}_X) \approx_k \mathfrak{Mod}_{coh}(\mathfrak{W}_X)$. Since $\Omega_X \simeq \mathcal{O}_X$ by the nth power of the symplectic form, one has $\mathbb{C}_{\Omega_X^{1/2}} \approx_{\mathbb{C}} \mathbb{C}_X^+$. As $q = \mathrm{id}$, the theorem thus reduces to the equivalence $\mathfrak{W}_X^{\mathrm{op}} \approx_k \mathfrak{W}_X$ given by the involution $*$.

If $V = \Lambda$ is Lagrangian, then $Z = \mathrm{pt}$. Hence $q^{-1}\mathfrak{W}_{\mathrm{pt}} \approx_k k_\Lambda^+$, and (5) asserts that

$$\mathfrak{S}_\Lambda \otimes_{\mathbb{C}} \mathbb{C}_{\Omega_\Lambda^{-1/2}} \approx_k k_\Lambda^+.$$

In other words, it asserts that $\mathfrak{S}_\Lambda \otimes_{\mathbb{C}} \mathbb{C}_{\Omega_\Lambda^{-1/2}} \subset \mathfrak{Mod}(\mathfrak{W}_X \otimes_{\mathbb{C}} \mathbb{C}_{\Omega_\Lambda^{1/2}})$ has a global object. This is a result of D'Agnolo-Schapira [6], obtained by adapting a similar theorem of Kashiwara [9] for microdifferential operators, along the techniques of [16] (see also [14] for the smooth case). As for (6), we recover the WKB analogue of [9], also stated in [6],

$$\mathfrak{Mod}_{\Lambda\text{-}reg}(\mathfrak{W}_X) \approx_k \mathfrak{Mod}_{loc\text{-}sys}(k_\Lambda^+ \otimes_{\mathbb{C}} \mathbb{C}_{\Omega_\Lambda^{-1/2}}),$$

where the right-hand side denotes the stack of twisted locally constant k-modules of finite rank.

Proof. [Proof of Theorem 4.1] Consider the two projections

$$X \xleftarrow[p_1]{} X \times Z \xrightarrow[p_2]{} Z.$$

By the graph embedding, V is identified with a Lagrangian submanifold V' of $X \times Z$. By [6] there exists a simple module $\mathcal{L}$ along V' in $\mathfrak{Mod}(\mathfrak{W}_{X\times Z} \otimes_{\mathbb{C}} \mathbb{C}_{\Omega_{V'}^{1/2}})$. Using [4], we get an integral transform k-functor with kernel $\mathcal{L}$

$$\mathfrak{Mod}(p_1^{-1}\mathfrak{W}_X) \longrightarrow \mathfrak{Mod}(p_2^{-1}\mathfrak{W}_Z^{\mathrm{op}} \otimes_{\mathbb{C}} \mathbb{C}_{\Omega_{V'}^{-1/2}}).$$

---

[e]We denote here by q^{-1} the stack-theoretical inverse image. In particular, $q^{-1}(\mathcal{A}^+) \approx_k (q^{-1}\mathcal{A})^+$, for a k-algebra $\mathcal{A}$ on Z.

Since q is identified with $p_2|_{V'}$ and the inclusion $V \subset X$ with $p_1|_{V'}$, we get an induced functor

$$\mathfrak{Mod}(\mathfrak{W}_X|_V) \longrightarrow \mathfrak{Mod}(q^{-1}\mathfrak{W}_Z^{\mathrm{op}} \otimes_{\mathbb{C}} \mathbb{C}_{\Omega_V^{-1/2}}).$$

This restricts to functors

$$\begin{array}{ccc} \mathfrak{Mod}_{V\text{-}reg}(\mathfrak{W}_X) & \longrightarrow & \mathfrak{Mod}_{coh}(q^{-1}\mathfrak{W}_Z^{\mathrm{op}} \otimes_{\mathbb{C}} \mathbb{C}_{\Omega_V^{-1/2}}) \\ \cup & & \cup \\ \mathfrak{S}_V & \longrightarrow & q^{-1}\mathfrak{W}_Z \otimes_{\mathbb{C}} \mathbb{C}_{\Omega_V^{1/2}}, \end{array}$$

which are local, and hence global, equivalences by the WKB analogue of [5] and by a direct computation, respectively. □

As a corollary, we get a sufficient condition for the existence of a globally defined twisted simple WKB module along V.

Corollary 4.1. *In the situation of the above theorem, assume that there exists a deformation quantization algebra $\mathcal{A}$ on Z such that $\mathcal{A}^+ \approx_k \mathfrak{W}_Z$. Then there exists a globally defined simple module along V in $\mathfrak{Mod}(\mathfrak{W}_X \otimes_{\mathbb{C}} \mathbb{C}_{\Omega_V^{1/2}})$.*

Proof. By Lemma 1.1, the k-algebroid $\mathfrak{W}_Z$ has a global object. Then its image by the adjunction functor $\mathfrak{W}_Z \to q_* q^{-1}\mathfrak{W}_Z$ gives a globally defined twisted simple WKB module along V. □

References

1. T. Aoki, T. Kawai, T. Koike and Y. Takei, *On the exact WKB analysis of operators admitting infinitely many phases*, Advances in Math. **181** (2004), 165–189.
2. L. Breen, *On the classification of 2-gerbes and 2-stacks*, Astérisque **225** (1994).
3. L. Breen and W. Messing, *Differential geometry of gerbes*, e-print (2003), **arXiv:math.AG/0106083 v3**.
4. A. D'Agnolo and P. Polesello, *Stacks of twisted modules and integral transforms*, in: Geometric Aspects of Dwork's Theory (A collection of articles in memory of Bernard Dwork), A. Adolphson, F. Baldassarri, P.Berthelot, N. Katz and F. Loeser eds., Walter de Gruyter, Berlin (2004), 461-505
5. A. D'Agnolo and P. Schapira, *Radon-Penrose transform for $\mathcal{D}$-modules*, J. Funct. Anal. **139** (1996), no. 2, 349–382.
6. A. D'Agnolo and P. Schapira, *Quantization of complex Lagrangian submanifolds*, preprint (2005).

7. J. Giraud, *Cohomologie non abélienne*, Grundlehren der Math. Wiss. **179**, Springer (1971).
8. M. Kashiwara, *Introduction to microlocal analysis,* L'Enseignement Mathématiques, **32** (1986), 5–37
9. M. Kashiwara, *Quantization of contact manifolds*, Publ. Res. Inst. Math. Sci. **32** no. 1 (1996), 1–7.
10. M. Kashiwara and P. Schapira, *Categories and sheaves*, preliminary version (2005) of a book in preparation.
11. M. Kashiwara and T. Oshima, *Systems of differential equations with regular singularities and their boundary value problems,* Ann. of Math. **106** (1977), 145–200.
12. M. Kontsevich, *Deformation quantization of algebraic varieties.* in: Euro-Conférence Moshé Flato, Part III (Dijon, 2000). Lett. Math. Phys. **56** no. 3 (2001), 271–294.
13. B. Mitchell. *Rings with several objects*, Advances in Math. **8** (1972), 1–161.
14. R. Nest and B. Tsygan, *Remarks on modules over deformation quantization algebras*, e-print (2004), **arXiv:math-ph/0411066**, to appear in: Moscow Math. Journal **4** (2004).
15. F. Pham, *Resurgence, quantized canonical transformations and multi-instantons expansions*, in: Algebraic Analysis, dedicated to Prof. M. Sato, Academic Press (1988), 699-726.
16. P. Polesello and P. Schapira, *Stacks of quantization-deformation modules on complex symplectic manifolds, Int. Math. Res. Notices **49** (2004), 2637–2664.*
17. *M. Sato, T. Kawai, and M. Kashiwara,* Microfunctions and pseudo-differential equations, *in: Komatsu (ed.),* Hyperfunctions and pseudo-differential equations, *Proceedings Katata 1971, Lecture Notes in Math. **287**, Springer (1973), 265–529.*

DEFORMATION QUANTIZATION ON A HILBERT SPACE

GIUSEPPE DITO
Institut de Mathématiques de Bourgogne
Université de Bourgogne
UMR CNRS 5584
B.P. 40780, 21078 Dijon Cedex, France
giuseppe.dito@u-bourgogne.fr

We study deformation quantization on an infinite-dimensional Hilbert space W endowed with its canonical Poisson structure. The standard example of the Moyal star-product is made explicit and it is shown that it is well defined on a subalgebra of $C^\infty(W)$. A classification of inequivalent deformation quantizations of exponential type, containing the Moyal and normal star-products, is also given.

1. Introduction

Deformation quantization provides an alternative formulation of Quantum Mechanics by interpreting quantization as a deformation of the commutative algebra of classical observables into a noncommutative algebra [1]. The quantum algebra is defined by a formal associative star-product $\star_\hbar$ which encodes the algebraic structure of the set of observables.

Deformation quantization has been applied with increasing generality to several areas of mathematics and physics. Most of these applications deal with star-products on finite-dimensional manifolds. See [3] for a recent review.

It is natural to consider an extension of deformation quantization to infinite-dimensional manifolds as it appears to be a good setting where quantum field theory of nonlinear wave equations can be formulated (e.g. in the sense of I. Segal [11]). In the star-product approach, the first steps in that direction are given in [4,5].

Recently, deformation quantization has become popular among field and string theorists. A generalization of Moyal star-product to infinite-dimensional spaces appears in several places in the literature. Let us just notice that the Witten star-product [12] appearing in string field theory is *heuristically* equivalent to an infinite-dimensional version of the Moyal star-product. A brute force generalization of Moyal star-product to field theory yields to some pathological and unpleasant features such as anomalies and breakdown of associativity. We think that it is worth writing down a mathematical study of the Moyal product in infinite dimension even if it is not an adequate product for field theory considerations.

In the finite-dimensional case, the existence of star-products on any (real) symplectic manifold has been established by DeWilde and Lecomte [2]. The general existence and classification problems for the deformation quantization of a Poisson manifold was solved by Kontsevich [9]. The very first problem that one faces when going over infinite-dimensional spaces, is to make sense of the star-product itself as a formal associative product. It contrasts with the finite-dimensional case where the deformation is defined on all of the smooth functions on the manifold. This is by far too demanding in the infinite-dimensional case even when the Poisson structure is well defined on all of the smooth functions (e.g. on Banach or Fréchet spaces). One should specify first an Abelian algebra of admissible functions which then can be deformed. For example, on $E = \mathcal{S}\times\mathcal{S}$, where $\mathcal{S}$ is the Schwartz space on $\mathbb{R}^n$, endowed with its canonical Poisson structure, one cannot expect to write down a star-product defined on all holomorphic functions on E, but has to restrict it to some subalgebra. For example, in [5] it is shown, that for such a simple star-product as the normal star-product, it is defined on the subalgebra of holomorphic functions of a and $\bar{a}$ (creation and annihilation 'operators') having semi-regular kernels. In [6], one can find a nice analysis for the normal star-product and the conditions on the kernel have been translated in terms of wave front set of the distributions.

After making precise what is a deformation quantization on a Hilbert space, we first present a study of Moyal product when the space-space is the direct sum of a Hilbert space with its dual. We identify a subalgebra of smooth functions, specified by conditions of Hilbert-Schmidt type on their derivatives, on which the Moyal product makes sense. We also define a family of star-products of exponential type, show that they are not all equivalent to each other and give the classification of their equivalence classes in terms of Hilbert-Schmidt operators.

2. Star-products on a Hilbert space

When infinite-dimensional spaces are involved, further conditions are needed to define a deformation quantization or a star-product. The algebra of functions on which the Poisson bracket and the star-product are defined should be specified along with the class of *admissible* cochains (especially when the issue of the equivalence of deformations is considered).

2.1. *Notations*

Let B be a Banach space over a field $\mathbb{K}$ ($\mathbb{R}$ or $\mathbb{C}$). The topological dual of B shall be denoted by B^*. The Banach space of bounded r-linear forms on B is denoted by $\mathcal{L}^r(B,\mathbb{K})$ and $\mathcal{L}^r_{\mathrm{sym}}(B,\mathbb{K})$ is the subspace of $\mathcal{L}^r(B,\mathbb{K})$ consisting of bounded symmetric r-linear forms on B. We shall denote by $C^\infty(B,\mathbb{K})$ the space of $\mathbb{K}$-valued functions on B that are smooth in the Fréchet sense. The Fréchet derivative of $F \in C^\infty(W,\mathbb{K})$ is denoted by DF and it is a smooth map from B to $\mathcal{L}^1(B,\mathbb{K}) = B^*$, i.e., $DF \in C^\infty(B,B^*)$. For $F \in C^\infty(B,\mathbb{K})$, the higher derivative $D^{(r)}F$ belongs to $C^\infty(B,\mathcal{L}^r_{\mathrm{sym}}(B,\mathbb{K}))$ and we shall use the following notation $D^{(r)}F(b).(b_1,\dots,b_r)$ for the rth-derivative of F evaluated at $b \in B$ in the direction of $(b_1,\dots,b_r) \in B^r$.

Let W be an infinite-dimensional separable Hilbert space over a field $\mathbb{K}$. For notational reasons, as it will become clear later, it would be convenient for us to not identify W^* with W. For any orthonormal basis $\{e_i\}_{i\geq 1}$ in W and corresponding dual basis $\{e_i^*\}_{i\geq 1}$ in W^*, we shall denote the partial derivative of $F \in C^\infty(W,\mathbb{K})$ evaluated at w in the direction of e_i by $\partial_i F(w) \in \mathbb{K}$, i.e., $\partial_i F(w) = DF(w).e_i$. Since F is differentiable in the Fréchet sense, we have $DF(w) = \sum_{i\geq 1} \partial_i F(w) e_i^*$ and thus, for any $w \in W$, that $\sum_{i\geq 1} |\partial_i F(w)|^2 < \infty$.

2.2. *Multidifferential operators*

Let us first make precise what we call a Poisson structure on W. In the following, we will consider a map P that sends W into a space of (not necessarily bounded) bilinear forms on W^* and a subalgebra $\mathcal{F}$ of $C^\infty(W,\mathbb{K})$. We define the subspace $\mathcal{D}_w^{\mathcal{F}} = \{DF(w) \mid F \in \mathcal{F}\}$ of W^*.

Definition 2.1. Let W be a Hilbert space. Let $\mathcal{F}$ be a subalgebra (for the pointwise product) of $C^\infty(W,\mathbb{K})$. A Poisson bracket on $(W,\mathcal{F})$ is a $\mathbb{K}$-bilinear map $\{\cdot,\cdot\}\colon \mathcal{F}\times\mathcal{F}\to\mathcal{F}$ such that:

i) there exists a map P from W to the space of bilinear forms on W^*, so that the domain of $P(w)$ contains $\mathcal{D}_w^{\mathcal{F}} \times \mathcal{D}_w^{\mathcal{F}}$ and $\forall F,G \in \mathcal{F}$, $\{F,G\}(w) = P(w).(DF(w),DG(w))$ where $w \in W$.

ii) $(\mathcal{F},\{\cdot,\cdot\})$ is a Poisson algebra, i.e., skew-symmetry, Leibniz rule, and Jacobi identity are satisfied.

The triple $(W, \mathcal{F}, \{\cdot,\cdot\})$ is called a Poisson space.

Let us give an example where $P(w)$ is an unbounded bilinear form on W^*.

Example 2.1. Consider a real Hilbert space W with orthonormal basis $\{e_i\}_{i\geq 0}$. We will realize the following subalgebra of the Witt algebra:

$$[L_m, L_n] = (m-n)L_{m+n}, \quad m,n \geq 0,$$

by functions on W. For $w \in W$, let $\phi_i(w) = \langle e_i^*, w\rangle$, $i \geq 0$, be the coordinate functions. The algebra $\mathcal{F}$ generated by the family of functions $\{\phi_i\}_{i\geq 0}$ is a subalgebra of $C^\infty(W, \mathbb{R})$ consisting of polynomial functions in a finite number of variables. The following expression:

$$\{F,G\}(w) = \sum_{m,n\geq 0} (m-n)\phi_{m+n}(w)\partial_m F(w)\partial_n G(w), \quad F, G \in \mathcal{F}, w \in W,$$

defines a Poisson bracket on $(W, \mathcal{F})$. Indeed the right-hand side is a finite sum and is a function in $\mathcal{F}$, and we have:

$$\{\phi_i, \phi_j\} = (i-j)\phi_{i+j},$$

from which Jacobi identity follows. The special case where $j = 0$ gives:

$$\{\phi_i, \phi_0\}(w) = P(w).(D\phi_i(w), D\phi_0(w)) = P(w).(e_i^*, e_0^*) = i\phi_i(w).$$

By choosing an appropriate w (e.g. $w = \sum_{i\geq 1} i^{-3/4} e_i$), then $P(w).(e_i^*, e_0^*) = i\phi_i(w)$ can become as large as desired by varying i. This shows that the bilinear form $P(w)$ cannot be bounded.

The generalization of Def. 2.1 to multidifferential operators on W is straightforward. Given a subalgebra $\mathcal{F}$ of $C^\infty(W, \mathbb{K})$, we define the following subspace of $\mathcal{L}^r_{\text{sym}}(W, \mathbb{K})$:

$$\mathcal{D}_w^{\mathcal{F}}(r) = \{D^{(r)}F(w) \mid F \in \mathcal{F}\}.$$

Definition 2.2. Let W be a Hilbert space. Let $\mathcal{F}$ be a subalgebra, for the pointwise product, of $C^\infty(W, \mathbb{K})$. Let $r \geq 1$, an r-differential operator $\mathbf{A}$ on $(W, \mathcal{F})$ is an r-linear map $\mathbf{A}\colon \mathcal{F}^r \to \mathcal{F}$ such that:

i) for $(n_1, \ldots, n_r) \in \mathbb{N}^r$, there exists a map $a^{(n_1,\ldots,n_r)}$ from W to a space of (not necessarily bounded) r-linear forms on $\mathcal{L}^{n_1}_{\text{sym}}(W, \mathbb{K}) \times \cdots \times \mathcal{L}^{n_r}_{\text{sym}}(W, \mathbb{K})$, i.e.,

$$a^{(n_1,\ldots,n_r)}(w)\colon \mathcal{D}_w^{(n_1,\ldots,n_r)} \subset \mathcal{L}^{n_1}_{\text{sym}}(W, \mathbb{K}) \times \cdots \times \mathcal{L}^{n_r}_{\text{sym}}(W, \mathbb{K}) \to \mathbb{K},$$

so that the domain $\mathcal{D}_w^{(n_1,\cdots,n_r)}$ of $a^{(n_1,\cdots,n_r)}(w)$ contains $\mathcal{D}_w^{\mathcal{F}}(n_1) \times \cdots \times \mathcal{D}_w^{\mathcal{F}}(n_r)$ and $a^{(n_1,\cdots,n_r)}$ is 0 except for finitely many $(n_1, \ldots, n_r)$;

ii) for any $F_1, \ldots, F_r \in \mathcal{F}$ and $w \in W$, we have

$$\mathbf{A}(F_1, \ldots, F_r)(w) = \sum_{n_1,\cdots,n_r \geq 0} a^{(n_1,\cdots,n_r)}(w).(D^{(n_1)}F_1(w), \cdots, D^{(n_r)}F_r(w)).$$

Notice that Poisson brackets as defined above are special cases of bidifferential operators in the sense of Def. 2.2 with $P = a^{(1,1)}$.

2.3. *Deformation quantization on W*

We now have all the ingredients to define what is meant by deformation quantization of a Poisson space $(W, \mathcal{F}, \{\cdot,\cdot\})$ when W is a Hilbert space.

Definition 2.3. Let W be a Hilbert space and $(W, \mathcal{F}, \{\cdot,\cdot\})$ be a Poisson space. A star-product on $(W, \mathcal{F}, \{\cdot,\cdot\})$ is a $\mathbb{K}[[\hbar]]$-bilinear product $\star_\hbar \colon \mathcal{F}[[\hbar]] \times \mathcal{F}[[\hbar]] \to \mathcal{F}[[\hbar]]$ given by $F \star_\hbar G = \sum_{r\geq 0} \hbar^r C_r(F, G)$ for $F, G \in \mathcal{F}$ and extended by $\mathbb{K}[[\hbar]]$-bilinearity to $\mathcal{F}[[\hbar]]$, and satisfying for any $F, G, H \in \mathcal{F}$:

i) $C_0(F, G) = FG$,

ii) $C_1(F, G) - C_1(G, F) = 2\{F, G\}$,

iii) for $r \geq 1$, $C_r \colon \mathcal{F} \times \mathcal{F} \to \mathcal{F}$ are bidifferential operators in the sense of Def. 2.2, vanishing on constants,

iv) $F \star_\hbar (G \star_\hbar H) = (F \star_\hbar G) \star_\hbar H$.

The triple $(W, \mathcal{F}[[\hbar]], \star_\hbar)$ is called a deformation quantization of the Poisson space $(W, \mathcal{F}, \{\cdot,\cdot\})$.

We also have a notion of equivalence of deformations adapted to our context:

Definition 2.4. Two deformation quantizations $(W, \mathcal{F}[[\hbar]], \star_\hbar^1)$ and $(W, \mathcal{F}[[\hbar]], \star_\hbar^2)$ of the same Poisson space $(W, \mathcal{F}, \{\cdot,\cdot\})$ are said to be equivalent if there exists a $\mathbb{K}[[\hbar]]$-linear map $T \colon \mathcal{F}[[\hbar]] \to \mathcal{F}[[\hbar]]$ expressed as a formal series $T = \mathrm{Id}_{\mathcal{F}} + \sum_{r\geq 1} \hbar^r T_r$ satisfying:

i) $T_r : \mathcal{F} \to \mathcal{F}$, $r \geq 1$, are differential operators in the sense of Def. 2.2, vanishing on constants,

ii) $T(F) \star_\hbar^1 T(G) = T(F \star_\hbar^2 G)$, $\forall F, G \in \mathcal{F}$.

3. Moyal product on a Hilbert space

We present an infinite-dimensional version of the Moyal product defined on a class of smooth functions specified by a Hilbert-Schmidt type of conditions on their derivatives.

3.1. *Poisson structure*

Let $\mathcal{H}$ be an infinite-dimensional separable Hilbert space. We consider the phase-space $W = \mathcal{H} \oplus \mathcal{H}^*$ endowed with its canonical strong symplectic structure $\omega((x_1, \eta_1), (x_2, \eta_2)) = \eta_1(x_2) - \eta_2(x_1)$, where $x_1, x_2 \in \mathcal{H}$ and $\eta_1, \eta_2 \in \mathcal{H}^*$.

Let $F\colon W \to \mathbb{C}$ be a C^∞ function (in the Fréchet sense). We shall denote by $D_1F(x,\eta)$ (resp. $D_2F(x,\eta)$) the first (resp. second) partial Fréchet derivative of F evaluated at point $(x,\eta) \in W$. With the identification $\mathcal{H}^{**} \sim \mathcal{H}$ we have $D_1F(x,\eta) \in \mathcal{H}^*$ and $D_2F(x,\eta) \in \mathcal{H}$. Let $\langle \cdot, \cdot \rangle\colon \mathcal{H}^* \times \mathcal{H} \to \mathbb{K}$ be the canonical pairing between $\mathcal{H}$ and $\mathcal{H}^*$.

With these notations, the bracket associated with the canonical symplectic structure on W takes the form:

$$\{F, G\}(x,\eta) = \langle D_1F(x,\eta), D_2G(x,\eta)\rangle - \langle D_1G(x,\eta), D_2F(x,\eta)\rangle, \qquad (1)$$

where $F, G \in C^\infty(W, \mathbb{K})$.

Proposition 3.1. *The space W endowed with the bracket (1) is an infinite-dimensional Poisson space or, equivalently, $(C^\infty(W,\mathbb{K}), \{\cdot,\cdot\})$ is a Poisson algebra.*

Proof. One has only to check that the map $(x,\eta) \mapsto \{F,G\}(x,\eta)$ belongs to $C^\infty(W,\mathbb{K})$ for any $F, G \in C^\infty(W,\mathbb{K})$. Then Leibniz property and Jacobi identity will follow. For $F, G \in C^\infty(W,\mathbb{K})$, the maps $(x,\eta) \mapsto (D_1F(x,\eta), D_2G(x,\eta))$ and $(\xi, y) \mapsto \langle \xi, y\rangle$ belong to $C^\infty(W, \mathcal{H}^* \times \mathcal{H})$ and $C^\infty(\mathcal{H}^* \times \mathcal{H}, \mathbb{K})$, respectively. The map $(x,\eta) \mapsto \{F,G\}(x,\eta)$, as composition of C^∞ maps, is therefore in $C^\infty(W,\mathbb{K})$. □

For any orthonormal basis $\{e_i\}_{i\geq 1}$ in $\mathcal{H}$ and dual basis $\{e_i^*\}_{i\geq 1}$ in $\mathcal{H}^*$, the complex number $\partial_i F(x,\eta)$ shall denote the partial derivative of F evaluated at (x,η) in the direction of e_i, i.e. $\partial_i F(x,\eta) = DF(x,\eta).(e_i, 0) = D_1F(x,\eta).e_i$, and, similarly, $\partial_{i^*} F(x,\eta) = DF(x,\eta).(0, e_i^*) = D_2F(x,\eta).e_i^*$ is the partial derivative in the direction of e_i^*. Notice that i^* should not be considered as a different index from i when sums are involved, it is merely a mnemonic notation to distinguish partial derivatives in $\mathcal{H}$ and in $\mathcal{H}^*$.

For $F \in C^\infty(W, \mathbb{K})$, we have for any $(x, \eta) \in W$ that $\sum_{i\geq 1} |\partial_i F(x,\eta)|^2 < \infty$ and $\sum_{i\geq 1} |\partial_{i^*} F(x,\eta)|^2 < \infty$, hence the Poisson bracket (1) admits an equivalent form in terms of an absolutely convergent series:

$$\{F,G\}(x,\eta) = \sum_{i\geq 1} \left(\partial_i F(x,\eta)\partial_{i^*}G(x,\eta) - \partial_i G(x,\eta)\partial_{i^*}F(x,\eta)\right). \quad (2)$$

3.2. *Functions of Hilbert-Schmidt type*

We now define a subalgebra of $C^\infty(W, \mathbb{K})$ suited for our discussion. Let us start with some definitions and notations.

For any $F \in C^\infty(W, \mathbb{K})$ and $(x,\eta) \in W$, the higher derivatives

$$D^{(r)}F(x,\eta)\colon W \times \cdots \times W \to \mathbb{K}, \quad r \geq 1,$$

are bounded symmetric r-linear maps and partial derivatives of F will be denoted $D^{(r)}_{\alpha_1\cdots\alpha_r}F(x,\eta)$ where $\alpha_1, \ldots, \alpha_r$ are taking values 1 or 2. Let us introduce:

$$\mathcal{H}^{(\alpha)} = \begin{cases} \mathcal{H}, & \text{if } \alpha = 1; \\ \mathcal{H}^*, & \text{if } \alpha = 2. \end{cases} \quad \alpha^\flat = \begin{cases} 2, & \text{if } \alpha = 1; \\ 1, & \text{if } \alpha = 2. \end{cases} \quad i^{(\alpha)} = \begin{cases} i, & \text{if } \alpha = 1; \\ i^*, & \text{if } \alpha = 2. \end{cases} \quad (3)$$

Also $i^\sharp$ will stand for either i or i^*. With these notations, partial derivatives of F are bounded r-linear maps:

$$D^{(r)}_{\alpha_1\cdots\alpha_r}F(x,\eta)\colon \mathcal{H}^{(\alpha_1)} \times \cdots \times \mathcal{H}^{(\alpha_r)} \to \mathbb{K}.$$

It is convenient to introduce new symbols such as ∂_{ij^*k} for higher partial derivatives, e.g., $\partial_{ij^*k}F(x,\eta) \in \mathbb{K}$ stands for $D^{(3)}F(x,\eta).((e_i,0),(0,e_j^*),(e_k,0)) = D^{(3)}_{121}F(x,\eta).(e_i,e_j^*,e_k)$, where $\{e_i\}_{i\geq 1}$ (resp. $\{e_i^*\}_{i\geq 1}$) is an orthonormal basis in $\mathcal{H}$ (resp. $\mathcal{H}^*$).

Definition 3.1. Let $\{e_i\}_{i\geq 1}$ be an orthonormal basis in $\mathcal{H}$ and $\{e_i^*\}_{i\geq 1}$ be the dual basis in $\mathcal{H}^*$. A function of Hilbert-Schmidt type is a function F in $C^\infty(W, \mathbb{K})$ such that

$$\sum_{i_1,\ldots,i_r\geq 1} |\partial_{i_1^\sharp\cdots i_r^\sharp}F(x,\eta)|^2 < \infty, \quad \forall r \geq 1, \forall (x,\eta) \in W. \quad (4)$$

The sums involved have to be interpreted in the sense of summable families. By Schwarz lemma for partial derivatives, it should be understood that Eq. (4) represents $r+1$ distinct sums corresponding to all of the choices

$i^\sharp = i$ or i^*. The set of functions of Hilbert-Schmidt type on W will be denoted by $\mathcal{F}_{HS}$.

The definition above is independent of the choice of the orthonormal basis.

Remark 3.1. Let $\mathbb{N}_*$ be the set of positive integers. For each $r \geq 1$, the set of families of elements $\{x_I\}_{I\in\mathbb{N}_*^r}$ in $\mathbb{K}$ such that $\sum_{I\in\mathbb{N}_*^r} |x_I|^2 < \infty$ is the Hilbert space $\ell^2(\mathbb{N}_*^r)$ for the usual operations and inner product. Then condition (4) can be equivalently stated in the following way: $F \in C^\infty(W, \mathbb{K})$ is of Hilbert-Schmidt type if and only if, for any $r \geq 1$ and any $(x, \eta) \in W$, the 2^r families $\{\partial_{i_1^\sharp \cdots i_r^\sharp} F(x, \eta)\}_{(i_1^\sharp,\ldots,i_r^\sharp)\in\mathbb{N}_*^r}$ belong to $\ell^2(\mathbb{N}_*^r)$.

Remark 3.2. The set $\mathcal{F}_{HS}$ does not contain all of the (continuous) polynomials on W. For example, the polynomial $P(y, \xi) = \langle \xi, y\rangle$ is not in $\mathcal{F}_{HS}$ as $\sum_{i,j\geq 1} |\partial_{ij^*} P(x, \eta)|^2 = \sum_{i,j\geq 1} \delta_{ij} = \infty$. In a quantum field theory context, the polynomial P corresponds to a free Hamiltonian in the holomorphic representation.

Proposition 3.2. *The set of functions of Hilbert-Schmidt type is a subalgebra of $C^\infty(W, \mathbb{K})$ for the pointwise product of functions.*

Proof. Let $a, b \in \mathbb{K}$ and $F, G \in \mathcal{F}_{HS}$. It is clear from Remark 1 that $aF + bG$ is in $F, G \in \mathcal{F}_{HS}$. The product FG belongs to $\mathcal{F}_{HS}$ as a consequence of the Leibniz rule for the derivatives and from: if $\{x_I\}_{I\in\mathbb{N}_*^r} \in \ell^2(\mathbb{N}_*^r)$ and $\{y_J\}_{J\in\mathbb{N}_*^s} \in \ell^2(\mathbb{N}_*^s)$, then $\{x_I y_J\}_{I\times J\in\mathbb{N}_*^{r+s}} \in \ell^2(\mathbb{N}_*^{r+s})$. □

Moreover the Poisson bracket (1) restricts to $\mathcal{F}_{HS}$ and we have:

Proposition 3.3. $(W, \mathcal{F}_{HS}, \{\cdot,\cdot\})$ *is a Poisson space.*

Proof. Let F and G be in $\mathcal{F}_{HS}$. According to the proof of Prop. 3.1, the map $\Phi\colon (x, \eta) \mapsto \langle D_1 F(x, \eta), D_2 G(x, \eta)\rangle$ is in $C^\infty(W, \mathbb{K})$ and splits as follows:

$$\Phi\colon W \xrightarrow{\ \Psi_1\ } \mathcal{H}^* \times \mathcal{H} \xrightarrow{\ \Psi_2\ } \mathbb{K}$$

$$(x, \eta) \mapsto (D_1 F(x, \eta), D_2 G(x, \eta)) \mapsto \langle D_1 F(x, \eta), D_2 G(x, \eta)\rangle,$$

where both Ψ_1 and Ψ_2 are C^∞ maps. We only need to check that Φ is of Hilbert-Schmidt type.

By applying the chain rule to $\Phi = \Psi_2 \circ \Psi_1$, it is easy to see that we can freely interchange partial derivatives with the sum sign and we get that the partial derivatives of Φ is a finite sum of terms of the form:

$$a_{j_1^\sharp \cdots j_r^\sharp, k_1^\sharp \cdots k_s^\sharp} \equiv \sum_{i \geq 1} \partial_{i j_1^\sharp \cdots j_r^\sharp} F(x, \eta) \partial_{i^* k_1^\sharp \cdots k_s^\sharp} G(x, \eta). \tag{5}$$

The Cauchy-Schwarz inequality implies that the family $\{a_{j_1^\sharp \cdots j_r^\sharp, k_1^\sharp \cdots k_s^\sharp}\}$ is in $\ell^2(\mathbb{N}_*^{r+s})$ and thus Φ belongs to $\mathcal{F}_{HS}$. Hence $\mathcal{F}_{HS}$ is closed under the Poisson bracket. □

3.3. *Moyal star-product on W*

We are now in position to define the Moyal star-product on W as an associative product on $\mathcal{F}_{HS}[[\hbar]]$.

For $F, G \in \mathcal{F}_{HS}$, $(x, \eta) \in W$, $r \geq 1$, $\alpha_1, \ldots, \alpha_r, \beta_1 \ldots, \beta_r$ equal to 1 or 2, and with the notations introduced previously, let us define:

$$\begin{aligned} &\langle\langle D^{(r)}_{\alpha_1 \cdots \alpha_r} F, D^{(r)}_{\beta_1 \cdots \beta_r} G \rangle\rangle (x, \eta) \\ &\quad = \sum_{i_1, \ldots, i_r \geq 1} \partial_{i_1^{(\alpha_1)} \cdots i_r^{(\alpha_r)}} F(x, \eta)\, \partial_{i_1^{(\beta_1)} \cdots i_r^{(\beta_r)}} G(x, \eta) \end{aligned} \tag{6}$$

Remark 3.3. The preceding definition does not depend on the choice of the orthonormal basis in $\mathcal{H}$ and the series is absolutely convergent as a consequence of the Cauchy-Schwarz inequality.

Let Λ be the canonical symplectic 2×2-matrix with $\Lambda^{12} = +1$. As in the finite-dimensional case, the powers of the Poisson bracket (1) are defined as:

$$C_r(F, G) = \sum_{\alpha_1, \ldots, \alpha_r = 1,2} \sum_{\beta_1, \ldots, \beta_r = 1,2} \Lambda^{\alpha_1 \beta_1} \cdots \Lambda^{\alpha_r \beta_r} \langle\langle D^{(r)}_{\alpha_1 \cdots \alpha_r} F, D^{(r)}_{\beta_1 \cdots \beta_r} G \rangle\rangle. \tag{7}$$

The next Proposition shows that the C_r are bidifferential operators in the sense of Def. 2.2 and they close on $\mathcal{F}_{HS}$. We shall use a specific version of the Hilbert tensor product $\otimes$ between Hilbert spaces (see e.g. Sect. 2.6 in [8]). Let $\mathcal{H}_1, \ldots, \mathcal{H}_r$ be Hilbert spaces with orthonormal bases $\{e_i^{(1)}\}_{i \geq 1}, \ldots, \{e_i^{(r)}\}_{i \geq 1}$. There exists a Hilbert space $\mathcal{T} = \mathcal{H}_1 \otimes \cdots \otimes \mathcal{H}_r$ and a bounded r-linear map $\Psi\colon (x_1,\dots,x_r)\mapsto x_1\otimes\cdots\otimes x_r$ from $\mathcal{H}_1\times\cdots\times\mathcal{H}_r$ to $\mathcal{H}_1\otimes\cdots\otimes\mathcal{H}_r$ satisfying:

$$\sum_{i_1,\dots,i_r\geq 1} |\langle\Psi(e^{(1)}_{i_1},\dots,e^{(r)}_{i_r}),x\rangle|^2<\infty,\quad \forall x\in\mathcal{T},$$

such that for any bounded r-linear form: $\Xi\colon \mathcal{H}_1\times\cdots\times\mathcal{H}_r\to\mathbb{K}$ satisfying:

$$\sum_{i_1,\dots,i_r\geq 1} |\Xi(e^{(1)}_{i_1},\dots,e^{(r)}_{i_r})|^2<\infty,$$

there exists a unique bounded linear form L on $\mathcal{T}$ so that $\Xi = L\circ\Psi$. This universal property allows to identify Ξ to an element of $\mathcal{T}^*$.

Proposition 3.4. *For $F,G\in\mathcal{F}_{HS}$ and $r\geq 1$, the map $(x,\eta)\mapsto C_r(F,G)(x,\eta)$ belongs to the space of functions of Hilbert-Schmidt type $\mathcal{F}_{HS}$.*

Proof. Each term in the finite sum (7) is of the form

$$\langle\langle D^{(r)}_{\alpha_1\cdots\alpha_r}F(x,\eta), D^{(r)}_{\alpha^\flat_1\cdots\alpha^\flat_r}G(x,\eta)\rangle\rangle, \tag{8}$$

where $\alpha_1,\dots,\alpha_r = 1$ or 2. From the definition of $\mathcal{F}_{HS}$, expression (8) is well defined for any $F,G\in\mathcal{F}_{HS}$ and thus defines a function on W:

$$\Phi\colon (x,\eta)\mapsto \langle\langle D^{(r)}_{\alpha_1\cdots\alpha_r}F(x,\eta), D^{(r)}_{\alpha^\flat_1\cdots\alpha^\flat_r}G(x,\eta)\rangle\rangle.$$

The case $r=1$ has been already proved in Prop. 3.3. For $r\geq 2$, we need to slightly modify the argument used in the proof of Prop. 3.3 since the bilinear map $\langle\langle\ ,\ \rangle\rangle$ defined by (6) is not a bounded bilinear form on the product of Banach spaces:

$$\mathcal{L}^r(\mathcal{H}^{(\alpha_1)},\dots,\mathcal{H}^{(\alpha_r)};\mathbb{K})\times\mathcal{L}^r(\mathcal{H}^{(\alpha^\flat_1)},\dots,\mathcal{H}^{(\alpha^\flat_r)};\mathbb{K}). \tag{9}$$

In order to show that Φ is in $C^\infty(W,\mathbb{K})$, we shall use the universal property of the Hilbert tensor product $\otimes$ mentioned above. For $F,G\in\mathcal{F}_{HS}$, we can consider the bounded r-linear maps $D^{(r)}_{\alpha_1\cdots\alpha_r}F(x,\eta)$ and $D^{(r)}_{\alpha^\flat_1\cdots\alpha^\flat_r}G(x,\eta)$ as bounded linear forms on $\mathcal{H}^{(\alpha)}\equiv\mathcal{H}^{(\alpha_1)}\otimes\cdots\otimes\mathcal{H}^{(\alpha_r)}$ and $\mathcal{H}^{(\alpha^\flat)}\equiv\mathcal{H}^{(\alpha^\flat_1)}\otimes\cdots\otimes\mathcal{H}^{(\alpha^\flat_r)}$, respectively. Then the unbounded bilinear map $\langle\langle\ ,\ \rangle\rangle$ on the product of spaces (9) restricts to the natural pairing of $\mathcal{H}^{(\alpha^\flat)}\sim\mathcal{H}^{(\alpha)*}$ and $\mathcal{H}^{(\alpha)}$ which is a smooth map. This shows that Ψ belongs to $C^\infty(W,\mathbb{K})$.

An argument similar to the one used in the proof of Prop. 3.3 shows that the partial derivatives of Φ involve a finite sum of terms of the form:

$$\sum_{i_1,\dots i_r\geq 1} \partial_{i_1^{(\alpha_1)}\cdots i_r^{(\alpha_r)}j_1^\sharp\cdots j_a^\sharp}F(x,\eta)\,\partial_{i_1^{(\alpha^\flat_1)}\cdots i_r^{(\alpha^\flat_r)}k_1^\sharp\cdots k_b^\sharp}G(x,\eta),$$

where we have used the notations introduced at the beginning of Subsection 3.2. A direct application of the Cauchy-Schwarz inequality gives:

$$\sum_{j_1,\dots,j_a\geq 1}\sum_{k_1,\dots,k_b\geq 1}\Big|\sum_{i_1,\dots i_r\geq 1}\partial_{i_1^{(\alpha_1)}\dots i_r^{(\alpha_r)}j_1^\sharp\dots j_a^\sharp}F\,\partial_{i_1^{(\alpha_1^\flat)}\dots i_r^{(\alpha_r^\flat)}k_1^\sharp\dots k_b^\sharp}G\Big|^2\leq$$
$$\sum_{\substack{i_1,\dots i_r\geq 1\\ j_1,\dots,j_a\geq 1}}\big|\partial_{i_1^{(\alpha_1)}\dots i_r^{(\alpha_r)}j_1^\sharp\dots j_a^\sharp}F\big|^2\sum_{\substack{i_1,\dots i_r\geq 1\\ k_1,\dots,k_b\geq 1}}\big|\partial_{i_1^{(\alpha_1^\flat)}\dots i_r^{(\alpha_r^\flat)}k_1^\sharp\dots k_b^\sharp}G\big|^2<\infty.$$

This shows that Φ is of Hilbert-Schmidt type and hence $(x,\eta)\mapsto C_r(F,G)(x,\eta)$ belongs to $\mathcal{F}_{HS}$. □

We summarize all the previous facts in the following:

Theorem 3.1. *Let the C_r's be given by (7), then the formula*

$$F\star_\hbar^M G = FG+\sum_{r\geq 1}\frac{\hbar^r}{r!}C_r(F,G), \tag{10}$$

defines an associative product on $\mathcal{F}_{HS}[[\hbar]]$ and hence a deformation quantization $(W,\mathcal{F}_{HS}[[\hbar]],\star_\hbar^M)$ of the Poisson space $(W,\mathcal{F}_{HS},\{\cdot,\cdot\})$.

Proof. That $\star_\hbar^M:\mathcal{F}_{HS}[[\hbar]]\times\mathcal{F}_{HS}[[\hbar]]\to\mathcal{F}_{HS}[[\hbar]]$ is a bilinear map is a direct consequence of Prop. 3.4 and Prop. 3.2. Associativity follows from the same combinatorics used in the finite-dimensional case and the fact that the derivatives distribute in the pairing $\langle\langle\ ,\ \rangle\rangle$ defining the C_r's. □

4. On the equivalence of deformation quantizations on W

We end this article by a discussion on the issue of equivalence of star-products on W. In the finite-dimensional case (i.e. on $\mathbb{R}^{2n}$ endowed with its canonical Poisson bracket), it is well known that all star-products are equivalent to each other. The situation we are dealing with here, although it is a direct generalization of the flat finite-dimensional case, allows inequivalent deformation quantizations. We will illustrate this fact on a family of star-products of exponential type containing the important case of the normal star-product.

4.1. *The Hochschild complex*

The space of functions of Hilbert-Schmidt type $\mathcal{F}_{HS}$ being an associative algebra over $\mathbb{K}$, we can consider the Hochschild complex $C^\bullet(\mathcal{F}_{HS},\mathcal{F}_{HS})$ and its cohomology $H^\bullet(\mathcal{F}_{HS},\mathcal{F}_{HS})$.

One has first to specify a class of cochains that would define the Hochschild complex. Here the cochains are simply r-differential operators in the sense of Def. 2.2 which vanishes on constants. The case where $\mathcal{F} = \mathcal{F}_{HS}$ in Def. 2.2 allows a more precise description of the r-differential operators. Consider an r-differential operator defined by:

$$\mathbf{A}(F_1, \dots, F_r)(w) = \sum_{n_1,\dots,n_r \geq 0} a^{(n_1,\cdots,n_r)}(w).(D^{(n_1)}F_1(w), \cdots, D^{(n_r)}F_r(w)),$$

where $F_1, \dots, F_r \in \mathcal{F}_{HS}$ and $w = (x, \eta) \in W$. For $F \in \mathcal{F}_{HS}$, the higher derivative $D^{(m)}F(x, \eta)$ defines an element of the mth tensor power of W (here we identify W^* with W). For a fixed $w = (x, \eta) \in W$ and $m \geq 0$, the linear map $F \mapsto D^{(m)}F(x, \eta)$ from $\mathcal{F}_{HS}$ to $W^{\overset{m}{\otimes}}$ is onto and we can look at the restriction of the r-linear form

$$a^{(n_1,\dots,n_r)}(w) \colon \mathcal{D}_w^{(n_1,\dots,n_r)} \subset \mathcal{L}_{\text{sym}}^{n_1}(W, \mathbb{K}) \times \cdots \times \mathcal{L}_{\text{sym}}^{n_r}(W, \mathbb{K}) \to \mathbb{K},$$

to the product $W^{\overset{n_1}{\otimes}} \times \cdots \times W^{\overset{n_r}{\otimes}}$ as a bounded r-linear form:

$$\tilde{a}^{(n_1,\dots,n_r)}(w) \colon W^{\overset{n_1}{\otimes}} \times \cdots \times W^{\overset{n_r}{\otimes}} \to \mathbb{K},$$

such that $w \mapsto \tilde{a}^{(n_1,\dots,n_r)}(w)$ is a smooth map from W to $\mathcal{L}^r(W^{\overset{n_1}{\otimes}} \times \cdots \times W^{\overset{n_r}{\otimes}}, \mathbb{K})$.

The Leibniz rule for the derivatives of a product can be written here for $F, G \in \mathcal{F}_{HS}$ as:

$$D^{(m)}(FG)(w).(w_1, \dots, w_m)$$

$$= \frac{1}{m!} \sum_{\sigma \in \mathfrak{S}_m} \sum_{k=0}^{k=m} \binom{m}{k} (D^{(k)}F(w) \otimes D^{(m-k)}G(w))(w_{\sigma_1} \otimes \cdots \otimes w_{\sigma_m}),$$

where $\mathfrak{S}_m$ is the symmetric group of degree m.

Let $\mathbf{A}(F) = \sum_{m \geq 0} \tilde{a}^{(m)}(w).(D^{(m)}F(w))$ be a differential operator with $\tilde{a}^{(m)} \in C^\infty(W, \mathcal{L}(W^{\overset{m}{\otimes}}, \mathbb{K}))$, then it follows from the above form for the Leibniz rule that for $F, G \in \mathcal{F}_{HS}$, $\mathbf{A}(FG)$ can be written as a finite sum

$$\sum_{m_1, m_2 \geq 0} \tilde{b}^{(m_1,m_2)}(w).(D^{(m_1)}F(w), D^{(m_2)}G(w))$$

for some $\tilde{b}^{(m_1,m_2)} \in C^\infty(W, \mathcal{L}^2(W^{\overset{m_1}{\otimes}} \times W^{\overset{m_2}{\otimes}}, \mathbb{K}))$, and thus $(F, G) \mapsto \mathbf{A}(FG)$ is a bidifferential operator. The generalization to r-differential operators of this fact is straightforward.

The Hochschild complex consists of multidifferential operators vanishing on constants, i.e., $C^\bullet(\mathcal{F}_{HS},\mathcal{F}_{HS}) = \oplus_{k\geq 0} C^k(\mathcal{F}_{HS},\mathcal{F}_{HS})$ where, for $k \geq 1$, the space of k-cochains is:

$$C^k(\mathcal{F}_{HS},\mathcal{F}_{HS}) \\ = \{\mathbf{A}\colon \mathcal{F}_{HS}^k \to \mathcal{F}_{HS} \mid \mathbf{A} \text{ is a } k\text{–differential operator vanishing on } \mathbb{K}\}.$$

The differential of a k-cochain $\mathbf{A}$ is the $(k+1)$-linear map $\delta\mathbf{A}$ given by:

$$\delta\mathbf{A}(F_0,\dots,F_k) = F_0\mathbf{A}(F_1,\dots,F_k) - \mathbf{A}(F_0F_1,F_2,\dots,F_k) + \cdots \\ + (-1)^k\mathbf{A}(F_0,F_1,\dots,F_{k-1}F_k) + (-1)^{k+1}\mathbf{A}(F_0,\dots,F_{k-1})F_k. \quad (11)$$

satisfies $\delta^2 = 0$, and according to the discussion above $\delta\mathbf{A}$ is a $(k+1)$-differential operator, vanishing on constants whenever $\mathbf{A}$ does. Thus $\delta\mathbf{A}$ is a cochain that belongs to $C^{k+1}(\mathcal{F}_{HS},\mathcal{F}_{HS})$, hence we indeed have a complex.

A k-cochain $\mathbf{A}$ is a k-cocycle if $\delta\mathbf{A} = 0$. We denote by $Z^k(\mathcal{F}_{HS},\mathcal{F}_{HS})$ the space of k-cocycles and by $B^k(\mathcal{F}_{HS},\mathcal{F}_{HS})$ the space of those k-cocycles which are coboundaries. The kth Hochschild cohomology space of $\mathcal{F}_{HS}$ valued in $\mathcal{F}_{HS}$ is defined as the quotient $H^k(\mathcal{F}_{HS},\mathcal{F}_{HS}) = Z^k(\mathcal{F}_{HS},\mathcal{F}_{HS})/B^k(\mathcal{F}_{HS},\mathcal{F}_{HS})$.

4.2. *Star-products of the exponential type*

Let $\mathcal{B}(\mathcal{H})$ denote the algebra of bounded operators on $\mathcal{H}$ and $\mathcal{B}_2(\mathcal{H})$, the two-sided $*$-ideal of Hilbert-Schmidt operators on $\mathcal{H}$. We shall describe a family of deformation quantizations $\{(W,\mathcal{F}_{HS}[[\hbar]],\star_\hbar^A)\}_{A\in\mathcal{B}(H)}$. Each star-product $\star_\hbar^A$ where $A \in \mathcal{B}(H)$ shall be the exponential of a Hochschild 2-cocycle, with the Moyal star-product corresponding to the case $A = 0$. It will turn out that the set of equivalence classes of star-products of this type is parameterized by $\mathcal{B}(\mathcal{H})/\mathcal{B}_2(\mathcal{H})$.

Let $A \in \mathcal{B}(\mathcal{H})$. For $F,G \in \mathcal{F}_{HS}$, the map

$$(x,\eta) \mapsto \langle D_1F(x,\eta), AD_2G(x,\eta)\rangle + \langle D_1G(x,\eta), AD_2F(x,\eta)\rangle \quad (12)$$

defines a smooth function on W and is symmetric in F and G. We denote by $E_A(F,G)(x,\eta)$ the right-hand side of (12). Moreover we have:

Proposition 4.1. *The bilinear map $(F,G) \mapsto E_A(F,G)$ is a Hochschild 2-cocycle.*

Proof. It is clear that $\delta E_A = 0$ and, since E_A vanishes on constants, it is sufficient to check that E_A is a bidifferential operator on $(W,\mathcal{F}_{HS})$, i.e., that the smooth function $(x, \eta) \mapsto E_A(F, G)(x, \eta)$ is of Hilbert-Schmidt type for any $F, G \in \mathcal{F}_{HS}$. Let $\{e_i\}_{i\geq 1}$ be an orthonormal basis of $\mathcal{H}$ and $\{e_i^*\}_{i\geq 1}$ the dual basis. Consider the basis $\{f_i\}_{i\geq 1}$ in $W = \mathcal{H} \oplus \mathcal{H}^*$ defined by $f_i = (e_{\frac{i+1}{2}}, 0)$ if i is odd, and $f_i = (0, e^*_{\frac{i}{2}})$ if i is even. To show that $E_A(F, G)$ is in $\mathcal{F}_{HS}$ is equivalent to show that

$$\sum_{i_1,\dots,i_n \geq 1} |D^{(n)} E_A(F, G)(x, \eta).(f_{i_1}, \dots, f_{i_n})|^2 < \infty. \tag{13}$$

holds for any $n \geq 1$ and $w = (x, \eta) \in W$.

The chain rule applied to

$$\Phi\colon W \xrightarrow{\Psi_1} \mathcal{H}^* \times \mathcal{H} \xrightarrow{\Psi_2} \mathbb{K}$$

$$(x, \eta) \mapsto (D_1 F(x, \eta), D_2 G(x, \eta)) \mapsto \langle D_1 F(x, \eta), A D_2 G(x, \eta)\rangle,$$

shows that the derivatives of E_A distributes in the pairing $\langle\ ,\ \rangle$.

The nth derivative of $E_A(F, G)$ in the direction of $(f_{i_1}, \dots, f_{i_n})$ is a finite sum of terms of the form:

$$\langle D^{(r)} D_1 F(x, \eta).(f_{k_1}, \dots, f_{k_r}), A D^{(s)} D_2 G(x, \eta).(f_{l_1}, \dots, f_{l_s})\rangle, \tag{14}$$

where $r + s = n$ and $(k_1, \dots, k_r, l_1, \dots, l_s)$ is a permutation of $(i_1, \dots, i_n)$, and similar terms with F and G inverted. It is worth noting that $D^{(r)} D_1 F(x, \eta).(f_{k_1}, \dots, f_{k_r})$ is the element of $\mathcal{H}^*$ defined by the bounded linear form

$$h \mapsto D^{(r+1)} F(x, \eta).((h, 0), f_{k_1}, \dots, f_{k_r})$$

on $\mathcal{H}$ and, similarly, $D^{(s)} D_2 G(x, \eta).(f_{l_1}, \dots, f_{l_s})$ is the element of $\mathcal{H}$ defined by the bounded linear form on $\mathcal{H}^*$:

$$\xi \mapsto D^{(s+1)} G(x, \eta).((0, \xi), f_{l_1}, \dots, f_{l_s}).$$

The modulus squared of (14) is bounded by the constant $||A||^2 \alpha_{k_1\dots k_r} \beta_{l_1\dots l_s}$, where

$$\alpha_{k_1\dots k_r} = \sum_{i\geq 1} |D^{(r+1)} F(x, \eta).((e_i, 0), f_{k_1}, \dots, f_{k_r})|^2,$$

$$\beta_{l_1\dots l_s} = \sum_{j\geq 1} |D^{(s+1)} G(x, \eta).((0, e_j^*), f_{l_1}, \dots, f_{l_s})|^2.$$

Since $F, G \in \mathcal{F}_{HS}$, we have $\displaystyle\sum_{k_1,\dots,k_r\geq 1} \alpha_{k_1\dots k_r} < \infty$ and $\displaystyle\sum_{l_1,\dots,l_s\geq 1} \beta_{l_1\dots l_s} < \infty$, from which inequality (13) follows. □

For any $A \in \mathcal{B}(\mathcal{H})$, let us define:

$$\begin{aligned} C_1^A(F,G) &= \{F,G\} + E_A(F,G) \\ &= \langle D_1F, (A+\mathbb{I})D_2G\rangle + \langle D_1G, (A-\mathbb{I})D_2F\rangle, \end{aligned} \tag{15}$$

where $\mathbb{I}$ is the identity operator on $\mathcal{H}$. Plainly, C_1^A is a 2-cocycle with constant coefficients, and one can define a star-product by taking the exponential of C_1^A:

$$F \star_\hbar^A G = \exp(\hbar C_1^A)(F,G) = FG + \sum_{r\geq 1} \frac{\hbar^r}{r!} C_r^A(F,G),$$

where $C_r^A = (C_1^A)^r$ in the sense of bidifferential operators. This formula defines an associative product on $\mathcal{F}_{HS}[[\hbar]]$, and we get a family of deformation quantizations $\{(W, \mathcal{F}_{HS}[[\hbar]], \star_\hbar^A)\}_{A\in\mathcal{B}(H)}$ of $(W, \mathcal{F}_{HS}, \{,\cdot,\cdot\})$. This family of star-products is easily described by their symbols. Let us consider the following family of smooth functions on W:

$$\Phi_{y,\xi}(x,\eta) = \exp(\langle \eta, y\rangle + \langle \xi, x\rangle), \quad x,y \in \mathcal{H},\ \eta,\xi \in \mathcal{H}^*. \tag{16}$$

The $\Phi_{y,\xi}$'s belong to $\mathcal{F}_{HS}$ and from (15) we deduce that:

$$C_r^A(\Phi_{y,\xi}, \Phi_{y',\xi'}) = (\langle \xi, (A+\mathbb{I})y'\rangle + \langle \xi', (A-\mathbb{I})y\rangle)^r \, \Phi_{y+y',\xi+\xi'},$$

and consequently:

$$\Phi_{y,\xi} \star_\hbar^A \Phi_{y',\xi'} = \exp\left(\hbar(\langle \xi, (A+\mathbb{I})y'\rangle + \langle \xi', (A-\mathbb{I})y\rangle)\right) \Phi_{y+y',\xi+\xi'}. \tag{17}$$

Example 4.1. Set $A = \mathbb{I}$ in (15), then $C_1^{\mathbb{I}}(F,G) = 2\langle D_1F, D_2G\rangle$ and the corresponding star-product reads

$$F \star_\hbar^{\mathbb{I}} G = FG + \sum_{r\geq 1} \frac{(2\hbar)^r}{r!} \langle D_{1\cdots 1}^{(r)} F, D_{2\cdots 2}^{(r)} G\rangle. \tag{18}$$

It is the well-known normal star-product (or Wick or standard depending on the interpretation of the variables $(x,\eta) \in W$). The cochains defining the normal star-product $C_r^{\mathbb{I}}$ correspond to only one term in the sum defining the rth cochain of the Moyal star-product, namely the term corresponding to $\alpha_1 = \cdots = \alpha_r = 1$ and $\beta_1 = \cdots = \beta_r = 2$ in the sum (7). One would expect that conditions (4) defining the functions of Hilbert-Schmidt type are not all needed in order to make sense of the normal star-product and would guess that this product can be defined on a wider class of functions. Actually, the normal star-product defines a deformation quantization on a larger space of functions (containing the free Hamiltonian) that we shall describe in a forthcoming paper.

At this stage, a natural question arises: are the deformation quantizations $\{(W, \mathcal{F}_{HS}[[\hbar]], \star_\hbar^A)\}_{A\in\mathcal{B}(H)}$ equivalent to each other? The answer is given in the:

Proposition 4.2. *Let $A \in \mathcal{B}(\mathcal{H})$. The Hochschild 2-cocycle E_A defined by (12) is a coboundary if and only if A is a Hilbert-Schmidt operator.*

Proof. Let A be a Hilbert-Schmidt operator on $\mathcal{H}$. The map from W to $\mathbb{K}$ defined by $(x, \eta) \mapsto \langle \eta, Ax\rangle$ is of Hilbert-Schmidt type and therefore there exists a bounded linear form $\tilde{A}\colon \mathcal{H}^* \otimes \mathcal{H} \to \mathbb{K}$ so that $\langle \eta, Ax\rangle = \langle \tilde{A}, \eta \otimes x\rangle$. The 2-cocycle E_A can then be written as:

$$E_A(F,G) = \langle \tilde{A}, D_1F \otimes D_2G + D_1G \otimes D_2F\rangle, \quad F, G \in \mathcal{F}_{HS}.$$

For $F \in \mathcal{F}_{HS}$, the mixed derivative $D_{12}^{(2)}F$ belongs to $C^\infty(W, \mathcal{L}(\mathcal{H} \otimes \mathcal{H}^*, \mathbb{K})) \sim C^\infty(W, \mathcal{H}^* \otimes \mathcal{H})$ and $T_A(F)(x, \eta) = -\langle \tilde{A}, D_{12}^{(2)}F(x, \eta)\rangle$ defines a differential operator on $(W, \mathcal{F}_{HS})$ vanishing on constants. For any two functions in $\mathcal{F}_{HS}$, the Leibniz rule reads:

$$D_{12}^{(2)}(FG) = FD_{12}^{(2)}(G) + GD_{12}^{(2)}(F) + D_1F \otimes D_2G + D_1G \otimes D_2F,$$

it belongs to $C^\infty(W, \mathcal{H}^* \otimes \mathcal{H})$ and a simple computation gives $\delta T_A = E_A$, therefore E_A is a coboundary if A is in the Hilbert-Schmidt class.

Conversely, if E_A is a coboundary, there exists a differential operator S on $(W, \mathcal{F}_{HS})$ vanishing on constants, so that $E_A = \delta S$. If a term of degree one occurs in S (a derivation) it can be subtracted without changing δS, hence we can assume that S has the form:

$$S(F)(x,\eta) = \sum_{m\geq 2} a^{(m)}(x,\eta).(D^{(m)}F(x,\eta)),$$

where $a^{(m)} \in C^\infty(W, \mathcal{L}(W^{\overset{m}{\otimes}}, \mathbb{K}))$ and only finitely many of them are nonzero. By computing δS and using $E_A = \delta S$, we find that only the term of degree 2 contributes:

$$\begin{aligned}&\langle D_1F(w)\, AD_2G(w)\rangle + \langle D_1G(w), AD_2F(w)\rangle \\ &= -a^{(2)}(w).(DF(w) \otimes DG(w) + DG(w) \otimes DF(w)), \quad w = (x,\eta) \in W.\end{aligned}$$

If we evaluate the equality above on $F(x, \eta) = \langle \xi, x\rangle$, $\xi \in \mathcal{H}^*$, and $G(x, \eta) = \langle \eta, y\rangle$, $y \in \mathcal{H}$, we find (with a slight abuse of notations):

$$\langle \xi, Ay\rangle = -a^{(2)}(w).(\xi \otimes y), \quad \forall \xi \in \mathcal{H}^*, \forall y \in \mathcal{H},$$

from which follows that $A \in \mathcal{B}(\mathcal{H})$ is a Hilbert-Schmidt operator on $\mathcal{H}$, as $a^{(2)}(w)$ is a bounded linear form on $\mathcal{H}^* \otimes \mathcal{H}$. □

As an immediate consequence of Prop. 4.2, we deduce a classification result for deformation quantizations of exponential type $\{(W, \mathcal{F}_{HS}[[\hbar]], \star_\hbar^A)\}_{A\in\mathcal{B}(H)}$.

Theorem 4.1. *Let $A, B \in \mathcal{B}(H)$. Two deformation quantizations $(W, \mathcal{F}_{HS}[[\hbar]], \star_\hbar^A)$ and $(W, \mathcal{F}_{HS}[[\hbar]], \star_\hbar^B)$ are equivalent if and only if $A - B$ is in the Hilbert-Schmidt class. Consequently, the set of equivalence classes of $\{(W, \mathcal{F}_{HS}[[\hbar]], \star_\hbar^A)\}_{A\in\mathcal{B}(H)}$ is parameterized by $\mathcal{B}(\mathcal{H})/\mathcal{B}_2(\mathcal{H})$.*

Proof. Suppose that $\star_\hbar^A$ and $\star_\hbar^B$ are equivalent, i.e., there exists a formal series of differential operators vanishing on constants: $T = \mathrm{Id}_{\mathcal{F}_{HS}} + \sum_{r\geq 1} \hbar^r T_r$, so that $T(F \star_\hbar^A G) = TF \star_\hbar^B TG$. Then it follows that $C_1^A = C_1^B + \delta T_1$ and, from the definitions (12) and (15) of E_A and C_1^A, we have $E_{A-B} = E_A - E_B = \delta T_1$, showing that E_{A-B} is a coboundary and hence $A - B$ is a Hilbert-Schmidt operator on $\mathcal{H}$.

Conversely, if $S \equiv A - B$ is a Hilbert-Schmidt operator, it defines a bounded linear form $\tilde{S}$ on $\mathcal{H}^* \otimes \mathcal{H}$ and a differential operator $T_1(F) = -\langle \tilde{S}, D_{12}^{(2)} F\rangle$ on $(W, \mathcal{F}_{HS})$ (cf. the proof of Prop. 4.2). Since the star-products $\star_\hbar^A$ and $\star_\hbar^B$ are defined by constant coefficient bidifferential operators, it is sufficient to establish the equivalence at the level of symbols. Now define the formal series of differential operators $T = \exp(\hbar T_1)$. Its symbol is given by $T(\Phi_{y,\xi}) = \exp(\hbar\langle \xi, (B - A)y\rangle)\Phi_{y,\xi}$, where $\Phi_{y,\xi}$ has been defined in (16). Using the symbol (17) associated to a star-product, we find:

$$T(\Phi_{y,\xi} \star_\hbar^A \Phi_{y',\xi'}) = T\Phi_{y,\xi} \star_\hbar^B T\Phi_{y',\xi'}, \quad y, y' \in \mathcal{H}, \xi, \xi' \in \mathcal{H}^*.$$

Therefore the deformation quantizations $(W, \mathcal{F}_{HS}[[\hbar]], \star_\hbar^A)$ and $(W, \mathcal{F}_{HS}[[\hbar]], \star_\hbar^B)$ are equivalent. □

The Moyal and normal star-products correspond to $A = 0$ and $A = \mathbb{I}$ in (15), respectively. Since the identity operator on the infinite-dimensional Hilbert space $\mathcal{H}$ is not in the Hilbert-Schmidt class, we have:

Corollary 4.1. *The Moyal and normal star-products are not equivalent deformations on $(W, \mathcal{F}_{HS}[[\hbar]])$.*

Remark 4.1. One can generalize the class of exponential type of star-products by allowing formal series with coefficients in $\mathcal{B}(\mathcal{H})$ in (15) or (17). The set of equivalence classes would then be $(\mathcal{B}(\mathcal{H})/\mathcal{B}_2(\mathcal{H}))[[\hbar]]$. Since we did not show that any star-product on $(W, \mathcal{F}_{HS}, \{\cdot,\cdot\})$ is equivalent to a star-product of the exponential type, $(\mathcal{B}(\mathcal{H})/\mathcal{B}_2(\mathcal{H}))[[\hbar]]$ can only be considered as a *lower bound* for the classification space of all the star-products on $(W, \mathcal{F}_{HS}, \{\cdot,\cdot\})$.

Remark 4.2. Recall that all the star-products on $\mathbb{R}^{2n}$ endowed with its canonical Poisson structure are equivalent to each other. This fact should be put in relation with Von Neumann's uniqueness theorem on the irreducible (continuous) representations of Weyl systems associated to the canonical commutation relations (CCR) $\{q_i, p_j\} = \delta_{ij}$. The inequivalent representations of Weyl systems associated to the infinite dimensional CCR have been described long time ago by Gårding and Wightman [7], and Segal [10]. Here the existence of inequivalent deformation quantizations conveys the idea that there should be a close link between the set of equivalence classes of star-products and representations of Weyl systems associated to the CCR. It might turn out that they actually are identical.

Acknowledgments

The author would like to thank Y. Maeda, N. Tose and S. Watamura, the organizers of the wonderful meeting *Noncommutative Geometry and Physics 2004* held at Keio University (February 2004), for their invitation to give a talk and warmest hospitality.

References

1. F. Bayen, M. Flato, C. Frønsdal, A. Lichnerowicz, and D. Sternheimer, *Deformation theory and quantization.* Ann. Phys., **111** (1978), 61–151.
2. M. DeWilde and P. B. A. Lecomte. *Existence of star-products and of formal deformations of the Poisson Lie algebra of arbitrary symplectic manifolds.* Lett. Math. Phys., **7** (1983), 487–496.
3. G. Dito and D. Sternheimer. *Deformation quantization: genesis, developments and metamorphoses.* In (G. Halbout, ed.), Deformation quantization, volume 1 of *IRMA Lectures in Mathematics and Theoretical Physics*, pages 9–54. Walter de Gruyter, Berlin, New York, 2002.
4. J. Dito. *Star-product approach to quantum field theory: The free scalar field.* Lett. Math. Phys., **20** (1990), 125–134.
5. J. Dito. *Star-products and nonstandard quantization for Klein-Gordon equation.* J. Math. Phys., **33** (1992), 791–801.
6. M. Dütsch and K. Fredenhagen. *Perturbative algebraic field theory, and deformation quantization.* Field Inst. Commun., **30** (2001), 151–160.
7. L. Gårding and A. S. Wightman. *Representations of the commutation relations.* Proc. Nat. Acad. Sci. USA., **40** (1954), 622–626.
8. R. V. Kadison and J. R. Ringrose. *Fundamentals of the Theory of Operator Algebras. Volume I: Elementary Theory*, volume 15 of Graduate Studies in Mathematics. American Mathematical Society, Providence, 1997.

9. M. Kontsevich. *Deformation quantization of Poisson manifolds.* In (G. Dito and D. Sternheimer, eds.), Proceedings of the Euroconference Moshé Flato 2000, *Lett. Math. Phys.*, **66** (2004), 157–216.
10. I. E. Segal. *Distributions in Hilbert space and canonical systems of operators.* Trans. Amer. Math. Soc., **88** (1958), 12–41.
11. I. E. Segal. *Symplectic structures and the quantization problem for wave equations.* Symposia Mathematica, **14** (1974), 79–117.
12. E. Witten. *Noncommutative geometry and string field theory.* Nucl. Phys. B, **268** (1986), 253–294.

SYMMETRIES AND MODULI SPACES OF THE SELF–DUAL YANG–MILLS EQUATIONS

JAMES D.E. GRANT *

Department of Mathematical Sciences
University of Aberdeen
Aberdeen
AB24 3UE
Scotland
james.grant@univie.ac.at

We review the construction of infinite-dimensional symmetry algebras of the self-dual Yang-Mills equations on $\mathbb{R}^4$ and the ADHM description of the moduli space of instantons on S^4. We report on recent work describing the action of the corresponding symmetry (pseudo)-group on the instanton moduli spaces.

1. Introduction

In Hamiltonian mechanics, one has a well defined notion of an integrable system. Such a system is defined by a $2n$-dimensional symplectic manifold (X, ω) with the dynamics being determined by a function $H : X \to \mathbb{R}$. The system is completely integrable in the sense of Liouville if there exist n functions preserved by the flow generated by the Hamiltonian that have vanishing Poisson brackets with one another and that are functionally independent on X, perhaps minus a set of measure zero, S. From a more global perspective, this implies that there is an action of an n-dimensional Abelian Lie group on $X \setminus S$ that, in the case that this group is compact, corresponds to a Hamiltonian torus action. Thus the space $X \setminus S$ is foliated by n-dimensional tori.

When one considers systems of partial differential equations, it is more difficult to define a suitable notion of integrability. One approach has been to study systems that admit infinite-dimensional symmetry algebras. In order to make contact with the finite-dimensional theory, however, one

*Address from 1 March 2005: Fakultät für Mathematik, Universität Wien, Nordbergstrasse 15, 1090 Wien, Austria.

would like to know whether there is a corresponding group action on the space of solutions of the system and, if so, what the corresponding orbit structure is. As an example for particular types of harmonic maps, the action of the corresponding symmetry group (the *dressing action*) on the space of solutions is well understood (see Ref. 14 and references therein).

A set of equations that is of particular interest from the point of view of integrable systems theory is the self-dual Yang-Mills equations, which is a system of partial differential equations defined on an arbitrary oriented four-manifold, X. If the four-manifold is self-dual then the self-dual Yang-Mills equations are generally considered to be integrable in the sense that there is a twistor-theoretic method of constructing solutions which is analogous to the inverse-scattering methods ubiquitous in standard integrable systems theory.[4,24,17] Most known integrable systems can be derived as symmetry reductions of the self-dual Yang-Mills equations on a self-dual manifold for particular choices of gauge group and manifold X.[17] It is therefore important to know whether there is a corresponding group action on the space of solutions to the self-dual Yang-Mills equations corresponding to the above symmetry algebra and, if so, what the orbits of this group action are.

The purpose of this article is twofold. Firstly to review the relevant literature and results on the various aspects of the self-dual Yang-Mills equations that this problem entails, and secondly to briefly present some preliminary results on the action of non-local symmetries on instanton moduli spaces (for full details see Ref. 12). The problem of understanding the orbit structure lies at the interface between local considerations (i.e. symmetry algebras of differential equations) and global ones (moduli spaces of instanton solutions). One of our aims at this point is give a unified notation appropriate to both points of view. For simplicity we restrict ourselves to the self-dual Yang-Mills equations on $\mathbb{R}^4$ and take the gauge group to be SU(2). In this case one can construct explicitly an infinite-dimensional algebra of non-local symmetries of the self-dual Yang-Mills equations.[7,9] Since for standard integrable systems (e.g. the KdV equation) it is necessary to restrict oneself to a suitable class of solutions in order for inverse-scattering methods to work (e.g. periodic solutions or solutions with rapid asymptotic fall-off), we restrict ourselves to instanton solutions of the self-dual Yang-Mills equations on $\mathbb{R}^4$ (i.e. solutions with curvature that is L^2). By conformal invariance and a theorem of Uhlenbeck[22] this means we may equivalently consider the instanton problem on S^4. In this case we have a full description of the moduli space of solutions of the self-dual Yang-Mills equations given by the ADHM construction.[3,1] Our main result (see Sec. 4) is that the tangent space to each instanton moduli space is generated by non-local symmetries of the form given in Refs. 7 and 9. As such, since the instanton moduli spaces are connected[10] the corresponding symmetry pseudo-group acts transitively on them.

We end by discussing some further lines of research that we are pursuing.

2. The self-dual Yang-Mills equations

Let $(X, \mathbf{g})$ be a connected, oriented, Riemannian four-manifold. Since X is oriented, we have a volume form $\boldsymbol{\nu} \in \Omega^4(X)$, and thus we may define a Hodge $*$ operation

$$* : \Omega^p(X) \to \Omega^{4-p}(X), \qquad p = 0, \ldots, 4,$$

by the relation

$$\boldsymbol{\alpha} \wedge *\boldsymbol{\beta} = \langle \boldsymbol{\alpha}, \boldsymbol{\beta} \rangle \, \boldsymbol{\nu}, \qquad \forall \boldsymbol{\alpha}, \boldsymbol{\beta} \in \Omega^p(X).$$

When restricted to $\Omega^2(X)$ on a Riemannian four-manifold, the map $*$ has two important properties:

- It is conformally invariant;
- $\left(*|_{\Omega^2(X)} \right)^2 = \mathrm{Id}|_{\Omega^2(X)}$. We therefore have a direct sum decomposition

$$\Omega^2(X) = \Omega^{2+}(X) \oplus \Omega^{2-}(X) \tag{1}$$

of the space of two-forms into self-dual and anti-self-dual two-forms.

Let $\pi : E \to X$ be a vector bundle over X with structure group G. A connection on E may be represented by a $\mathfrak{g}$-valued one-form $\mathbf{A} \in \Omega^1(X, \mathfrak{g})$, with curvature $\mathbf{F_A} \in \Omega^2(X, \mathfrak{g})$. Using the decomposition of $\Omega^2(X)$ we may therefore write

$$\mathbf{F_A} = \mathbf{F_A^+} + \mathbf{F_A^-},$$

where $\mathbf{F_A^\pm} \in \Omega^{2\pm}(X, \mathfrak{g})$ are the self-dual and anti-self-dual parts of the curvature.

Definition 2.1. A connection on a vector bundle $\pi : E \to X$ is a solution of the *self-dual Yang-Mills equations* if its curvature obeys the condition

$$*\mathbf{F_A} = \mathbf{F_A}.$$

(Equivalently $\mathbf{F_A^-} = 0$.)

If we take E to be TX, the tangent bundle of X, with the Levi-Civita connection then the curvature of the connection is the Riemann tensor, $\mathbf{R}$. Viewing this as an element of $S^2\left(\Omega^2(X)\right)$ we may decompose it in block diagonal form relative to the decomposition (1) to yield

$$\mathbf{R} = \begin{pmatrix} W^+ + \frac{s}{12} & \mathrm{Ric}_0 \\ \mathrm{Ric}_0 & W^- + \frac{s}{12} \end{pmatrix}$$

where $W^{\pm}$ denote the self-dual and anti-self-dual parts of the Weyl tensor, Ric_0 denotes the trace-free part of the Ricci tensor, and s denotes the scalar curvature of the metric $\mathbf{g}$.

Definition 2.2. A Riemannian manifold $(X, \mathbf{g})$ is *self-dual* if the anti-self-dual part of the Weyl tensor vanishes:

$$W^- = 0.$$

From now on we fix X to be S^4 with its standard (self-dual) conformal structure. For simplicity we take $G = \mathrm{SU}(2) \cong \mathrm{Sp}(1)$, although it is straightforward to generalise our results to other groups.

2.1. *Instanton numbers and index theorem results*

Lemma 2.1. *[Refs. 5, 4] Let $\pi : E \to S^4$ be a rank-2 Hermitian complex vector bundle associated to a principal* SU(2) *bundle $P \to S^4$. Such bundles are characterised by the second Chern class*

$$c_2(E) := -\frac{1}{8\pi^2} \int_X \|\mathbf{F}_\mathbf{A}\|^2 \, \mathbf{dvol}_X,$$

where $\mathbf{A}$ is any connection on E and $\mathbf{F}_\mathbf{A} \in \Omega^2(S^4, \mathrm{End}(E))$ is its curvature. If $\int_X \|\mathbf{F}_\mathbf{A}\|^2 \, \mathbf{dvol}_X < \infty$, then $p_2(E) = -k$ where k is an integer, referred to as the instanton number. *If the connection $\mathbf{A}$ is a self-dual, then $k \geq 0$, with equality if and only if the connection is flat.*

There is a natural action of the group of gauge transformations (i.e. maps $S^4 \to \mathrm{SU}(2)$) on the space of connections. We therefore define the moduli space of k-instanton solutions:

$$\mathcal{M}_k := \frac{\{(E, \mathbf{A}) : \mathbf{F}_\mathbf{A} = *\mathbf{F}_\mathbf{A}, c_2(E) = -k\}}{\text{Gauge transformations}}.$$

Theorem 2.1. *[Ref. 4] $\mathcal{M}_k$ is a manifold (possibly with singularities corresponding to reducible connections) of dimension $8k - 3$.*

2.2. *The ADHM construction*

The ADHM construction[3,1] allows us to construct the $(8k-3)$-parameter family of solutions of the self-dual Yang-Mills equations on S^4 using quaternionic linear algebra. Viewing S^4 as $\mathbb{HP}^1$ then a point $p \in S^4$ corresponds to a quaternionic line $\Sigma \subset \mathbb{H}^2$. Choosing a complex structure on $\mathbb{H}^2$, we can identify it with $\mathbb{C}^4$, with Σ corresponding to a complex surface. Since, as a subspace of $\mathbb{H}^2$, Σ is invariant under right multiplication by the unit quaternion j, this implies that as a subspace of $\mathbb{C}^4$ it is invariant under the corresponding anti-linear anti-involution

$$\sigma : \mathbb{C}^4 \to \mathbb{C}^4 : (z_1, z_2, z_3, z_4) \mapsto (-\overline{z_2}, \overline{z_1}, -\overline{z_4}, \overline{z_3}).$$

Under the natural projection $\mathbb{C}^4 \to \mathbb{C}P^3$ the surface Σ projects to a rational curve which we denote by $\sigma(p) \cong \mathbb{C}P^1 \subset \mathbb{C}P^3$. Curves that arise in this way are called *real lines* in $\mathbb{C}P^3$. The involution on $\mathbb{C}P^3$ induced by σ leaves the real lines invariant, and acts as the anti-podal map on each real $\mathbb{C}P^1$.

Theorem 2.2. *[**Ref.** 23] There is a bijective correspondence between a). Solutions of the self-dual Yang-Mills equations on S^4 and b). holomorphic vector bundles on $\mathbb{C}P^3$ that are (holomorphically) trivial when restricted to each real line. We refer to this correspondence as the* Ward correspondence.

Remark 2.1. The holomorphic bundle over $\mathbb{C}P^3$ will carry additional structures depending on the particular group G that we are considering. For the case $G = \mathrm{SU}(2)$, we require that the determinant bundle $\det E$ is trivial and that E admits a positive real form.

The ADHM construction uses methods from algebraic geometry to construct holomorphic vector bundles on $\mathbb{C}P^3$ and therefore, via the Ward correspondence, self-dual Yang-Mills connections on S^4. For each $z = (z_1, z_2, z_3, z_4) \in \mathbb{C}^4$, we define a linear map

$$A(z) : W \to V,$$

where W, V are complex vector spaces of dimension $k, 2k+2$ respectively, which is of the form

$$A(z) = \sum_{i=1}^{4} z_i A_i.$$

The space W is assumed to admit an anti-linear involution $\sigma_W : W \to W$. The space V is assumed to have a non-degenerate, skew-symmetric form

$(\cdot, \cdot)$ and an anti-linear anti-involution $\sigma_V : V \to V$ that is compatible with the form in the sense that

$$(\sigma_V u, \sigma_V v) = \overline{(u, v)}, \qquad \forall u, v \in V.$$

We then require that the map $A(z)$ satisfies the compatibility condition that

$$\sigma_V (A(z)w) = A(\sigma(z))\sigma_W(w), \qquad \forall w \in W \tag{2}$$

and impose the following conditions:

- For all $z \in \mathbb{C}^4$, the space $U_z := A(z)(W) \subset V$ is of dimension k;
- For all $z \in \mathbb{C}^4$, U_z is isotropic with respect to $(\cdot, \cdot)$ i.e. $U_z \subseteq U_z^\perp$, where $^\perp$ denotes the complement with respect to the form $(\cdot, \cdot)$.

If we then define the quotient $E_z := U_z^\perp / U_z$, then the dimension and isotropy constraints on U_z imply that the collection of E_z defines a holomorphic, rank-2 complex vector bundle $E \to \mathbb{C}P^3$ with structure group $\mathrm{SL}(2, \mathbb{C})$. The reality condition (2) then imply that the bundle is trivial on restriction to any real line and that the structure group reduces from $\mathrm{SL}(2, \mathbb{C})$ to $\mathrm{SU}(2)$.

The power of the ADHM construction is that all k-instanton solutions of the self-dual Yang-Mills equations arise in this fashion.

2.3. *Patching matrix description*

Given a finite point $\mathbf{x} := (x^1, x^2, x^3, x^4) \in \mathbb{R}^4 \subset S^4$, then we may view

$$x := x^1 + ix^2 + jx^3 + kx^4 = u + jv \in \mathbb{H}$$

as an affine coordinate on $\mathbb{R}^4 \cong \mathbb{H} \subset \mathbb{H}P^1$, where

$$u := x^1 + ix^2, \qquad v := x^3 - ix^4.$$

As mentioned above, the point $\mathbf{x}$ therefore defines a quaternionic line $\Sigma(\mathbf{x}) \subset \mathbb{H}^2$ and thence a real line $\sigma(\mathbf{x}) \cong \mathbb{C}P^1 \subset \mathbb{C}P^3$ which takes the form

$$\sigma(\mathbf{x}) = \left\{ [z_1, z_2, z_1\, u - z_2\, \overline{v}, z_1\, v + z_2\, \overline{u}] \,:\, (z_1, z_2) \in \mathbb{C}^2 \setminus \{(0,0)\} \right\}.$$

On a fixed real line $\sigma(\mathbf{x})$, $\mathbf{x} \in \mathbb{R}^4$, we introduce the affine coordinate $z = z_2/z_1 \in \mathbb{C}P^1$. The image of point $\infty \in S^4$ is the real line

$$l_\infty := \{[0, 0, z_3, z_4] \,:\, (z_3, z_4) \in \mathbb{C}^2 \setminus \{(0,0)\}\}.$$

Given an instanton solution on S^4, we may consider the restriction of the solution to $\mathbb{R}^4 \subset S^4$. (Uhlenbeck's theorem[22] implies that there is no loss of information in doing so.) We may therefore consider the corresponding restriction of the bundle $E|_{\mathbb{C}P^3 \setminus l_\infty}$ which, for convenience, we denote by $E \to \mathbb{C}P^3 \setminus l_\infty$. We may split the region $\mathbb{C}P^3 \setminus l_\infty$ into coordinate regions

$$\mathcal{S}_0 := \{((u,v),z) \in \mathbb{C}^2 \times \mathbb{C}P^1 : z \neq \infty\} = \mathbb{C}^2 \times U_0,$$
$$\mathcal{S}_\infty := \{((u,v),z) \in \mathbb{C}^2 \times \mathbb{C}P^1 : z \neq 0\} = \mathbb{C}^2 \times U_\infty,$$

where

$$U_0 := \{[z_1, z_2] \in \mathbb{C}P^1 : z_1 \neq 0\}, \qquad U_\infty := \{[z_1, z_2] \in \mathbb{C}P^1 : z_2 \neq 0\}.$$

Since $\mathcal{S}_0, \mathcal{S}_\infty \cong \mathbb{C}^3$, a result of Grauert (see e.g. Ref. 18) implies that the bundle E restricted to either of these regions is trivial. Therefore the bundle E is characterised by the transition functions $G : \mathcal{S}_0 \cap \mathcal{S}_\infty \to \mathrm{SL}(2,\mathbb{C})$.

We may construct the map G directly from the ADHM data. With respect to bases on V, W, $A(z)$ can be viewed as a $(2k+2) \times k$ matrix with complex coefficients. The columns of $A(z)$ then define a set of k vectors $v_1(z), \ldots, v_k(z) \in \mathbb{C}^{2k+2}$ that span the space U_z. These vectors obey the reality condition

$$\sigma_V(v_i(z)) = v_i(\sigma(z)), \qquad i = 1, \ldots, k.$$

Since U_z is isotropic with respect to the symplectic form $(\cdot,\cdot)$, we deduce that

$$(v_i(z), v_j(z)) = 0.$$

We now restrict to a real line $\sigma(\mathbf{x}) = \{\mathbf{x}\} \times (U_0 \cup U_\infty) \subset \mathbb{C}P^3 \setminus l_\infty$. On the subset $\{\mathbf{x}\} \times U_0$ the annihilator $U_z^\perp$ is spanned by $\{v_i(z)\}$ along with two vectors $\{e_A(z) : A = 1, 2\}$ that span $U_z^\perp / U_z$. We therefore have

$$(v_i(z), e_A(z)) = 0,$$

and may, without loss of generality, assume that

$$(e_1(z), e_2(z)) = -(e_2(z), e_1(z)) = 1. \tag{3}$$

We may also define a basis $\{f_A(z) : A = 1, 2\}$ for $U_z^\perp / U_z$ on the region $\{\mathbf{x}\} \times U_\infty$ by the relations

$$f_1(z) := -\sigma_V(e_2(\sigma(z))), \qquad f_2(z) := \sigma_V(e_1(\sigma(z))).$$

This basis has the property that

$$(f_1(z), f_2(z)) = 1, \qquad (v_i(z), f_A(z)) = 0.$$

Given that $\{e_A(z)\}$ and $\{f_A(z)\}$ are both bases for $U_z^{\perp}/U_z$ for $z \in U_0 \cap U_\infty$, there exist functions $G_A{}^B(z), \lambda_A{}^i(z)$ with the property that

$$f_A(z) = G_A{}^B(z) e_B(z) + \lambda_A{}^i(z) v_i(z). \tag{4}$$

The matrix G is then the transition function of our bundle E. The definition of the vectors $\{f_A\}$ in terms of the $\{e_A\}$ implies that

$$G^* = G,$$
$$\det G = 1,$$

where $G^*(z) := \overline{{}^tG(\sigma(z))}$.

Finally note that since the bundle E is holomorphic over $\mathbb{C}P^3 \setminus l_\infty$ the transition functions are holomorphic with respect to the natural complex structure on $\mathbb{C}P^3$ restricted to $\mathbb{C}P^3 \setminus l_\infty$. In terms of the coordinates introduced above this implies that

$$G = G\left(u - z\overline{v}, v + z\overline{u}, z\right).$$

3. Symmetries of the self-dual Yang-Mills equations

In terms of the complex coordinates u, v introduced on $\mathbb{C}^2 \cong \mathbb{R}^4$ above, the standard flat metric on $\mathbb{R}^4$ takes the form

$$\mathbf{g} = \frac{1}{2}\left(du \otimes d\overline{u} + d\overline{u} \otimes du + dv \otimes d\overline{v} + d\overline{v} \otimes dv\right).$$

The corresponding volume form is then

$$\epsilon = dt \wedge dx \wedge dy \wedge dz = \frac{1}{4} du \wedge d\overline{u} \wedge dv \wedge d\overline{v}.$$

In terms of these coordinates, the self-dual Yang-Mills equations for a connection, $\mathbf{A}$, correspond to the following conditions on the components of the curvature tensor, $\mathbf{F_A}$, of the connection:

$$F_{uv} = 0; \tag{5a}$$
$$F_{u\overline{u}} + F_{v\overline{v}} = 0; \tag{5b}$$
$$F_{\overline{u}\overline{v}} = 0. \tag{5c}$$

Introducing the vector fields

$$\mathbf{X}(z) := \partial_{\overline{v}} + z\partial_u, \qquad \mathbf{Y}(z) := \partial_{\overline{u}} - z\partial_v,$$

which depend on an arbitrary parameter $z \in \mathbb{C} \cup \infty = \mathbb{C}P^1$, then the self-dual Yang-Mills equations (5) are equivalent to the condition that

$$\mathbf{F_A}\left(\mathbf{X}(z), \mathbf{Y}(z)\right) = 0, \qquad \forall z \in \mathbb{C}P^1.$$

3.1. *Non-local symmetries of the self-dual Yang-Mills equations*

An important property of the self-dual Yang-Mills equations is that they are the compability condition for the following overdetermined equations[7,6,21,8]

$$(\partial_{\overline{v}} + z\partial_u)\, \psi(x,z) = -\,(A_{\overline{v}} + zA_u)\, \psi(x,z), \tag{6a}$$

$$(\partial_{\overline{u}} - z\partial_v)\, \psi(x,z) = -\,(A_{\overline{u}} - zA_v)\, \psi(x,z), \tag{6b}$$

for a map $\psi : \mathbb{R}^4 \times U \to \mathrm{SL}(2,\mathbb{C})$, where $U \subset \mathbb{C}P^1$ is a suitable domain. In particular we may find a solution $\psi_0 : \mathbb{R}^4 \times U_0 \to \mathrm{SL}(2,\mathbb{C})$ that is analytic in z for $z \neq \infty$. Given such a solution we may then construct a solution $\psi_\infty(x,z) : \mathbb{R}^4 \times U_\infty \to \mathrm{SL}(2,\mathbb{C})$ that is analytic in z for $z \neq 0$ by taking

$$\psi_\infty(x,z) = \left(\psi_0(x,\sigma(z))^{-1}\right)^\dagger ,$$

where

$$\sigma(z) := -\frac{1}{\overline{z}}.$$

Note that σ may be viewed as the anti-podal map on $\mathbb{C}P^1$ viewed in terms of affine coordinates.

Equations (6) for ψ_0 and ψ_∞ imply that we may write the components of the connection in the form

$$A_u = -\,(\partial_u \psi_\infty(\infty))\, \psi_\infty(\infty)^{-1}, \qquad A_v = -\,(\partial_v \psi_\infty(\infty))\, \psi_\infty(\infty)^{-1}, \tag{7}$$

$$A_{\overline{u}} = -\,(\partial_{\overline{u}} \psi_0(0))\, \psi_0(0)^{-1}, \qquad A_{\overline{v}} = -\,(\partial_{\overline{v}} \psi_0(0))\, \psi_0(0)^{-1}. \tag{8}$$

If we define $J := \psi_\infty(\infty)^{-1} \cdot \psi_0(0)$ then the remaining part of the self-dual Yang-Mills equations imply that J obeys the Yang-Pohlmeyer equation

$$\partial_u \left(J_{\overline{u}} J^{-1}\right) + \partial_v \left(J_{\overline{v}} J^{-1}\right) = 0. \tag{9}$$

It is known that the only local symmetries of the self-dual Yang-Mills equations on flat $\mathbb{R}^4$ are gauge transformations and those generated by the action of the conformal group. On the other hand, there exists a non-trivial family of non-local symmetries of the self-dual Yang-Mills equations.[7,9] Let $J(t)$ denote a 1-parameter family of solutions of (9) that is assumed to depend smoothly on the parameter $t \in I$, where I is a open subinterval of the real line that contains the origin. Taking the derivative of (9) with respect to t we find that we require

$$\partial_u \left(J\partial_{\overline{u}} \left(J^{-1}\frac{\partial J}{\partial t} \right) J^{-1} \right) + \partial_v \left(J\partial_{\overline{v}} \left(J^{-1}\frac{\partial J}{\partial t} \right) J^{-1} \right) = 0. \tag{10}$$

The construction of Refs. 7 and 9 proceeds as follows. Let $\chi(z)$ be a solution of the system

$$\begin{aligned}\left[(\partial_{\overline{v}} - J_{\overline{v}}J^{-1}) + z\partial_u\right]\chi(x,z) = 0,\\ \left[(\partial_{\overline{u}} - J_{\overline{u}}J^{-1}) - z\partial_v\right]\chi(x,z) = 0.\end{aligned}$$

If we then define

$$\frac{\partial J}{\partial t} = \chi(x,z)T(x,z)\chi(x,z)^{-1}\cdot J, \tag{11}$$

where the function T obeys the relations

$$(\partial_{\overline{v}} + z\partial_u)\,T = (\partial_{\overline{u}} - z\partial_v)\,T = 0, \tag{12}$$

then $\partial J/\partial t$ is a solution of the linearisation (10). Without loss of generality, we may take

$$\chi(x,z) = \psi_\infty(\infty)^{-1}\cdot\psi_\infty(z),$$

which is analytic for $z \neq 0$, and has the property that $\chi(x,\infty) = \mathrm{Id}$. If we expand the right-hand-side of (11) as a Laurent series in z then the coefficients in the expansion define a (generally infinite) family of solutions of the linearisation equations for J. Moving from one coefficient in this expansion to the next defines a map between solutions of the linearisation equations which, in integrable systems terminology, defines the recursion operator of the self-dual Yang-Mills equations and its related integrable hierarchy.[17]

Such symmetries of the self-dual Yang-Mills equations with gauge group $G = \mathrm{GL}(n,\mathbb{C})$ have been studied within the Sato/Segal-Wilson approach to integrable systems by Takasaki[21] and thus have the interpretation of an infinite-dimensional family of projective transformations on an infinite-dimensional Grassmannian. It is not clear, however, how to implement this approach for $G = \mathrm{SU}(n)$, as the transformations generally do not preserve the unitarity of field J. Takasaki's approach has, however, been used to study dressing actions on harmonic maps, where the reality conditions are more straightforward.[16]

In order to maintain unitarity of J, we will consider the symmetry

$$\begin{aligned}\frac{\partial J}{\partial t} &= \chi(x,z)T(x,z)\chi(x,z)^{-1}\cdot J + J\cdot\left(\chi(x,z)^{-1}\right)^\dagger T(x,z)^\dagger\chi(x,z)^\dagger\\ &= \psi_\infty(\infty)^{-1}\left[\psi_\infty(z)T(x,z)\psi_\infty(z)^{-1}\right.\\ &\qquad\left.+\psi_0(\sigma(z))T(x,z)^\dagger\psi_0(\sigma(z))^{-1}\right]\psi_0(0),\end{aligned} \tag{13}$$

where $T(x,z)$ obeys the condition (12).

If we now consider the symmetry (13) with $T(x, z)$ a constant element of $\mathfrak{g} \otimes \mathbb{C}$, then the algebra of such symmetries is isomorphic to the Kac-Moody algebra of $\mathfrak{g} \otimes \mathbb{C}$. The natural question is whether there is a corresponding group action on the space of solutions. A partial solution to this problem was given by Crane[8] who showed that taking T to be constant, one could define an action of the (analytic) loop group of $G_{\mathbb{C}}$ (i.e. the group of analytic maps from S^1 to $G_{\mathbb{C}}$) on the space of solutions of the self-dual Yang-Mills equations. As Crane showed by example, however, this action does not preserve the instanton condition that the curvature of the connection be L^2. More generally, for a holomorphic bundle on $\mathbb{C}P^3 \setminus l_\infty$ defined by the patching matrix $G(u - z\overline{v}, v + z\overline{u}, z)$, a general $T(u - z\overline{v}, v + z\overline{u}, z)$ generates a transformation of the form

$$G(u - z\overline{v}, v + z\overline{u}, z) \mapsto g(u - z\overline{v}, v + z\overline{u}, z) \cdot G(u - z\overline{v}, v + z\overline{u}, z) \cdot g^*(u - z\overline{v}, v + z\overline{u}, z),$$

where

$$g^*(u - z\overline{v}, v + z\overline{u}, z) := g\left(u + \frac{\overline{v}}{\overline{z}}, v - \frac{\overline{u}}{\overline{z}}, -\frac{1}{\overline{z}}\right)^\dagger .$$

The function g is an arbitrary holomorphic functions of its arguments. If g is analytic for $z \neq \infty$, this transformation is simply a holomorphic change of basis on the bundle over $\mathbb{C}P^3$, and leaves the self-dual Yang-Mills field unchanged. However, if g has poles at finite values of z, it will have a non-trivial effect on the connection and generates a distinct solution of the self-dual Yang-Mills equations. This action has been given a cohomological description by Park (see Ref. 19 and references therein), which has been further investigated by Popov and Ivanova (see Refs. 20, 15 and references therein).

4. One-parameter families of ADHM data

We now consider a one-parameter family of ADHM data $A(t, z) : W \to V$, where $t \in I$ is a parameter with values in an open subset $I \subseteq \mathbb{R}$ that contains the origin. We assume that $A(t, z)$ is a smooth, continuous function of t. Such a one-parameter family of data defines a one-parameter family of holomorphic vector bundles $E(t) \to \mathbb{C}P^3$ and hence a one-parameter family of instanton solutions of the self-dual Yang-Mills equations on S^4. We now wish to investigate how the elements of the explicit constructions of the previous sections depend on $A(t, z)$.

The image $A(t,z)(W)$ is now spanned by the vectors $\{v_i(t,z)\}$. On a fixed real line the vector space $U_z^{\perp}/U_z$ is spanned by $\{e_A(t,z)\}$ for each $z \in U_0$, which we assume are normalised such that (3) is satisfied for each $t \in I$. Constructing the vectors $\{f_A(t,z)\}$, we then define the patching matrix $G(t,z)$ and the functions $\lambda_A{}^i(t,z)$ as in equation (4).

A short calculation (see Ref. 12) implies that the t-derivative of the patching matrix obeys the relation

$$\frac{\partial}{\partial t} G(t,z) = d(t,z)G(t,z) + G(t,z)d(t,z)^*, \tag{14}$$

where $d(t,z)$ is a matrix with components

$$d_A{}^B(t,z) = \sum_{C=1}^{2} \epsilon^{BC} \left(\frac{\partial}{\partial t} f_A(t,z), f_C(t,z) - \sum_{i=1}^{k} \lambda_C{}^i(t,z) v_i(t,z) \right),$$

where $\epsilon^{BC} = -\epsilon^{CB}$ and $\epsilon^{12} = 1$ are the components of the volume form on the bundle and we have defined $d^*(t,z) = d(t,\sigma(z))^{\dagger}$.

At the formal level, it follows from equation (14) that

$$G(t,z) = \alpha(t,z)\, G(0,z)\, (\alpha(t,z))^* , \tag{15}$$

where $\alpha(t,z)$ satisfies the first order ordinary differential equation

$$\dot{\alpha}(t,z) = d(t,z)\alpha(t,z), \qquad \alpha(0:z) = \mathrm{Id}.$$

The form of $G(t,z)$ given in equation (15) is then precisely of the form given in Ref. 8. More precisely, the differential equation (14) satisfied by $G(t,z)$ is a flow generated by symmetries of the form (13). Therefore given any smooth path in $\gamma : I \to \mathcal{M}_k$, where I is an open sub-interval of $\mathbb{R}$, then the tangent vector to γ at any point is the fundamental vector field corresponding to a symmetry of the form given in Refs. 7 and 9.

At the global level the transformations generated by (13) are generally only locally defined and therefore form a pseudo-group (or groupoid) rather than a Lie group. Since the moduli spaces $\mathcal{M}_k$ are connected[10] the above calculations imply that this pseudo-group acts transitively on each $\mathcal{M}_k$. Since the moduli spaces are finite-dimensional, the tangent space at each point will be generated by a finite-dimensional sub-algebra of the algebra of symmetries (13) at each point. An implicit description of this sub-algebra may be derived from the ADHM data via equation (14). Whether this finite-dimensional sub-algebra gives rise to a transitive action of a finite-dimensional group on each moduli space $\mathcal{M}_k$, and whether the transitive action of a pseudo-group on the $\mathcal{M}_k$ leads to any information concerning the global structure of the moduli spaces is currently under investigation.

5. Open problems

Outside of clarifying some of the technical issues surrounding our results, we mention some natural ways in which the current work could be extended. A reformulation of the ADHM construction due to Atiyah[2] and Donaldson[10] shows that there is a $1-1$ correspondence between (framed) instantons solutions of instanton number k and based holomorphic maps $f : \mathbb{C}P^1 \to \Omega G$ of degree k. Given that such maps are in many ways analogous to harmonic maps from $\mathbb{C}P^1 \to \mathbb{C}P^N$, it seems likely that the dressing action on instanton moduli spaces may be most clearly understood within this formalism.

Although the generalisation from $G = \mathrm{SU}(2)$ to arbitrary G is essentially trivial, a natural question is whether the results we have found hold for other four-manifolds. In the case where the base manifold is $\mathbb{C}P^2$ with $G = \mathrm{SU}(2)$ we can analyse the one-instanton moduli space, $\mathcal{M}_1(\mathbb{C}P^2)$. This moduli space is isomorphic to a cone on $\mathbb{C}P^2$, with the vertex of the cone corresponding to the unique, homogeneous, reducible SU(2) connection on the bundle $L \oplus L^{-1}$. Preliminary investigations[13] indicate that in this case the corresponding symmetry group action has two, distinct orbits, the first consisting of the irreducible connections and the second consisting of the reducible connection which is a fixed point of the group action. Given that on a general four-manifold the space of reducible connections is central to Donaldson's theory of four-manifolds[11] this might suggest an interesting correspondence between topological field theory and group actions in integrable systems.

Acknowledgements

This work was partially supported by the European Contract Human Potential Programme, Research Training Network HPRN-CT-2000-00101 (EDGE), the Keio University COE program on Integrative Mathematical Science: Progress in Mathematics Motivated by Natural and Social Phenomena, and by START-project Y237–N13 of the Austrian Science Fund. The author is grateful to the Mathematics departments in the Università dell'Aquila, the University of Aberdeen, Keio University and the Universität Wien for their support and hospitality.

References

1. M.F. Atiyah, *The Geometry of Yang-Mills fields.* Lezioni Fermiane, Accademia Nazionale dei Lincei et Scuola Normale Superiore, Pisa, 1979.

2. M.F. Atiyah, *Instantons in two and four dimensions.* Commun. Math. Phys. **93** (1984), 437–451.
3. M.F. Atiyah, V.G. Drinfeld, N.J. Hitchin and Yu.I. Manin, *Construction of instantons.* Phys. Lett. A **65** (1978), 185–187.
4. M.F. Atiyah, N.J. Hitchin and I.M. Singer, *Self-duality in four-dimensional Riemannian geometry.* Proceedings of the Royal Society of London **A362** (1978), 425–461.
5. A. Belavin , A. Polyakov, A. Schwartz, and Y. Tyupkin, *Pseudoparticle solutions of the Yang-Mills equations.* Phys. Lett. B **59** (1975), 85–87.
6. A.A. Belavin and V.E. Zakharov, *Yang-Mills equations as inverse scattering problem.* Phys. Lett. B **73** (1978), 53–57.
   L.L. Chau, M.L. Ge and A. Sinha and Y.S. Wu, *Hidden-symmetry algebra for the self-dual Yang-Mills equation.* Phys. Lett. B **121** (1983), 391–396.
7. L.L. Chau, M.L. Ge and Y.S. Wu, *Kac-Moody algebra in the self-dual Yang-Mills system. Phys. Rev.* **D25** (1982), 1086–1094.
   L.L. Chau and Y.S. Wu, *More about hidden-symmetry algebra for self-dual Yang-Mills system. Phys. Rev.* **D26** (1982), 3581–3592.
8. L. Crane, *Action of the loop group on the self dual Yang-Mills equation.* Commun. Math. Phys. **110** (1987), 391–414.
9. L. Dolan, *A new symmetry group of real self-dual Yang-Mills theory.* Phys. Lett. B **113** (1982), 387–390.
   L. Dolan, *Kac-Moody algebras and exact solvability in hadronic physics.* Phys. Rep. **109** (1984), 1–94.
10. S.K. Donaldson *Instantons and geometric invariant theory.* Commun. Math. Phys. **93** (1984), 453–460.
11. S.K. Donaldson and P.B. Kronheimer, *The Geometry of Four Manifolds.* Oxford University Press, Oxford, 1990.
12. J.D.E. Grant, *The ADHM construction and non-local symmetries of the self-dual Yang-Mills equations.* In preparation.
13. J.D.E. Grant, *The action of non-local symmetries on the moduli space of solutions of the self-dual Yang-Mills equations:* $\mathbb{C}P^2$. In preparation.
14. M.A. Guest, *Harmonic Maps, Loop Groups, and Integrable Systems.* London Mathematical Society Student Texts **38** Cambridge University Press, Cambridge, 1997.
15. T.A. Ivanova, *Moduli space of self-dual gauge fields, holomorphic bundles and cohomology sets.* in Moduli spaces in mathematics and physics (Hindawi Publ. Corp., Cairo, 2001) 79–90. Preprint **math-ph/9902015**.
16. M. Jacques and Y. Saint-Aubin, *Infinite-dimensional Lie algebras acting on the solution space of various σ models.* J. Math. Phys. **28** (1987), 2463–2479.
   G. Arsenault, M. Jacques and Y. Saint-Aubin, *Collapse and exponentiation of infinite symmetry algebras of Euclidean projective and Grassmannian σ models.* J. Math. Phys. **29** (1988), 1465–1471.
   G. Arsenault and Y. Saint-Aubin, *The hidden symmetry of $U(n)$ principal σ models revisited: I. Explicit expressions for the generators.* Nonlinearity **2** (1989), 571–591.
   G. Arsenault and Y. Saint-Aubin, *The hidden symmetry of $U(n)$ principal σ*

models revisited: II. The algebraic structure. Nonlinearity **2** (1989), 593–607.

17. L.J. Mason L.J. and N.M.J. Woodhouse, *Integrability, Self-Duality, and Twistor Theory.* Oxford University Press, Oxford, 1996.
18. C. Okonek, M. Schneider and H. Spindler, *Vector Bundles on Complex Projective Spaces,* Progress in Mathematics **3** Birkhauser, Boston, 1980.
19. Q-Han Park, *2D sigma model approach to 4D instantons.* Internat. J. Modern Phys. A **7** (1992), 1415–1447.
20. A.D. Popov, *Self-dual Yang-Mills: symmetries and moduli space.* Rev. Math. Phys. **11** (1999), 1091–1149. Preprint **hep-th/9803183**.
21. K. Takasaki, *A new approach to the self-dual Yang-Mills equations.* Commun. Math. Phys. **94** (1984), 35–59.
22. K.K. Uhlenbeck, *Removable singularities in Yang-Mills fields.* Commun. Math. Phys. **83** (1982), 11–29.
23. R.S. Ward, *On self-dual gauge fields.* Phys. Lett. A **61**, 81–82 (1977).
24. R.S. Ward and R.O. Wells, *Twistor Geometry and Field Theory* Cambridge University Press, Cambridge, 1990.

NONCOMMUTATIVE SOLITONS AND INTEGRABLE SYSTEMS *

MASASHI HAMANAKA
Graduate School of Mathematics, Nagoya University, Chikusa-ku, Nagoya, 464-8602, JAPAN
hamanaka@math.nagoya-u.ac.jp

We review recent developments of soliton theories and integrable systems on noncommutative spaces. The former part is a brief review of noncommutative gauge theories focusing on ADHM construction of noncommutative instantons. The latter part is a report on recent results of existence of infinite conserved densities and exact multi-soliton solutions for noncommutative Gelfand-Dickey hierarchies. Some examples of noncommutative Ward's conjecture are also presented. Finally, we discuss future directions on noncommutative Sato's theories and twistor theories.

1. Introduction

Non-Commutative (NC) extension of field theories has been studied intensively for the last several years. NC gauge theories are equivalent to ordinary gauge theories in the presence of background magnetic fields and succeeded in revealing various aspects of them. (For reviews, see e.g. [20,35,86,97,152].) NC solitons especially play important roles in the study of D-brane dynamics, such as the confirmation of Sen's conjecture on tachyon condensation. (For reviews, see e.g. [70].) One of the distinguished features of NC theories is resolution of singularities. This gives rise to various new physical objects such as U(1) instantons and makes it possible to analyze singular configurations as usual. (For a review, see my Ph.D thesis [60].)

NC extension of integrable equations such as the KdV equation is also one of the hot topics. These equations imply no gauge field and NC extension of them perhaps might have no physical picture or no good property on integrabilities. To make matters worse, NC extension of $(1+1)$-dimensional

*This work was supported in part by JSPS Research Fellowships for Young Scientists (#0310363) and the Daiko Foundation (#9095).

equations introduces infinite number of time derivatives, which makes it hard to discuss or define the integrability. However, some of them actually possess integrable properties, such as the existence of infinite number of conserved quantities and the linearizability which are widely accepted as definition of complete integrability of equations. Furthermore, a few of them can be derived from NC (anti-)self-dual Yang-Mills (ASDYM) equations by suitable reductions. This fact may give some physical meanings and good properties to the lower-dimensional NC field equations and makes us expect that Ward's conjecture [164] still holds on NC spaces. So far, however, those equations have been examined one by one. Now it is time to discuss the geometrical and physical origin of the special properties and integrabilities, in more general framework.

That is why we propose the following programs as future directions:

- Construction of NC twistor theory
- Confirmation of NC Ward's conjecture
- Completion of NC Sato's theory

Twistor theory [139] is the most essential framework in the study of integrability of ASD Yang-Mills(-Higgs) equations. (See, e.g. [118,165].) NC extension of twistor theories are already discussed by several authors, e.g. [69,83,92,106,153]. This would lays the geometrical foundation of integrabilities of ASD YM(H) equations.

NC Ward conjecture is very important to give physical pictures to lower-dimensional integrable equations and to make it possible to apply analysis of NC solitons to that of the corresponding D-branes. Origin of the integrable-like properties would be also revealed from the viewpoints of NC twistor theory and preserved supersymmetry in the D-brane systems.

Sato's theory is known to be one of the most beautiful theories of solitons and reveals essential aspects of the integrability, such as, the construction of exact multi-soliton solutions, the structure of the solution space, the existence of infinite conserved quantities, and the hidden symmetry of them, on commutative spaces. So it is reasonable to extend Sato's theory onto NC spaces in order to clarify various integrable-like aspects directly.

In this article, we report recent developments of NC extension of soliton theories and integrable systems focusing on NC ADHM construction and NC Sato's theory. As recent results [61,62], we prove the existence of infinite conserved quantities and exact multi-soliton solutions for Gelfand-Dickey hierarchies on NC spaces and give the explicit representations with both space-space and space-time noncommutativities.

2. NC Instantons and BPS Monopoles

2.1. *NC Gauge Theories*

NC spaces are defined by the noncommutativity of the coordinates:

$$[x^i, x^j] = i\theta^{ij}, \tag{1}$$

where θ^{ij} are real constants and called the *NC parameters.* This relation looks like the canonical commutation relation in quantum mechanics and leads to "space-space uncertainty relation." Hence the singularity which exists on commutative spaces could resolve on NC spaces. This is one of the prominent features of NC field theories and yields various new physical objects.

NC field theories are given by the exchange of ordinary products in the commutative field theories for the star-products and realized as deformed theories from the commutative ones. In this context, they are often called the *NC-deformed theories.*

The star-product is defined for ordinary fields on commutative spaces. For Euclidean spaces, it is explicitly given by

$$\begin{aligned} f \star g(x) &:= \exp\left(\frac{i}{2}\theta^{ij}\partial_i^{(x')}\partial_j^{(x'')}\right) f(x')g(x'')\Big|_{x'=x''=x} \\ &= f(x)g(x) + \frac{i}{2}\theta^{ij}\partial_i f(x)\partial_j g(x) + O(\theta^2), \end{aligned} \tag{2}$$

where $\partial_i^{(x')} := \partial/\partial x'^i$ and so on. This explicit representation is known as the *Moyal product* [54,124].

The star-product has associativity: $f \star (g \star h) = (f \star g) \star h$, and returns back to the ordinary product in the commutative limit: $\theta^{ij} \to 0$. The modification of the product makes the ordinary spatial coordinate "noncommutative," that is, $[x^i, x^j]_\star := x^i \star x^j - x^j \star x^i = i\theta^{ij}$.

We note that the fields themselves take c-number values as usual and the differentiation and the integration for them are well-defined as usual. A nontrivial point is that NC field equations contain infinite number of derivatives in general. Hence the integrability of the equations are not so trivial as commutative cases.

2.2. *ADHM Construction of Instantons*

In this subsection, we treat NC instantons by Atiyah-Drinfeld-Hitchin-Manin (ADHM) construction [6]. ADHM construction is a strong method to generate instanton solutions with arbitrary instanton number for

$SU(N), SO(N)$ and $Sp(N)$. This is based on a duality, that is, one-to-one correspondence between the instanton moduli space and the moduli space of ADHM-data which are specified by the ASD equation and ADHM equation, respectively. The concrete steps are as follows (For reviews on commutative spaces, see e.g [34,58,60].):

- Step (i): Solving ADHM equation:

$$\begin{aligned} &[B_1, B_1^\dagger] + [B_2, B_2^\dagger] + II^\dagger - J^\dagger J = -[z_1, \bar{z}_1] - [z_2, \bar{z}_2] = 0, \\ &[B_1, B_2] + IJ = -[z_1, z_2] = 0. \end{aligned} \tag{3}$$

  We note that the coordinates $z_{1,2}$ always appear in pair with the matrices $B_{1,2}$ and that is why we see the commutator of the coordinates in the RHS. These terms, of course, vanish on commutative spaces, however, they cause nontrivial contributions on NC spaces, which is seen later soon.
- Step (ii): Solving "0-dimensional Dirac equation" in the background of the ADHM date:

$$\nabla^\dagger V = 0, \tag{4}$$

  with the normalization condition:

$$V^\dagger V = 1. \tag{5}$$

- Step (iii): By using the solution V, we can construct the corresponding instanton solution as

$$A_\mu = V^\dagger \partial_\mu V, \tag{6}$$

  which actually satisfies the ASD equation:

$$\begin{aligned} &F_{z_1\bar{z}_1} + F_{z_2\bar{z}_2} = [D_{z_1}, D_{\bar{z}_1}] + [D_{z_2}, D_{\bar{z}_2}] = 0, \\ &F_{z_1 z_2} = [D_{z_1}, D_{z_2}] = 0. \end{aligned} \tag{7}$$

In this subsection, we give some examples of the explicit instanton solutions focusing on BPST instanton solution.

BPST instanton solution (1-instanton, $\dim M_{2,1}^{\mathrm{BPST}} = 5$)

This solution is the most basic and important and is constructed almost trivially by ADHM procedure.

- Step (i): ADHM equation is a $k \times k$ matrix-equation and in the present $k = 1$ case, it is trivially solved. The commutator part of $B_{1,2}$ is automatically dropped out and the matrices $B_{1,2}$ can be taken as arbitrary complex number. The remaining part I, J are also easily solved:

$$B_1 = \alpha_1, \; B_2 = \alpha_2, \; I = (\rho, 0), \; J = \begin{pmatrix} 0 \\ \rho \end{pmatrix}, \quad \alpha_{1,2} \in \mathbf{C}, \; \rho \in \mathbf{R}. \tag{8}$$

Here the real and imaginary part of α are denoted by $\alpha_1 = b_2 + ib_1$, $\alpha_2 = b_4 + ib_3$, respectively.

- Step (ii): The "0-dimensional Dirac equation" is also easily solved in this case. (See, e.g. [60].)
- Step (iii): The instanton solution is constructed as

$$A_\mu = V^\dagger \partial_\mu V = \frac{i(x-b)^\nu \eta_{\mu\nu}^{(-)}}{(x-b)^2 + \rho^2}. \tag{9}$$

The field strength $F_{\mu\nu}$ is calculated from this gauge field as

$$F_{\mu\nu} = \frac{2i\rho^2}{(|z-\alpha|^2 + \rho^2)^2} \eta_{\mu\nu}^{(-)}. \tag{10}$$

Now let us take the zero-size limit. Then the distribution of the field strength $F_{\mu\nu}$ converses into the singular, delta-functional configuration. Instantons have smooth configurations by definition and hence the zero-size instanton does not exists, which corresponds to the singularity of the (complete) instanton moduli space which is called the *small instanton singularity*. On NC spaces, the singularity is resolved and new class of instantons appear.

2.3. *ADHM Construction of NC Instantons*

In this subsection, we construct some typical NC instanton solutions by using ADHM method in the operator formalism. In NC ADHM construction, the self-duality of the NC parameter is important, which reflects the properties of the instanton solutions.

The steps are all the same as the commutative one:

- Step (i): ADHM equation is deformed by the noncommutativity of the coordinates as we mentioned in the previous subsection:

$$\begin{aligned} &(\mu_{\mathbf{R}} :=) \;\; [B_1, B_1^\dagger] + [B_2, B_2^\dagger] + II^\dagger - J^\dagger J = -2(\theta_1 + \theta_2) =: \zeta, \\ &(\mu_{\mathbf{C}} :=) \;\; [B_1, B_2] + IJ = 0. \end{aligned} \tag{11}$$

We note that if the NC parameter is ASD, that is, $\theta_1+\theta_2=0$, then the RHS of the first equation of ADHM equation becomes zero.[a]

- Step (ii): Solving the NC "0-dimensional Dirac equation"

$$\hat{\nabla}^\dagger \hat{V} = 0 \tag{12}$$

with the normalization condition.

- Step (iii): the ASD gauge fields are constructed from the zero-mode V,

$$\hat{A}_\mu = \hat{V}^\dagger \partial_\mu \hat{V}, \tag{13}$$

which actually satisfies the NC ASD equation:

$$(\hat{F}_{z_1\bar{z}_1} + \hat{F}_{z_2\bar{z}_2} =)\ [\hat{D}_{z_1}, \hat{D}_{\bar{z}_1}] + [\hat{D}_{z_2}, \hat{D}_{\bar{z}_2}] - \frac{1}{2}\left(\frac{1}{\theta_1} + \frac{1}{\theta_2}\right) = 0,$$
$$(\hat{F}_{z_1 z_2} =)\ [\hat{D}_{z_1}, \hat{D}_{z_2}] = 0. \tag{14}$$

There is seen to be a beautiful duality between (11) and (14) We note that when the NC parameter is ASD, then the constant terms in both (11) and (14) disappear.

In this way, NC instantons are actually constructed. Here we have to take care about the inverse of the operators.

Comments on instanton moduli spaces

Instanton moduli spaces are determined by the value of $\mu_{\mathbf{R}}$ [126,127]. Namely,

- In $\mu_{\mathbf{R}} = 0$ case, instanton moduli spaces contain small instanton singularities, (which is the case for commutative $\mathbf{R}^4$ and special NC $\mathbf{R}^4$ where θ : ASD).
- In $\mu_{\mathbf{R}} \neq 0$ case, small instanton singularities are resolved and new class of smooth instantons, U(1) instantons exist, (which is the case for general NC $\mathbf{R}^4$)

Since $\mu_{\mathbf{R}} = \zeta = -2(\theta_1+\theta_2)$ as Eq. (11), the self-duality of the NC parameter is important.

When the NC parameter is ASD, that is, $\theta_1+\theta_2 = 0$, instanton moduli space implies the singularities. The origin of the "phase

[a] When we treat SD gauge fields, then the RHS is proportional to $(\theta_1-\theta_2)$. Hence the relative self-duality between gauge fields and NC parameters is important.

diagram" corresponds to commutative instantons. The θ-axis represents instantons on $\mathbf{R}^2_{\rm NC} \times \mathbf{R}^2_{\rm Com}$. The other instantons basically have the same properties, hence let us fix the NC parameter θ self-dual. This type of instantons are just discussed first by Nekrasov and Schwarz [131]. The ASD-SD instantons (the combination of self-dualities of gauge fields and NC parameters is ASD-SD) are discussed in e.g. [23,39,43,44,45,81,84,88,93,95,104,110,112,128,131,138,155,157]. The ASD-ASD instantons [5] are constructed by ADHM construction in [46,59,167], and ADHM construction of instantons on $\mathbf{R}^2_{\rm NC} \times \mathbf{R}^2_{\rm Com}$ are discussed in [94]. Witten's ansatz [168] for NC instantons are studied in [22,24,147]. Geometrical origin of instanton number of NC instantons is also discussed in e.g. [45,71,72,85,119,144,158,156]. For comprehensive discussion on ADHM construction, see e.g. [21,60,166]. Instantons in Born-Infeld actions in the background of B-fields are discussed in [79,99,123,149,154].

$U(2)$, $k = 1$ *solution* (NC BPST, θ: SD)

This solution is also obtained by ADHM procedure with the "Furuuchi's Method" [43,44]. The solution of NC ADHM equation is

$$B_{1,2} = 0, \quad I = (\sqrt{\rho^2 + \zeta}, 0), \quad J = \begin{pmatrix} 0 \\ \rho \end{pmatrix}. \tag{15}$$

Comparing the solution of commutative ADHM equation, the date I is deformed by the noncommutativity of the coordinates, which shows that the size of instantons becomes larger than that of commutative one because of the existence of ζ. In fact, in the $\rho \to 0$ limit, the configuration is still smooth and the U(1) part is alive. This is essentially the same as a $U(1)$ instanton solution.

BPST instantons on commutative and NC spaces are summarized as follows.

BPST instanton		NC BPST instanton
$\mu_{\mathbf{R}} = 0$, $\mu_{\mathbf{C}} = 0$	ADHM equation	$\mu_{\mathbf{R}} = \zeta$, $\mu_{\mathbf{C}} = 0$
$B_{1,2} = \alpha_{1,2}$, $I = (\rho, 0)$, $J^t = (0, \rho)$	ADHM data	$B_{1,2} = \alpha_{1,2}$, $I = (\sqrt{\rho^2 + \zeta}, 0)$, $J^t = (0, \rho)$
$\mathbf{R}^4\times$ orbifold $\mathbf{C}^2/\mathbf{Z}_2$ (singular)	moduli space	$\mathbf{R}^4\times$ Eguchi-Hanson $\widetilde{\mathbf{C}^2/\mathbf{Z}_2}$ (regular)
$F_{\mu\nu} \to$ delta function (singular)	zero-size limit	$F_{\mu\nu} \to U(1)$ instanton (regular)

More detailed discussion are presented in e.g. [45,60,98,101,129,147,166].

Some other BPS solitons

There are many works on the study of other NC BPS solitons as follows:

- NC monopoles: [8,50,55,56,57,59,65,73,74,77,76,78,122,129,140,141]
- NC vortices in abelian Higgs models: [9,10,90,114]
- NC solitons in $CP(n)$ models: [38,41,49,51,109,125,136]
- Higher-dim. NC instantons: [40,64,80,87,96,108,120,130,133,142,145,160,169]

3. Towards NC Sato's Theories

3.1. *NC Gelfand-Dickey's Hierarchies*

In this section, we derive various NC soliton equations in terms of pseudo-differential operators which include negative powers of differential operators.

An N-th order (monic) pseudo-differential operator A is represented as follows

$$A = \partial_x^N + a_{N-1}\partial_x^{N-1} + \cdots + a_0 + a_{-1}\partial_x^{-1} + a_{-2}\partial_x^{-2} + \cdots. \tag{16}$$

Here we introduce useful symbols:

$$A_{\geq r} := \partial_x^N + a_{N-1}\partial_x^{N-1} + \cdots + a_r\partial_x^r, \tag{17}$$

$$A_{\leq r} := A - A_{\geq r+1} = a_r\partial_x^r + a_{r-1}\partial_x^{r-1} + \cdots, \tag{18}$$

$$\mathrm{res}_r A := a_r. \tag{19}$$

The symbol $\mathrm{res}_{-1}A$ is especially called the *residue* of A.

The action of a differential operator ∂_x^n on a multiplicity operator f is formally defined as the following generalized Leibniz rule:

$$\partial_x^n \cdot f := \sum_{i\geq 0} \binom{n}{i} (\partial_x^i f)\partial_x^{n-i}, \tag{20}$$

where the binomial coefficient is given by

$$\binom{n}{i} := \frac{n(n-1)\cdots(n-i+1)}{i(i-1)\cdots 1}. \tag{21}$$

We note that the definition of the binomial coefficient (21) is applicable to the case for negative n, which just define the action of negative power of differential operators. The examples are,

$$\begin{aligned}\partial_x^{-1}\cdot f &= f\partial_x^{-1} - f'\partial_x^{-2} + f''\partial_x^{-3} - \cdots, \\ \partial_x^{-2}\cdot f &= f\partial_x^{-2} - 2f'\partial_x^{-3} + 3f''\partial_x^{-4} - \cdots, \\ \partial_x^{-3}\cdot f &= f\partial_x^{-3} - 3f'\partial_x^{-4} + 6f''\partial_x^{-5} - \cdots,\end{aligned} \tag{22}$$

where $f' := \partial f/\partial x, f'' := \partial^2 f/\partial x^2$ and so on, and ∂_x^{-1} in the RHS acts as an integration operator $\int^x dx$.

The composition of pseudo-differential operators is also well-defined and the total set of pseudo-differential operators forms an operator algebra. For more on pseudo-differential operators and Sato's theory, see e.g. [7,16,27,121,134].

Let us introduce a Lax operator as the following first-order pseudo-differential operator:

$$L = \partial_x + u_2\partial_x^{-1} + u_3\partial_x^{-2} + u_4\partial_x^{-3} + \cdots, \tag{23}$$

where the coefficients u_k $(k = 2, 3, \ldots)$ are functions of infinite variables $(x^1, x^2, \ldots)$ with $x^1 \equiv x$:

$$u_k = u_k(x^1, x^2, \ldots). \tag{24}$$

The noncommutativity is arbitrarily introduced for the variables $(x^1, x^2, \ldots)$ as Eq. (1) here.

The NC KP hierarchy is defined in Sato's framework as

$$\partial_m L = [B_m, L]_\star, \quad m = 1, 2, \ldots, \tag{25}$$

where the action of ∂_m on the pseudo-differential operator L should be interpreted to be coefficient-wise, that is, $\partial_m L := [\partial_m, L]$ or $\partial_m \partial_x^k = 0$. The operator B_m is given by

$$B_m := (\underbrace{L\star\cdots\star L}_{m \text{ times}})_{\geq 0} =: (L^m)_{\geq 0}. \tag{26}$$

The KP hierarchy gives rise to a set of infinite differential equations with respect to infinite kind of fields from the coefficients in Eq. (25) for a fixed m. Hence it contains huge amount of differential (evolution) equations for all m. The LHS of Eq. (25) becomes $\partial_m u_k$ which shows a flow in the x^m direction.

If we put the constraint $L^l = B_l$ on the NC KP hierarchy (25), we get infinite set of l-reduced NC KP hierarchies. We can easily show

$$\frac{\partial u_k}{\partial x^{Nl}} = 0, \tag{27}$$

for all N, k because

$$\frac{dL^l}{dx^{Nl}} = [B_{Nl}, L^l]_\star = [(L^l)^N, L^l]_\star = 0, \tag{28}$$

which implies Eq. (27). The reduced NC KP hierarchy is called the *l-reduction* of the NC KP hierarchy. This time, the constraint $L^l = B_l$ gives simple relationships which make it possible to represent infinite kind of fields $u_{l+1}, u_{l+2}, u_{l+3}, \ldots$ in terms of $(l-1)$ kind of fields $u_2, u_3, \ldots, u_l$. (cf. Appendix A in [61].)

Let us see explicit examples.

- NC KP hierarchy

  The coefficients of each powers of (pseudo-)differential operators in the NC KP hierarchy (25) yield a series of infinite NC "evolution equations," that is, for $m = 1$

$$\partial_x^{1-k})\quad \partial_1 u_k = u_k', \quad k = 2, 3, \ldots \quad \Rightarrow \quad x^1 \equiv x, \tag{29}$$

for $m = 2$

$$\begin{aligned}
\partial_x^{-1})\quad & \partial_2 u_2 = u_2'' + 2u_3', \\
\partial_x^{-2})\quad & \partial_2 u_3 = u_3'' + 2u_4' + 2u_2 \star u_2' + 2[u_2, u_3]_\star, \\
\partial_x^{-3})\quad & \partial_2 u_4 = u_4'' + 2u_5' + 4u_3 \star u_2' - 2u_2 \star u_2'' + 2[u_2, u_4]_\star, \\
\partial_x^{-4})\quad & \partial_2 u_5 = \cdots,
\end{aligned} \tag{30}$$

and for $m = 3$

$$\begin{aligned}
\partial_x^{-1})\quad \partial_3 u_2 &= u_2''' + 3u_3'' + 3u_4' + 3u_2' \star u_2 + 3u_2 \star u_2', \\
\partial_x^{-2})\quad \partial_3 u_3 &= u_3''' + 3u_4'' + 3u_5' + 6u_2 \star u_3' + 3u_2' \star u_3 \\
&\quad +3u_3 \star u_2' + 3[u_2, u_4]_\star, \\
\partial_x^{-3})\quad \partial_3 u_4 &= u_4''' + 3u_5'' + 3u_6' + 3u_2' \star u_4 + 3u_2 \star u_4' + 6u_4 \star u_2' \\
&\quad -3u_2 \star u_3'' - 3u_3 \star u_2'' + 6u_3 \star u_3' \\
&\quad +3[u_2, u_5]_\star + 3[u_3, u_4]_\star, \\
\partial_x^{-4})\quad \partial_3 u_5 &= \cdots.
\end{aligned} \tag{31}$$

These just imply the $(2+1)$-dimensional NC KP equation [137,100] with $2u_2 \equiv u$, $x^2 \equiv y$, $x^3 \equiv t$ and $\partial_x^{-1} = \int^x dx$:

$$\frac{\partial u}{\partial t} = \frac{1}{4}\frac{\partial^3 u}{\partial x^3} + \frac{3}{4}\frac{\partial (u \star u)}{\partial x} + \frac{3}{4}\partial_x^{-1}\frac{\partial^2 u(x')}{\partial y^2} - \frac{3}{4}\left[u, \partial_x^{-1}\frac{\partial u(x')}{\partial y}\right]_\star \tag{32}$$

Important point is that infinite kind of fields $u_3, u_4, u_5, \ldots$ are represented in terms of one kind of field $2u_2 \equiv u$ as is seen in Eq. (30).

This guarantees the existence of NC KP hierarchy which implies the existence of reductions of the NC KP hierarchy. The order of nonlinear terms are determined in this way.

- NC KdV Hierarchy (2-reduction of the NC KP hierarchy)

  Taking the constraint $L^2 = B_2 =: \partial_x^2 + u$ for the NC KP hierarchy, we get the NC KdV hierarchy. This time, the following NC Lax hierarchy

$$\frac{\partial u}{\partial x^m} = \left[B_m, L^2\right]_\star, \tag{33}$$

  include neither positive nor negative power of (pseudo-)differential operators for the same reason as commutative case and gives rise to the m-th KdV equation for each m. For example, the NC KdV hierarchy (33) becomes the $(1+1)$-dimensional NC KdV equation [30] for $m = 3$ with $x^3 \equiv t$

$$\dot{u} = \frac{1}{4}u''' + \frac{3}{4}\left(u' \star u + u \star u'\right), \tag{34}$$

  and the $(1+1)$-dimensional 5-th NC KdV equation [159] for $m = 5$ with $x^5 \equiv t$

$$\dot{u} = \frac{1}{16}u''''' + \frac{5}{16}(u \star u''' + u''' \star u) + \frac{5}{8}(u' \star u' + u \star u \star u)' \tag{35}$$

- NC Boussinesq Hierarchy (3-reduction of the NC KP hierarchy)

  The 3-reduction $L^3 = B_3$ yields the NC Boussinesq hierarchy which includes the $(1+1)$-dimensional NC Boussinesq equation [159] with $t \equiv x^2$:

$$\ddot{u} = \frac{1}{3}u'''' + (u \star u)'' + ([u, \partial_x^{-1}\dot{u}]_\star)', \tag{36}$$

  where $\ddot{u} := \partial^2 u/\partial t^2$.

In this way, we can generate infinite set of the l-reduced NC KP hierarchies. (This is called the *NC Gelfand-Dickey hierarchies* which reduce to the ordinary Gelfand-Dickey hierarchies [47] in the commutative limit.) The present discussion is also applicable to the matrix Sato theory where the fields u_k $(k = 1, 2, \ldots)$ are $N \times N$ matrices. For $N = 2$, the Lax hierarchy includes the Ablowitz-Kaup-Newell-Segur (AKNS) system [4], the Davey-Stewarson equation, the NLS equation and so on. (For commutative discussions, see e.g. [16,27].) NC Bogoyavlenskii-Calogero-Schiff (BCS) equation [159] is also derived.

3.2. *Conservation Laws*

Here we prove the existence of infinite conservation laws for the wide class of NC soliton equations. The existence of infinite number of conserved quantities would lead to infinite-dimensional hidden symmetry from Nöther's theorem.

First we would like to comment on conservation laws of NC field equations [67]. The discussion is basically the same as commutative case because both the differentiation and the integration are the same as commutative ones in the Moyal representation.

Let us suppose the conservation law

$$\frac{\partial\sigma(t,x^i)}{\partial t} = \partial_i J^i(t,x^i), \tag{37}$$

where $\sigma(t,x^i)$ and $J^i(t,x^i)$ are called the *conserved density* and the *associated flux*, respectively. The conserved quantity is given by spatial integral of the conserved density:

$$Q(t) = \int_{\text{space}} d^D x \sigma(t,x^i), \tag{38}$$

where the integral $\int_{\text{space}} dx^D$ is taken for spatial coordinates. The proof is straightforward:

$$\begin{aligned}\frac{dQ}{dt} &= \frac{\partial}{\partial t}\int_{\text{space}} d^D x\sigma(t,x^i) = \int_{\text{space}} d^D x \partial_i J_i(t,x^i)\\ &= \int_{\substack{\text{spatial}\\ \text{infinity}}} dS^i J_i(t,x^i) = 0,\end{aligned} \tag{39}$$

unless the surface term of the integrand $J_i(t,x^i)$ vanishes. The convergence of the integral is also expected because the star-product naively reduces to the ordinary product at spatial infinity due to: $\partial_i \sim O(r^{-1})$ where $r := |x|$.

Here let us return back to NC hierarchy. In order to discuss the conservation laws, we have to specify for what equations the conservation laws are. The specified equations possess space and time coordinates in the infinite coordinates $x_1, x_2, x_3, \cdots$. Identifying $t \equiv x^m$, we can get conserved densities for the NC Lax hierarchies as follows ($n = 1, 2, \dots$) [61]:

$$\sigma_n = \text{res}_{-1} L^n + \theta^{im} \sum_{k=0}^{m-1}\sum_{l=0}^{k} \binom{k}{l} \partial_x^{k-l}\text{res}_{-(l+1)} L^n \diamond \partial_i \text{res}_k L^m, \tag{40}$$

where the suffices i must run in the space-time directions only. The symbol

"$\diamond$" is called the *Strachan product* [151] and defined by

$$f(x) \diamond g(x) := \sum_{s=0}^{\infty} \frac{(-1)^s}{(2s+1)!} \left(\frac{1}{2} \theta^{ij} \partial_i^{(x')} \partial_j^{(x'')} \right)^{2s} f(x')g(x'') \Big|_{x'=x''=x} . \quad (41)$$

This is a commutative and non-associative product.

We can easily see that deformation terms appear in the second term of Eq. (40) in the case of space-time noncommutativity. On the other hand, in the case of space-space noncommutativity, the conserved density is given by the residue of L^n as commutative case.

For examples, explicit representation of the NC KP equation with space-time noncommutativity, the NC KdV equation is

$$\sigma_n = \text{res}_{-1} L^n - 3\theta \left((\text{res}_{-1} L^n) \diamond u_3' + (\text{res}_{-2} L^n) \diamond u_2' \right). \quad (42)$$

We have a comment on conserved densities for one-soliton configuration. One soliton solutions can always reduce to the commutative ones because $f(t-x) \star g(t-x) = f(t-x)g(t-x)$ [30,67]. Hence the conserved densities are not deformed in the NC extension.

The present discussion is applicable to the NC matrix Sato theory, including the NC AKNS system, the NC Davey-Stewarson equation, the NC NLS equation, and the NC BCS equation.

3.3. *Some Exact Solutions*

Here we show the existence of exact (multi-soliton) solutions by giving the explicit formula.

First, let us comment on 1-soliton solutions. Defining $z := x + vt, \bar{z} := x - vt$, we easily see

$$f(z) \star g(z) = f(z)g(z) \quad (43)$$

because the star-product (2) is rewritten in terms of $(z, \bar{z})$ as

$$f(z,\bar{z}) \star g(z,\bar{z}) = e^{iv\theta(\partial_{\bar{z}'}\partial_{z''} - \partial_{z'}\partial_{\bar{z}''})} f(z',\bar{z}')g(z'',\bar{z}'') \Big|_{\substack{z'=z''=z \\ \bar{z}'=\bar{z}''=\bar{z}.}} \quad (44)$$

Hence NC one soliton-solutions are essentially the same as the commutative ones.

Next, we prove that NC Burgers equations derived from NC Gelfand-Dickey hierarchies are integrable in the sense that they are linearizable.

NC Burgers equation is obtained by a special reduction of NC mKP hierarchies [67,61]:

$$\dot{u} - u'' - 2u \star u' = 0. \quad (45)$$

The solutions of the following NC diffusion equations

$$\dot{\psi} = \psi'', \tag{46}$$

solve Eq.(45) via the NC Cole-Hopf transformation: $u = \psi^{-1} \star \psi'$. The naive solution of the NC diffusion equation (46) is

$$\psi(t,x) = 1 + \sum_{i=1}^{N} h_i e^{ak_i^2 t} \star e^{\pm k_i x} = 1 + \sum_{i=1}^{N} h_i e^{\frac{i}{2}ak_i^3\theta} e^{ak_i^2 t \pm k_i x}, \tag{47}$$

where h_i, k_i are complex constants. The final form in (47) shows that the naive solution on commutative space is deformed by $e^{\frac{i}{2}ak_i^3\theta}$ due to the noncommutativity. This reduces to the N-shock wave solution in fluid dynamics. Hence we call it the *NC N-shock wave solution.* Exact solutions for $N = 1, 2$ are obtained by L. Martina and O. Pashaev [115] in terms of u and nontrivial effects of the NC-deformation are actually reported.

This is a very interesting result. The NC Burgers equation contains infinite number of time derivatives in the nonlinear term and integrability would be naively never expected. Initial value problems are hard to define. Nevertheless, the NC Burgers equation is linearizable and the linearized equation is a differential equation of first order with respect to time and the initial value problem is well-defined. This shows that the NC Burgers equation is completely integrable.

General arguments for NC hierarchies are possible. Exact solutions for them are already given by Etingof, Gelfand and Retakh [37] as explicit forms in terms of quasi-determinants [48]. In Moyal deformations, the solutions are actually *multi-soliton* solutions, which can be seen in the asymptotic behavior. In scattering process, the soliton configurations are stable and never decay. Noncommutativity affects the phase shifts only. Exact solutions for NC KP eq. would coincide with those by Paniak [137]. More detailed discussion will be reported later soon [62]. Exact solutions are also discussed in [33,143,161].

3.4. *Some Examples of NC Ward's Conjecture*

In this subsection, we present some examples of NC Ward's conjecture, including NC NLS eq., NC Burgers eq., NC KdV eq. and so on. (For commutative discussions, see e.g. [1,2,3,19,89,118].)

- NC NLS equation

  Let us consider the following NC ASDYM equation with $G = U(2)$, which is dimensionally reduced to 2-dimensional space time

(The convention is the same as [118].):

$$Q' = 0, \quad \dot{Q} + \Phi'_w + [\Phi_z, Q]_\star = 0,$$
$$\Phi'_z - \dot{\Phi}_w + [\Phi_w, \Phi_z]_\star = 0. \tag{48}$$

where Q, Φ_w and Φ_z denote the original gauge fields.

Now let us take a further reduction on the gauge fields in the ASDYM eq. (48) as follows [113]:

$$Q = \frac{i}{2}\begin{pmatrix} 1 & 0 \\ 0 & -1 \end{pmatrix}, \ \Phi_w = \begin{pmatrix} 0 & \psi \\ -\bar{\psi} & 0 \end{pmatrix}, \ \Phi_z = i\begin{pmatrix} \psi \star \bar{\psi} & 0 \\ 0 & -\bar{\psi} \star \psi \end{pmatrix} \tag{49}$$

Then the NC ASDYM (48) reduces to

$$i\dot{\psi} = \psi'' + 2\psi \star \bar{\psi} \star \psi. \tag{50}$$

This is just the NC NLS equation [29].

We note that the gauge group is not $SU(2)$ but $U(2)$ on NC spaces because the matrix Φ_z is not traceless. This is a very consistent result because in the original NC Yang-Mills theories, $U(1)$ part of the gauge group is essential [35,60,152]. NC KdV equation is also derived in a similar way [63], which is new.

- NC Burgers equation

  Let us consider the following NC ASDYM equation with $G = U(1)$ (Eq. (3.1.2) in [118]):

$$\partial_w A_z - \partial_z A_w + [A_w, A_z]_\star = 0, \quad \partial_{\tilde{w}} A_{\tilde{z}} - \partial_{\tilde{z}} A_{\tilde{w}} + [A_{\tilde{w}}, A_{\tilde{z}}]_\star = 0,$$
$$\partial_z A_{\tilde{z}} - \partial_{\tilde{z}} A_z + \partial_{\tilde{w}} A_w - \partial_w A_{\tilde{w}} + [A_{\tilde{z}}, A_z]_\star + [A_w, A_{\tilde{w}}]_\star = 0. \tag{51}$$

where $(z, \tilde{z}, w, \tilde{w})$ and $A_{z,\tilde{z},w,\tilde{w}}$ denote the coordinates of the original $(2+2)$-dimensional space and the gauge fields, respectively. We note that the commutator part should remain though the gauge group is $U(1)$ because the elements of the gauge group could be operators and the gauge group could be considered to be non-abelian: $U(\infty)$. This commutator part actually plays an important role as usual in NC theories.

Now let us take the simple dimensional reduction $\partial_{\tilde{z}} = \partial_{\tilde{w}} = 0$ and put the following constraints (with $w \equiv t, \ z \equiv x$):

$$A_{\tilde{z}} = A_{\tilde{w}} = 0, \quad A_z = u, \quad A_w = u' + u \star u. \tag{52}$$

Then the NC ASDYM equation (51) reduces to :

$$\dot{u} - u'' - 2u \star u' = 0. \tag{53}$$

This is just the NC Burgers equation which is linearizable and hence completely integrable in this sense [67,115]. We note that without the commutator part $[A_w, A_z]_\star$, the nonlinear term should be symmetric: $u' \star u + u \star u'$, which leads to neither the Lax representation nor linearized equations via a NC Cole-Hopf transformation [67]. This shows that the special feature in the original NC gauge theories plays a crucial role in integrability for the lower-dimensional equation. Therefore the NC Burgers equation is expected to have some non-trivial property special to NC spaces such as the existence of $U(1)$ instantons.

- NC KdV equation

Finally Let us consider another reduction of NC ASDYM eq. onto NC KdV eq. which is different from that by Legaré [113]. Let us start with the following NC ASDYM equation with $G = SL(2, \mathbf{R})$, which is dimensionally reduced to 2-dimensional space-time (The convention is the same as [12]):

$$[P, B] = 0, \quad P' - Q' + [P, Q]_\star + [H, B]_\star = 0,$$
$$\dot{Q} - H' + [Q, H]_\star = 0. \tag{54}$$

where B, H, P and Q denote the original gauge fields.

Now let us take further reduction on the gauge fields as follows:

$$B = \begin{pmatrix} 0 & 0 \\ -1 & 0 \end{pmatrix}, \quad P = \frac{1}{2}\begin{pmatrix} 0 & 0 \\ -u & 0 \end{pmatrix},$$
$$Q = \begin{pmatrix} 0 & 1 \\ -u & 0 \end{pmatrix}, \quad H = \frac{1}{4}\begin{pmatrix} -u' & 2u \\ -u'' - 2u \star u & u' \end{pmatrix}. \tag{55}$$

Then the NC ASDYM equation (54) reduces to

$$\dot{u} + \frac{1}{4}u''' + \frac{3}{4}(u' \star u + u \star u') = 0. \tag{56}$$

This is just the NC KdV equation (34). This time, however, it is naively hard to interpret the corresponding D-brane configuration because the gauge group is $SL(2, \mathbf{R})$.

NC KP equation is also derived in a similar way [63]. These results are new.

In this way, we can derive various integrable equations from NC ASDYM eqs. by reductions. Existence of these reductions guarantees the lower-dimensional integrable equations actually have the corresponding physical situations and could be applied to analysis of D-brane dynamics in special reduced situations.

An (incomplete) list of works on NC integrable equations

- NC Burgers eqs: [67,115]
- NC Fordy-Kulish systems: [31]
- NC KdV eqs: [30]
- NC KP eqs: [37,137,100,161]
- NC Non-Linear Schrödinger eqs: [29,162]
- NC Liouville, sine-Gordon, sinh-Gordon and Toda field eqs: [15,17,18,28,52,53,102,103,111,171]
- NC hierarchies etc.: [33,61,68,143,163]
- NC dressing and splitting methods: [14,82,83,104,105,106,107,170]
- NC mini-twistor spaces: [106]
- NC twistor theories: [69,92,153]

4. Conclusion and Discussion

In the present article, we reported recent developments of NC extension of soliton theories and integrable systems focusing on ADHM construction of NC instantons and NC Sato's theories. In the former part, we saw how and ADHM constructions work and the small instanton singularities are resolved on NC spaces. In the latter part, we proved the existence of infinite number of conserved densities and exact multi-soliton solutions for wide class of NC soliton equations. This suggests that NC soliton equations could be completely integrable in some sense and an infinite-dimensional symmetry would be hidden.

As a next step, completion of NC Sato's theory is the most worth keeping to investigate. In order to reveal what the hidden symmetry is, we have to construct theories of tau-functions which play crucial roles in Sato's theories. (See also [33,162,143].) The symmetry would be represented in terms of some kind of deformed infinite-dimensional affine Lie algebras.

From the original motivation, confirmation of NC Ward's conjecture would be the most important via the construction of NC twistor theories. Some aspects of NC Twistor theories have been already discussed by many authors e.g. [69,83,92,106,153]. This would clarifies integrability of NC ASDYM equations.

In reductions of ASDYM equations, we mainly should take metric of $(2,2)$-type signature which is called the *split signature*. ASDYM theories with the split signature can be embedded [107] in $N = 2$ string theories [135]. Simple reductions of them are studied intensively by Lechtenfeld's group [14,83,104,170]. This guarantees that NC integrable equations would have physical meanings and might lead to various successful applications to the corresponding D-brane dynamics and so on. It is also very interesting to clarify what symmetries in reductions guarantee integrabilities in lower dimensional integrable equations. One approach is from the viewpoint of Lagrangian formalism in supersymmetric Yang-Mills theories. The BPS equations just correspond to integrable equations and preserved supersymmetries would relate to their integrability [63]. Various BPS D-brane configurations (e.g. [36]) might have a relation to our studies.

Supersymmetric extension (e.g [113,132]) and higher dimensional extension (e.g. [159]) would be interesting and straightforwardly possible. Extension to non(-anti)commutative superspaces is also considerable. We also expect special properties would still survive in these extensions.

For space-time noncommutativity, we have to consider foundation of Hamiltonian formalism from the beginning in order to establish what integrability for them is, especially, symplectic structures, Poisson brackets, Liouville's theorem, Nöther's theorem, action-angle variables, initial value problems and so on. Geometrical interpretations of them must be also clarified.

Though our program is going well now, there are still many things worth studying to be seen.

Acknowledgements

It is a great pleasure to thank the organizers for invitation and hospitality during the COE workshop on NC Geometry and Physics, 26 February - 3 March, 2004 at Keio university. He would be grateful to H. Awata, A. Dimakis, S. Kakei, M. Kato, I. Kishimoto, O. Lechtenfeld, K. Lee, L. Mason, M-y. Mo, A. Mukherjee, F. Müller-Hoissen, A. Nakamula, S. Odake, D. Popov, K. Takasaki and K. Toda for useful comments. Thanks are also due to organizers and audiences during the workshops YITP-W-03-07 on "QFT 2003" and YITP-W-04-03 on "QFT 2004," for hospitality and discussion, and to L. Mason, O. Lechtenfeld, F. Müller-Hoissen and K. Lee for financial support and hospitality during the stays at Oxford university, Hannover university, MPI for Flow Research, Göttingen and KIAS on January 2004, August 2004 and March 2005. This work was partially supported by JSPS Research Fellowships for Young Scientists (#0310363) and the Daiko Foundation (#9095).

References

1. M. J. Ablowitz, S. Chakravarty and R. G. Halburd, J. Math. Phys. **44** (2003) 3147.
2. M. J. Ablowitz, S. Chakravarty and L. A. Takhtajan, Commun. Math. Phys. **158** (1993) 289.
3. M. J. Ablowitz and P. A. Clarkson, *Solitons, Nonlinear Evolution Equations and Inverse Scattering*, (Cambridge UP, 1991) [ISBN/0-521-38730-2].
4. M. J. Ablowitz, D. J. Kaup, A. C. Newell and H. Segur, Problems," tiyahStud. Appl. Math. **53** (1974) 249.
5. M. Aganagic, R. Gopakumar, S. Minwalla and A. Strominger, JHEP **0104** (2001) 001 [hep-th/0009142].
6. M. F. A, N. J. Hitchin, V. G. Drinfeld and Y. I. Manin, Phys. Lett. A **65** (1978) 185.
7. O. Babelon, D. Bernard and M. Talon, *Introduction to classical integrable systems*, (Cambridge UP, 2003) [ISBN/0-521-82267-X].
8. D. Bak, Phys. Lett. B **471** (1999) 149 [hep-th/9910135].
9. D. Bak, Phys. Lett. B **495** (2000) 251 [hep-th/0008204].
10. D. Bak, K. M. Lee and J. H. Park, Phys. Rev. D **63** (2001) 125010 [hep-th/0011099].
11. D. s. Bak, K. y. Lee and J. H. Park, Phys. Rev. D **66** (2002) 025021 [hep-th/0204221].
12. I. Bakas and D. A. Depireux, Mod. Phys. Lett. A **6** (1991) 399.
13. A. A. Belavin, A. M. Polyakov, A. S. Schwartz and Y. S. Tyupkin, Phys. Lett. B **59** (1975) 85.
14. S. Bieling, J. Phys. A **35** (2002) 6281 [hep-th/0203269].
15. H. Blas, H. L. Carrion and M. Rojas, JHEP **03** (2005) 037 [hep-th/0502051].
16. M. Błaszak, *Multi-Hamiltonian Theory of Dynamical Systems* (Springer, 1998) [ISBN/3-540-64251-X].
17. I. Cabrera-Carnero, hep-th/0503147.
18. I. Cabrera-Carnero and M. Moriconi, Nucl. Phys. B **673** (2003) 437 [hep-th/0211193]; hep-th/0303168.
19. S. Chakravarty, S. L. Kent and E. T. Newman, J. Math. Phys. **36** (1995) 763.
20. C. S. Chu, "Non-commutative geometry from strings," hep-th/0502167.
21. C. S. Chu, V. V. Khoze and G. Travaglini, Nucl. Phys. B **621** (2002) 101 [hep-th/0108007].
22. D. H. Correa, P. Forgacs, E. F. Moreno, F. A. Schaposnik and G. A. Silva, JHEP **0407** (2004) 037 [hep-th/0404015].
23. D. H. Correa, G. S. Lozano, E. F. Moreno and F. A. Schaposnik, Phys. Lett. B **515** (2001) 206 [hep-th/0105085].
24. D. H. Correa, E. F. Moreno and F. A. Schaposnik, Phys. Lett. B **543** (2002) 235 [hep-th/0207180].
25. T. Curtright, D. Fairlie and C. K. Zachos, Phys. Lett. B **405** (1997) 37 [hep-

th/9704037]; C. K. Z, D. Fairlie and T. Curtright, "Matrix membranes and integrability," hep-th/9709042.

26. E. Date, M. Jimbo, M. Kashiwara and T. Miwa, "Transformation groups for soliton equations," in *Nonlinear integrable systems —classical theory and quantum theory* (World Sci., 1983) 39 [ISBN/9971950324] and references therein.
27. L. A. Dickey, *Soliton equations and Hamiltonian systems (2nd Ed.)*, Adv. Ser. Math. Phys. **26** (2003) 1 [TSBN/9812381732].
28. A. Dimakis and F. Müller-Hoissen, Int. J. Mod. Phys. B **14** (2000) 2455 [hep-th/0006005].
29. A. Dimakis and F. Müller-Hoissen, hep-th/0007015. Czech. J. Phys. **51** (2001) 1285.
30. A. Dimakis and F. Müller-Hoissen, Phys. Lett. A **278** (2000) 139 [hep-th/0007074].
31. A. Dimakis and F. Müller-Hoissen, J. Phys. A **34** (2001) 2571 [nlin.si/0008016].
32. A. Dimakis and F. Müller-Hoissen, Lett. Math. Phys. **54** (2000) 123 [hep-th/0007160]; J. Phys. A **34** (2001) 9163 [nlin.si/0104071].
33. A. Dimakis and F. Müller-Hoissen, J. Phys. A **37** (2004) 4069 [hep-th/0401142]; J. Phys. A **37** (2004) 10899 [hep-th/0406112]; nlin.si/0408023; nlin.si/0501003.
34. N. Dorey, T. J. Hollowood, V. V. Khoze and M. P. Mattis, Phys. Rept. **371** (2002) 231 [hep-th/0206063].
35. M. R. Douglas and N. A. Nekrasov, Rev. Mod. Phys. **73** (2002) 977 [hep-th/0106048].
36. M. Eto, Y. Isozumi, M. Nitta, K. Ohashi, K. Ohta and N. Sakai, hep-th/0412024; M. Eto, Y. Isozumi, M. Nitta, K. Ohashi and N. Sakai, hep-th/0412048; M. Eto, Y. Isozumi, M. Nitta, K. Ohashi, K. Ohta, N. Sakai and Y. Tachikawa, hep-th/0503033 and references therein.
37. P. Etingof, I. Gelfand and V. Retakh, Math. Res. Lett. **4** (1997) 413 [q-alg/9701008].
38. O. Foda, I. Jack and D. R. T. Jones, Phys. Lett. B **547** (2002) 79 [hep-th/0209111].
39. F. Franco-Sollova and T. Ivanova, J. Phys. A **36** (2003) 4207 [hep-th/0209153].
40. A. Fujii, Y. Imaizumi and N. Ohta, Nucl. Phys. B **615** (2001) 61 [hep-th/0105079].
41. K. Furuta, T. Inami, H. Nakajima and M. Yamamoto, Phys. Lett. B **537** (2002) 165 [hep-th/0203125]; JHEP **0208** (2002) 009 [hep-th/0207166].
42. K. Furuta, T. Inami and M. Yamamoto, hep-th/0211129.
43. K. Furuuchi, Prog. Theor. Phys. **103** (2000) 1043 [hep-th/9912047].
44. K. Furuuchi, Commun. Math. Phys. **217** (2001) 579 [hep-th/0005199].
45. K. Furuuchi, hep-th/0010006.
46. K. Furuuchi, JHEP **0103** (2001) 033 [hep-th/0010119].
47. I. M. Gelfand and L. A. Dikii, Funct. Anal. Appl. **10** (1976) 13 (Russian), 259 (English).

48. I. Gelfand and V. Retakh, Funct. Anal. Appl. **25** (1991) 91; Funct. Anal. Appl. **2** (1992) 1.
49. S. Ghosh, Nucl. Phys. B **670** (2003) 359 [hep-th/0306045].
50. C. Gomez and J. J. Manjarin, hep-th/0111169.
51. T. R. Govindarajan and E. Harikumar, Phys. Lett. B **602** (2004) 238 [hep-th/0406273].
52. M. T. Grisaru, L. Mazzanti, S. Penati and L. Tamassia, JHEP **0404** (2004) 057 [hep-th/0310214].
53. M. T. Grisaru and S. Penati, Nucl. Phys. B **655** (2003) 250 [hep-th/0112246].
54. H. J. Groenewold, Physica **12** (1946) 405.
55. D. J. Gross and N. A. Nekrasov, JHEP **0007** (2000) 034 [hep-th/0005204].
56. D. J. Gross and N. A. Nekrasov, JHEP **0010** (2000) 021 [hep-th/0007204].
57. D. J. Gross and N. A. Nekrasov, JHEP **0103** (2001) 044 [hep-th/0010090].
58. M. Hamanaka, "ADHM/Nahm construction and its duality," Soryushiron Kenkyu (Kyoto) **106** (2002) 1 (Japanese).
59. M. Hamanaka, Phys. Rev. D **65** (2002) 085022 [hep-th/0109070].
60. M. Hamanaka, "Noncommutative solitons and D-branes," Ph. D thesis (University of Tokyo, 2003) hep-th/0303256.
61. M. Hamanaka, J. Math. Phys. **46** (2005) 052701 [hep-th/0311206].
62. M. Hamanaka, to appear.
63. M. Hamanaka, work in progress.
64. M. Hamanaka, Y. Imaizumi and N. Ohta, Phys. Lett. B **529** (2002) 163 [hep-th/0112050].
65. M. Hamanaka and S. Terashima, JHEP **0103** (2001) 034 [hep-th/0010221].
66. M. Hamanaka and K. Toda, Phys. Lett. A **316** (2003) 77 [hep-th/0211148].
67. M. Hamanaka and K. Toda, J. Phys. A **36** (2003) 11981 [hep-th/0301213].
68. M. Hamanaka and K. Toda, Proc. Inst. Math. NAS Ukraine (2004) 404 [hep-th/0309265].
69. K. C. Hannabuss, Lett. Math. Phys. **58** (2001) 153 [hep-th/0108228].
70. J. A. Harvey, [hep-th/0102076].
71. J. A. Harvey, [hep-th/0105242].
72. J. A. Harvey and G. W. Moore, J. Math. Phys. **42** (2001) 2765 [hep-th/0009030].
73. K. Hashimoto, JHEP **0012** (2000) 023 [hep-th/0010251].
74. K. Hashimoto, Phys. Rev. D **65** (2002) 065014 [hep-th/0107226].
75. A. Hashimoto and K. Hashimoto, JHEP **9911** (1999) 005 [hep-th/9909202].
76. K. Hashimoto, H. Hata and S. Moriyama, JHEP **9912** (1999) 021 [hep-th/9910196].
77. K. Hashimoto and T. Hirayama, Nucl. Phys. B **587** (2000) 207 [hep-th/0002090].
78. K. Hashimoto, T. Hirayama and S. Moriyama, JHEP **0011** (2000) 014 [hep-th/0010026].
79. K. Hashimoto and H. Ooguri, Phys. Rev. D **64** (2001) 106005 [hep-th/0105311].
80. Y. Hiraoka, Phys. Lett. B **536** (2002) 147 [hep-th/0203047]; hep-th/0205283; Phys. Rev. D **67** (2003) 105025 [hep-th/0301176].

81. P. M. Ho, hep-th/0003012.
82. Z. Horváth, O. Lechtenfeld and M. Wolf JHEP **0212** (2002) 060 [hep-th/0211041].
83. M. Ihl and S. Uhlmann, Int. J. Mod. Phys. B **18** (2003) 4889 [hep-th/0211263].
84. T. Ishikawa, S. I. Kuroki and A. Sako, JHEP **0112** (2001) 000 [hep-th/0109111].
85. T. Ishikawa, S. I. Kuroki and A. Sako, JHEP **0208** (2002) 028 [hep-th/0201196].
86. K. Ito, "Field theory in noncommutative spacetime and superstring theory," the Bulletin of the Physical Society of Japan **59** (2004) 856 (Japanese).
87. T. A. Ivanova and O. Lechtenfeld, hep-th/0502117.
88. T. A. Ivanova, O. Lechtenfeld and H. Muller-Ebhardt, Mod. Phys. Lett. A **19** (2004) 2419 [hep-th/0404127].
89. T. A. Ivanova and A. D. Popov, Theor. Math. Phys. **102** (1995) 280 [Teor. Mat. Fiz. **102** (1995) 384].
90. D. P. Jatkar, G. Mandal and S. R. Wadia, JHEP **0009** (2000) 018 [hep-th/0007078].
91. M. Jimbo and T. Miwa, Publ. Res. Inst. Math. Sci. Kyoto **19** (1983) 943.
92. A. Kapustin, A. Kuznetsov and D. Orlov, Commun. Math. Phys. **221** (2001) 385 [hep-th/0002193].
93. K. Y. Kim, B. H. Lee and H. S. Yang, J. Korean Phys. Soc. **41** (2002) 290 [hep-th/0003093].
94. K. Y. Kim, B. H. Lee and H. S. Yang, Phys. Lett. B **523** (2001) 357 [hep-th/0109121].
95. K. Y. Kim, B. H. Lee and H. S. Yang, Phys. Rev. D **66** (2002) 025034 [hep-th/0205010].
96. C. j. Kim, K. M. Lee and S. H. Yi, Phys. Lett. B **543** (2002) 107 [hep-th/0204109].
97. A. Konechny and A. Schwarz, Phys. Rept. **360** (2002) 353 [hep-th/0012145].
98. A. Konechny and A. Schwarz, Phys. Rept. **360** (2002) 353 [hep-th/0107251].
99. P. Kraus and M. Shigemori, JHEP **0206** (2002) 034 [hep-th/0110035].
100. B. Kupershmidt, *KP or mKP* (AMS, 2000) [ISBN/0821814001].
101. O. Lechtenfeld, Fortsch. Phys. **52** (2004) 596 [hep-th/0401158].
102. O. Lechtenfeld, Czech. J. Phys. **54** (2004) 1351 [hep-th/0409108].
103. O. Lechtenfeld, L. Mazzanti, S. Penati, A. D. Popov and L. Tamassia, Nucl. Phys. B **705** (2005) 477 [hep-th/0406065].
104. O. Lechtenfeld and A. D. Popov, JHEP **0111** (2001) 040 [hep-th/0106213]; Phys. Lett. B **523** (2001) 178 [hep-th/0108118].
105. O. Lechtenfeld and A. D. Popov, JHEP **0203** (2002) 040 [hep-th/0109209].
106. O. Lechtenfeld and A. D. Popov, JHEP **0401** (2004) 069 [hep-th/0306263].
107. O. Lechtenfeld, A. D. Popov and B. Spendig, Phys. Lett. B **507** (2001) 317 [hep-th/0012200]; JHEP **0106** (2001) 011 [hep-th/0103196].
108. O. Lechtenfeld, A. D. Popov and R. J. Szabo, JHEP **0312** (2003) 022 [hep-th/0310267].
109. B. H. Lee, K. M. Lee and H. S. Yang, Phys. Lett. B **498** (2001) 277 [hep-

th/0007140].
110. B. H. Lee and H. S. Yang, Phys. Rev. D **66** (2002) 045027 [hep-th/0206001].
111. K. M. Lee, JHEP **0408** (2004) 054 [hep-th/0405244].
112. K. M. Lee, D. Tong and S. Yi, Phys. Rev. D **63** (2001) 065017 [hep-th/0008092].
113. M. Legare, hep-th/0012077; J. Phys. A **35** (2002) 5489.
114. G. S. Lozano, E. F. Moreno and F. A. Schaposnik, Phys. Lett. B **504** (2001) 117 [hep-th/0011205].
115. L. Martina and O. K. Pashaev, hep-th/0302055.
116. M. Marino, R. Minasian, G. Moore and A. Strominger, JHEP **0001** (2000) 005 [hep-th/9911206].
117. L. J. Mason and G. A. J. Sparling, Phys. Lett. A **137** (1989) 29.
118. L. J. Mason and N. M. Woodhouse, *Integrability, Self-Duality, and Twistor Theory* (Oxford UP, 1996) [ISBN/0-19-853498-1].
119. Y. Matsuo, Phys. Lett. B **499** (2001) 223 [hep-th/0009002].
120. M. Mihailescu, I. Y. Park and T. A. Tran, Phys. Rev. D **64** (2001) 046006 [hep-th/0011079].
121. T. Miwa, M. Jimbo and E. Date, (translated by M. Reid), "Solitons : differential equations, symmetries and infinite dimensional algebras," (Cambridge UP, 2000) [ISBN/0521561612].
122. S. Moriyama, Phys. Lett. B **485** (2000) 278 [hep-th/0003231].
123. S. Moriyama, JHEP **0008** (2000) 014 [hep-th/0006056].
124. J. E. Moyal, Proc. Cambridge Phil. Soc. **45** (1949) 99;
125. J. Murugan and R. Adams, JHEP **0212** (2002) 073 [hep-th/0211171].
126. H. Nakajima, "Resolutions of moduli spaces of ideal instantons on $\mathbf{R}^4$," *Topology, Geometry and Field Theory* (World Sci., 1994) 129 [ISBN/981-02-1817-6].
127. H. Nakajima, *Lectures on Hilbert Schemes of Points on Surfaces* (AMS, 1999) [ISBN/0-8218-1956-9].
128. N. A. Nekrasov, Commun. Math. Phys. **241** (2003) 143 [hep-th/0010017].
129. N. A. Nekrasov, hep-th/0011095.
130. N. A. Nekrasov, hep-th/0203109.
131. N. Nekrasov and A. Schwarz, Commun. Math. Phys. **198** (1998) 689 [hep-th/9802068].
132. H. Nishino and S. Rajpoot, Phys. Lett. B **572** (2003) 91 [hep-th/0306290].
133. K. Ohta, Phys. Rev. D **64** (2001) 046003 [hep-th/0101082].
134. Y. Ohta, J. Satsuma, D. Takahashi and T. Tokihiro, Prog. Theor. Phys. Suppl. **94** (1988) 210.
135. H. Ooguri and C. Vafa, Mod. Phys. Lett. A **5** (1990) 1389; Nucl. Phys. B **361** (1991) 469; Nucl. Phys. B **367** (1991) 83.
136. H. Otsu, T. Sato, H. Ikemori and S. Kitakado, JHEP **0307** (2003) 054 [hep-th/0303090]; JHEP **0406** (2004) 006 [hep-th/0404140]; hep-th/0503118.
137. L. D. Paniak, hep-th/0105185.
138. S. Parvizi, Mod. Phys. Lett. A **17** (2002) 341 [hep-th/0202025].
139. R. Penrose, Gen. Rel. Grav. **7** (1976) 31.
140. A. P. Polychronakos, Phys. Lett. B **495** (2000) 407 [hep-th/0007043].

141. A. D. Popov, A. G. Sergeev and M. Wolf, J. Math. Phys. **44** (2003) 4527 [hep-th/0304263].
142. A. D. Popov and R. J. Szabo, hep-th/0504025.
143. M. Sakakibara, J. Phys. A **37** (2004) L599 [nlin.si/0408002].
144. A. Sako, JHEP **0304** (2003) 023 [hep-th/0209139].
145. A. Sako and T. Suzuki, hep-th/0503214.
146. M. Sato and Y. Sato, "Soliton equations as dynamical systems on infinite dimensional Grassmann manifold," in *Nonlinear Partial Differential Equations in Applied Sciences* (North-Holland, 1982) 259.
147. F. A. Schaposnik, Braz. J. Phys. **34** (2004) 1349 [hep-th/0310202]; hep-th/0408132.
148. A. Schwarz, Commun. Math. Phys. **221** (2001) 433 [hep-th/0102182].
149. N. Seiberg and E. Witten, JHEP **9909** (1999) 032 [hep-th/9908142].
150. K. Shigechi, M. Wadati and N. Wang, Nucl. Phys. B **706** (2005) 518 [hep-th/0404249].
151. I. A. B. Strachan, J. Geom. Phys. **21** (1997) 255 [hep-th/9604142].
152. R. J. Szabo, Phys. Rept. **378** (2003) 207 [hep-th/0109162].
153. K. Takasaki, J. Geom. Phys. **37** (2001) 291 [hep-th/0005194].
154. S. Terashima, Phys. Lett. B **477** (2000) 292 [hep-th/9911245].
155. Y. Tian, hep-th/0307264.
156. Y. Tian, Mod. Phys. Lett. A **19** (2004) 1315 [hep-th/0404128].
157. Y. Tian and C. J. Zhu, Commun. Theor. Phys. **38** (2002) 691 [hep-th/0205110]. Phys. Rev. D **67** (2003) 045016 [hep-th/0210163].
158. Y. Tian, C. J. Zhu and X. C. Song, Mod. Phys. Lett. A **18** (2003) 1691 [hep-th/0211225].
159. K. Toda, Proceedings of workshop on Integrable Theories, Solitons and Duality, Sao Paulo, Brazil, 1-6 July 2002 [JHEP PRHEP-unesp2002/038].
160. P. Valtancoli, Int. J. Mod. Phys. A **18** (2003) 1125 [hep-th/0209118].
161. N. Wang and M. Wadati, J. Phys. Soc. Jap. **72** (2003) 1366; J. Phys. Soc. Jap. **72** (2003) 1881; J. Phys. Soc. Jap. **72** (2003) 3055;
162. N. Wang and M. Wadati, J. Phys. Soc. Jap. **73** (2004) 1689.
163. N. Wang and M. Wadati, J. Phys. Soc. Jap. **73** (2004) 1689.
164. R. S. Ward, Phil. Trans. Roy. Soc. Lond. A **315** (1985) 451; Lect. Notes Phys. **280** (Springer, 1986) 106; "Integrable systems in twistor theory," in *Twistors in Mathematics and Physics* (Cambridge UP, 1990) 246.
165. R. S. Ward and R. O. Wells, *Twistor Geometry and Field Theory* (Cambridge UP, 1990) [ISBN/0-521-42268-X].
166. S. Watamura, "Gauge theories on noncommutative spaces," Proceedings of workshop on Mathematical Physics 2002, Tokyo, Japan, 21-23 September 2002 (Japanese).
167. R. Wimmer, hep-th/0502158.
168. E. Witten, Phys. Rev. Lett. **38** (1977) 121.
169. E. Witten, JHEP **0204** (2002) 012 [hep-th/0012054].
170. M. Wolf, JHEP **0206** (2002) 055 [hep-th/0204185].
171. A. Zuevsky, J. Phys. A **37** (2004) 537.

ON THE CONNECTEDNESS OF THE SPACES OF SURFACE GROUP REPRESENTATIONS

NAN-KUO HO
National Cheng-Kung University,
Department of Mathematics,
Tainan, Taiwan
nho@fields.utoronto.ca, nankuo@mail.ncku.edu.tw

CHIU-CHU MELISSA LIU
Harvard University,
Department of Mathematics,
Cambridge, MA, USA
ccliu@math.harvard.edu

Let G be a compact, connected, semisimple Lie group. It is known that for a closed compact connected orientable surface Σ the order of the group $H^2(\Sigma, \pi_1(G))$ is equal to the number of connected components of the space $\mathrm{Hom}(\pi_1(\Sigma), G)/G$ which can also be identified with the moduli space of gauge equivalence classes of flat G-bundles over Σ. If G is any compact connected Lie group, [1] implies that the number of connected components of the space $\mathrm{Hom}(\pi_1(\Sigma), G)/G$ is equal to the order of the the group $\pi_1(G_{ss})$ where G_{ss} is the maximal semisimple subgroup of G. For a closed compact connected nonorientable surface, we show that the number of connected components of the space $\mathrm{Hom}(\pi_1(\Sigma), G)/G$ is equal to the order of the group $\pi_1(G)/2\pi_1(G)$ for any compact connected Lie group G .

1. introduction

Let Σ be a closed compact surface, and let $\pi_1(\Sigma)$ be its fundamental group. For any connected Lie group G, we denote $\mathrm{Hom}(\pi_1(\Sigma), G)$ the space of all homomorphism from $\pi_1(\Sigma)$ to G. In [7], W. M. Goldman computed the number of connected components of spaces $\mathrm{Hom}(\pi_1(\Sigma), G)$ for $G = PSL(2, \mathbb{C})$ or $PSL(2, \mathbb{R})$, and Σ is a Riemann surface of genus bigger than 1. In the same paper, he also made a conjecture for the general connected complex semisimple Lie group G that there is a bijection between $\pi_0(\mathrm{Hom}(\pi_1(\Sigma), G)/G)$ and $H^2(\Sigma, \pi_1(G)) \cong \pi_1(G)$, where the G-action on $\mathrm{Hom}(\pi_1(\Sigma), G)$ is induced by the conjugation action of G on itself. This conjecture was proved by J. Li in [14] as well as the following corollary:

Theorem 1. *Let Σ be a closed, orientable Riemann surface of genus $\ell > 1$. Let G be a connected, compact, semisimple Lie group. Then there is a bijection*

$$\pi_0(\mathrm{Hom}(\pi_1(\Sigma), G)/G) \to H^2(\Sigma; \pi_1(G)) \cong \pi_1(G),$$

where G acts on $\mathrm{Hom}(\pi_1(\Sigma), G)$ by conjugation.

Our first result is an analogue of Theorem 1 (cf:[8]).

Theorem 2. *Let Σ be a closed, compact, nonorientable surface which is homeomorphic to the connected sum of m copies of the real projective plane, where $m \neq 1, 2, 4$. Let G be a connected, compact, semisimple Lie group. Then there is a bijection*

$$\pi_0(\mathrm{Hom}(\pi_1(\Sigma), G)/G) \to H^2(\Sigma; \pi_1(G)) \cong \pi_1(G)/2\pi_1(G),$$

where $2\pi_1(G)$ denote the subgroup $\{k^2 \mid k \in \pi_1(G)\}$ of the finite abelian group $\pi_1(G)$.

Geometrically, $\mathrm{Hom}(\pi_1(\Sigma), G)/G$ can be identified with the moduli space of gauge equivalence classes of flat G-bundles over Σ, where a flat G-bundle is a principal G-bundle together with a flat connection. It is known that there is a one-to-one correspondence between topological principal G-bundles over Σ and elements in $\pi_1(G)$ or $\pi_1(G)/2\pi_1(G)$ depending on the surface being orientable or nonorientable (respectively). The geometric interpretation of Theorem 1 and Theorem 2 is that the moduli space of gauge equivalence classes of flat connections on a fixed underlying topological principal G-bundle over Σ is always nonempty and connected when the surface is compact closed and G is compact, connected, semisimple. In fact, this was the idea of J. Li's proof for Goldman's conjecture. More details of this geometric interpretation will be given in Section 6.

When the compact Lie group fails to be semisimple, the geometry of the space $\mathrm{Hom}(\pi_1(\Sigma), G)/G$ becomes more complicated. Let us first recall the structure theorem of compact connected Lie group [[13], Theorem 4.29]:

Lemma 3. *Let G be a compact connected Lie group with center $Z(G)$, and let H be the identity component of $Z(G)$. Let $\mathfrak{g}$ be the Lie algebra of G, and let G_{ss} be the analytic subgroup of G with Lie algebra $[\mathfrak{g}, \mathfrak{g}]$. Then G_{ss} has finite center, H and G_{ss} are closed subgroups, and G is the commuting product $G = G_{ss}H$.*

In Lemma 3, G_{ss} is semisimple, and the map $G_{ss} \times H \to G = G_{ss}H$ given by $(g,h) \mapsto gh$ is a finite cover which is also a group homomorphism. With the notation given above, we are able to state the theorems of connected components of $\mathrm{Hom}(\pi_1(\Sigma), G)/G$ for general compact connected Lie group G:

Theorem 4. *Let Σ be a closed, orientable Riemann surface with genus bigger than zero. Let G be a compact connected Lie group, and let G_{ss} be the maximal connected semi-simple subgroup of G. Then there is a bijection*

$$\pi_0(\mathrm{Hom}(\pi_1(\Sigma), G)/G) \to \pi_1(G_{ss}).$$

The group $\pi_1(G_{ss})$ in Theorem 4 is a finite abelian group. This theorem shows that not every principal G-bundle is equipped with flat structure (unless G is semisimple).

A proof of Theorem 4 can be extracted from [1]. However, here we will give another proof of Theorem 4, and derive the following result for non-orientable surfaces (cf:[10]):

Theorem 5. *Let Σ be a closed, compact, nonorientable surface which is homeomorphic to the connected sum of m copies of the real projective plane, where $m \neq 1, 2, 4$. Let G be a compact connected Lie group. Then there is a bijection*

$$\pi_0(\mathrm{Hom}(\pi_1(\Sigma), G)/G) \to \pi_1(G)/2\pi_1(G)$$

where $2\pi_1(G)$ denote the subgroup $\{k^2 \mid k \in \pi_1(G)\}$ of the abelian group $\pi_1(G)$.

The group $\pi_1(G)/2\pi_1(G)$ in Theorem 5 fits in the following short exact sequence of abelian groups:

$$1 \to \pi_1(G_{ss})/2\pi_1(G_{ss}) \to \pi_1(G)/2\pi_2(G) \to (\mathbb{Z}/2\mathbb{Z})^{\dim H} \to 1$$

where $G_{ss} = [G, G]$ and H is the connected component of the identity of the center of G. In particular, $\pi_1(G)/2\pi_1(G)$ is a finite abelian group so that it makes sense to count the number of connected components of $\mathrm{Hom}(\pi_1(\Sigma), G)/G$.

In fact, with Theorem 4 and Theorem 5 given, Theorem 1 and Theorem 2 are just corollary in the sense that $G_{ss} = G$. Nevertheless, since the proof for G semisimple is very fundamental and easy to understand, we keep the proof in this note.

The key ingredient of the proofs is the following theorem:

Lemma 6. *Let G be a compact, semisimple, connected and simply connected Lie group. Let ℓ be a positive integer. Then the commutator map $\mu_G^\ell : G^{2\ell} \to G$ defined by*

$$\mu_G^\ell(a_1, b_1, \ldots, a_\ell, b_\ell) = a_1 b_1 a_1^{-1} b_1^{-1} \cdots a_\ell b_\ell a_\ell^{-1} b_\ell^{-1} \tag{1}$$

is surjective, and $(\mu_G^\ell)^{-1}(g)$ is connected for all $g \in G$.

The surjectivity follows from Goto's commutator theorem [[11], Theorem 6.55]. The commutator map μ_G^ℓ in Lemma 6 is a group valued moment map of the quasi-Hamiltonian G-space $G^{2\ell}$ in the sense of Alekseev, Malkin, and Meinrenken [3]. By Theorem 7.2 in [3], all the fibers of the moment map of a connected q-Hamiltonian G-space are connected if G is a compact, connected and simply connected Lie group. We will explained later in Remark 8 that why this Lemma can not be directly applied to the nonorientable surface case as to the orientable surface case.

This note is organized as follows: in Section 2 we give the proof for compact semisimple Lie group G; in Section 3, we give a brief summary of the structure theorem for compact Lie group; in Section 4 we give the proof for general compact Lie group G, and in Section 5 we give some examples for classical Lie groups and what happened when $m = 1, 2$ or 4. In Section 6, we give algebraic and geometric interpretations of the obstruction map used in our proof.

Throughout this note, our surface is closed without boundary. For those who are interested in the case when the surface has markings, please see [10].

2. Compact semisimple Lie group

2.1. *Orientable surfaces*

In this section, Σ is a closed Riemann surface with genus $\ell > 0$. Note that $\mathrm{Hom}(\pi_1(S^2), G)/G$ is just a single point even if G is not simply connected, so Theorem 1 is not true for the genus zero case.

Let G be a connected, compact, semisimple Lie group, and let $\rho : \tilde{G} \to G$ be the universal covering map which is also a group homomorphism. Then $\tilde{G}$ is a compact, semisimple, connected and simply connected Lie group, and $\mathrm{Ker}\rho$ is a subgroup of the center $Z(\tilde{G})$ of $\tilde{G}$ which is abelian.

Let $\mu_G^\ell : G^{2\ell} \to G$ be the commutator map defined as in (1), and let e be the identity element of G. Then $\mathrm{Hom}(\pi_1(\Sigma), G)$ can be identified with

$$(\mu_G^\ell)^{-1}(e) = \{(a_1, b_1, \ldots, a_\ell, b_\ell) \in G^{2\ell} \mid a_1 b_1 a_1^{-1} b_1^{-1} \cdots a_\ell b_\ell a_\ell^{-1} b_\ell^{-1} = e\}.$$

There is an obstruction map (see e.g.[7])

$$o_2 : \mathrm{Hom}(\pi_1(\Sigma), G) \cong (\mu_G^\ell)^{-1}(e) \to \mathrm{Ker}\rho$$

defined by

$$(a_1, b_1, \ldots, a_\ell, b_\ell) \in (\mu_G^\ell)^{-1}(e) \mapsto \tilde{a}_1\tilde{b}_1\tilde{a}_1^{-1}\tilde{b}_1^{-1} \cdots \tilde{a}_\ell\tilde{b}_\ell\tilde{a}_\ell^{-1}\tilde{b}_\ell^{-1} \in \mathrm{Ker}\rho,$$

where $(\tilde{a}_1, \tilde{b}_1 \ldots, \tilde{a}_\ell, \tilde{b}_\ell)$ is a preimage of $(a_1, b_1, \ldots, a_\ell, b_\ell)$ under $\rho^{2\ell} : \tilde{G}^{2\ell} \to G^{2\ell}$. It is easily checked that the definition does not depend on the choice of $(\tilde{a}_1, \tilde{b}_1, \ldots, \tilde{a}_\ell, \tilde{b}_\ell)$ and o_2 descends to a continuous map

$$\bar{o}_2 : \mathrm{Hom}(\pi_1(\Sigma), G)/G \to \mathrm{Ker}\rho.$$

For each $k \in \mathrm{Ker}\rho$, there is a surjective continuous map

$$(\mu_{\tilde{G}}^\ell)^{-1}(k) \to \bar{o}_2^{-1}(k)$$

which is the restriction of $\tilde{G}^{2\ell} \xrightarrow{\rho^{2\ell}} G^{2\ell} \to G^{2\ell}/G$, where G acts on $G^{2\ell}$ by diagonal conjugation. By Lemma 6, $(\mu_{\tilde{G}}^\ell)^{-1}(k)$ is nonempty and connected for each $k \in \mathrm{Ker}\rho$, so $\bar{o}_2$ is surjective, and $\bar{o}_2^{-1}(k)$ is connected for each $k \in \mathrm{Ker}\rho \cong \pi_1(G)$. This gives Theorem 1 as well as the genus one case.

2.2. *Nonorientable surfaces*

Let G be a connected, compact, semisimple Lie group, and let $\rho : \tilde{G} \to G$ be the universal covering map. Let Σ be a closed compact nonorientable surface. Then Σ is homeomorphic to the connected sum of m copies of $\mathbb{RP}^2$, where m is a positive integer. We will only give the proof for $m = 2\ell + 1$ which is homeomorphic to the connected sum of a Riemann surface of genus ℓ and $\mathbb{RP}^2$. The proof for $m = 2\ell + 2$ is similiar and is omitted here(see [8] for the details.).

Let Σ be homeomorphic to the connected sum of a Riemann surface of genus ℓ and $\mathbb{RP}^2$. The space $\mathrm{Hom}(\pi_1(\Sigma), G)$ can be identified with

$$X_{2\ell+1} = \{(a_1, b_1, \ldots, a_\ell, b_\ell, c) \in G^{2\ell+1} \mid a_1b_1a_1^{-1}b_1^{-1} \cdots a_\ell b_\ell a_\ell^{-1} b_\ell^{-1} c^2 = e\},$$

where e is the identity element of G.

Let $K = \mathrm{Ker}\rho \cong \pi_1(G)$, and let $2K$ be the subgroup $\{k^2 \mid k \in K\}$ of the finite abelian group K. There is an obstruction map

$$o_2 : \mathrm{Hom}(\pi_1(\Sigma), G) \cong X_m \to K/2K$$

defined by

$$(a_1, b_1, \ldots, a_\ell, b_\ell, c) \in X \mapsto [\tilde{a}_1\tilde{b}_1\tilde{a}_1^{-1}\tilde{b}_1^{-1} \cdots \tilde{a}_\ell\tilde{b}_\ell\tilde{a}_\ell^{-1}\tilde{b}_\ell^{-1}\tilde{c}^2] \in K/2K,$$

where $(\tilde{a}_1, \tilde{b}_1, \ldots, \tilde{a}_\ell, \tilde{b}_\ell, \tilde{c})$ is a preimage of $(a_1, b_1, \ldots, a_\ell, b_\ell, c)$ under $\rho^{2\ell+1} : \tilde{G}^{2\ell+1} \to G^{2\ell+1}$. It is easily checked that the definition does not depend on the choice of $(\tilde{a}_1, \tilde{b}_1, \ldots, \tilde{a}_\ell, \tilde{b}_\ell, \tilde{c})$ and o_2 descends to a continuous map

$$\bar{o}_2 : \mathrm{Hom}(\pi_1(\Sigma), G)/G \to K/2K.$$

For each $k \in K$ and positive integer m, we define $\tilde{X}_m(k) \subset \tilde{G}^m$ for $m = 2\ell + 1$ as follows.

$$\tilde{X}_{2\ell+1}(k) = \{(a_1, b_1, \ldots, a_\ell, b_\ell, c) \in \tilde{G}^{2\ell+1} \mid a_1 b_1 a_1^{-1} b_1^{-1} \cdots a_\ell b_\ell a_\ell^{-1} b_\ell^{-1} c^2 = k\}$$

Let $[k]$ denote the image of k under $K \to K/2K$. Then there is a surjective continuous map

$$\tilde{X}_m(k) \to \bar{o}_2^{-1}([k])$$

which is the restriction of $\tilde{G}^m \xrightarrow{\rho^m} G^m \to G^m/G$, where G acts on G^m by diagonal conjugation. For each $k \in K$, let $q_k \in \tilde{T}$ be a square root of $k \in Z(\tilde{G})$, where we fix a maximal torus $\tilde{T}$ of $\tilde{G}$. Let $\tilde{e}$ denote the identity element of $\tilde{G}$. Then $\tilde{X}_m(k)$ contains $(\tilde{e}, \ldots, \tilde{e}, q_k)$, so it is nonempty. Therefore,

$$\bar{o}_2 : \mathrm{Hom}(\pi_1(\Sigma), G)/G \to K/2K$$

is surjective for any closed compact nonorientable surface Σ.

It remains to show that when $m \neq 1, 2, 4$, $\tilde{X}_m(k)$ is connected for all $k \in K$(more generally for all $k \in Z(\tilde{G})$), which will imply $\bar{o}_2^{-1}([k])$ is connected for all $[k] \in K/2K$.

Let $\ell > 0$ and $Q' : \tilde{G}^{2\ell+1} \to \tilde{G}$ be the projection to the last factor. The restriction of Q' to $\tilde{X}_{2\ell+1}(k)$ gives a map $Q : \tilde{X}_{2\ell+1}(k) \to \tilde{G}$, $(a_1, b_1, \ldots, a_\ell, b_\ell, c) \mapsto c$. Lemma 6 implies that Q is surjective and and $Q^{-1}(c)$ is connected for all $c \in \tilde{G}$. To prove that $\tilde{X}_{2\ell+1}(k)$ is connected, it suffices to show that for any $c \in \tilde{G}$, there is a path $\gamma : [0, 1] \to \tilde{X}_{2\ell+1}(k)$ such that $\gamma(0) \in Q^{-1}(q_k)$ and $\gamma(1) \in Q^{-1}(c)$.

For any $c \in \tilde{G}$, there exists $g \in \tilde{G}$ such that $g^{-1}cg \in \tilde{T}$. Let $\tilde{\mathfrak{g}}$ and $\tilde{\mathfrak{t}}$ be the Lie algebras of $\tilde{G}$ and $\tilde{T}$, respectively. Let $\exp : \tilde{\mathfrak{g}} \to \tilde{G}$ be the exponential map. Then $g^{-1}cgq_k^{-1} = \exp \xi$ for some $\xi \in \tilde{\mathfrak{t}}$. Let $\tilde{W}$ be the Weyl group of $\tilde{G}$, and let $w \in \tilde{W}$ be a Coxeter element. The linear map $w : \tilde{\mathfrak{t}} \to \tilde{\mathfrak{t}}$ has no eigenvalue equal to 1 [12, Section 3.16], so there exists $\xi' \in \tilde{\mathfrak{t}}$ such that $w \cdot \xi' - \xi' = \xi$. Recall that $\tilde{W} = N(\tilde{T})/\tilde{T}$, where $N(\tilde{T})$ is the normalizer of $\tilde{T}$ in $\tilde{G}$, so $w = a\tilde{T} \in N(\tilde{T})/\tilde{T}$ for some $a \in \tilde{G}$. We have

$$a \exp(t\xi') a^{-1} \exp(-t\xi') = \exp(t\xi)$$

for any $t \in \mathbb{R}$.

The group $\tilde{G}$ is connected, so there exists a path $\tilde{g} : [0,1] \to \tilde{G}$ such that $\tilde{g}(0) = \tilde{e}$ and $\tilde{g}(1) = g$. Define $\gamma : [0,1] \to \tilde{G}^{2\ell+1}$ by

$$\gamma(t) = (a(t), b(t), \tilde{e}, \ldots, \tilde{e}, c(t)),$$

where

$$a(t) = \tilde{g}(t) a \tilde{g}(t)^{-1},\ b(t) = \tilde{g}(t)\exp(-2t\xi')\tilde{g}(t)^{-1},\ c(t) = \tilde{g}(t) q_k \exp(t\xi)\tilde{g}(t)^{-1}.$$

Then the image of γ lies in $\tilde{X}_{2\ell+1}(k)$, $\gamma(0) = (a, \tilde{e}, \tilde{e}, \ldots, \tilde{e}, q_k) \in Q^{-1}(q_k)$, and

$$\gamma(1) = (gag^{-1}, g\exp(-2\xi')g^{-1}, \tilde{e}, \ldots, \tilde{e}, c) \in Q^{-1}(c).$$

This gives Theorem 2.

Remark 7.

One can see the isomorphism $H^2(\Sigma; \pi_1(G)) \cong \pi_1(G)/2\pi_1(G)$ as follows. By the universal coefficient theorem of cohomology, we have a short exact sequence

$$0 \to \mathrm{Ext}(H_1(\Sigma;\mathbb{Z}), \pi_1(G)) \to H^2(\Sigma; \pi_1(G)) \to \mathrm{Hom}(H_2(\Sigma;\mathbb{Z}), \pi_1(G)) \to 0,$$

where $H_1(\Sigma, \mathbb{Z}) \cong \mathbb{Z}^{m-1} \oplus \mathbb{Z}/2\mathbb{Z}$, and $H_2(\Sigma, \mathbb{Z}) = 0$. So we have

$$H^2(\Sigma, \pi_1(G)) \cong \mathrm{Ext}(\mathbb{Z}/2\mathbb{Z}, \pi_1(G)) \cong \pi_1(G)/2\pi_1(G).$$

3. Structure theorem

Let G be a compact connected Lie group. Let $G_{ss} = [G, G]$ be its commutator group. Then G_{ss} is the maximal connected semisimple subgroup of G. Let H be the connected component of the identity of the center $Z(G)$ of G. Then H is a compact torus. The map $\phi : H \times G_{ss} \to G = HG_{ss}$ given by $(h, g) \mapsto hg$ is a finite cover which is also a group homomorphism. The kernel of ϕ is isomorphic to $D = H \cap G_{ss} \subset Z(G_{ss})$, which is a finite abelian group. Note that G_{ss} is a normal subgroup of G, and the quotient

$$G/G_{ss} \cong H/D$$

is a compact torus.

Let $\rho_{ss} : \tilde{G}_{ss} \to G_{ss}$ be the universal covering map which is also a group homomorphism. Then $\tilde{G}_{ss}$ is a compact, connected, simply connected Lie group, and $\mathrm{Ker}(\rho_{ss})$ is a subgroup of $Z(\tilde{G}_{ss})$.

Let $\mathfrak{g}$ and $\mathfrak{h}$ be the Lie algebras of G and H respectively, and let $\exp_H : \mathfrak{h} \to H$ be the exponential map. The map

$$\rho : \tilde{G} = \mathfrak{h} \times \tilde{G}_{ss} \to G$$

given by

$$(X, g) \to \exp_H(X)\rho_{ss}(g)$$

is the universal covering map which is also a group homomorphism. Here we give a group structure for $\mathfrak{h}$ the addition operation in the vector space $\mathfrak{h}$ and 0 as the identity element for the group structure. Notice that $\tilde{G}$ is not compact. Let $\pi_1 : H \to H/D$ and $\pi_2 : G_{ss} \to G_{ss}/D$ be natural projections. Then

$$\pi_1 \circ \exp_H : \mathfrak{h} \to H/D \cong G/G_{ss}$$

is the universal covering map which is also a group homomorphism, and

$$\check{\Lambda} = \mathrm{Ker}(\pi_1 \circ \exp_H) \cong \pi_1(G/G_{ss}) \cong \mathbb{Z}^{\dim H}.$$

We have

$$\begin{aligned}\mathrm{Ker}(\rho) &= \{(X, g) \in \mathfrak{h} \times \tilde{G}_{ss} \mid \exp_H(X)\rho_{ss}(g) = e\} \\ &\subset \check{\Lambda} \times \mathrm{Ker}(\pi_2 \circ \rho_{ss}) \subset \check{\Lambda} \times Z(\tilde{G}_{ss}) \subset \mathfrak{h} \times Z(\tilde{G}_{ss}) = Z(\tilde{G}).\end{aligned}$$

The map $(X, g) \mapsto X$ defines a surjective group homomorphism $p : \mathrm{Ker}(\rho) \to \check{\Lambda}$. The kernel of p is $\{0\} \times \mathrm{Ker}(\rho_{ss}) \cong \mathrm{Ker}(\rho_{ss})$. So we have an exact sequence of abelian groups

$$1 \to \mathrm{Ker}(\rho_{ss}) \to \mathrm{Ker}(\rho) \to \check{\Lambda} \to 1, \tag{2}$$

which can be rewritten as

$$1 \to \pi_1(G_{ss}) \to \pi_1(G) \to \pi_1(G/G_{ss}) \to 1. \tag{3}$$

where $\pi_1(G_{ss})$ is a finite abelian group, and

$$\pi_1(G/G_{ss}) = \pi_1(H/D) \cong \mathbb{Z}^{\dim H}.$$

4. Compact Lie group

4.1. *Orientable surfaces*

In this section, Σ is a closed Riemann surface with genus $\ell > 0$. Note that $\mathrm{Hom}(\pi_1(S^2), G)/G$ consists of a point even if $\pi_1(G_{ss})$ is nontrivial, so Theorem 4 is not true for the genus zero case.

Let G, H, G_{ss}, D, $\tilde{G}$, $\mathfrak{h}$, ρ, ρ_{ss} be as in Section 3, and let $\mu_G^\ell : G^{2\ell} \to G$ be the commutator map defined by (1). Then $\mathrm{Hom}(\pi_1(\Sigma), G)$ can be identified with

$$(\mu_G^\ell)^{-1}(e) = \{(a_1, b_1, \ldots, a_\ell, b_\ell) \in G^{2\ell} \mid a_1 b_1 a_1^{-1} b_1^{-1} \cdots a_\ell b_\ell a_\ell^{-1} b_\ell^{-1} = e\}.$$

There is an obstruction map (see e.g. [7])

$$o_2 : \mathrm{Hom}(\pi_1(\Sigma), G) \cong (\mu_G^\ell)^{-1}(e) \to \mathrm{Ker}(\rho)$$

defined by

$$(a_1, b_1, \ldots, a_\ell, b_\ell) \in (\mu_G^\ell)^{-1}(e) \mapsto \mu_{\tilde{G}}^\ell(\tilde{a}_1, \tilde{b}_1, \ldots, \tilde{a}_\ell, \tilde{b}_\ell) \in \mathrm{Ker}(\rho),$$

where $(\tilde{a}_1, \tilde{b}_1, \ldots, \tilde{a}_\ell, \tilde{b}_\ell)$ is a preimage of $(a_1, b_1, \ldots, a_\ell, b_\ell)$ under $\rho^{2\ell} : \tilde{G}^{2\ell} \to G^{2\ell}$. The definition does not depend on the choice of $(\tilde{a}_1, \tilde{b}_1, \ldots, \tilde{a}_\ell, \tilde{b}_\ell)$ because $\mathrm{Ker}(\rho) \subset Z(\tilde{G})$. It is easily checked that o_2 descends to a continuous map

$$\bar{o}_2 : \mathrm{Hom}(\pi_1(\Sigma), G)/G \to \mathrm{Ker}(\rho)$$

whose image is contained in $\mathrm{Ker}(\rho_{ss})$. Here $\mathrm{Ker}(\rho_{ss})$ is identified with

$$\{0\} \times \mathrm{Ker}(\rho_{ss}) \subset \mathfrak{h} \times \tilde{G}_{ss} = \tilde{G}.$$

For each $k \in \mathrm{Ker}(\rho_{ss})$, there is a surjective map

$$(\mu_{\tilde{G}}^\ell)^{-1}(k) \to \bar{o}_2^{-1}(k)$$

which is the restriction of $\tilde{G}^{2l} \xrightarrow{\rho^{2\ell}} G^{2\ell} \to G^{2\ell}/G$, where G acts on $G^{2\ell}$ by diagonal conjugation. We have

$$(\mu_{\tilde{G}}^\ell)^{-1}(k) \cong \mathfrak{h}^{2\ell} \times (\mu_{\tilde{G}_{ss}}^\ell)^{-1}(k).$$

By Lemma 6, $(\mu_{\tilde{G}_{ss}}^\ell)^{-1}(k)$ is nonempty and connected for each $k \in \mathrm{Ker}(\rho_{ss})$, so $(\mu_{\tilde{G}}^\ell)^{-1}(k)$ is also nonempty and connected for each $k \in \mathrm{Ker}(\rho_{ss})$. Therefore, the image of $\bar{o}_2$ is $\mathrm{Ker}(\rho_{ss})$ and $\bar{o}^{-1}(k)$ is connected for each $k \in \mathrm{Ker}(\rho_{ss}) \cong \pi_1(G_{ss})$. This completes the proof of Theorem 4.

4.2. *Nonorientable surfaces*

Let G be a compact connected Lie group, and let $\rho : \tilde{G} \to G$ be the universal covering map, defined as in Section 3. Let Σ be a closed compact nonorientable surface. Then Σ is homeomorphic to the connected sum of m copies of $\mathbb{RP}^2$, where m is a positive integer. We will only give the proof when m odd since the proof for m even is similiar.

Let Σ be homeomorphic to the connected sum of a Riemann surface of genus ℓ and $\mathbb{RP}^2$. The space $\mathrm{Hom}(\pi_1(\Sigma), G)$ can be identified with

$$X_{2\ell+1} = \{(a_1, b_1, \ldots, a_\ell, b_\ell, c) \in G^{2\ell+1} \mid a_1 b_1 a_1^{-1} b_1^{-1} \cdots a_\ell b_\ell a_\ell^{-1} b_\ell^{-1} c^2 = e\},$$

where e is the identity element of G.

Let $K = \mathrm{Ker}(\rho) \cong \pi_1(G)$, and let $2K$ be the subgroup $\{k^2 \mid k \in K\}$ of the abelian group K. There is an obstruction map

$$o_2 : \mathrm{Hom}(\pi_1(\Sigma), G) \cong X_m \to K/2K$$

defined by

$$(a_1, b_1, \ldots, a_\ell, b_\ell, c) \in X \mapsto [\tilde{a}_1 \tilde{b}_1 \tilde{a}_1^{-1} \tilde{b}_1^{-1} \cdots \tilde{a}_\ell \tilde{b}_\ell \tilde{a}_\ell^{-1} \tilde{b}_\ell^{-1} \tilde{c}^2] \in K/2K,$$

as given in section 2.2, where $(\tilde{a}_1, \tilde{b}_1, \ldots, \tilde{a}_\ell, \tilde{b}_\ell, \tilde{c})$ is a preimage of $(a_1, b_1, \ldots, a_\ell, b_\ell, c)$ under $\rho^{2\ell+1} : \tilde{G}^{2\ell+1} \to G^{2\ell+1}$. It is easily checked that the definition does not depend on the choice of $(\tilde{a}_1, \tilde{b}_1, \ldots, \tilde{a}_\ell, \tilde{b}_\ell, \tilde{c})$ and o_2 descends to a continuous map

$$\bar{o}_2 : \mathrm{Hom}(\pi_1(\Sigma), G)/G \to K/2K. \tag{4}$$

For each $k \in K$ and each positive integer m, we define $\tilde{X}_m(k) \subset \tilde{G}^m$ for $m = 2\ell + 1$ as follows:

$$\tilde{X}_{2\ell+1}(k) = \{(a_1, b_1, \ldots, a_\ell, b_\ell, c) \in \tilde{G}^{2\ell+1} \mid a_1 b_1 a_1^{-1} b_1^{-1} \cdots a_\ell b_\ell a_\ell^{-1} b_\ell^{-1} c^2 = k\}$$

Let $[k]$ denote the image of k under $K \to K/2K$. Then there is a surjective continuous map

$$\tilde{X}_m(k) \to \bar{o}_2^{-1}([k])$$

which is the restriction of $\tilde{G}^m \overset{\rho^m}{\to} G^m \to G^m/G$, where G acts on G^m by diagonal conjugation.

It remains to show that when $m \neq 1, 2, 4$, $\tilde{X}_m(k)$ is connected for all $k \in K$, which implies that $\bar{o}_2^{-1}([k])$ is connected for all $[k] \in K/2K$.

Recall that $\tilde{G} = \mathfrak{h} \times \tilde{G}_{ss}$, and $K = \mathrm{Ker}(\rho) \subset \check{\Lambda} \times Z(\tilde{G}_{ss})$, where we use the notation in Section 3. Given $k \in K$, write $k = (v, q)$, where $v \in \check{\Lambda} \subset \mathfrak{h}$, and $q \in Z(\tilde{G}_{ss}) \subset \tilde{G}_{ss}$.

We have

$$\begin{aligned}\tilde{X}_{2\ell+1}(v, q) = \Big\{&((x_1, \bar{a}_1), (y_1, \bar{b}_1), \ldots (x_\ell, \bar{a}_\ell), (y_\ell, \bar{b}_\ell), (z, \bar{c})) \in (\mathfrak{h} \times \tilde{G}_{ss})^{2\ell+1} \mid \\ &2z = v, \bar{a}_1 \bar{b}_1 \bar{a}_1^{-1} \bar{b}_1^{-1} \cdots \bar{a}_\ell \bar{b}_\ell \bar{a}_\ell^{-1} \bar{b}_\ell^{-1} \bar{c}^2 = q\} \\ \cong\, &\mathfrak{h}^{2\ell} \times \bar{X}_{2\ell+1}(q),\end{aligned}$$

where

$$\bar{X}_{2\ell+1}(q)=\{(\bar{a}_1,\bar{b}_1,\dots,\bar{a}_\ell,\bar{b}_\ell,\bar{c})\in \tilde{G}_{ss}^{2\ell+1}|\bar{a}_1\bar{b}_1\bar{a}_1^{-1}\bar{b}_1^{-1}\cdots\bar{a}_\ell\bar{b}_\ell\bar{a}_\ell^{-1}\bar{b}_\ell^{-1}\bar{c}^2=q\}.$$

It is shown in Section 2.2 that $\bar{X}_{2\ell+1}(q)$ is nonempty and connected for $\ell\geq 1$ and $q\in Z(\tilde{G}_{ss})$, so $\tilde{X}_{2\ell+1}(v,q)$ is nonempty and connected for $\ell\geq 1$.

This completes the proof of Theorem 5.

Remark 8. Let M be a manifold, and let G be a compact Lie group. By Definition 2.2 (B1) in [3], a necessary condition for a map $\mu: M\to G$ to be a group valued moment map is $\mu^*\chi$ being exact, where χ is the canonical closed bi-invariant 3-form on G.

Let $\tilde{G}$ be its universal covering, define $\tilde{\mu}_m:\tilde{G}^m\to\tilde{G}$ by

$$(a_1,b_1,\dots,a_\ell,b_\ell,c)\mapsto a_1b_1a_1^{-1}b_1^{-1}\cdots a_\ell b_\ell a_\ell^{-1}b_\ell^{-1}c^2,\quad \text{if } m=2\ell+1,$$
$$(a_1,b_1,\dots,a_\ell,b_\ell,c_1,c_2)\mapsto a_1b_1a_1^{-1}b_1^{-1}\cdots a_\ell b_\ell a_\ell^{-1}b_\ell^{-1}c_1^2c_2^2,\quad \text{if } m=2\ell+2.$$

Let $Q_j:\tilde{G}^m\to\tilde{G}$ be the projection to the j-th factor. We have

$$\tilde{\mu}_{2\ell+1}^*\chi=2Q_{2\ell+1}^*\chi,\quad \tilde{\mu}_{2\ell+2}^*\chi=2Q_{2\ell+1}^*\chi+2Q_{2\ell+2}^*\chi,$$

which represent nontrivial cohomology classes. So $\tilde{\mu}_m$ cannot be a group valued moment map, and we cannot apply Theorem 7.2 in [3] directly to $\tilde{\mu}_m$ to conclude that $\tilde{X}_m(k)=\tilde{\mu}_m^{-1}(k)$ is connected.

5. Examples and The cases $m=1,2,4$

Here we give some classical Lie groups as examples of our theorem:

For compact connected semisimple Lie group: if $G=SU(n)$, or $Spin(n)$, or $Sp(n)$, the number of connected components of the space $\mathrm{Hom}(\pi_1(\Sigma),G)/G$ is one since they are all simply connected; and if $G=SO(n)$ then the number of connected components of the space $\mathrm{Hom}(\pi_1(\Sigma),G)/G$ is two. For compact connected Lie group where its semisimple part is simply connected: if $G=U(n)$, then the number of connected components of $\mathrm{Hom}(\pi_1(\Sigma),G)/G$ is one if Σ is orientable and is two if Σ is nonorientable.

For $m=1$, we have

$$\mathrm{Hom}(\pi_1(\mathbb{RP}^2),G)/G=\{g\in G\mid g^2=e\}/G,$$

where G acts by conjugation. It is not hard to see that this is just a set of finite points for $G=SU(n)$, or $Sp(n)$, so Theorem 2 is not true for $m=1$.

It is not clear to us if Theorem 2 and Theorem 5 are true for $m=2$. We expect them to hold for $m=4$, but we do not know how to prove it using the approach in this paper.

6. Obstruction class

Let Σ be a connected closed compact surface. Let G be a connected, compact, semisimple Lie group. Let $\rho : \tilde{G} \to G$ be the universal covering map which is also a group homomorphism. Then $\mathrm{Ker}\rho$ is contained in the center $Z(\tilde{G})$ of $\tilde{G}$. Let $K = \mathrm{Ker}\rho \cong \pi_1(G)$ which is a finite abelian group.

In this section, we will give three interpretations of the obstruction map

$$o_2 : \mathrm{Hom}(\pi_1(\Sigma), G) \to H^2(\Sigma; K).$$

If Σ is homeomorphic to a 2-sphere, then $\mathrm{Hom}(\pi_1(\Sigma), G)$ consists of a single point e. We extend the definition in Section 2.1 to the genus zero case by defining $o_2(e)$ to be the identity element of $H^2(\Sigma; K)$. So the obstruction map o_2 is defined for any closed compact surfaces Σ. Note that in the genus zero case, the map o_2 is not necessary to be surjective while for any other compact closed surface, the map o_2 is surjective.

6.1. *Non-abelian Čech cohomology*

For an account of of non-abelian Čech cohomology, see Appendix A in [15]. Here we focus on the interpretation over topological bundles.

Given a group A, let $\underline{A}$ denote the sheaf on Σ for which $\underline{A}(U)$ is the group of continuous functions from U to A, where U is any open subset of Σ. Note that $\underline{K} = K$ since K is discrete. The short exact sequence of sheaves

$$1 \to K \to \underline{\tilde{G}} \to \underline{G} \to 1$$

gives rise to the following long exact sequence of Čech cohomology

$$\begin{aligned} 1 \to H^0(\Sigma; K) \to H^0(\Sigma; \underline{\tilde{G}}) \to H^0(\Sigma, \underline{G}) \\ \xrightarrow{\delta'} H^1(\Sigma; K) \to H^1(\Sigma; \underline{\tilde{G}}) \to H^1(\Sigma; \underline{G}) \xrightarrow{\Delta'} H^2(\Sigma; K). \end{aligned} \tag{5}$$

The pointed sets $H^1(\Sigma; \underline{\tilde{G}})$ and $H^1(\Sigma; \underline{G})$ can be identified with the set of equivalence classes of topological principal $\tilde{G}$-bundles and G-bundles, respectively. Note that $H^1(\Sigma; K) \cong \mathrm{Hom}(\pi_1(\Sigma), K)$. The exact sequence (5) can be rewritten as

$$\begin{aligned} 1 \to \quad K \quad \to \mathrm{Map}(\Sigma, \tilde{G}) \to \mathrm{Map}(\Sigma, G) \\ \xrightarrow{\delta'} \mathrm{Hom}(\pi_1(\Sigma), K) \to \mathrm{Prin}_{\tilde{G}}(\Sigma) \to \mathrm{Prin}_G(\Sigma) \xrightarrow{\Delta'} H^2(\Sigma, K). \end{aligned} \tag{6}$$

The map δ' sends $f : \Sigma \to G$ to $f_* : \pi_1(\Sigma) \to \pi_1(G) \cong K$, and Δ' sends a principal G-bundle to the obstruction to lifting it to a principal $\tilde{G}$-bundle.

For example, if $G = SO(n)$ $(n > 2)$, then $K = \mathbb{Z}/2\mathbb{Z}$, and $\Delta'(P)$ is the second Stiefel-Whitney class $w_2(P) \in H^2(\Sigma, \mathbb{Z}/2\mathbb{Z})$.

We have a commutative diagram of sheaves

$$\begin{array}{ccccccccc} 1 & \longrightarrow & K & \longrightarrow & \tilde{G} & \longrightarrow & G & \longrightarrow & 1 \\ & & \downarrow & & \downarrow & & \downarrow & & \\ 1 & \longrightarrow & K & \longrightarrow & \underline{\tilde{G}} & \longrightarrow & \underline{G} & \longrightarrow & 1 \end{array}$$

In particular, we have the following commutative diagram

$$\begin{array}{ccc} H^1(\Sigma; G) = \mathrm{Hom}(\pi_1(\Sigma), G)/G & \xrightarrow{\Delta} & H^2(\Sigma; K) \\ {\scriptstyle i}\downarrow & & \downarrow{\scriptstyle id} \\ H^1(\Sigma; \underline{G}) = \quad \mathrm{Prin}_G(\Sigma) & \xrightarrow{\Delta'} & H^2(\Sigma; K) \end{array}$$

where i sends a flat G-bundle to the underlying topological principal G-bundle. Therefore, if P is the underlying topological principal G-bundle of a flat G-bundle associated to $p \in \mathrm{Hom}(\pi_1(\Sigma), G)$, then $\Delta'(P) = o_2(p) \in H^2(\Sigma; K)$.

6.2. *Obstruction theory*

For an account of obstruction theory of fiber bundles, see [19].

The first obstruction to the triviality of a principal G-bundle P over Σ, or equivalently, the first obstruction of existence of a cross-section, lies in $H^1(\Sigma; \pi_0(G))$, which vanishes because G is connected. So P is trivial over the 1-skeleton. The second obstruction lies in $H^2(\Sigma; \pi_1(G)) = H^2(\Sigma; K)$ and coincides with $\Delta'(P)$, the obstruction to lifting P to a principal $\tilde{G}$-bundle. For a surface Σ, $\Delta'(P)$ is the only obstruction to the triviality of P. So the topological principal G-bundles over Σ are classified by $\Delta'(P) \in H^2(\Sigma; K)$. The set $\mathrm{Prin}_{\tilde{G}}(\Sigma)$ consists of a single point corresponding to the trivial $\tilde{G}$-bundle, and Δ' in (5) is a bijection.

Note that if G is not simply connected, then there exists nontrivial topological G-bundles over S^2, but the underlying topological principal G-bundle of a flat G-bundle over S^2 must be trivial. So Δ sends the only point in $H^1(S^2; G) = H^1(\pi_1(S^2), G))$ to the identity element of $H^2(S^2; K) = H^2(\pi_1(S^2), K)$.

6.3. *Elementary approach*

There is an elementary approach to the classification of topological principal G-bundles over Σ which we learned from E. Meinrenken. Let D be a disc around a point $x \in \Sigma$. Any principal G-bundle P is trivial over D and over $\Sigma\backslash\{x\}$, and the topological type of P is determined by the homotopy class of the transition function $\psi : D\backslash\{x\} \to G$, or equivalently, an element in $K = \pi_1(G)$. This gives a surjective map $K \to \mathrm{Prin}_G(\Sigma)$ which is injective if Σ is orientable and induces a bijection $K/2K \to \mathrm{Prin}_G(\Sigma)$ if Σ is nonorientable. Therefore, we have a bijection

$$\phi : H^2(\Sigma; K) \to \mathrm{Prin}_G(\Sigma).$$

The argument in [14] Section 6 shows that if P is the underlying topological principal G-bundle of a flat G-bundle associated to $p \in \mathrm{Hom}(\pi_1(\Sigma), G)$, then $\phi(o_2(p)) = P$.

Based on above discussions, Theorem 1 and Theorem 2 can be reformulated as follows.

Theorem 9. *Let Σ be a closed compact connected orientable surface of genus $\ell > 0$, or a closed compact connected nonorientable surface which is homeomorphic to m copies of the real projective plane, where $m \neq 1, 2, 4$. Let G be a connected, compact, semisimple Lie group, and let P be a topological principal G-bundle over Σ. Then the moduli space of gauge equivalence classes of flat connections on P is nonempty and connected.*

In the notation of the above theorem, if $\Sigma = S^2$ and P is nontrivial, which can occur when G is not simply connected, then the moduli space of gauge equivalence classes of flat connections on P is empty. If Σ is homeomorphic to m copies of RP^2, where $m = 1, 2, 4$, then the moduli space of gauge equivalence classes of flat connections on P is nonempty and can be disconnected for $m = 1$.

From the above discussion, our proof of Theorem 4 gives the following statement:

Theorem 10. *Let Σ be a closed compact connected orientable surface with genus $\ell > 0$. Let G be a compact connected Lie group, and let P be a principal G-bundle over Σ. Then the moduli space of gauge equivalence classes of flat connections on P is nonempty if and only if the obstruction class $\Delta'(P) \in H^2(\Sigma; \pi_1(G)) \cong \pi_1(G)$ is a torsion element. In this case, the moduli space of gauge equivalence classes of flat connections on P is connected.*

Similarly, our proof of Theorem 5 gives the following statement:

Theorem 11. *Let Σ be a closed compact connected nonorientable surface which is homeomorphic to m copies of the real projective plane. Let G be* *a compact connected Lie group, and let P be a principal G-bundle over Σ. Then the moduli space of gauge equivalence classes of flat connections on P is nonempty. Moreover, it is connected if $m \neq 1, 2, 4$.*

Acknowledgements

The first author was supported by Grant NSC 92-2119-M-006-006 and NSERC Postdoc Fellowship. Both authors would like to thank Lisa Jeffrey and Eckhard Meinrenken for many useful suggestions and kind help.

References

1. M.F. Atiyah and R. Bott, *Yang-Mills equations over Riemann surfaces*, Philos. Trans. Roy. Soc. London Ser. A **308** (1983), no. 1505, 523–615.
2. M.F. Atiyah, R. Bott, and A. Shapiro, *Clifford modules*, Topology **3** (1964) suppl. 1, 3–38.
3. A. Alekseev, A. Malkin, and E. Meinrenken, *Lie group valued moment maps*, J. Differential Geom. **48** (1998), no. 3, 445–495.
4. A. Alekseev, E. Meinrenken, and C. Woodward, *Duistermaat-Heckman measures and moduli spaces of flat bundles over surfaces*, Geom. Funct. Anal. **12** (2002), no. 1, 1–31.
5. Theodor Bröcker and Tammo tom Dieck, *Representations of compact Lie groups*, Graduate Texts in Mathematics, 98. Springer-Verlag, New York, 1985.
6. W.M. Goldman, *The symplectic nature of fundamental groups of surfaces*, Adv. in Math. **54** (1984), no. 2, 200–225.
7. W.M. Goldman, *Topological components of spaces of representations*, Invent. Math. **93** (1988), no. 3, 557–607.
8. N.-K. Ho and C.-C.M. Liu, *Connected Components of the Space of Surface Group Representations* IMRN (2003), no. 44, 2359–2372.
9. N.-K. Ho and C.-C.M. Liu, *On the Connectedness of Moduli Spaces of Flat Connections over Compact Surfaces*, Canad. J. Math. Vol.56(6) (2004), 1228-1236.
10. N.-K. Ho and C.-C.M. Liu, *Connected Components of the Space of Surface Group Representations II*, **arXiv:math.SG/0406069**. To appear in IMRN.
11. K.H. Hofmann and S.A. Morris, *The structure of compact groups. A primer for the student—a handbook for the expert*, de Gruyter Studies in Mathematics, 25. Walter de Gruyter & Co., Berlin, 1998.
12. J.E. Humphreys, *Reflection Groups and Coxeter Groups*, Cambridge Studies in Advanced Mathematics, vol.29, Cambridge University Press, Cambridge, 1990.
13. A.W. Knapp, *Lie Groups Beyond an Introduction*, Progress in Mathematics, 140, Birkhäuser Boston, Inc., Boston, MA, 1996.
14. J. Li, *The space of surface group representations*, Manuscripta Math. **78** (1993), no. 3, 223–243.

15. H.B. Lawson and M. Michelsohn, *Spin Geometry*, Princeton Mathematical Series 38, Princeton University Press, Princeton, NJ, 1989.
16. A. Ramanathan, *Stable principal bundles on a compact Riemann surface*, Math. Ann. **213** (1975), 129–152.
17. A. Ramanathan, *Moduli for principal bundles*, Algebraic geometry (Proc. Summer Meeting, Univ. Copenhagen, Copenhagen, 1978), pp. 527–533, Lecture Notes in Math., 732, Springer, Berlin, 1979.
18. J.-P. Serre, *Local Fields*, translated from the French by Marvin Jay Greenberg, Graduate Texts in Mathematics, 67. Springer-Verlag, New York-Berlin, 1979.
19. N. Steenrod, *The Topology of Fibre Bundles*, Princeton Mathematical Series, vol. 14. Princeton University Press, Princeton, N. J., 1951.

WITTEN'S DEFORMED LAPLACIAN AND ITS CLASSICAL MECHANICS

ATSUSHI INOUE
Department of Mathematics,
Tokyo Institute of Technology,
Tokyo, 152-0033, Japan
inoue@math.titech.ac.jp

Though the brilliant idea of supersymmetry is introuduced in mathematics, there seems no well-established mathematical theorems on classical mechanics corresponding to the supersymmetric quantum mechanics. Here, we give such theorems corresponding to Witten's deformed Laplacian.

1. Introduction

1.1. *Getzler's interpretation of Witten's idea*

Let (M^d, g) be a d-dimensional compact Riemannian manifold. For the fundamental solution $E(t, q, q')$ of the heat equation $\dfrac{\partial}{\partial t} - \dfrac{1}{2}\Delta_g$, Weyl Theorem on (M^d, g) for $e^{(t/2)\Delta_g}$ states roughly that when $t \to 0$, we have

$$\operatorname{tr}\exp\left[\frac{t}{2}\Delta_g\right] = \int_M d_g q\, E(t, q, q) \sim (2\pi t)^{-d/2}\mathrm{vol}_g M.$$

Here, under suitable conditions on (M^d, g), we have approximately

$$E(t, \bar{q}, \underline{q}) \sim A(t, \bar{q}, \underline{q}) \exp\left[-S(t, \bar{q}, \underline{q})\right],$$

with $S(t, \bar{q}, \underline{q})$ is a solution of the Hamilton-Jacobi equation, and $A(t, \bar{q}, \underline{q})$ is a solution of the continuity equation. Moreover, for the classical path $\gamma_c(\tau)$ from $(0, \underline{q})$ to $(t, \bar{q})$, we have

$$S(t, \bar{q}, \underline{q}) = \int_0^t L(\gamma_c(\tau), \dot{\gamma}_c(\tau))d\tau.$$

Atiyah-Singer Index Theorem may be considered as an extension of Weyl Theorem on (M, g), to the one on $(\widetilde{M}, \tilde{g})$ where $\widetilde{M}$ stands for a "supermanifold" with "superRiemannian" metric $\tilde{g}$. This Getzler's interpretation is very attractive but it is only asserted in the introduction of his paper and is left unfixed mathematically. From our point of view, we try (but not yet completed) to prove

$$\mathrm{str}\ \exp\left[\frac{t}{2}\widetilde{\Delta}_{\tilde{g}}\right] \sim \int_{\tilde{M}} d_{\tilde{g}}Q\,\mathcal{E}(t,Q,Q) = \text{topological index} \sim \text{analytical index},$$

directly by constructing a parametrix $\mathcal{E}(t,\bar{Q},\underline{Q})$ of $\dfrac{\partial}{\partial t} - \dfrac{1}{2}\widetilde{\Delta}_{\tilde{g}}$ where $\widetilde{\Delta}_{\tilde{g}}$ is the supersymmetric extension of Δ_g. Here, str stands for super-trace, $\bar{Q} = (\bar{x},\bar{\theta})$, $\underline{Q} = (\underline{x},\underline{\pi})$,

$$\mathcal{E}(t,\bar{Q},\underline{Q}) \sim \mathcal{A}(t,\bar{Q},\underline{Q}) \exp\left[-\mathcal{S}(t,\bar{Q},\underline{Q})\right],$$

with $\mathcal{S}(t,\bar{Q},\underline{Q})$ is a solution of the Hamilton-Jacobi equation, and $\mathcal{A}(t,\bar{Q},\underline{Q})$ is a solution of the continuity equation on the supermanifold $\widetilde{M}$. See, appendix as the simplest realization of these.

1.2. *Things to be fixed*

1.2.1. *What is susyQM?*

What is supersymmetric quantum mechanics(=susyQM)?

1.2.2. *What is quantization?*

How to associate a "classical mechanics" to a given susyQM and how to recover it from that classical mechanics? This procedure is sometimes called"quantization".

1.2.3. *Necessity for Superanalysis*

In order to get "classical mechanics" corresponding to "system of PDEs", we need analytical tools based on Grassmann Algebra [a]. This analysis is called Superanalysis(=Analysis treating boson and fermion on equal footing, so said by Berezin-Marinov). Moreover, in order to get superextended Hamiltonian from the classical scalar one, we need superspace. In our case, usage of Fréchet-Grassmann Algebra based on countable Grassmann generators, is necessary which should be compared with the analysis based on Banach-Grassmann Algebra.

[a] Here, we use the fact that Clifford algebra has a representation on Grassmann algebra and that every matrix is decomposed by Clifford algebra

1.2.4. *What is explained in this paper?*

Preparing superanalysis with countably infinite Grassmann generators from scratch, we explain the classical mechanics corresponding to the system of PDEs, that is, the Hamilton flow for Witten's deformed Laplacian (cf. Inoue[20]). We mention the method for extending supersymmetrically the classical mechanics with Hamiltonian $H(x,\xi)$ to $\mathcal{H}(x,\theta,\xi,\pi)$. In the appendix, taking the harmonic oscillator as the simplest one, we explain the quantization by heat and Schrödinger from classical mechanics. Moreover, after extending it supersymmetrically, we calculate its action and the kernel of the super heat operator having supertrace with value 1, whose interpretation is not clear for the time being.

2. Witten's deformed Laplacian

2.1. *susyQM*

Following description is taken from Cycon et al. [7]:

Definition 2.1. Let $\mathbb{H}$ be a Hilbert space and let $\mathbf{H}$ and $\mathbf{Q}$ be selfadjoint operators, and $\mathbf{P}$ be a bounded self-adjoint operator on $\mathbb{H}$ such that

$$\begin{gathered}\mathbf{H}=\mathbf{Q}^2\geq 0,\quad \mathbf{P}^2=I,\\ [\mathbf{Q},\mathbf{P}]_+=\mathbf{QP}+\mathbf{PQ}=0.\end{gathered}\tag{1}$$

Then, we say that the system $(\mathbf{H},\mathbf{P},\mathbf{Q})$ has supersymmetry or it defines a susyQM. $\mathbf{Q}$ is called supercharge of $\mathbf{H}$.

Under this circumstance, we may decompose

$$\begin{gathered}\mathbb{H}=\mathbb{H}_{\mathrm{b}}\oplus\mathbb{H}_{\mathrm{f}},\\ \mathbb{H}_{\mathrm{f}}=\{u\in\mathbb{H}\,|\,\mathbf{P}u=-u\},\quad \mathbb{H}_{\mathrm{b}}=\{u\in\mathbb{H}\,|\,\mathbf{P}u=u\}.\end{gathered}$$

Using this decomposition and identifying an element $u=u_{\mathrm{b}}+u_{\mathrm{f}}\in\mathbb{H}$ as a vector $\begin{pmatrix}u_{\mathrm{b}}\\ u_{\mathrm{f}}\end{pmatrix}$, we have a representation

$$\mathbf{P}=\begin{pmatrix}I_{\mathrm{b}} & 0\\ 0 & -I_{\mathrm{f}}\end{pmatrix}=\begin{pmatrix}1 & 0\\ 0 & -1\end{pmatrix}.$$

Since $\mathbf{P}$ and $\mathbf{Q}$ anti-commute and $\mathbf{Q}$ is self-adjoint, $\mathbf{Q}$ has always the form

$$\mathbf{Q}=\begin{pmatrix}0 & A^*\\ A & 0\end{pmatrix},\ \mathbf{H}=\begin{pmatrix}A^*A & 0\\ 0 & AA^*\end{pmatrix},$$

where A, called the annihilation operator, is an operator which maps $\mathbb{H}_b$ into $\mathbb{H}_f$, and its adjoint A^*, called the creation operator, maps $\mathbb{H}_f$ into $\mathbb{H}_b$. Thus, $\mathbf{P}$ commutes with $\mathbf{H}$, and $\mathbb{H}_b$ and $\mathbb{H}_f$ are invariant under $\mathbf{H}$, i.e. $\mathbf{H}\mathbb{H}_b \subset \mathbb{H}_b$ and $\mathbf{H}\mathbb{H}_f \subset \mathbb{H}_f$. That is, there is a one-to-one correspondence between densely defined closed operators A and self-adjoint operators $\mathbf{Q}$ (supercharges) of the above form.

Definition 2.2. We define a supersymmetric index of H if it exists by

$$\mathrm{ind}_s(H) \equiv \dim(\mathrm{Ker}\,(H|\mathbb{H}_b)) - \dim(\mathrm{Ker}\,(H|\mathbb{H}_f)) \in \bar{\mathbb{Z}} = \mathbb{Z} \cup \{\pm\infty\}.$$

On the other hand, we give

Definition 2.3. Let X, Y be two Banach spaces and let $\mathcal{C}(X,Y)$ be a set of densely defined closed operators from X to Y. $T \in \mathcal{C}(X,Y)$ is called Fredholm iff the range of T, $R(T)$, is closed in Y and both $\ker T$ and $Y/R(T)$ are finite-dimensional. $T \in \mathcal{C}(X,Y)$ is called semi-Fredholm iff $R(T)$ is closed in Y and at least one of Ker T and $Y/R(T)$ is finite-dimensional. If the operator is semi-Fredholm, then the Fredholm index $\mathrm{ind}_F(T) = \dim(\mathrm{Ker}\ T) - \dim(Y/R(T))$ exists in $\bar{\mathbb{Z}}$.

If the operator A is semi-Fredholm, we have the relation

$$\mathrm{ind}_s(\mathbf{H}) = \mathrm{ind}_F(A) \equiv \dim(\mathrm{Ker}\ A) - \dim(\mathrm{Ker}\ A^*).$$

In order to check whether the supersymmetry is broken or unbroken, E. Witten introduced the so-called Witten index.

Definition 2.4. Let $(\mathbf{H}, \mathbf{P}, \mathbf{Q})$ be susyQM with (1). Putting, for $t > 0$

$$\Delta_t(\mathbf{H}) = \mathrm{tr}\,(e^{-tA^*A} - e^{-tAA^*}) = \mathrm{str}\ e^{-t\mathbf{H}},$$

we define, if the limit exists, the (heat kernel regulated) Witten index W_H of $(\mathbf{H}, \mathbf{P}, \mathbf{Q})$ by

$$W_H = \lim_{t\to\infty} \Delta_t(\mathbf{H}).$$

We define also the (heat kernel regulated) axial anomaly $\mathcal{A}_H$ of $(\mathbf{H}, \mathbf{P}, \mathbf{Q})$ by

$$\mathcal{A}_H = \lim_{t\to 0} \Delta_t(\mathbf{H}).$$

We have

Theorem 2.1. *Let* $\mathbf{Q}$ *be a supercharge of* $\mathbf{H}$. *If* $\exp(-t\mathbf{Q}^2)$ *is trace class for some* $t > 0$, *then* $\mathbf{Q}$ *is Fredholm and*

$$\mathrm{ind}_t(\mathbf{Q})(\textit{independent of } t) = \mathrm{ind}_F(\mathbf{Q}) = \mathrm{ind}_s(\mathbf{H}).$$

2.2. *An example*

Let (M, g), $g = \sum_{i,j=1}^{d} g_{ij}(q)dq^i dq^j$ be a d-dimensional smooth Riemannian manifold. We put

$$\Lambda(M) = \bigoplus_{k=0}^{d} \Lambda^k(M), \quad \Lambda_0(M) = \bigoplus_{k=0}^{d} \Lambda_0^k(M),$$

where

$$\begin{aligned}\Lambda^k(M) =& \{\sum_{|I|=k} \omega_I dq^I \mid \omega_I \in C^\infty(M : \mathbb{C})\},\\ \Lambda_0^k(M) =& \{\omega \in \Lambda^k(M) \mid \omega_I \in C_0^\infty(M : \mathbb{C})\},\\ \tilde{\Lambda}^k(M) =& \{\omega \in \Lambda^k(M) \mid \|\omega\| < \infty\},\end{aligned}$$

with

$$\begin{aligned}I = (i_1 \cdots i_k),\ 1 \le i_1 < \cdots < i_k \le m,\\ dq^I = dq^{i_1} \wedge \cdots \wedge dq^{i_k},\ |I| = k.\end{aligned}$$

Let d be an exterior differential acting on $\omega_I(q)dq^I \in \Lambda^k(M)$ as

$$d\omega = \sum_{j=1}^{d} \frac{\partial \omega_I(q)}{\partial q^j} dq^j \wedge dq^I.$$

$\mathbf{P}$ is defined by $\mathbf{P}\omega = (-1)^k \omega$ for $\omega \in \Lambda^k(M)$.

Put $\mathbb{H} = \overline{\Lambda(M)}$ where $\overline{\Lambda(M)} = \cup_{k=0}^{d} \overline{\Lambda^k(M)}$ with $\overline{\Lambda^k(M)}$ is the closure of $\tilde{\Lambda}^k(M)$ in L^2-norm $\|\cdot\|$. Denoting the adjoint of d in $\overline{\Lambda(M)}$ by d^* and putting

$$\begin{aligned}\mathbf{Q}_1 = d + d^*, \quad \mathbf{Q}_2 = i(d - d^*),\\ \mathbf{H} = \mathbf{Q}_1^2 = \mathbf{Q}_2^2 = dd^* + d^* d,\end{aligned}$$

we have that $(\mathbf{H}, \mathbf{Q}_\alpha, \mathbf{P})$ has the supersymmetry on $\mathbb{H}$ for each $\alpha = 1, 2$.

2.3. *Witten's deformed Laplacian*

For any real-valued function ϕ on M, we put

$$d_\lambda = e^{-\lambda\phi} d e^{\lambda\phi}, \quad d_\lambda^* = e^{\lambda\phi} d^* e^{-\lambda\phi}$$

where λ is a real parameter. We have $d_\lambda^2 = 0 = {d_\lambda^*}^2$.

$$\begin{aligned}\mathbf{Q}_{1\lambda} = d_\lambda + d_\lambda^*, \quad \mathbf{Q}_{2\lambda} = i(d_\lambda - d_\lambda^*),\\ \mathbf{H}_\lambda = d_\lambda d_\lambda^* + d_\lambda^* d_\lambda.\end{aligned}$$

Defining $\mathbf{P}$ as before, we have the supersymmetric system $(\mathbf{H}_\lambda, \mathbf{Q}_{\alpha\lambda}, \mathbf{P})$ on $\mathbb{H}$ for each $\alpha = 1, 2$.

We calculate $\mathbf{H}_\lambda$ more explicitly:

$$\mathbf{H}_\lambda = dd^* + d^*d + \lambda^2(d\phi)^2 + \sum_{i,j=1}^{d} \lambda \frac{D^2\phi}{Dq^iDq^j}[a^{i*}, a^j]_-$$

Here, the annihilation and creation operators a^j and a^{j*}, respectively, are defined as follows: For any $0 \le \ell \le d$ and $q \in M$,

$$\begin{cases} a_q^{i*}dq^J = dq^i \wedge dq^J, \\ a_q^i dq^J = \displaystyle\sum_{k=1}^{\ell}(-1)^k g^{ij_k}(q)dq^{\check{J}_k} \end{cases}$$

where

$$J = (j_1, \cdots, j_\ell),\ \check{J}_k = (j_1, \cdots, j_{k-1}, j_{k+1}, \cdots, j_\ell).$$

Then, these give mappings from

$$\Lambda(T^*M) \to \Lambda(T^*M),$$

and we get

$$[a_q^i, a_q^j][a_q^i, a_q^j]_+ = 0, \quad [a_q^i, a_q^{j*}]_+ = g^{ij}(q), \quad [a_q^{i*}, a_q^{j*}]_+ = 0.$$

Moreover,

$$(d\phi)^2 = g^{ij}\frac{\partial\phi}{\partial q^i}\frac{\partial\phi}{\partial q^j}, \quad \frac{D^2\phi}{Dq^iDq^j} = \nabla_i\nabla_j\phi,$$
$$\frac{D\psi^i}{Dt} = \frac{d\psi^i}{dt} - \Gamma^i_{jk}\dot{q}^j\psi^k.$$

[Notation]: For $\omega = \sum_I \omega_I(q)dq^I$, we put

$$(\nabla_j\omega)_I = \frac{\partial\omega_I}{\partial q^j} - \sum_{r,l}\Gamma^l_{ji_r}\omega_{I^r_l}.$$

$I^r_l = (i_1, \cdots, i_{r-1}, l, i_{r+1}, \cdots, i_k)$

Then, we have

$$\nabla_j = \partial_{q^j} - \sum_{k,l,m}\Gamma^k_{jl}g_{km}a^{l*}a^m,$$

$$d = \sum_{l=1}^{d} a^{l*}\nabla_l = -\sum_{l=1}^{d}\nabla_l^* a^{l*}, \quad d^* = -\sum_{l=1}^{d} a^l\nabla_l,$$

$$d^*\omega = \sum_{j,r}(-1)^{r-1} g^{ji_r}\left[\frac{\partial \omega_I(q)}{\partial q^j} - \sum_{l,s} \Gamma^l_{j\,i_s}\omega_{I^l_s}(q)\right] dq^{\check{I}_r}.$$

What we should clarify in the forthcoming papers, are the following two things:

(1) The most important thing is to **consider the operator $\mathbf{H}_\lambda$ as the quantized one from the action**[b]

$$\begin{aligned}\mathcal{L}_\lambda =& \frac{1}{2}\int dt \left[g_{ij}\left(\frac{dq^i}{dt}\frac{dq^j}{dt} + i\bar{\psi}^i \frac{D\psi^j}{Dt}\right)\right. \\ &\left. + \frac{1}{4}R_{ijkl}\bar{\psi}^i\psi^k\bar{\psi}^j\psi^l - \lambda^2 g^{ij}\frac{\partial\phi}{\partial q^i}\frac{\partial\phi}{\partial q^j} - \lambda\frac{D^2\phi}{Dq^iDq^j}\bar{\psi}^i\psi^j\right].\end{aligned} \tag{2}$$

Here, we used the summation convention and ψ^i and $\bar{\psi}^i$ are anti-commuting fields tangent to M, which becomes **the creation and annihilation operators after quantization.**

Remark. This problem will be solved if we may introduce the super heat operator $\partial_t - \mathbf{H}_\lambda$ from the classical quantity derived from $\mathcal{L}_\lambda$. We need to define supermanifold, super-diffeomorphism, super-pseudodifferential calculus. Some part of this project is under construction from our point of view.

(2) How to recognize **the effect of instanton or tunneling paths** corresponding to $\mathcal{L}_\lambda$ to the quantum mechanics governed by the above Lagrangian? Here, tunneling paths satisfy the classical mechanics defined by

$$\begin{aligned}\bar{\mathcal{L}}_\lambda &= \frac{1}{2}\int dt\left(g_{ij}\frac{dq^i}{dt}\frac{dq^j}{dt} + \lambda^2 g^{ij}\frac{\partial\phi}{\partial q^i}\frac{\partial\phi}{\partial q^j}\right) \\ &= \frac{1}{2}\int dt \left|\frac{dq^i}{dt} \pm \lambda g^{ij}\frac{\partial\phi}{\partial q^j}\right|^2 \mp \lambda\int dt\,\frac{d\phi}{dt}.\end{aligned}$$

Remark. Though physicists may use freely Feynman's integral and Laplace method under Feynmann measure, these are not admitted mathematically. In Witten's work, he used Feynman's quantization method using path integral and apply to it Laplace's method for asymptotic behavior with respect to a parameter, which leads to count the tunneling effect rather directly.

---

[b]in physics, they call this as Lagrangian, but in mathematics, we call the integrand as Lagrangian

3. Main ingredients

3.1. *Fundamentals of superanalysis*

[**Superspace**] For letters $\{\sigma_j\}_{j=1}^{\infty}$ satisfying the Grassmann relation

$$\sigma_j\sigma_k + \sigma_k\sigma_j = 0,\ j,k = 1,2,\cdots,$$

we put

$$\begin{cases} \mathfrak{C} = \{X = \sum_{I\in\mathcal{I}} X_I\sigma^I \mid X_I \in \mathbb{C}\}, \quad \mathfrak{C} \cong \mathfrak{C}_{\rm ev} \oplus \mathfrak{C}_{\rm od} \cong \mathfrak{C}_{\rm ev} \times \mathfrak{C}_{\rm od}, \\ \mathfrak{C}_{\rm ev} = \{X = \sum_{I\in\mathcal{I},|I|={\rm even}} X_I\sigma^I \mid X_I \in \mathbb{C}\}, \\ \mathfrak{C}_{\rm od} = \{X = \sum_{I\in\mathcal{I},|I|={\rm odd}} X_I\sigma^I \mid X_I \in \mathbb{C}\}, \end{cases}$$

where

$$\mathcal{I} = \{I = (i_k) \in \{0,1\}^{\mathbf{N}} \mid |I| = \sum_k i_k < \infty\},$$

$$\sigma^I = \sigma_1^{i_1}\sigma_2^{i_2}\cdots,\ I = (i_1, i_2, \cdots),\ \tilde{0} = (0,0,\cdots).$$

Besides trivially defined linear operations of sums and scalar multiplications, we have a product operation in $\mathfrak{C}$: For $X = \sum_{J\in\mathcal{J}} X_J\sigma^J$, $Y = \sum_{K\in\mathcal{I}} Y_K\sigma^K$, we put $XY = \sum_{I\in\mathcal{I}} (XY)_I\sigma^I$ with

$$(XY)_I = \sum_{I=J+K} (-1)^{\tau(I;J,K)} X_J Y_K.$$

Here, $\tau(I;J,K)$ is an integer defined by

$$\sigma^J\sigma^K = (-1)^{\tau(I;J,K)}\sigma^I, \quad I = J+K.$$

Proposition 3.1. $\mathfrak{C}$ *forms a* ∞*-dimensional Fréchet-Grassmann algebra over* $\mathbb{C}$*, that is, an associative, distributive and non-commutative ring with degree, which is endowed with the Fréchet topology.*

Remarks. (0) Degree in $\mathfrak{C}$ *is defined by introducing subspaces*

$$\mathfrak{C}_{[j]} = \{X = \sum_{I\in\mathcal{I},|I|=j} X_I\sigma^I\} \text{ for } j = 0,1,\cdots$$

which satisfy

$$\mathfrak{C} = \oplus_{j=0}^{\infty}\mathfrak{C}_{[j]}, \quad \mathfrak{C}_{[j]}\cdot\mathfrak{C}_{[k]} \subset \mathfrak{C}_{[j+k]}.$$

(1) Define $\mathrm{proj}_I(X) = X_I$ *for* $X = \sum_{I\in\mathcal{I}} X_I\sigma^I \in \mathfrak{C}$. *The topology in* $\mathfrak{C}$ *is given by;* $X \to 0$ *in* $\mathfrak{C}$ *iff for any* $I \in \mathcal{I}$, $\mathrm{proj}_I(X) \to 0$ *in* $\mathbb{C}$.

This topology is equivalent to the one introduced by the metric $\mathrm{dist}(X, Y) = \mathrm{dist}(X - Y)$ *where* $\mathrm{dist}(X)$ *is defined by*

$$\mathrm{dist}(X) = \sum_{I\in\mathcal{I}} \frac{1}{2^{r(I)}} \frac{|\,\mathrm{proj}_I(X)|}{1 + |\,\mathrm{proj}_I(X)|}$$

with

$$r(I) = 1 + \frac{1}{2}\sum_{k=1}^{\infty} 2^k i_k \textit{ for } I \in \mathcal{I}.$$

(2) We introduce parity in $\mathfrak{C}$ *by setting*

$$p(X) = \begin{cases} 0 & \textit{if } X = \sum_{I\in\mathcal{I},|I|=\mathrm{even}} X_I\sigma^I, \\ 1 & \textit{if } X = \sum_{I\in\mathcal{I},|I|=\mathrm{odd}} X_I\sigma^I, \\ \textit{undefined} & \textit{if otherwise.} \end{cases}$$

$X \in \mathfrak{C}$ *is called homogeneous if it satisfies* $p(X) = 0$ *or* $= 1$. *We put*

$$\begin{cases} \mathfrak{C}_{\mathrm{ev}} = \{X \in \mathfrak{C} \,|\, p(X) = 0\}, \\ \mathfrak{C}_{\mathrm{od}} = \{X \in \mathfrak{C} \,|\, p(X) = 1\}, \\ \mathfrak{C} \cong \mathfrak{C}_{\mathrm{ev}} \oplus \mathfrak{C}_{\mathrm{od}} \cong \mathfrak{C}_{\mathrm{ev}} \times \mathfrak{C}_{\mathrm{od}}. \end{cases}$$

Analogous to $\mathfrak{C}$, *we define*

$$\begin{cases} \mathfrak{R} = \{X \in \mathfrak{C} \,|\, \pi_{\mathrm{B}}X \in \mathbb{R}\}, \\ \mathfrak{R}_{\mathrm{ev}} = \mathfrak{R} \cap \mathfrak{C}_{\mathrm{ev}}, \quad \mathfrak{R}_{\mathrm{od}} = \mathfrak{R} \cap \mathfrak{C}_{\mathrm{od}} = \mathfrak{C}_{\mathrm{od}}, \\ \mathfrak{R} \cong \mathfrak{R}_{\mathrm{ev}} \oplus \mathfrak{R}_{\mathrm{od}} \cong \mathfrak{R}_{\mathrm{ev}} \times \mathfrak{R}_{\mathrm{od}}. \end{cases}$$

We introduced the body (projection) map π_{B} *by*

$$\pi_{\mathrm{B}}X = \mathrm{proj}_{\tilde{0}}(X) = X_{\tilde{0}} = X_{\mathrm{B}} \textit{ for any } X \in \mathfrak{C}.$$

Introducing the body (projection) map π_{B} *by*

$$\pi_{\mathrm{B}}X = \mathrm{proj}_{\tilde{0}}(X) = X_{\tilde{0}} = X_{\mathrm{B}} \textit{ for any } X \in \mathfrak{C},$$

we put

$$\begin{cases} \mathfrak{R} = \{X \in \mathfrak{C} \,|\, \pi_{\mathrm{B}}X \in \mathbb{R}\}, \\ \mathfrak{R}_{\mathrm{ev}} = \mathfrak{R} \cap \mathfrak{C}_{\mathrm{ev}}, \quad \mathfrak{R}_{\mathrm{od}} = \mathfrak{R} \cap \mathfrak{C}_{\mathrm{od}} = \mathfrak{C}_{\mathrm{od}}, \\ \mathfrak{R} \cong \mathfrak{R}_{\mathrm{ev}} \oplus \mathfrak{R}_{\mathrm{od}} \cong \mathfrak{R}_{\mathrm{ev}} \times \mathfrak{R}_{\mathrm{od}}. \end{cases}$$

We define the (real) superspace $\mathfrak{R}^{m|n}$ by

$$\mathfrak{R}^{m|n} = \mathfrak{R}^m_{\mathrm{ev}} \times \mathfrak{R}^n_{\mathrm{od}} \ni (x,\theta) = X,$$

with $x = (x_1, \cdots, x_m)$, $\theta = (\theta_1, \cdots, \theta_n)$ where $x_j \in \mathfrak{R}_{\mathrm{ev}}$, $\theta_k \in \mathfrak{R}_{\mathrm{od}}$.

[**Supersmooth functions**] For $f(q) \in C^\infty(\mathbb{R}^m : \mathbb{C})$, we put, for $x = x_{\mathrm{B}} + x_{\mathrm{S}}$,

$$f(x) = \sum_{|\alpha|=0}^{\infty} \frac{1}{\alpha!} \partial_q^\alpha f(x_{\mathrm{B}}) x_{\mathrm{S}}^\alpha,$$

which is called the Grassmann continuation of $f(q)$. We define a class of supersmooth functions $\mathcal{C}_{SS}(\mathfrak{R}^{m|n})$ by

$$u(X) = u(x,\theta) = \sum_{|a|\le n} u_a(x)\theta^a,$$

with $u_a(x)$ is the Grassmann continuation of $u_a(q) \in C^\infty(\mathbb{R}^m : \mathbb{C})$.
[**Derivations**] For a given supersmooth function $u(X)$ on $\mathfrak{R}^{m|n}$, we define its derivatives as follows: For $j = 1, 2, \cdots, m$ and $k = 1, 2, \cdots, n$, we put

$$\begin{cases} U_j(X) = \sum_{|a|\le n} \partial_{x_j} u_a(x)\theta^a, \\ U_{k+m}(X) = \sum_{|a|\le n} (-1)^{l_k(a)} u_a(x) \theta_1^{a_1} \cdots \theta_k^{a_k - 1} \cdots \theta_n^{a_n} \end{cases}$$

where $l_k(a) = \sum_{j=1}^{k-1} a_j$ and $\theta_k^{-1} = 0$. $U_\kappa(X)$ are called the partial derivatives of u with respect to X_κ at $X = (x,\theta)$ and are denoted by

$$U_j(X) = \frac{\partial}{\partial x_j} u(x,\theta) = \partial_{x_j} u(x,\theta) \text{ for } j = 1, 2, \cdots, m,$$

$$U_{m+s}(X) = \frac{\partial}{\partial \theta_s} u(x,\theta) = \partial_{\theta_s} u(x,\theta) \text{ for } s = 1, 2, \cdots, n$$

or simply by

$$U_\kappa(X) = \partial_{X_\kappa} u(X) \text{ for } \kappa = 1, \cdots, m+n.$$

For

$$\mathfrak{a} = (\alpha, a),\ \ \alpha = (\alpha_1, \cdots, \alpha_m) \in \mathbf{N}^m,\ \ a = (a_1, \cdots, a_n) \in \{0,1\}^n,$$

$$|\alpha| = \sum_{j=1}^m \alpha_j,\ \ |a| = \sum_{k=1}^n a_k,\ \ |\mathfrak{a}| = |\alpha| + |a|,$$

we put

$$\partial_X^{\mathfrak{a}} = \partial_x^{\alpha}\partial_\theta^{a} \text{ with } \partial_x^{\alpha} = \partial_{x_1}^{\alpha_1}\cdots\partial_{x_n}^{\alpha_m}, \qquad \partial_\theta^{a} = \partial_{\theta_1}^{a_1}\cdots\partial_{\theta_n}^{a_n}.$$

Though Taylor expansion and Implicit function theorem hold for super-smooth functions, we don't mention them here.

[**Integration**] Formally

$$\begin{aligned}\int_{\mathfrak{R}^{m|n}} dx d\theta\, u(x,\theta) &= \int_{\mathfrak{R}^{m|0}} dx \left\{\int_{\mathfrak{R}^{0|n}} d\theta\, u(x,\theta)\right\} \\ &= \int_{\mathfrak{R}^{0|n}} d\theta \left\{\int_{\mathfrak{R}^{m|0}} dx\, u(x,\theta)\right\} = \int_{\mathfrak{R}^{m|n}} d\theta dx\, u(x,\theta).\end{aligned}$$

Especially, the integration w.r.t. odd variables, there exists a well-known formula (Berezin integral):

$$\int_{\mathfrak{R}^{0|n}} d\theta_n\cdots d\theta_1\, \theta_1\cdots\theta_n = 1, \quad \int_{\mathfrak{R}^{0|n}} d\theta_n\cdots d\theta_1\, 1 = 0.$$

[**Fourier transform**] Putting

$$\langle x|\xi\rangle = \sum_{j=1}^{m} x_j\xi_j,\ \langle\theta|\pi\rangle = \sum_{k=1}^{n}\theta_k\pi_k,\ \iota_n = e^{-\frac{\pi i}{4}n(n-2)},$$

we define

$$\begin{aligned}(F_e v)(\xi) &= (2\pi\hbar)^{-m/2}\int_{\mathfrak{R}^{m|0}} dx\, e^{-i\hbar^{-1}\langle x|\xi\rangle} v(x), \\ (\bar{F}_e w)(x) &= (2\pi\hbar)^{-m/2}\int_{\mathfrak{R}^{m|0}} d\xi\, e^{i\hbar^{-1}\langle x|\xi\rangle} w(\xi), \\ (F_o v)(\pi) &= \hbar^{n/2}\iota_n \int_{\mathfrak{R}^{0|n}} d\theta\, e^{-i\hbar^{-1}\langle\theta|\pi\rangle} v(\theta), \\ (\bar{F}_o w)(\theta) &= \hbar^{n/2}\iota_n \int_{\mathfrak{R}^{0|n}} d\pi\, e^{i\hbar^{-1}\langle\theta|\pi\rangle} w(\pi).\end{aligned}$$

Putting also,

$$\langle X|\Xi\rangle = \langle x|\xi\rangle + \langle\theta|\pi\rangle \in \mathfrak{R}_{\mathrm{ev}},\ c_{m,n} = (2\pi\hbar)^{-m/2}\hbar^{n/2}\iota_n,$$

we rewrite

$$\begin{aligned}(\mathcal{F}u)(\xi,\pi) &= c_{m,n}\int_{\mathfrak{R}^{m|n}} dX\, e^{-i\hbar^{-1}\langle X|\Xi\rangle} u(X) = \sum_a [(F_e u_a)(\xi)][(F_o\theta^a)(\pi)], \\ (\bar{\mathcal{F}}v)(x,\theta) &= c_{m,n}\int_{\mathfrak{R}^{m|n}} d\Xi\, e^{i\hbar^{-1}\langle X|\Xi\rangle} v(\Xi) = \sum_a [(\bar{F}_e v_a)(x)][(\bar{F}_o\pi^a)(\theta)].\end{aligned}$$

Remark. There are so many super-papers using Banach-Grassmann algebras. It is because that there is no implicit function theorems in general Fréchet spaces. We have the implicit function theorems on our superspace based on Fréchet space with grading, like ILH(inverse limit of Hilbert spaces).

3.2. *New look at matrix structure*

The fact "Clifford algebra is represented on Grassmann algebra" is embodied as follows (here, we restrict ourselves to 2×2-matrices):

We decompose a matrix $\mathbf{A} = \begin{pmatrix} a & b \\ c & d \end{pmatrix}$ as

$$\begin{aligned}\mathbf{A} &= \frac{a+d}{2}\begin{pmatrix} 1 & 0 \\ 0 & 1 \end{pmatrix} + \frac{a-d}{2}\begin{pmatrix} 1 & 0 \\ 0 & -1 \end{pmatrix} + \frac{b+c}{2}\begin{pmatrix} 0 & 1 \\ 1 & 0 \end{pmatrix} + \frac{b-c}{2}\begin{pmatrix} 0 & 1 \\ -1 & 0 \end{pmatrix} \\ &= \frac{a+d}{2}\mathbf{I}_2 + \frac{a-d}{2}\boldsymbol{\sigma}_3 + \frac{b+c}{2}\boldsymbol{\sigma}_1 + \frac{b-c}{2}i\boldsymbol{\sigma}_2.\end{aligned}$$

We identify a vector ${}^t(u_0, u_1)$ with a function $u(\theta) = u_0 + u_1\theta_1\theta_2$, then, for example,

$$\Big(\theta_1\theta_2 - \frac{\partial^2}{\partial\theta_1\partial\theta_2}\Big)u(\theta) = u_1 + u_0\theta_1\theta_2.$$

This means a vector ${}^t(u_0, u_1)$ is transformed to ${}^t(u_1, u_0)$, that is, the operation of matrix $\boldsymbol{\sigma}_1$ appears. Therefore, the action of $\mathbf{A}$ to a vector ${}^t(u_0, u_1)$ is regarded as the action of a differential operator

$$A\Big(\theta, \frac{\partial}{\partial\theta}\Big) = \frac{a+d}{2} + \frac{a-d}{2}\Big(1 - \theta_1\frac{\partial}{\partial\theta_1} - \theta_2\frac{\partial}{\partial\theta_2}\Big) + c\theta_1\theta_2 - b\frac{\partial^2}{\partial\theta_1\partial\theta_2}$$

to a function $u(\theta) = u_0 + u_1\theta_1\theta_2$.

Remarks. (1) Correspondences between a vector and a function, are defined by

$$\sharp\begin{pmatrix} u_0 \\ u_1 \end{pmatrix} = u_0 + u_1\theta_1\theta_2 = u(\theta), \qquad \flat u(\theta) = \begin{pmatrix} v_0 \\ v_1 \end{pmatrix} \tag{3}$$

where

$$v_0 = u(\theta)\big|_{\theta_1=\theta_2=0}, \quad v_1 = \frac{\partial^2 u(\theta)}{\partial\theta_2\partial\theta_1}\Big|_{\theta_1=\theta_2=0} = \int_{\mathfrak{R}^{0|2}} u(\theta)d\theta.$$

Therefore,

$$\sharp\begin{pmatrix} a & b \\ c & d \end{pmatrix}\begin{pmatrix} u_0 \\ u_1 \end{pmatrix} = \sharp\begin{pmatrix} a & b \\ c & d \end{pmatrix}\flat\cdot\sharp\begin{pmatrix} u_0 \\ u_1 \end{pmatrix}$$
$$= \left[\frac{a+d}{2} + \frac{a-d}{2}\left(1 - \theta_1\frac{\partial}{\partial\theta_1} - \theta_2\frac{\partial}{\partial\theta_2}\right) + c\theta_1\theta_2 - b\frac{\partial^2}{\partial\theta_1\partial\theta_2}\right](u_0 + u_1\theta_1\theta_2).$$

(2) Though not only $-i(1 - 2\theta_1\partial_{\theta_1})$ but also $-i(1 - 2\theta_2\partial_{\theta_2})$ yields the operator $\sharp\boldsymbol{\sigma}_3\flat$ when it acts on $u_0 + u_1\theta_1\theta_2$, we take the average of these following the belief that symmetric one should be more preferable. Moreover, under this choice, the operators $\sigma_j(\theta, \partial_\theta)$ $(j = 1, 2, 3)$ annihilate the set $\{v_1\theta_1 + v_2\theta_2 \mid v_1, v_2$ are independent of $\theta_1, \theta_2\}$.

(3) We may regard that

$$\theta_1,\ \theta_2 \sim dz_1,\ dz_2,$$
$$\theta_j(u_0 + u_1\theta_1\theta_2) \sim dz_j\wedge(u_0 + u_1 dz_1\wedge dz_2) = u_0 dz_j,$$
$$\frac{\partial}{\partial\theta_j}(u_0 + u_1\theta_1\theta_2) \sim \frac{d}{dz_j}\lfloor(u_0 + u_1 dz_1\wedge dz_2) = \begin{cases} u_1 dz_2 & \text{if } j = 1, \\ -u_1 dz_1 & \text{if } j = 2. \end{cases}$$

4. A simple example

We consider operators $\hat{p}$, $\hat{q}$, $\hat{\theta}$ and $\hat{\pi}$ satisfying

$$[\hat{p}, \hat{q}]_- = -i, \quad [\hat{\theta}, \hat{\pi}]_+ = -i \text{ with } \hat{\theta}^2 = 0,\ \hat{\pi}^2 = 0.$$

A representation of these operators are given by

$$\hat{q} = q\times,\ \hat{p} = \frac{1}{i}\frac{\partial}{\partial q} \text{ and } \hat{\theta} = \theta\times,\ \hat{\pi} = \frac{1}{i}\frac{\partial}{\partial\theta}$$

acting on functions $u(q, \theta) = u_0(q) + u_1(q)\theta$.

For selfadjoint operators H_1 and H_2, we put

$$\mathbf{H} = \begin{bmatrix} H_1 & 0 \\ 0 & H_2 \end{bmatrix}$$

and

$$\mathbf{Q} = \begin{bmatrix} 0 & 0 \\ A & 0 \end{bmatrix},\ \mathbf{Q}^\dagger = \begin{bmatrix} 0 & A^\dagger \\ 0 & 0 \end{bmatrix}$$

and

$$[\mathbf{H}, \mathbf{Q}]_- = [\mathbf{H}, \mathbf{Q}^\dagger]_- = 0,\ [\mathbf{Q}, \mathbf{Q}^\dagger]_+ = \mathbf{H},$$
$$[\mathbf{Q}, \mathbf{Q}]_+ = 0,\ [\mathbf{Q}^\dagger, \mathbf{Q}^\dagger]_+ = 0.$$

Here,

$$H_j = -\frac{\hbar^2}{2m}\frac{d^2}{dq^2} + V_j,\ j = 1, 2.$$

Putting

$$A = \frac{\hbar}{\sqrt{2m}}\frac{d}{dq} + W(q),\ A^\dagger = -\frac{\hbar}{\sqrt{2m}}\frac{d}{dq} + W(q),$$

we have

$$\begin{aligned} H_1 &= A^\dagger A, \\ V_1(q) &= W^2(q) - \frac{\hbar}{\sqrt{2m}} W'(q) \\ V_2(q) &= W^2(q) + \frac{\hbar}{\sqrt{2m}} W'(q). \end{aligned}$$

In matrix notation,

$$\begin{aligned} \mathbf{H} &= \Big(-\frac{\hbar^2}{2m}\frac{d^2}{dq^2} + W^2(q) \Big)\mathbf{I}_2 - \frac{\hbar}{\sqrt{2m}} W'(q)\boldsymbol{\sigma}_3, \\ \hat{\mathcal{H}} = \sharp\mathbf{H}\flat &= \Big(-\frac{\hbar^2}{2m}\frac{d^2}{dq^2} + W^2(q) \Big) + i\frac{\hbar}{\sqrt{2m}} W'(q)[\hat{\theta}, \hat{\pi}]_+ \end{aligned}$$

is called Witten model. In this case,

$$\mathbf{P} = (-1)^F = 1 - 2i\hat{\theta}\hat{\pi} \sim \boldsymbol{\sigma}_3.$$

5. Hamilton and Lagrange mechanics on superspace

5.1. *Superextension of a given Lagrangian*

For a given Lagrangian

$$L(q, \dot{q}) = \frac{1}{2} g_{ij}(q)\dot{q}^i\dot{q}^j + A_j(q)\dot{q}^j - V(q) \in C^\infty(T\mathbb{R}^d : \mathbb{R}),$$

using Legendre transformation, we associate a Hamiltonian

$$H(q, p) = \frac{1}{2} g^{ij}(q)(p_i - A_i(q))(p_j - A_j(q)) + V(q) \in C^\infty(T^*\mathbb{R}^d : \mathbb{R}). \quad (4)$$

To such a Hamiltonian, via extending formally that Lagrangian when

$$V(q) = (1/2)g^{ij}(q)W_{,i}(q)W_{,j}(q),$$

we may associate a supersymmetric extension

$$\begin{aligned}\mathcal{H}(x,\xi,\theta,\pi) = \frac{1}{2}g^{ij}\,(\xi_i - \frac{\sqrt{-1}}{2}(g_{ik,l}-g_{il,k})\theta^k\pi^l - A_i) \\ \times\,(\xi_j - \frac{\sqrt{-1}}{2}(g_{jm,n}-g_{jn,m})\theta^m\pi^n - A_j) \\ + \frac{1}{2}R_{ikjl}\theta^j\theta^l\pi^i\pi^k + \frac{1}{2}g^{jk}W_{,j}W_{,k} - W_{;ij}\theta^i\pi^j\end{aligned} \tag{5}$$

which belongs to $\mathcal{C}_{\mathrm{SS}}(\mathfrak{R}^{2d|2d}:\mathfrak{R}_{\mathrm{ev}})$. Here, the functions $g^{ij}=g^{ij}(x)$ of $x\in\mathfrak{R}^{d|0}$ etc. appeared above are Grassmann continuations of the corresponding ones $g^{ij}=g^{ij}(q)$ of $q\in\mathbb{R}^d$, etc.

Remark. It is not easy to say what part of the Hamiltonian of (5) *corresponds to*

$$-\lambda^2 g^{ij}\frac{\partial\phi}{\partial q^i}\frac{\partial\phi}{\partial q^j} - \lambda\frac{D^2\phi}{Dq^iDq^j}\bar{\psi}^i\psi^j,$$

the part of Lagrangian (2). *But, roughly speaking, for* $A_i=0$,

$$\frac{1}{2}g^{jk}W_{,j}W_{,k} - W_{;ij}\theta^i\pi^j$$

corresponding to above terms with $\lambda=1$.

Using the coordinate (x,ξ,θ,π), we consider the following symplectic 2-form on $\mathfrak{R}^{2d|2d}$

$$\Omega = \sum_{j=1}^{d} d\xi_j\wedge dx_j + \sum_{k=1}^{d} d\pi_k\vee d\theta_k.$$

Here, $\wedge$ and $\vee$ are anti-symmetric and symmetric tensor products, respectively, and $\mathfrak{R}^{2d|2d}$ is regarded as the cotangent space $\mathcal{T}^*\mathfrak{R}^{d|d}$ of $\mathfrak{R}^{d|d}$.

Then, associated with a Hamiltonian function $\mathcal{H}$, we define a Hamiltonian vector field $X_{\mathcal{H}}$ as follows

$$\Omega\rfloor X_{\mathcal{H}} = \vec{d}\mathcal{H}.$$

Using the integral curve of this vector field, we do a routine work as in the usual symplectic geometry.

5.2. *Super Hamiltonian mechanics*

We start this section by formulating the super classical mechanics defined by a super Hamiltonian function on $\mathfrak{R}^{2d|2d}$.

5.2.1. *Super Hamiltonian flows*

Let a function $\mathcal{H}(x,\xi,\theta,\pi)$ on $\mathfrak{R}^{2d|2d}$ be given by (5). For the sake of simplicity, we assume;

Assumption. $g_{ij}, A_i, W \in \mathcal{B}(\mathbb{R}^d)$.

We consider the following initial value problem:

$$\begin{cases} \dfrac{d}{dt}q(t) = \partial_p H(q(t),p(t)), \\ \dfrac{d}{dt}p(t) = -\partial_q H(q(t),p(t)) \end{cases} \tag{6}$$

with the initial condition at $t=0$ given by

$$(q(0),p(0)) = (\underline{q},\underline{p}) \in \mathbb{R}^d \times \mathbb{R}^d = T^*\mathbb{R}^d. \tag{7}$$

Our assumption gurantees that

Proposition 5.1. *For $T>0$ and $t \in [-T,T]$, there exists a unique solution of (6) with (7) which is C^∞ in $(t,\underline{q},\underline{p})$. Moreover, there exists a constant $\delta_0(T)>0$ with the following properties : If $|t|<\delta_0(T)$, there exist positive constants C_0 and $C^{(0)}_{\alpha,\beta}$ for $|\alpha|+|\beta|\geq 1$, depending on $(t,\underline{q},\underline{p})$ such that $\lim_{t\to 0} C_*(t,\underline{q},\underline{p}) = 0$ and*

$$\begin{cases} |q(t,\underline{q},\underline{p}) - \underline{q}| \leq C_0(t,\underline{q},\underline{p}), \\ |p(t,\underline{q},\underline{p}) - \underline{p}| \leq C_0(t,\underline{q},\underline{p}), \end{cases}$$

$$\begin{cases} |\partial_{\underline{q}}^\alpha \partial_{\underline{p}}^\beta (q(t,\underline{q},\underline{p}) - \underline{q})| \leq C^{(0)}_{\alpha,\beta}(t,\underline{q},\underline{p}), \\ |\partial_{\underline{q}}^\alpha \partial_{\underline{p}}^\beta (p(t,\underline{q},\underline{p}) - p)| \leq C^{(0)}_{\alpha,\beta}(t,\underline{q},\underline{p}), \end{cases}$$

$$\begin{cases} |(\partial_t q)(t,\underline{q},\underline{p}) + \partial_p H(\underline{q},\underline{p})| \leq C_0(t,\underline{q},\underline{p}), \\ |(\partial_t p)(t,\underline{q},\underline{p}) - \partial_q H(\underline{q},\underline{p})| \leq C_0(t,\underline{q},\underline{p}). \end{cases}$$

For $t \in [-T,T]$, we want to construct a solution $(x(t),\xi(t),\theta(t),\pi(t))$ of the super Hamiltonian equation given by

$$\begin{cases} \dfrac{d}{dt}x(t) = \ \ \partial_\xi \mathcal{H}(x(t),\xi(t),\theta(t),\pi(t)), \\ \dfrac{d}{dt}\xi(t) = -\partial_x \mathcal{H}(x(t),\xi(t),\theta(t),\pi(t)), \\ \dfrac{d}{dt}\theta(t) = -\partial_\pi \mathcal{H}(x(t),\xi(t),\theta(t),\pi(t)), \\ \dfrac{d}{dt}\pi(t) = -\partial_\theta \mathcal{H}(x(t),\xi(t),\theta(t),\pi(t)) \end{cases} \tag{8}$$

with the initial condition at $t = \underline{t}$ given by

$$(x(0), \xi(0), \theta(0), \pi(0)) = (\underline{x}, \underline{\xi}, \underline{\theta}, \underline{\pi}) \in \mathfrak{R}^{d|d} \times \mathfrak{R}^{d|d}. \tag{9}$$

We first observe the body part of (8).

Proposition 5.2. *Under Assumption, there exists a unique solution of* (8) *with* (9), *for any* $T > 0$.

Next, we investigate the differentiability of $(x(t), \xi(t), \theta(t), \pi(t))$ with respect to the initial data $(\underline{x}, \underline{\xi}, \underline{\theta}, \underline{\pi})$.

Proposition 5.3. *Fix* $T > 0$ *arbitrarily. Under Assumption, the solution* $(x(t), \xi(t), \theta(t), \pi(t))$ *of* (8) *is smooth in* $(t, \underline{x}, \underline{\xi}, \underline{\theta}, \underline{\pi})$.

Moreover, for $k = |\alpha| + |\beta| + |a| + |b| \geq 1$, *there exists* $1 \geq \delta_0(T) > 0$ *such that the following properties hold: If* $|t| \leq \delta_0(T)$, *there exist* C_1 *and* $C_k^{(0)}$ *such that* $\lim_{t\to 0} C_*^*(t, \underline{q}, \underline{p}) = 0$ *and*

$$|\pi_{\mathrm{B}}\big(\partial_{\underline{x}}^{\alpha}\partial_{\underline{\xi}}^{\beta}\partial_{\underline{\theta}}^{a}\partial_{\underline{\pi}}^{b}(x(t) - \underline{x})\big)| \leq C_k^{(0)}(t, \underline{q}, \underline{p}),$$

$$|\pi_{\mathrm{B}}\big(\partial_{\underline{x}}^{\alpha}\partial_{\underline{\xi}}^{\beta}\partial_{\underline{\theta}}^{a}\partial_{\underline{\pi}}^{b}(\xi(t) - \underline{\xi})\big)| \leq C_k^{(0)}(t, \underline{q}, \underline{p}),$$

for $|a| + |b|$ *=even. Analogously, when* $|a| + |b|$ *=odd, we have*

$$|\pi_{\mathrm{B}}\big(\partial_{\underline{x}}^{\alpha}\partial_{\underline{\xi}}^{\beta}\partial_{\underline{\theta}}^{a}\partial_{\underline{\pi}}^{b}(\theta(t) - \underline{\theta})\big)| \leq C_k^{(0)}(t, \underline{q}, \underline{p}),$$

$$|\pi_{\mathrm{B}}\big(\partial_{\underline{x}}^{\alpha}\partial_{\underline{\xi}}^{\beta}\partial_{\underline{\theta}}^{a}\partial_{\underline{\pi}}^{b}(\pi(t) - \underline{\pi})\big)| \leq C_k^{(0)}(t, \underline{q}, \underline{p}).$$

Moreover, for any fixed $t, \underline{\xi}, \underline{\pi}$, the mapping

$$\begin{pmatrix} \underline{x} \\ \underline{\theta} \end{pmatrix} \mapsto \begin{pmatrix} \bar{x} = x(t, \underline{x}, \underline{\xi}, \underline{\theta}, \underline{\pi}) \\ \bar{\theta} = \theta(t, \underline{x}, \underline{\xi}, \underline{\theta}, \underline{\pi}) \end{pmatrix}$$

is a C^∞-diffeomorphism in $\mathfrak{R}^{d|d}$. We write the inverse C^∞-diffeomorphism as

$$\begin{pmatrix} \bar{x} \\ \bar{\theta} \end{pmatrix} \mapsto \begin{pmatrix} \underline{x} = y(t, \bar{x}, \underline{\xi}, \bar{\theta}, \underline{\pi}) \\ \underline{\theta} = \omega(t, \bar{x}, \underline{\xi}, \bar{\theta}, \underline{\pi}) \end{pmatrix}.$$

This mapping is supersmooth in $(\bar{x}, \underline{\xi}, \bar{\theta}, \underline{\pi})$ for fixed t and has appropriate estimates.

5.3. *Action integral*

We construct the action integral along the Hamiltonian flow given above. First, we remark

Lemma 5.1. *Let $(x(t), \xi(t), \theta(t), \pi(t))$ be the Hamiltonian flow defined by (8). Then we have*

$$\frac{d}{dt}\mathcal{H}(x(t), \xi(t), \theta(t), \pi(t)) = 0.$$

Now, we define

$$\begin{aligned}\mathcal{S}_0(t) &= \mathcal{S}_0(t, \underline{x}, \underline{\xi}, \underline{\theta}, \underline{\pi}) \\ &= \langle \underline{x} | \underline{\xi} \rangle + \langle \underline{\theta} | \underline{\pi} \rangle + \int_0^t \mathcal{L}(x(\tau), \theta(\tau), \xi(\tau), \pi(\tau))\, d\tau\end{aligned}$$

where

$$\mathcal{L}(x, \xi, \theta, \pi) = \langle \partial_\xi \mathcal{H}(x, \xi, \theta, \pi) | \xi \rangle - \langle \partial_\pi \mathcal{H}(x, \xi, \theta, \pi) | \pi \rangle - \mathcal{H}(x, \xi, \theta, \pi).$$

Putting

$$\mathcal{S}(t, s, \bar{x}, \underline{\xi}, \bar{\theta}, \underline{\pi}) = \mathcal{S}_0(t, s, y(t, s, \bar{x}, \underline{\xi}, \bar{\theta}, \underline{\pi}), \underline{\xi}, \omega(t, s, \bar{x}, \underline{\xi}, \bar{\theta}, \underline{\pi}), \underline{\pi})),$$

we have:

Proposition 5.4. *(Hamilton-Jacobi equation) Let $|t - s| < \delta_1$, then, on $(-T, T) \times \mathfrak{R}^{2d|2d}$, $\mathcal{S}(t, s, \bar{x}, \underline{\xi}, \bar{\theta}, \underline{\pi})$ is smooth, and*

$$\begin{cases} \partial_t \mathcal{S}(t, s, \bar{x}, \underline{\xi}, \bar{\theta}, \underline{\pi}) + \mathcal{H}(\bar{x}, \partial_{\bar{x}} \mathcal{S}, \bar{\theta}, \partial_{\bar{\theta}} \mathcal{S}) = 0, \\ \mathcal{S}(s, s, \bar{x}, \underline{\xi}, \bar{\theta}, \underline{\pi}) = \langle \underline{x} | \underline{\xi} \rangle + \langle \underline{\theta} | \underline{\pi} \rangle. \end{cases}$$

5.3.1. *Continuity equation*

Put

$$\mathcal{D}(t, s, \bar{x}, \underline{\xi}, \bar{\theta}, \underline{\pi}) = \operatorname{sdet} \mathbf{D},$$

where

$$\mathbf{D} = \begin{bmatrix} \partial_{\bar{x}} y(t, s, \bar{x}, \underline{\xi}, \bar{\theta}, \underline{\pi}) & \partial_{\bar{\theta}} y(t, s, \bar{x}, \underline{\xi}, \bar{\theta}, \underline{\pi}) \\ \partial_{\bar{x}} \omega(t, s, \bar{x}, \underline{\xi}, \bar{\theta}, \underline{\pi}) & \partial_{\bar{\theta}} \omega(t, s, \bar{x}, \underline{\xi}, \bar{\theta}, \underline{\pi}) \end{bmatrix}$$

which is well-defined for $|t - s| \le \delta_1(T)$, $t, s \in [-T, T]$.

Proposition 5.5. *(Continuity equation)* $\mathcal{D}(t,s,\bar{x},\underline{\xi},\bar{\theta},\underline{\pi})$ *satisfies the following for* $|t-s| \le \delta_1(T)$*:*

$$\begin{cases} \partial_t \mathcal{D}(t,s,\bar{x},\underline{\xi},\bar{\theta},\underline{\pi}) = -\sum_{j=1}^{m} \partial_{\bar{x}_j}\{\mathcal{D}\mathcal{H}_{\underline{\xi}_j}(\bar{x},\partial_{\bar{x}}S,\bar{\theta},\partial_{\bar{\theta}}S)\} \\ \qquad\qquad\qquad\qquad\qquad -\sum_{r=1}^{n} \partial_{\bar{\theta}_r}\{\mathcal{D}\mathcal{H}_{\underline{\pi}_r}(\bar{x},\partial_{\bar{x}}S,\theta,\partial_{\theta}S)\}, \\ \mathcal{D}(s,s,\bar{x},\underline{\xi},\bar{\theta},\underline{\pi}) = 1. \end{cases}$$

Appendix A. A simple calculable example

In order to clarify the physicist's claim, we take the simplest example, the harmonic oscillator on $\mathbb{R}$, though $\mathbb{R}$ is non-compact, which is extended to the one on $\mathfrak{R}^{1|1}$. Moreover, we calculate the supertrace of the super-heat operator.

Appendix A.1. *Ordinary case*

We consider the heat equation with quadratic potential:

$$\frac{\partial u}{\partial t} = H(q,D)u \text{ with } H(q,D) = \frac{1}{2}\frac{\partial^2}{\partial q^2} - \frac{1}{2}\omega^2 q^2, \tag{A.1}$$

with symbol $H(q,p)$ of $H(q,D)$ is defined by

$$H(q,p) = e^{-qp}H(q,D)e^{qp} = \frac{1}{2}p^2 - \frac{1}{2}\omega^2 q^2. \tag{A.2}$$

Here, we should remark that we don't use $e^{-iqp}H(q,D)e^{iqp}$ as in QM.

The classical mechanics corresponding to $H(q,p)$ is given by

$$\dot{q} = p, \quad \dot{p} = \omega^2 q \text{ with } (q(0),p(0)) = (\underline{q},\underline{p}),$$

which is solved as

$$q(s) = \underline{q}\cosh(\omega t) + \frac{\underline{p}}{\omega}\sinh(\omega t), \quad p(s) = \underline{p}\cosh(\omega t) + \omega\underline{q}\sinh(\omega t).$$

Then, we put

$$\begin{aligned} S_0(t,\underline{q},\underline{p}) &= \int_0^t ds[\dot{q}(s)p(s) - H(q(s),p(s))] \\ &= \frac{\sinh(\omega t)\cosh(\omega t)}{2\omega}(\underline{p}^2 + \omega^2\underline{q}^2) + \frac{\cosh(2\omega t) - 1}{2}\underline{q}\underline{p}. \end{aligned} \tag{A.3}$$

Appendix A.1.1. *L-formulation*

Substituting $\underline{p} = \dfrac{\omega}{\sinh(\omega t)}(\bar{q} - \cosh(\omega t)\underline{q})$ into $S_0(t, \underline{q}, \underline{p})$, we have

$$\begin{aligned} S(t, \bar{q}, \underline{q}) &= \frac{\omega \cosh(\omega t)}{2\sinh(\omega t)}(\bar{q}^2 + \underline{q}^2) - \frac{\omega}{\sinh(\omega t)}\bar{q}\underline{q}, \\ D(t, \bar{q}, \underline{q}) &= -\frac{\partial^2 S(t, \bar{q}, \underline{q})}{\partial \bar{q} \partial \underline{q}} = \frac{\omega}{\sinh(\omega t)}. \end{aligned} \tag{A.4}$$

They satisfy

$$\begin{aligned} S_t + H(\bar{q}, S_{\bar{q}}) = 0 \text{ with } \lim_{t \to 0} tS(t, \bar{q}, \underline{q}) = \frac{(\bar{q} - \underline{q})^2}{2}, \\ D_t + \partial_{\bar{q}}(DH_p) = 0 \text{ with } \lim_{t \to 0} tD(t, \bar{q}, \underline{q}) = 1 \text{ where } H_p = H_p(\bar{q}, S_{\bar{q}}). \end{aligned} \tag{A.5}$$

For the future use, we explain the **quantization by heat**: Define

$$\begin{aligned} H_t^L\, \underline{u}(\bar{q}) &= \frac{1}{\sqrt{2\pi}} \int_{\mathbb{R}} d\underline{q}\, D^{1/2}(t, \bar{q}, \underline{q}) e^{-S(t, \bar{q}, \underline{q})} \underline{u}(\underline{q}) \\ &= \sqrt{\frac{\omega}{2\pi \sinh(\omega t)}} \int_{\mathbb{R}} d\underline{q}\, e^{-S(t, \bar{q}, \underline{q})} \underline{u}(\underline{q}). \end{aligned} \tag{A.6}$$

It is well-known that $u(t, q) = H_t^L\, \underline{u}(q)$ satisfies (A.1) with $u(0, q) = \underline{u}(q)$. Therefore, for a Hamiltonian $H(q, p)$, we may associate an operator

$$H(q, D) = \frac{\partial}{\partial t}(H_t^L \underline{u})(q)\Big|_{t=0}$$

by Feynman's procedure of heat type.

Moreover, since $S(t, \underline{q}, \underline{q}) = \dfrac{\omega}{\sinh(\omega t)}(\cosh(\omega t) - 1)\underline{q}^2$, we get

$$\begin{aligned} \mathrm{tr}\, H_t^L &= \sqrt{\frac{\omega}{2\pi \sinh(\omega t)}} \int_{\mathbb{R}} d\underline{q}\, e^{-S(t, \underline{q}, \underline{q})} \\ &= \sqrt{\frac{1}{2(\cosh(\omega t) - 1)}} = (e^{\omega t/2} - e^{-\omega t/2})^{-1} \sim (\omega t)^{-1}\ (t \to 0). \end{aligned}$$

The reason why we don't have $t^{-1/2}$ as $t \to 0$ is not clear, presumablly, that comes from the "boundary points" $\{+\infty\} \cup \{-\infty\}$.

Appendix A.1.2. *H-formulation*

On the other hand, since $\underline{q} = \underline{q}(\bar{q}, \underline{p}) = \frac{1}{\cosh(\omega t)}(\bar{q} - \frac{\sinh(\omega t)}{\omega}\underline{p})$, we put

$$\begin{aligned} S(t, \bar{q}, \underline{p}) &= [\underline{q}\underline{p} + S_0(t, \underline{q}, \underline{p})]\Big|_{\underline{q}=\underline{q}(\bar{q},\underline{p})} \\ &= -\frac{\sinh(\omega t)}{2\omega\cosh(\omega t)}(\underline{p}^2 - \omega^2\bar{q}^2) + \frac{1}{\cosh(\omega t)}\bar{q}\underline{p}, \\ D(t, \bar{q}, \underline{p}) &= \frac{\partial^2 S(t, \bar{q}, \underline{p})}{\partial\bar{q}\partial\underline{p}} = \frac{1}{\cosh(\omega t)}. \end{aligned} \tag{A.7}$$

They satisfy

$$\begin{gathered} S_t + H(\bar{q}, S_{\bar{q}}) = 0 \text{ with } \lim_{t\to 0} S(t, \bar{q}, \underline{p}) = \bar{q}\underline{p}, \\ D_t + \partial_{\bar{q}}(DH_p) = 0 \text{ with } \lim_{t\to 0} D(t, \bar{q}, \underline{p}) = 1 \text{ where } H_p = H_p(\bar{q}, S_{\bar{q}}). \end{gathered} \tag{A.8}$$

Then, we define

$$\begin{aligned} H_t^H\,\underline{u}(\bar{q}) &= \frac{1}{\sqrt{2\pi}}\int_{\mathbb{R}} d\underline{p}\, D(t, \bar{q}, \underline{p})^{1/2} e^{-S(t,\bar{q},\underline{p})}\hat{\underline{u}}(\underline{p}) \\ &= \frac{1}{\sqrt{2\pi}}\int_{\mathbb{R}} d\underline{p}\sqrt{\frac{1}{\cosh(\omega t)}}e^{-S(t,\bar{q},\underline{p})}\hat{\underline{u}}(\underline{p}) \\ &= \frac{1}{2\pi}\iint_{\mathbb{R}^2} d\underline{p}\, d\underline{q}\sqrt{\frac{1}{\cosh(\omega t)}}e^{-S(t,\bar{q},\underline{p})-i\underline{q}\underline{p}}\underline{u}(\underline{q}). \end{aligned} \tag{A.9}$$

Remark. In the last expression of (A.9), we may change the order of integration from $d\underline{p}\,d\underline{q}$ to $d\underline{q}\,d\underline{p}$, if we regard the integral as oscillatory one. Moreover, from this, we have $H_t^H = H_t^L$.

Appendix A.2. *An example of supersymmetric extension*

Let

$$H(q, p) = \frac{1}{2}(p - A(q))^2 + V(q) \in C^\infty(T^*\mathbb{R} : \mathbb{R})$$

be given with $A(q), V(q) \in C^\infty(\mathbb{R} : \mathbb{R})$. Using the Legendre transformation as

$$\dot{q} = \frac{\partial H}{\partial p} = p - A(q) \quad L(q, \dot{q}) = \dot{q}p - H(q, p),$$

we get a Lagrangian

$$L(q, \dot{q}) = \frac{1}{2}\dot{q}^2 + A(q)\dot{q} - V(q) \in C^\infty(T\mathbb{R} : \mathbb{R}).$$

Instead of a path $q : \mathbb{R} \ni t \to q(t) \in \mathbb{R}$, we consider a generalized path $\Phi : \mathbb{R} \ni t \to \Phi(t) = x(t) + \sqrt{-1}(\rho_1\psi_2(t) - \rho_2\psi_1(t)) + \sqrt{-1}\rho_1\rho_2 F(t) \in \mathfrak{R}_{\mathrm{ev}}$ with $\rho_\alpha \in \mathfrak{R}_{\mathrm{od}}$ $(\alpha = 1, 2)$ being odd parameters and

$$\begin{cases} x : \mathbb{R} \ni t \to x(t) \in \mathfrak{R}_{\mathrm{ev}}, \\ \psi_\alpha : \mathbb{R} \ni t \to \psi_\alpha(t) \in \mathfrak{R}_{\mathrm{od}} \text{ with } \alpha = 1, 2, \\ F : \mathbb{R} \ni t \to F(t) \in \mathfrak{R}_{\mathrm{ev}}. \end{cases}$$

Introducing operators

$$\mathcal{D}_\alpha = \frac{\partial}{\partial \rho_\alpha} - \sqrt{-1}\rho_\alpha \frac{\partial}{\partial t} \text{ with } \alpha = 1, 2,$$

and $\epsilon_{\alpha\beta} = -\epsilon_{\beta\alpha}$, $\epsilon_{12} = 1$, we extend $L(q, \dot{q})$ as

$$\tilde{\mathcal{L}}_0 = -\frac{1}{4}(\mathcal{D}_\alpha\Phi)\epsilon_{\alpha\beta}(\mathcal{D}_\beta\Phi) + \frac{\sqrt{-1}}{2}A\rho_\alpha\epsilon_{\alpha\beta}\mathcal{D}_\beta\Phi - \sqrt{-1}W(\Phi).$$

In the above, $A(q)$ is extended from a function of $q \in \mathbb{R}$ to that of $\Phi = x + i(\rho_1\psi_2 - \rho_2\psi_1) + i\rho_1\rho_2 F \in \mathfrak{R}_{\mathrm{ev}}$ as

$$A(\Phi) = A(x) + \sqrt{-1}A'(x)(\rho_1\psi_2 - \rho_2\psi_1 + \rho_1\rho_2 F) + A''(x)\rho_1\rho_2\psi_1\psi_2,$$

called the Grassmann continuation. $W(\Phi)$ is analogously extended from $W(q)$ whose relation to $V(q)$ will be given later.

Now, we have

$$\begin{aligned} \mathcal{L}'_0 &:= \int d\rho_2 d\rho_1\, \tilde{\mathcal{L}}_0 \\ &= \frac{1}{2}\dot{x}^2 + A(x)\dot{x} + \frac{1}{2}F^2 + W'(x)F \qquad \text{(A.10)} \\ &\qquad + \frac{\sqrt{-1}}{2}\left(\psi_2\dot{\psi}_2 - \dot{\psi}_1\psi_1\right) - \sqrt{-1}W''(x)\psi_1\psi_2. \end{aligned}$$

Assuming that the "auxilliary field F" should satisfy

$$0 = \frac{\delta \mathcal{L}'_0}{\delta F} = F + W', \qquad \text{(A.11)}$$

we arrived at

$$\begin{aligned} \mathcal{L}_0 &= \frac{1}{2}\dot{x}^2 + A(x)\dot{x} \\ &\qquad + \frac{\sqrt{-1}}{2}\left(\psi_2\dot{\psi}_2 - \dot{\psi}_1\psi_1\right) - \frac{1}{2}W'(x)^2 - \sqrt{-1}W''(x)\psi_1\psi_2. \end{aligned} \qquad \text{(A.12)}$$

This is the desired Lagrangian with variables $x, \dot{x}, \psi_\alpha, \dot{\psi}_\alpha$, but variables $\psi_\alpha, \dot{\psi}_\alpha$ are not independent each other. In fact, they satisfy

$$\{\psi_\alpha, \psi_\beta\} = \psi_\alpha\psi_\beta + \psi_\beta\psi_\alpha = 0, \quad \{\psi_\alpha, \dot{\psi}_\beta\} = 0 \text{ and } \{\dot{\psi}_\alpha, \dot{\psi}_\beta\} = 0.$$

To find out "independent odd variables" in (A.12), we introduce new variables by the following two methods:

(I) Defining new variables as

$$\begin{cases} \xi = \dfrac{\delta\mathcal{L}_0}{\delta\dot{x}} = \dot{x} + ax, \\ \phi_\alpha = \dfrac{\delta\mathcal{L}_0}{\delta\dot{\psi}_\alpha} = -\dfrac{\sqrt{-1}}{2}\psi_\alpha \text{ for } \alpha = 1, 2, \end{cases} \tag{A.13}$$

we put

$$\begin{aligned} \mathcal{H}(x, \xi, \psi_1, \psi_2) &= \dot{x}\xi + \dot{\psi}_\alpha\phi_\alpha - \mathcal{L}_0 \\ &= \frac{1}{2}(\xi - A(x))^2 + \frac{1}{2}W'(x)^2 + \frac{\sqrt{-1}}{2}W''(x)\psi_\alpha\epsilon_{\alpha\beta}\psi_\beta. \end{aligned} \tag{A.14}$$

Rewriting the variables ψ_1, ψ_2 as θ, π, respectively, we get

$$\mathcal{H}(x, \xi, \theta, \pi) = \frac{1}{2}(\xi - A(x))^2 + \frac{1}{2}W'(x)^2 + \sqrt{-1}W''(x)\theta\pi. \tag{A.15}$$

(II) In the above, we use the "real" odd variables ψ_α. We "complexify" these variables by putting

$$\begin{aligned} &\psi = \frac{1}{\sqrt{2}}(\psi_1 + \sqrt{-1}\psi_2), \quad \bar{\psi} = \frac{1}{\sqrt{2}}(\psi_1 - \sqrt{-1}\psi_2), \\ &i.e. \quad \psi_1 = \frac{1}{\sqrt{2}}(\psi + \bar{\psi}), \quad \psi_2 = \frac{1}{\sqrt{2}\sqrt{-1}}(\psi - \bar{\psi}), \end{aligned} \tag{A.16}$$

and then we rewrite $\mathcal{L}_0$ as

$$\bar{\mathcal{L}}_0 = \frac{1}{2}\dot{x}^2 + A(x)\dot{x} + \frac{\sqrt{-1}}{2}(\psi\dot{\bar{\psi}} + \bar{\psi}\dot{\psi}) - \frac{1}{2}W'(x)^2 - W''(x)\bar{\psi}\psi. \tag{A.17}$$

Introducing new variables as

$$\begin{aligned} \xi &= \frac{\delta\bar{\mathcal{L}}_0}{\delta\dot{x}} = \dot{x} + A(x), \\ \phi = \frac{\delta\bar{\mathcal{L}}_0}{\delta\dot{\psi}} &= -\frac{\sqrt{-1}}{2}\bar{\psi}, \quad \bar{\phi} = \frac{\delta\bar{\mathcal{L}}_0}{\delta\dot{\bar{\psi}}} = -\frac{\sqrt{-1}}{2}\psi, \end{aligned} \tag{A.18}$$

we put

$$\begin{aligned} \mathcal{H}(x, \xi, \psi, \bar{\psi}) &:= \dot{x}\xi + \dot{\psi}\phi + \dot{\bar{\psi}}\bar{\phi} - \bar{\mathcal{L}}_0 \\ &= \frac{1}{2}(\xi - A(x))^2 + \frac{1}{2}W'(x)^2 + W''(x)\bar{\psi}\psi. \end{aligned}$$

Rewriting ψ and $\bar{\psi}$ by θ and π, respectively, we get finally a function

$$\mathcal{H}(x,\xi,\theta,\pi)=\frac{1}{2}(\xi-A(x))^2+\frac{1}{2}W'(x)^2 - W''(x)\theta\pi \in \mathcal{C}_{SS}(\mathfrak{R}^{2|2}:\mathfrak{R}_{\mathrm{ev}}). \tag{A.19}$$

Here, $(x,\theta)\in\mathfrak{R}^{1|1}$, $(\xi,\pi)\in\mathfrak{R}^{1|1}$.

Remarks. (0) The difference between (A.15) *and* (A.19) *is the existence of* $\sqrt{-1}$ *in front of the term* $W''(x)\theta\pi$*. This difference is rather significant when we consider Witten index for supersymmetric quantum mechanics using the kernel representation of the corresponding evolution operator.*
(1) As there is no preference at this stage to take π *and* θ *instead of* θ *and* π*, there is no significance of the sign* $\pm$ *in front of the terms* $\sqrt{-1}W''(x)\theta\pi$ *in* (A.15) *or* $W''(x)\theta\pi$ *in* (A.19).
(2) We may regard $\mathcal{H}(x,\xi,\theta,\pi)$ *as a Hamiltonian in* $\mathcal{C}_{SS}(T^*\mathfrak{R}^{1|1}:\mathfrak{R}_{\mathrm{ev}})$.
(3) These Hamiltonians (A.15) *and* (A.19) *are called supersymmetric extensions of (3.1) because they give supersymmetric quantum mechanics after quantization (see §4). The procedure above is author's unmatured understanding of amalgam of physics papers such as Cooper and Freedman* [6]*, Davis, Macfarlane, Popat and van Holten* [8] *etc. But supersymmetry in superspace* $\mathfrak{R}^{m|n}$ *will be studied separately.*
(4) On the other hand, using the identification (1.13), we have

$$\mathcal{H}^{\hbar}_{\pm}(x,\partial_x,\theta,\partial_\theta)=\#\begin{pmatrix} H^{\hbar}_{\pm}-\frac{\hbar}{2}b & 0\\ 0 & H^{\hbar}_{\pm}+\frac{\hbar}{2}b\end{pmatrix}\flat=\#\mathbb{H}^{\hbar,b}_{\pm}\flat.$$

Moreover, in this case, the "complete Weyl symbol of the above $\mathcal{H}^{\hbar}(x,\partial_x,\theta,\partial_\theta)$*" is calculated by*

$$\begin{aligned}\mathcal{H}_{\pm}(x,\xi,\theta,\pi)&=(e^{-i\hbar^{-1}(x\xi+\theta\pi)}\mathcal{H}^{\hbar}_{\pm}(x,\partial_x,\theta,\partial_\theta)e^{i\hbar^{-1}(x\xi+\theta\pi)}\Big|_{\hbar=0}\\ &=\frac{1}{2}(\xi-ax)^2\pm\frac{1}{2}b^2x^2+\sqrt{-1}b\theta\pi.\end{aligned}$$

$\mathcal{H}_+$ *equals to* (A.15) *when* $A(q)=aq$ *and* $W(x)=bx^2/2$*, and* $\mathcal{H}_-$ *is obtained from* (A.19) *with* $A(q)=aq$ *and* $W(x)=-\sqrt{-1}bx^2/2$*. These give the relation between* $W(q)$ *and* $V(q)$*. (See susyQM defined in §1.)*
(5) Witten [29] *considered as a quantum mechanical operator*

$$\mathbb{H}(q,\partial_q)=\left(-\frac{1}{2}\partial_q^2+v(q)\right)\begin{pmatrix}1&0\\0&1\end{pmatrix}-\frac{1}{2}w(q)\begin{pmatrix}1&0\\0&-1\end{pmatrix}.$$

This operator is supersymmetric when there exists a function $u(q)$ *such that*

$$v(q)=\frac{1}{2}u'(q)^2,\quad w(q)=u''(q).$$

Appendix A.3. *Super case*

As a supersymmetric extension of $H(q,p)$ above, we consider

$$\mathcal{H}(x,\xi,\theta,\pi) = \frac{1}{2}\xi^2 - \frac{1}{2}\omega^2 x^2 - \omega\theta\pi. \tag{A.20}$$

Corresponding to this Hamiltonian function, we may construct a fundamental solution of heat type evolution equation.

Appendix A.3.1. *LH-formulation*

In order not to use the Fourier transformation in bosonic direction, we put

$$\begin{aligned} \mathcal{S}(t,\bar{x},\bar{\theta},\underline{x},\underline{\pi}) &= [\underline{\theta}\,\underline{\pi} + S_0(t,\underline{x},\underline{\xi})]\Big|_{\underline{\xi}=\underline{\xi}(\bar{x},\bar{\theta},\underline{x},\underline{\pi}),\underline{\theta}=\underline{\theta}(\bar{x},\bar{\theta},\underline{x},\underline{\pi})} \\ &= e^{-\omega t}\bar{\theta}\underline{\pi} + S(t,\bar{x},\underline{x}) \end{aligned} \tag{A.21}$$

with

$$S(t,\bar{x},\underline{x}) = \frac{\omega\cosh(\omega t)}{2\sinh(\omega t)}(\bar{x}^2 + \underline{x}^2) - \frac{\omega}{\sinh(\omega t)}\bar{x}\underline{x}.$$

We define

$$\mathcal{D}(t,\bar{x},\bar{\theta},\underline{\xi},\underline{\pi}) = \operatorname{sdet}\begin{pmatrix} -\frac{\partial^2\mathcal{S}}{\partial\bar{x}\partial\underline{x}} & 0 \\ 0 & -\frac{\partial^2\mathcal{S}}{\partial\bar{\theta}\partial\underline{\pi}} \end{pmatrix} = \frac{\omega}{\sinh(\omega t)}e^{\omega t}$$

and we have

$$\mathcal{D}^{1/2}e^{-\mathcal{S}} = e^{\omega t/2}\sqrt{\frac{\omega}{\sinh(\omega t)}}\,e^{-e^{-\omega t}\bar{\theta}\underline{\pi} - S(t,\bar{x},\underline{x})}$$

Define the Fourier transformations by

$$\begin{cases} \int_{\mathfrak{R}^{0|1}} d\theta\, e^{-\theta\pi}(v_0 + v_1\theta) = v_1 - v_0\pi, \\ \int_{\mathfrak{R}^{0|1}} d\pi\, e^{\theta\pi}(v_1 - v_0\pi) = -v_0 - v_1\theta. \end{cases}$$

Remarking $e^{-e^{-\omega t}\bar{\theta}\underline{\pi}} = 1 - e^{-\omega t}\bar{\theta}\underline{\pi}$ and

$$-\int d\underline{\pi}\, e^{-e^{-\omega t}\bar{\theta}\underline{\pi}}(\underline{v}_1(\underline{x}) - \underline{v}_0(\underline{x})\underline{\pi}) = \underline{v}_0(\underline{x}) - e^{-\omega t}\underline{v}_1(\underline{x})\bar{\theta},$$

we have

$$\begin{aligned} \mathcal{H}_t^{LH}\underline{v}(\bar{q},\bar{\theta}) = v(t,\bar{q},\bar{\theta}) &= \sqrt{\frac{1}{2\pi}}\int d\underline{x}\,d\underline{\pi}\,\mathcal{D}^{1/2}e^{-\mathcal{S}}(\underline{v}_1(\underline{x}) - \underline{v}_0(\underline{x})\underline{\pi}) \\ &= \sqrt{\frac{\omega}{2\pi\sinh(\omega t)}}\int d\underline{x}\,e^{-S(t,\bar{x},\underline{x})}(e^{\omega t/2}\underline{v}_0(\underline{x}) - e^{-\omega t/2}\underline{v}_1(\underline{x})\bar{\theta}). \end{aligned}$$

On the other hand, since

$$\int d\underline{\theta}\,(e^{\omega t/2}\underline{\theta}+e^{-\omega t/2}\bar{\theta})(\underline{v}_0(\underline{x})+\underline{v}_1(\underline{x})\underline{\theta}) = e^{\omega t/2}\underline{v}_0(\underline{x}) - e^{-\omega t/2}\underline{v}_1(\underline{x})\bar{\theta},$$

we have

$$[\mathrm{Ker}\,(Pe^{-t\hat{\mathcal{H}}})](\bar{x},\underline{x},\bar{\theta},\underline{\theta}) = (1-2\bar{\theta}\partial_{\bar{\theta}})(e^{\omega t/2}\underline{\theta}+e^{-\omega t/2}\bar{\theta}) \times \sqrt{\frac{\omega}{2\pi\sinh(\omega t)}}e^{-\mathcal{S}(t,\bar{x},\underline{x})},$$

which implies

$$\begin{aligned}\mathrm{str}\,\exp[-t\hat{\mathcal{H}}] &= \int d\underline{x}\,d\underline{\theta}\,[\mathrm{Ker}\,(Pe^{-t\hat{\mathcal{H}}})](\underline{x},\underline{x},\underline{\theta},\underline{\theta})\\ &= (e^{\omega t/2}-e^{-\omega t/2})\sqrt{\frac{\omega}{2\pi\sinh(\omega t)}}\sqrt{\frac{\pi\sinh(\omega t)}{\omega(\cosh(\omega t)-1)}}\\ &= \frac{e^{\omega t/2}-e^{-\omega t/2}}{\sqrt{2(\cosh(\omega t)-1)}} = 1.\end{aligned}$$

Appendix A.3.2. *LH-formulation for general dimension d*

Let $u(t,x,\theta)=\sum_{|a|\le d}u_a(t,x)\theta^a$. Since we have

$$\prod_{j=1}^{d}(\theta_j\pi_j)^{b_j} = (-1)^{|b|(|b|-1)/2}\theta^b\pi^b,$$

$$\theta^a\theta^b = (-1)^{\tau(a,b)}\theta^{a+b},\ \text{with}\ \tau(b,a)\equiv|a||b|+\tau(a,b) \mod 2,$$

we get, with $b_j=1-a_j$, $a+b=\tilde{1}=(1,\cdots,1)$, $d=|a|+|b|$,

$$\begin{aligned}\int d\theta\,e^{\langle\theta|\pi\rangle}\theta^a = \int d\theta\prod_{j=1}^{d}(1+\theta_j\pi_j)\theta^a &= \int d\theta\,\theta^a\prod_{j=1}^{d}(\theta_j\pi_j)^{b_j}\\ &= (-1)^{|b|(|b|-1)/2+\tau(a,b)}\pi^b,\end{aligned}$$

$$\int d\pi\,e^{-\langle\theta|\pi\rangle}\pi^b = (-1)^{|a|(|a|-1)/2+\tau(b,a)}\theta^a,$$

$$\int d\pi\,e^{-\langle\theta|\pi\rangle}\Big(\int d\theta'\,e^{\langle\theta'|\pi\rangle}\theta'^a\Big) = (-1)^{d(d-1)/2}\theta^a.$$

Therefore

$$\int d\theta\, e^{\langle\theta|\pi\rangle} \sum_{|a|\le d} u_a \theta^a = \int d\theta \sum_{|a|\le d} u_a \theta^a \prod_{j=1}^{d} (\theta_j \pi_j)^{b_j}$$
$$= \sum_{|a|\le d} (-1)^{|b|(|b|-1)/2+\tau(a,b)} u_a \pi^b,$$

$$\int d\pi\, e^{-\langle\theta|\pi\rangle} \sum_{|a|\le d} (-1)^{|b|(|b|-1)/2+\tau(a,b)} u_a \pi^b = (-1)^{d(d-1)/2} \sum_{|a|\le d} u_a \theta^a.$$

On the other hand, as

$$\int d\underline{\pi}\, e^{\Sigma_{j=1}^d e^{-\omega_j t}\bar{\theta}_j \underline{\pi}_j} (-1)^{|b|(|b|-1)/2+\tau(a,b)} \underline{\pi}^b$$
$$= \int d\underline{\pi}\, (-1)^{|b|(|b|-1)/2+\tau(a,b)} \underline{\pi}^b \prod_{j=1}^{d} (e^{-\omega_j t} \bar{\theta}_j \underline{\pi}_j)^{a_j}$$
$$= \prod_{j=1}^{d} e^{-a_j \omega_j t} (-1)^{d(d-1)/2} (-1)^{|a|+\tau(a,b)} \bar{\theta}^a,$$

we have

$$\exp[-t\hat{\mathcal{H}}]\underline{u}(\bar{x},\bar{\theta}) = \frac{(-1)^{d(d-1)/2}}{(2\pi)^{d/2}} \int d\underline{x}\, d\underline{\pi}\, \mathcal{D}^{1/2}(t,\bar{x},\underline{x},\bar{\theta},\underline{\pi}) e^{-\mathcal{S}(t,\bar{x},\underline{x},\bar{\theta},\underline{\pi})}$$
$$\times \sum_{|a|\le d} (-1)^{|b|(|b|-1)/2+\tau(a,b)} \underline{u}_a(\underline{x}) \underline{\pi}^b$$
$$= \int d\underline{x} \prod_{j=1}^{d} \left(\frac{\omega_j}{2\pi \sinh(\omega_j t)} \right)^{1/2} e^{-S(t,\bar{x},\underline{x})} e^{\omega t/2}$$
$$\times \sum_{|a|\le d} \prod_{j=1}^{d} e^{-a_j \omega_j t} (-1)^{|a|+\tau(a,b)} \underline{u}_a(\underline{x}) \bar{\theta}^a.$$

Since

$$\int d\underline{\theta} \sum_{|a'|+|b'|=d} e^{\Sigma_{j=1}^d (a'_j - b'_j)\omega_j t/2} \underline{\theta}^{a'} \bar{\theta}^{b'} \sum_{|a|\le d} u_a \underline{\theta}^a$$
$$= \sum_{|a|\le d} \prod_{j=1}^{d} e^{(b_j - a_j)\omega_j t/2} (-1)^{|a|+\tau(a,b)} \underline{u}_a(\underline{x}) \bar{\theta}^a.$$

and

$$Pu = \prod_{j=1}^{d} \left(1 - 2\theta_j \frac{\partial}{\partial\theta_j}\right) \sum_{|a|\le d} u_a \theta^a = \sum_{|a|\le d} (-1)^{|b|} u_a \theta^a,$$

we have the corresponding susyQM as before.

Moreover, we have

$$\begin{aligned}\operatorname{str}\exp[-t\hat{\mathcal{H}}] &= \int d\underline{x}\,d\underline{\theta}\prod_{j=1}^{d}(e^{\omega_j t/2}-e^{-\omega_j t/2})\underline{\theta}_1\cdots\underline{\theta}_d e^{-S(t,\underline{x},\underline{x})}\\ &= \prod_{j=1}^{d}(e^{\omega_j t/2}-e^{-\omega_j t/2})\frac{1}{\sqrt{\prod_{j=1}^{d}2(\cosh(\omega_j t)-1)}} = 1.\end{aligned}$$

On the other hand, putting $v(t,x,\theta)=\exp[-t\hat{\mathcal{H}}]\underline{v}(x,\theta)$, we have

$$\begin{aligned}&\frac{\partial}{\partial t}v(t,x,\theta)\\ &= \Big(\frac{1}{2}\sum_{j=1}^{d}\frac{\partial^2}{\partial x_j^2}-\frac{1}{2}\sum_{j=1}^{d}\omega_j^2x_j^2+\sum_{j=1}^{d}\frac{\omega_j}{2}\big[\frac{\partial}{\partial\theta_j},\theta_j\big]_-\Big)v(t,x,\theta)\end{aligned}\tag{A.22}$$

with $v(0,x,\theta)=\underline{v}(x,\theta)$. This equation (A.22) is the quantized one of the superextension of harmonic oscillator in $\mathbb{R}^d$ and also the explicit form corresponding to the one in §4.

References

1. L. Alvarez-Gaumé, *Supersymmetry and the Atiyah–Singer index theorem*, Commun.Math.Phys. **90** (1983), 161-173.
2. R. Abraham and J.E. Marsden, *Foundations of Mechanics*, Second edition, Massachusetts, Benjamin,1980.
3. F.A. Berezin and M.S. Marinov, *Particle spin dynamics as the Grassmann variant of classical mechanics*, Annals of Physics **104** (1977), 336-362.
4. R. Casalbuoni, *On the quantization of systems with anticommuting variables*, Nuovo Cimento **A33** (1976), 115-125.
5. R. Casalbuoni, *The classical mechanics for Bose-Fermi systems*, Nuovo Cimento **A33** (1976), 389-431.
6. F. Cooper and B. Freedman, *Aspects of supersymmetric quantum mechanics*, Annals of Physics **146** (1983), 262-288.
7. H.L. Cycon, R.G. Froese, W. Kirsh and B. Simon, *Schrödinger Operators with Application to Quantum Mechanics and Global Geometry*, New York, Springer-Verlag, GMT, 1987.
8. A.C. Davis, A.J. Macfarlane, P.C. Popat and J.W. van Holten, *The quantum mechnics of the supersymmetric nonlinear σ-model*, J.Phys.A:Math.Gen. **17** (1984), 2945-2954.
9. B. DeWitt, *Dynamical theory in curved spaces I. A review of the classical and quantum action principles*, Reviews of modern physics **29** (1984), 377-397.
10. D. Fujiwara, *A construction of the fundamental solution for the Schrödinger equation*, J.D'Analyse Math. **35** (1979), 41-96.

11. D. Fujiwara, *Remarks on convergence of the Feynman path integrals*, Duke Math.J. **47** (1980), 559-600.
12. D. Friedan and P. Windey, *Supersymmetric derivation of the Atiyah–Singer index and the chiral anomaly*, Nuclear Phys.B(FS11) **253** (1984), 395-416.
13. D. Fujiwara, *A construction of the fundamental solution for the Schrödinger equation*, J. D'Analyse Math. **35** (1979), 41-96.
14. D. Fujiwara, *Remarks on convergence of the Feynman path integrals*, Duke Math. J. **47** (1980), 559-600.
15. E. Getzler, *Pseudodifferential operators on supermanifolds and the Atiyah-Singer index theorem*, Commun.Math.Phys. **92** (1983), 163-178.
16. A. Inoue, *On a construction of the fundamental solution for the free Weyl equation by Hamiltonian path-integral method —an exactly solvable case with "odd variable coefficients"*, Tôhoku J.Math. **50** (1998), 91-118.
17. A. Inoue, *On a construction of the fundamental solution for the free Dirac equation by Hamiltonian path-integral method —the classical counterpart of Zitterbewegung*, Japanese J.Math. **24** (1998), 297-334.
18. A. Inoue, *A partial solution for Feynman's problem –a new derivation of the Weyl equation*, Mathematical Physics and Quantum Field Theory, Electron.J.Diff.Eqns., Conf. **04** (2000), 121-145.
19. A. Inoue, *An extension of the method of characteristic to a system of Partial Differential Operators – an application to the Weyl equation with external field by "Super Hamiltonian path-integral method"* **arXiv:math-ph/0212065**.
20. A. Inoue, *On a supersymmetric Hamiltonian over* $\mathbf{R}^{2d|2d}$ *and its classical orbits*, not yet published.
21. A. Inoue and Y. Maeda, *On integral transformations associated with a certain Lagrangian–as a prototype of quantization*, J.Math.Soc.Japan **37** (1985), 219-244.
22. A. Inoue, *On a construction of a good parametrix for the Pauli equation by Hamiltonian path-integral method — an application of superanalysis*, Japanese J. Math.**29** (2003), 27-107.
23. H. Kitada, *On a construction of the fundamental solution for Schrödinger equations*, J. Fac. Sci. Univ. Tokyo **27** (1980), 193-226.
24. H. Kitada and H. Kumano-go, *A family of Fourier integral operators and the fundamental solution for a Schrödinger equation*, Osaka J. Math. **18** (1981), 291-360.
25. A. Klein, L.J. Landau and J.F. Perez, *Supersymmetry and the Parisi-Sourlas dimensional reduction: a rigorous proof*, Commun.Math.Phys. **94** (1984), 459-482.
26. J. Mañes and B. Zumino, *WKB method, susy quantum mechanics and the index theorem*, Nuclear Physics B(FS16) **270** (1986), 651-686.
27. V.P. Maslov, *Théorie desperturbations et méthodes asymptotiques*, Dunod, Gauthier-Villars (1972).
28. A.V. Smilga, *How to quantize supersymmetric theories*, Nuclear Physics **B292** (1987), 363-380.
29. E. Witten, *Constraints on supersymmetry breaking*, Nuclear Physics **B202**

(1982) 253-316.

30. E. Witten, *Supersymmetry and Morse theory*, J.Diff.Geom. **17** (1982), 661-692.

THE VANISHING PROBLEM FOR COHOMOLOGY OF SUPERSPACES

DAISUKE KATO*

Deartment of Mathematics, Keio University
3-14-1 Hiyoshi ,
Yokohama , Japan
y11535@educ.cc.keio.ac.jp

In this paper, we discuss the vanishing problem in two types of cohomology theory of supermanifolds, Vi-cohomology and VZ-cohomology, which are generalizations of deRham cohomology.

1. Introduction

Supermanifolds are a type of noncommutative manifolds, whose ring of functions is locally a direct sum of commutative algebra and anti-commutative algebra. The model spaces are called superspaces (which correspond to Euclidean space in the ordinary setting).

The notion of supermanifolds was first introduced by Berezin-Leites[2], and a simple generalization of deRham complex was introduced by Leites[3] and it was used in the study of invariant differential operators with respect to the action of the group of superdiffeomorphisms (Bernstein and Leites[3]). However, their cohomology coincides with deRham cohomology of the underlying manifold, so it has no information about the odd part of supermanifolds.

Voronov and Zorich[13] introduced a new type of differential forms as densities which satisfy additional conditions. We call these forms VZ–forms. The VZ–forms seem to be deeply related to the integration theory on supermanifolds and the theory of subsupermanifolds (cf. Voronov[11]).

The definition of VZ–forms seems nice, but the explicit form of them is complicated even for low dimension. In order to simplify the treatment of

*This work is partially supported by COE Program " Integrative Mathematical Science".

VZ–forms, Victoria[9] introduced new definition of differential forms. However, his definition is different from that of VZ–forms, although he claimed that they are equivalent. Victoria's differential forms is called by Vi–forms in this article.

Calculation of VZ-cohomology was done by Voronov[11], [12], Voronov-Zorich[13], though cohomology groups have not been completely determined even for superspaces.

All proofs are omitted in this article. See Kato[6] for the proofs.

2. Preliminaries

We treat real supermanifolds in this article. For the foundation of real supermanifolds, see Berezin[1], Leites[8], and so on.

We fix notations. Supermanifolds is denoted by $M := (M_{\mathrm{red}}, \mathcal{O}_M)$. M_{red} is the underlying manifold of the supermanifold M, and $\mathcal{O}_M$ denotes the structure sheaf of M. $C^\infty(M)$ denotes the superalgebra of global sections of $\mathcal{O}_M$ and $\mathfrak{X}(M)$ for the $C^\infty(M)$-supermodule of superderivations of $C^\infty(M)$. We call $X \in \mathfrak{X}(M)$ a vector field on M.

We define Vi–forms in the first half of this section.

Denote $\Omega_{Vi}^{1|0}(\mathbb{R}^{m|n}) := \mathfrak{X}(\mathbb{R}^{m|n})^*$ and $\Omega_{Vi}^{0|1}(\mathbb{R}^{m|n}) := \Pi\,\mathfrak{X}(\mathbb{R}^{m|n})^*$, where $\mathfrak{X}(\mathbb{R}^{m|n})^*$ is the $C^\infty(\mathbb{R}^{m|n})$-dual of $\mathfrak{X}(\mathbb{R}^{m|n})$ and Π is the parity changing functor.

We will define $\Omega_{Vi}^{p|q}(\mathbb{R}^{m|n})$, which is the $C^\infty(\mathbb{R}^{m|n})$-supermodule of VZ-$p|q$-forms on $\mathbb{R}^{m|n}$, as a subsupermodule of

$$\mathrm{T}^{p|q}(\mathfrak{X}(\mathbb{R}^{m|n})^*)$$

$$:= \overbrace{\Omega_{Vi}^{1|0}(\mathbb{R}^{m|n}) \otimes \cdots \Omega_{Vi}^{1|0}(\mathbb{R}^{m|n})}^{\text{p-tuples}} \otimes \overbrace{\Omega_{Vi}^{0|1}(\mathbb{R}^{m|n}) \otimes \cdots \Omega_{Vi}^{0|1}(\mathbb{R}^{m|n})}^{\text{q-tuples}},$$

which is tensored as a $C^\infty(\mathbb{R}^{m|n})$-supermodule.

Denote $\mathrm{d}^0 x_i := \left(\frac{\partial}{\partial x_i}\right)^*$ and $\mathrm{d}^1 x_i := \Pi\left(\frac{\partial}{\partial x_i}\right)^*$. $\{\mathrm{d}^0x_1, \cdots, \mathrm{d}^0x_{m+n}\}$ (resp. $\{\mathrm{d}^1x_1, \cdots, \mathrm{d}^1x_{m+n}\}$) is a $C^\infty(\mathbb{R}^{m|n})$-basis of $\Omega_{Vi}^{1|0}(\mathbb{R}^{m|n})$ (resp. $\Omega_{Vi}^{0|1}(\mathbb{R}^{m|n})$).

We use the notion of supersignature of permutations (see Victoria[10]) in order to define $\Omega_{Vi}^{p|q}(\mathbb{R}^{m|n})$. We denote $\mathbf{Sgn}_{p|q}$ for the supersignature and sgn for the ordinary signature. We define

$$\mathrm{d}^0x_{i_1}\wedge\cdots\wedge\mathrm{d}^0x_{i_p}\wedge\mathrm{d}^1x_{i_{p+1}}\wedge\cdots\wedge\mathrm{d}^1x_{i_{p+q}} := \sum_{\sigma\in\mathfrak{S}_{p+q}} (-1)^{\mathrm{sgn}(\sigma)}(-1)^{\mathbf{Sgn}_{p|q}(\sigma)}$$
$$\mathrm{d}^0x_{\sigma(i_1)}\otimes\cdots\otimes\mathrm{d}^0x_{\sigma(i_p)}\otimes\mathrm{d}^1x_{\sigma(i_{p+1})}\otimes\cdots\otimes\mathrm{d}^1x_{\sigma(i_{p+q})}.$$

From now on, we omit $\wedge$s.

$\Omega^{p|q}_{Vi}(\mathbb{R}^{m|n})$ is the $C^\infty(\mathbb{R}^{m|n})$-subsupermoule of $\mathrm{T}^{p|q}\left(\mathfrak{X}(\mathbb{R}^{m|n})^*\right)$, generated by

$$\{\mathrm{d}^0x_{i_1}\cdots\mathrm{d}^1x_{i_{p+q}}; 1\le i_1\le\cdots\le i_{p+q}\le m+n\}.$$

In order to regard $\Omega^{p|q}_{Vi}(\mathbb{R}^{m|n})$ as a (noncommutative) superalgebra, we define also:

$$\mathrm{d}^aA\mathrm{d}^bB := (-1)^{(a+\tilde{A})(b+\tilde{B})}\mathrm{d}^bB\mathrm{d}^aA.$$

Any $\alpha\in\Omega^{p|q}_{Vi}(\mathbb{R}^{m|n})$ can be uniquely written by

$$\alpha = \sum_{1\le i_k\le m+n} \mathrm{d}^{a_1}x_{i_1}\cdots\mathrm{d}^{a_{p+q}}x_{i_{p+q}}f_{i_1\cdots i_{p+q}},$$

with $f_{i_1\cdots i_{p+q}}\in C^\infty(\mathbb{R}^{m|n})$. The differentials $(D_{Vi})_0 : \Omega^{p|q}_{Vi}(\mathbb{R}^{m|n})\longrightarrow\Omega^{p+1|q}_{Vi}(\mathbb{R}^{m|n})$ and $(D_{Vi})_1 : \Omega^{p|q}_{Vi}(\mathbb{R}^{m|n})\longrightarrow\Omega^{p|q+1}_{Vi}(\mathbb{R}^{m|n})$ are defined by

$$(D_{Vi})_0 := \sum_{k=1}^{m+n}\mathrm{d}^0x_k\frac{\partial}{\partial x_k}$$
$$(D_{Vi})_1 := \sum_{k=1}^{m+n}\mathrm{d}^1x_k\frac{\partial}{\partial x_k}$$

as left operators.

In the ordinary setting, the functor $f\mapsto f^*$ (f^* is the pull-back of f) from the category of C^∞-manifolds to the category of differential complexes plays an important role. In the super setting, one can consider a similar functor. For a morphism $\varphi : \mathbb{R}^{m_1|n_1}\longrightarrow\mathbb{R}^{m_2|n_2}$, the pull–back $\varphi^*_{Vi} : \Omega^{p|q}_{Vi}(\mathbb{R}^{m_2|n_2})\longrightarrow\Omega^{p|q}_{Vi}(\mathbb{R}^{m_1|n_1})$ is defined by:

$$\varphi^*_{Vi}(\mathrm{d}^ay_i\,f_i) := \sum_{k_1}^{m_1+n_1}\mathrm{d}^ay_k\frac{\partial\varphi^\#(y_i)}{\partial x_k}\varphi^\#(f_i)$$

($x_1,\cdots,x_{m_1+n_1}$ is a coordinate system of $\mathbb{R}^{m_1|n_1}$ and $y_1,\cdots,y_{m_2+n_2}$ is a coordinate system of $\mathbb{R}^{m_2|n_2}$) and

$$\varphi^*_{Vi}(\alpha_1\cdots\alpha_{p+q}) := \varphi^*_{Vi}(\alpha_1)\cdots\varphi^*_{Vi}(\alpha_{p+q})$$

(each α_i is an element of $\Omega^{1|0}_{Vi}(\mathbb{R}^{m|n})$ or $\Omega^{0|1}_{Vi}(\mathbb{R}^{m|n})$).

VZ–$p|q$–forms are defined as $p|q$–densities (see Shander[4]) satisfying some conditions. $p|q$-densities on M are defined as C^∞-functions on the $p|q$–frame superbundle $\mathrm{E}^{p|q}(M)$. For the definition of $\mathrm{E}^{p|q}(M)$, see Shander[4].

We denote by $x_1, \cdots, x_{m+n}$ a local coordinate system of the base supermanifold M and by $\begin{pmatrix} X_1^1 & \cdots & X_1^{m+n} \\ \vdots & & \vdots \\ X_{p+q}^1 & \cdots & X_{p+q}^{m+n} \end{pmatrix}$ a coordinate system of the fiber part, induced by $x_1, \cdots, x_{m+n}$.

Let $\mathrm{GL}^+_{m|n}\left(C^\infty(\mathbb{R}^{m|n})\right)$ denote a Lie supergroup consisting of $m|n$–square supermatrices over $C^\infty(\mathbb{R}^{m|n})$ such that $\det(X_{11}) > 0, \det(X_{22}) > 0$.

Remark 1 As Shander[4] and Voronov[12] pointed out, any orientable supermanifold admits four different orientations. Indeed, Voronov[12] defined four types of superforms corresponding to these orientations.

Although we should treat these four orientations equally, only the orientation identified with $\mathrm{GL}^+_{m|n}$ is discussed in this article, because the similar argument with that in this article is possible for other orientations.

We consider a superaction (see Boyer, Sánchez-Valenzuela[5]) A_P of $P \in \mathrm{GL}^+_{p|q}\left(C^\infty(\mathbb{R}^{m|n})\right)$ on $\mathrm{E}^{p|q}(M)$ defined by

$$A_P^\#(x_i) := x_i \qquad A_P^\#(X_i^j) := P_i^k X_k^j,$$

where $P = (P_i^j)_{ij}$.

Let $\alpha \in C^\infty\left(\mathrm{E}^{p|q}(M)\right)$ satisfy the following conditions:

(1) For any $P \in \mathrm{GL}^+_{p|q}\left(C^\infty(\mathbb{R}^{m|n})\right)$,

$$\alpha_P^\#(\alpha) = \mathrm{Ber}(P) \cdot \alpha.$$

Here $\mathrm{Ber}(P)$ denotes the Berezinian (superdeterminant) of P.

(2)

$$\left(\frac{\partial^2}{\partial X_b^a \partial X_d^c} + (-1)^{\tilde{a}\tilde{c}}(-1)^{\tilde{d}(\tilde{a}+\tilde{c})} \frac{\partial^2}{\partial X_d^a \partial X_b^c}\right)(\alpha) = 0$$

We denote the $C^\infty(M)$-supermodule of VZ–$p|q$–forms on M by $\Omega^{p|q}_{VZ}(M)$.

The differential $D_{VZ} : \Omega^{p|q}_{VZ}(M) \longrightarrow \Omega^{p+1|q}_{VZ}(M)$ is defined by

$$D_{VZ} := (-1)^p \sum_{i,j,k} X^i_{p+q+1}\left(\frac{\partial}{\partial x_i} - (-1)^{\tilde{j}\tilde{k}} X_k^j \frac{\partial^2}{\partial x_j \partial X_k^i}\right)$$

as a left operator.

Similarly with Vi–forms, a functor $\varphi \mapsto \varphi^*_{VZ}$ is defined by

$$\varphi^*_{VZ}(x_i) := \varphi^{\#}(x_i), \qquad \varphi^*_{VZ}(X^j_i) := \sum_{k=1}^{m+n} Y^k_i \frac{\partial \varphi^{\#}(x_j)}{\partial y_k}.$$

3. The difference between Ω_{VZ} and Ω_{Vi}

Before investigating the difference between VZ–forms and Vi–forms, we recall the notion of Berezinian. Let $\mathcal{A}$ be a superalgebra and let $\mathrm{GL}_{p|q}(\mathcal{A})$ denote the group of invertible $p|q$–square supermatrices with coefficients in $\mathcal{A}$. For

$$A = \begin{pmatrix} A_{11} & A_{12} \\ A_{21} & A_{22} \end{pmatrix} \in \mathrm{GL}_{p|q}(\mathcal{A}),$$

Ber : $\mathrm{GL}_{p|q}(\mathcal{A}) \longrightarrow \mathcal{A}_0$ ($\mathcal{A}_0$ is the even part of $\mathcal{A}$) is defined by

$$\mathrm{Ber}(A) = \frac{\det(A_{11} - A_{12}A_{22}^{-1}A_{21})}{\det(A_{22})}.$$

Since Berezinians are rational functions, $p|q$–densities on M are rational functions of the fiber part of $\mathrm{E}^{p|q}(M)$, for $q \neq 0$. This is also the reason why $p|q$–densities are defined on the $p|q$–frame bundle, not the $p|q$–tangent bundles (the $p|q$–tangent bundle means the Whitney sum of p–multi Whitney sum of the tangent bundle and q–multi Whitney sum of the parity changed tangent bundle).

On the other hand, $\Omega^{p|q}_{Vi}(M)$ is a $C^\infty(M)$–subsupermodule of $\mathrm{T}^{p|q}(\mathfrak{X}(M)^*)$, and thus any Vi–form is a polynomial of the fiber part of $\mathrm{T}^{p|q}(M)$. This situation causes the difference of VZ–forms and Vi–forms. An example is in Kato[6].

4. About Künneth formula for Vi–cohomology

Künneth formula for ordinary deRham cohomology is:

Theorem 1 *Let M, N be ordinary C^∞–manifold. Then,*

$$\mathrm{H}^p_{dR}(M \times N) \cong \bigoplus_{k_1+k_2=p} \mathrm{H}^{k_1}_{dR}(M) \otimes \mathrm{H}^{k_2}_{dR}(N)$$

However, the direct generalization to Vi–cohomology (also to VZ–cohomology) seems to be impossible even for superspaces (needless to say curved supermanifolds).

Victoria[9] tried to prove Künneth–like formula

Claim 1

$$\mathrm{H}^{p|q}_{(D_{Vi})_0}(\mathbb{R}^{m|n}) \cong \bigoplus_{q_1+q_2=q} \mathrm{H}^{0|q_1}_{(D_{Vi})_0}(\mathbb{R}^{m|0}) \otimes \mathrm{H}^{p|q_2}_{(D_{Vi})_0}(\mathbb{R}^{0|n})$$
$$\mathrm{H}^{p|q}_{(D_{Vi})_1}(\mathbb{R}^{m|n}) \cong \bigoplus_{p_1+p_2=p} \mathrm{H}^{p_1|q}_{(D_{Vi})_1}(\mathbb{R}^{m|0}) \otimes \mathrm{H}^{p_2|q}_{(D_{Vi})_1}(\mathbb{R}^{0|n})$$

by using double complexes.

For brevity, we denote

$$\overline{\Omega}^{p|q}(\mathbb{R}^{m|n}) := \bigoplus_{\substack{k_1+k_2=p \\ l_1+l_2=q}} \Omega^{k_1|l_1}_{Vi}(\mathbb{R}^{m|0}) \otimes \Omega^{k_2|l_2}_{Vi}(\mathbb{R}^{0|n}).$$

For any single term in $\Omega^{p|q}_{Vi}(\mathbb{R}^{m|n})$, there exists $\alpha_0 \in \Omega^{k_1|l_1}_{Vi}(\mathbb{R}^{m|0})$ and $\alpha_1 \in \Omega^{k_2|i_2}_{Vi}(\mathbb{R}^{0|n})$ satisfying $\alpha = \alpha_0 \otimes \alpha_1$ and they are unique up to multiplications by real numbers, thus it defines an isomorphism $\iota_{p|q} : \Omega^{p|q}_{Vi}(\mathbb{R}^{m|n}) \longrightarrow \overline{\Omega}^{p|q}(\mathbb{R}^{m|n})$.

Victoria defined the differential of the complex

$$\cdots \longrightarrow \overline{\Omega}^{p|q}(\mathbb{R}^{m|n}) \longrightarrow \overline{\Omega}^{p+1|q}(\mathbb{R}^{m|n}) \longrightarrow \cdots \tag{1}$$

by $(D_T)_0 := \bigoplus ((D_{Vi})_0 \otimes \mathrm{id} \oplus \mathrm{id} \otimes (D_{Vi})_0)$ and that of the complex

$$\cdots \longrightarrow \overline{\Omega}^{p|q}(\mathbb{R}^{m|n}) \longrightarrow \overline{\Omega}^{p|q+1}(\mathbb{R}^{m|n}) \longrightarrow \cdots \tag{2}$$

by $(D_T)_1 := \bigoplus ((D_{Vi})_1 \otimes \mathrm{id} \oplus \mathrm{id} \otimes (D_{Vi})_1)$, and claimed that the ι is a cochain isomorphism from (1) to $(D_{Vi})_0$–complex and (2) to $(D_{Vi})_1$–complex.

After the above preliminary, he tried to prove (1) by using the double complex. Although the argument about the double complex is correct, ι is not cochain isomorphism unfortunately (an example is the case of $\mathbb{R}^{1|1}$). The reason why the direct generalization of Künneth formula does not go well seems to be that differentials D_{Vi} mix the $\Omega^{p|0}_{Vi}$ part with $\Omega^{0|q}_{Vi}$ part. The similar situation seems to occur for VZ–cohomology.

5. The vanishing of Vi–cohomology

In the ordinary setting, the vanishing of deRham cohomology group of Euclidean spaces is proved by constructing homotopy operators, that is, $\mathbb{R}$–linear mapping $K : \Omega^{p}_{dR}(\mathbb{R}^{m+1}) \longrightarrow \Omega^{p+1}_{dR}(\mathbb{R}^{m+1})$ which satisfies

$$d \circ K \pm K \circ d = (s \circ \pi)^*.$$

Here d is the differential, $\pi : \mathbb{R}^{m+1} \longrightarrow \mathbb{R}^m$ is the projection, $s : \mathbb{R}^m \longrightarrow \mathbb{R}^{m+1}$ is the inclusion map, $(s \circ \pi)^*$ is the pull-back of $s \circ \pi$.

A similar proof is applicable for Vi–cohomology of superspaces, and we have:

Proposition 1

$$\mathrm{H}^{p|q}_{(D_{Vi})_0}(\mathbb{R}^{m|n}) = 0 \quad \textit{for} \quad p \neq 0 \tag{3}$$

$$\mathrm{H}^{p|q}_{(D_{Vi})_1}(\mathbb{R}^{m|n}) = 0 \quad \textit{for} \quad q \neq 0 \tag{4}$$

$\mathrm{H}^{0|q}_{D_{Vi}}(\mathbb{R}^{m|n})$ and $\mathrm{H}^{p|0}_{D_{Vi}}(\mathbb{R}^{m|n})$ seem to be infinite–dimensional vector spaces for most $p$or q.

6. The vanishing problem for VZ–cohomology

For VZ–cohomology of superspaces, Voronov[11] proved

Proposition 2

$$\mathrm{H}^{p|q}_{D_{VZ}}(\mathbb{R}^{m|n}) = 0 \qquad \textit{for } p > q,$$
$$\mathrm{H}^{p|q}_{D_{VZ}}(\mathbb{R}^{m|n}) \cong \mathrm{H}^{p|q}_{D_{VZ}}(\mathbb{R}^{0|n})\,.$$

Furthermore he proved for a general supermanifold M:

Theorem 2

$$\mathrm{H}^{p|0}_{D_{VZ}}(M) \cong \mathrm{H}^p(M_{\mathrm{red}}, \mathbb{R}), \qquad \mathrm{H}^{p|n}_{D_{VZ}}(M) \cong \mathrm{H}^p(M_{\mathrm{red}}, \mathbb{R})\,,$$

where $\mathrm{H}^*(\cdot, \mathbb{R})$ *is Cêch cohomology with* $\mathbb{R}$*–coefficients and* n *is the odd dimension of* M.

Proposition 2 means it is sufficient that we know VZ–cohomology of purely odd superspaces $\mathbb{R}^{0|n}$ in order to know VZ–cohomology of superspaces. However, it does not seem that $\mathrm{H}^{p|q}_{D_{VZ}}(\mathbb{R}^{0|n})$ have been determined explicitly for $p \leq q$ up to now.

Voronov[12] generalized VZ–cohomology and defined new cohomology of supermanifolds, named "stable cohomology" by him. This is a $\mathbb{Z}_2$-graded cohomology (which means the complex is graded by positive and negative integers, not only non-negative integers).

In the positive degree, stable cohomology coincides with VZ–cohomology for the positive degree (Voronov[12]), so it is sufficient that we consider VZ–cohomology. On the other hand, the determination of the negative half of stable cohomology of superspaces (and also supermanifolds) seems to be an interesting problem.

Fortunately, super version of Cartan's homotopy formula was found by Voronov[12], so the generalization of Proposition 2 and Theorem 2 might be possible by using this formula (actually Proposition 2 and Theorem 2 was proved by using VZ–cohomology version of Cartan's formula (Voronov[11])).

Acknowledgements

This work was essentially done while the author was a graduate student at Shinshu University. I would like to thank Prof Kojun Abe for his advice and encouragement. I also would like to thank Prof Dimitry Leites for many useful discussions.

References

1. F. A. Berezin, *Introduction to superanalysis.* D.Reidel Publishing Co., 1987.
2. F. A. Berezin and D. A. Leites, *Supermanifolds.* Soviet Math. Dokl. **16** (1975), 1218-1222.
3. J. N. Bernstein and D. A. Leites, *Invariant differential operators and irreducible representations of Lie superalgebras of vector fields.* Selecta Math. Sov. **1** (1981), 143–160.
4. J. N. Bernstein, D. A. Leites and V. S. Shander, *Seminar on supersymmetries,* Reports of Dept. Math. Stockholm Univ., Stockholm Univ.
5. C. P. Boyer and O. M. Sánchez-Valenzuela, *Lie supergroup actions on supermanifolds.* Trans. Amer. Math. Soc. **323** (1991), 151–175.
6. D. Kato, *The Poincarè type lemma for superspaces,* to appear.
7. D. A. Leites, *Cohomology of Lie superalgebras.* Funct. Anal. Appl. **9** (1975), 75–76.
8. D. A. Leites, *Introduction to the theory of supermanifolds.* Russ. Math. Surveys **35** (1980), 3–53.
9. C. M. Victoria, *Cohomology ring of supermanifolds.* New Zealand J. Math. **27** (1988), 123–144.
10. C. M. Victoria, *Symmetric and antisymmetric tensors over free supermodules.* New Zealand J. Math. **27** (1998), 277–291.
11. T. Voronov, *Geometric integration theory on supermanifolds.* Harwood Academic Publishers, 1991.
12. T. Voronov, *Dual forms on supermanifolds and Cartan calculus.* Comm. Math. Phys. **228** (2002), 1–16.
13. T. Voronov and A. V. Zorich, *Complex of forms on a supermanfiold.* Funct. Anal. Appl. **20** (1986), 132–133.

HIGHER DIMENSIONAL SPHERICAL D-BRANES AND MATRIX MODEL

YUSUKE KIMURA

Department of Physics, Kyoto University,
Kyoto, 606-8502, Japan
ykimura@gauge.scphys.kyoto-u.ac.jp

Some aspects of higher dimensional fuzzy spheres are investigated with the emphasis on their algebraic structure and dual descriptions based on [9].

1. Introduction

Several papers have been devoted to the study of relationships between noncommutative geometry and string theories. One of the most interesting facts is the appearance of matrix-valued coordinates of D-branes. Transverse coordinates of N D-branes are expressed by $U(N)$ matrices. When the transverse coordinates do not commute, the D-branes show characteristics of higher dimensional objects and are described by noncommutative geometry.

Let us consider a system composed of D0-branes in the flat background. When the coordinates satisfy the following noncommutative relationship

$$[X_1, X_2] = -iC1, \tag{1}$$

the D0-branes form a two-dimensional object (D2-brane). The size of the matrices have to be infinity in this case. This is a classical solution of a low energy effective action of D0-branes in the flat background. On the other hand, the same configuration can be obtained by considering a D2-brane in the presence of the strong NS-NS $B_{\mu\mu}$ field background. In this case, world-volume coordinates on the D2-brane are described by noncommutative geometry. The magnetic charge can be regarded as D0-branes. These two viewpoints are actually the same descriptions and represent a bound state of D0-branes and a D2-brane. In general, the above descriptions are summarized as follows: *Higher dimensional D-branes are obtained by imposing a noncommutative relationship to coordinates of lower dimen-*

sional D-branes. On the other hand, lower dimensional D-branes bounded on higher dimensional D-branes are described by a nontrivial gauge field configuration. The first description is used in matrix models. We can understand the origin of the two descriptions by investigating the Chern-Simons action:

$$S_{CS} = \mu_p \int_{p+1} Tr\left(P[e^{i\lambda^{-1} i_X i_X} \sum C^{(n)}] e^{\lambda F}\right), \qquad (2)$$

where $i_X i_X C^{(2)} = \frac{1}{2}[X^j, X^i] C^{(2)}_{ij}$ and $\lambda = 2\pi l_s^2$. X is a transverse coordinate of Dp-branes and $C^{(n)}$ is an R-R field. This action shows a coupling between D-branes with different dimensions. Let us see the typical two cases of $p = 0$ and $p = 2$ with some conditions:

- $p = 0$ with conditions $[X, X] \neq 0$ and $F = 0$;

$$S_{CS} = \mu_0 \int_{0+1} Tr\left(P[C^{(1)} + \frac{i}{2\lambda}[X_j, X_i] C^{(3)}_{ij} + \cdots]\right). \qquad (3)$$

  A commutator $Tr[X, X]$ provides the charge of D2-branes.

- $p = 2$ with $[X, X] = 0$ and $F \neq 0$;

$$S_{CS} = \mu_2 \int_{2+1} Tr\left(C^{(3)} + \lambda C^{(1)} \wedge F + \cdots\right). \qquad (4)$$

  The Chern class TrF provides the charge of D0-branes.

As these examples show, we can say that noncommutative geometrical description of D-branes is closely related to a gauge field configuration on D-branes.

We would like to study this correspondence for curved D-branes to understand the structure of curved noncommutative spaces. They have some characteristics which can not seen for the flat noncommutative space.

- Compact noncommutative spaces are described by matrices whose sizes are finite. Field theories on such spaces may be regarded as regularized theories.
- Curved noncommutative D-branes are constructed by identifying coordinates with generators of Lie algebra. The dynamics of the D-branes depends on the properties of the associated algebra.
- The construction of higher dimensional curved noncommutative spaces is difficult since such they do not always have symplectic structures.

In this paper, we study two descriptions of higher dimensional curved D-branes. We study spherical noncommutative spaces called fuzzy spheres as examples of curved noncommutative spaces. A fuzzy two-sphere has a dual description in terms of an abelian gauge field on a spherical D2-brane. On the other hand, we need a nonabelian gauge field in the case of a higher dimensional system. This difference makes higher dimensional noncommutative spaces complicated and interesting.

2. Fuzzy Sphere

In this section, we review the algebraic structure of higher dimensional fuzzy spheres [2,14,7,8]. We begin with the $SO(2k+1)$ algebra since it is a symmetry of a $2k$-sphere,

$$[\hat{G}_{\mu\nu}, \hat{G}_{\lambda\rho}] = 2\left(\delta_{\nu\lambda}\hat{G}_{\mu\rho} + \delta_{\mu\rho}\hat{G}_{\nu\lambda} - \delta_{\mu\lambda}\hat{G}_{\nu\rho} - \delta_{\nu\rho}\hat{G}_{\mu\lambda}\right). \tag{5}$$

We consider a representation whose highest weight state is labelled by $[0, \cdots, 0, n]$, where n is a positive integer. We have used the Dynkin index to label the representation. We denote the size of the matrix representation by N_k. It is calculated as

$$\begin{aligned} N_1 &= n+1, \quad N_2 = \frac{1}{6}(n+1)(n+2)(n+3), \\ N_3 &= \frac{1}{360}(n+1)(n+2)(n+3)^2(n+4)(n+5), \\ N_4 &= \frac{1}{302400}(n+1)(n+2)(n+3)^2(n+4)^2(n+5)^2(n+6)(n+7). \end{aligned} \tag{6}$$

A big difference between a fuzzy two-sphere and a fuzzy $2k$-sphere ($k \neq 1$) is that N_1 can take any positive integers while N_k ($k \neq 1$) cannot.

It is convenient to introduce $\hat{G}_\mu$ which satisfy

$$\begin{aligned} [\hat{G}_\mu, \hat{G}_\nu] &= 2\hat{G}_{\mu\nu}, \\ [\hat{G}_\mu, \hat{G}_{\nu\lambda}] &= 2\left(\delta_{\mu\nu}\hat{G}_\lambda - \delta_{\mu\lambda}\hat{G}_\nu\right). \end{aligned} \tag{7}$$

When $n = 1$, $\hat{G}_\mu$ becomes the $(2k+1)$-dimensional gamma matrix. Other relationships for $\hat{G}_\mu$ and $\hat{G}_{\mu\nu}$ are summarized in Appendix A. We can construct coordinates of a fuzzy $2k$-sphere as $X_\mu = \alpha\hat{G}_\mu$ because of the following relationship

$$\hat{G}_\mu\hat{G}_\mu = n(n+2k)1_{N_k}. \tag{8}$$

The radius of a fuzzy $2k$-sphere is then given by

$$r^2 = \alpha^2 n(n+2k). \tag{9}$$

A noncommutative scale on the fuzzy sphere can be defined as

$$l_{nc}^2 \equiv \alpha^2 n \simeq r^2/n. \tag{10}$$

We easily find that $\hat{G}_\mu$ and $\hat{G}_{\mu\nu}$ form the $SO(2k+2)$ algebra. When we define $\Sigma_{\mu\nu} \equiv \hat{G}_{\mu\nu}$ and $\Sigma_{2k+2,\mu} \equiv i\hat{G}_\mu$, Σ_{ab} $(a, b = 1, \cdots, 2k+2)$ satisfy the $SO(2k+2)$ algebra and belong to the irreducible spinor representation of $SO(2k+2)$.

We now make a comment on the number of independent matrices. $\hat{G}_\mu$ and $\hat{G}_{\mu\nu}$ have $(2k+1)$ and $k(2k+1)$ components respectively. Although there are totally $(k+1)(2k+1)$ components, the dimension of the corresponding noncommutative space is less than $(k+1)(2k+1)$ because of some constraints between them. The number of independent components is really $k(k+1)$. It is shown in [7] that the fuzzy $2k$-sphere is given by the coset space $SO(2k+1)/U(k)$. A fuzzy sphere is also shown to have a bundle structure over an usual sphere. This is a reason why it is called fuzzy sphere though the coset space is not given by $SO(2k+1)/SO(2k)$. In the remainder of this paper, we clarify this distinctive property from the viewpoint of the dual description of the fuzzy sphere.

3. D0-branes in the R-R field strength background

In this section, we consider a collection of D0-branes in a constant R-R field strength background. A coupling of D0-branes with the R-R field strength causes the so-called dielectric effect [12]. We will see that their transverse coordinates form the fuzzy sphere at an extremum of a low energy effective action of the D0-branes. The tension and the charge of a Dp-brane are defined as

$$T_p = \mu_p = \frac{2\pi}{g(2\pi l_s)^{p+1}}. \tag{11}$$

We also define $\lambda \equiv 2\pi l_s^2$. In this notation, the charge of a Dp-brane is related to that of a D0-brane as $\mu_0 = (2\pi\lambda)^{p/2}\mu_p$. Basically we follow the notation of [12].

The dynamics of a world-volume theory of D-branes is described by the Born-Infeld action [10] and the Chern-Simons action [11,5]. The case of multiple D-branes was further developed in [15,12]. The Born-Infeld action for N D0-branes in a flat space, with all other fields except transverse scalar fields vanishing, is given by

$$S_{BI} = -T_0 \int dt \left(N - \frac{\lambda^2}{4} Tr[\Phi_i, \Phi_j][\Phi_i, \Phi_j] \right), \tag{12}$$

where we have assumed the condition $\lambda[\Phi_i, \Phi_j] \ll 1$. Φ_i is an $N \times N$ matrix-valued coordinate whose dimension is $length^{-1}$, representing a transverse motion of N D0-branes. We define a coordinate whose dimension is $length$ as $X_i = \lambda\Phi_i$. The second term in (12) is also obtained form the dimensional reduction of the ten-dimensional $U(N)$ super Yang-Mills action.

We next consider the Chern-Simons term. A coupling of N D0-branes to the R-R potential is given in [15,12] by

$$
\begin{aligned}
S_{CS} &= \mu_0 \int Tr\left(P[e^{i\lambda i_\Phi i_\Phi} \sum C^{(n)}]\right) \\
&= \mu_0 \int Tr\left(P\left[C^{(1)} + i\lambda i_\Phi i_\Phi C^{(3)} - \frac{\lambda^2}{2}(i_\Phi i_\Phi)^2 C^{(5)} \right.\right. \\
&\qquad \left.\left. - i\frac{\lambda^3}{3!}(i_\Phi i_\Phi)^3 C^{(7)} + \frac{\lambda^4}{4!}(i_\Phi i_\Phi)^4 C^{(9)}\right]\right), \quad (13)
\end{aligned}
$$

where we have considered a situation that the NS-NS two-form field B and the gauge field strength F vanish. We now assume that only a constant R-R $(2k+2)$-form field strength is nonzero

$$
F^{(2k+2)}_{ti_1\ldots i_{2k+1}} = f_k \epsilon_{i_1\ldots i_{2k+1}}, \quad (14)
$$

where f_k is a constant and determined later. The leading order interaction term in this case is given by

$$
S_{CS} = -\frac{1}{(2k+1)}\frac{i^k}{k!}\lambda^{k+1} f_k \mu_0 \int dt Tr(\Phi_{i_1}\cdots\Phi_{i_{2k+1}})\epsilon_{i_1\ldots i_{2k+1}}. \quad (15)
$$

By combining (12) with (15), a low energy effective scalar potential of N D0-branes in a constant R-R $(2k+2)$-form field strength background is given by

$$
\begin{aligned}
V = \lambda^2 T_0 \left(-\frac{1}{4}Tr[\Phi_i, \Phi_j][\Phi_i, \Phi_j] \right. \\
\left. + \frac{1}{(2k+1)}\frac{i^k}{k!}\lambda^{k-1} f_k Tr(\Phi_{i_1}\cdots\Phi_{i_{2k+1}})\epsilon_{i_1\ldots i_{2k+1}}\right), \quad (16)
\end{aligned}
$$

where we have ignored the rest energy of N D0-branes, NT_0, which is not important in this discussion. The indices i, j run over $1, 2, \cdots, 2k+1$. f_k is determined by requiring the condition that a fuzzy $2k$-sphere becomes a classical solution of this matrix model. The equation of motion of the matrix model (16) is

$$
[\Phi_j, [\Phi_i, \Phi_j]] - \frac{i^k}{k!}\lambda^{k-1} f_k \Phi_{i_2}\cdots\Phi_{i_{2k+1}}\epsilon_{ii_2\ldots i_{2k+1}} = 0. \quad (17)
$$

We substitute the following ansatz

$$X_i = \lambda\Phi_i = \alpha\hat{G}_i^{(k)} \tag{18}$$

into the above equation. We then find that f_k should be given by

$$f_1 = \frac{4\alpha}{\lambda}, \quad f_2 = \frac{4}{\alpha(n+2)}, \quad f_3 = \frac{3\lambda}{\alpha^3(n+2)(n+4)},$$
$$f_4 = \frac{2\lambda^2}{\alpha^5(n+2)(n+4)(n+6)}. \tag{19}$$

We can calculate the value of the potential (16) for this solution as

$$V_k = 2k\left(1 - \frac{4}{2k+1}\right) T_0 \frac{\alpha^4}{\lambda^2} n(n+2k)N_k. \tag{20}$$

It must be noted that V_1 is negative while $V_k(k = 2, 3, 4)$ are positive.

Let us investigate another classical solution. A set of commuting matrices

$$[\Phi_i, \Phi_j] = 0 \tag{21}$$

is also a classical solution. Since Φ_i commute each other, they are simultaneously diagonalized and represent a set of N separated D0-branes. The potential energy (16) for this solution is zero. Since V_k in (20) is positive when $k = 2, 3, 4$, a fuzzy $2k$-sphere is classically unstable and is expected to collapse into the solution (21). This situation is completely opposite to the case of the fuzzy two-sphere. Since the potential energy of a fuzzy two-sphere is lower than that of the solution (21), a fuzzy two-sphere is classically stable[12]. This is one of differences between a fuzzy two-sphere and a fuzzy $2k$-sphere ($k \neq 2$). Another difference can be found by noticing that f_k depends on n when k takes $2, 3, 4$. Due to the n dependence of f_k, reducible representations cannot be classical solutions. (If we require a reducible representation to be a classical solution of the model, the irreducible representation cannot.) Taking account of these differences, the classical dynamics of the higher dimensional fuzzy sphere is completely different from that of the fuzzy two-sphere.

4. Dual description of fuzzy sphere

In the previous section, we realized a higher dimensional fuzzy sphere as a classical solution of a matrix model of D0-branes. Noncommutative geometry is fully used in this description. The classical solution really represents a bound state of spherical D$2k$-branes and D0-branes. From the viewpoint of a world-volume theory on D$2k$-branes, D0-branes are expressed by a nontrivial gauge field configuration. The Chern character provides the charge of D0-branes. In this section, we study this dual description of the higher dimensional fuzzy spheres [a]. We consider a system of spherical D$2k$-branes and a monopole gauge field in a constant R-R $(2k+2)$-form field strength background, and compare the potential energy with (20). We restrict the analysis to the case of $k=3$. Other cases are also studied in [9].

We begin by introducing an $SO(6)$ gauge field on a six-sphere:

$$A_\alpha = \frac{-i}{2r(r+x_7)}\Sigma^N_{\beta\alpha}x_\beta, \quad A_7 = 0 \tag{22}$$

where $\Sigma^N_{\alpha\beta} = (\Sigma^N_{ij}, \Sigma^N_{6i}) \equiv (\gamma_{ij}, i\gamma_i) = (\frac{1}{2}[\gamma_i,\gamma_j], i\gamma_i)$, γ_i $(i = 1,\cdots,5)$ is the five-dimensional gamma matrix. Notation of indices is as follows: $\mu,\nu = 1,2,\cdots,7$ and $\alpha,\beta = 1,2,\cdots,6$. a,b are used for the world-volume indices. $\Sigma^N_{\alpha\beta}$ transforms in the spinor representation of $SO(6)$. To relate the analysis in this section with that in the previous section, we need to consider a higher dimensional representation of $SO(6)$. Such a representation was already obtained in Section 2, the $k=2$ case being relevant in the present case. We replace the five-dimensional gamma matrices γ_i and γ_{ij} with the higher dimensional representation $\hat{G}^{(2)}_i$ and $\hat{G}^{(2)}_{ij}$. Accordingly $\Sigma^N_{\alpha\beta}$ is replaced with $\hat{\Sigma}^N_{\alpha\beta} = (\hat{\Sigma}^N_{ij}, \hat{\Sigma}^N_{2k,i}) \equiv (\hat{G}^{(2)}_{ij}, i\hat{G}^{(2)}_i)$. The index (2) is added to emphasize that the matrices are related to the $k=2$ case in Section 2. Since these matrices are realized as the size N_2, the gauge field becomes an $N_2 \times N_2$ matrix. What is interesting is these matrices construct a fuzzy four-sphere.

Before we begin calculations, let us explain the meaning of the index N in $\hat{\Sigma}^N_{\alpha\beta}$. Let us consider the N_3-dimensional irreducible representation of $SO(7)$, which is denoted by $\hat{G}^{(3)}_{\mu\nu}$ $(\mu,\nu = 1,\cdots,7)$. As is explained in Section 2, it is associated with a fuzzy six-sphere. We now consider an $SO(6)$ subalgebra $\hat{G}^{(3)}_{\alpha\beta}$ $(\alpha,\beta = 1,\cdots,6)$. The representation of $\hat{G}^{(3)}_{\alpha\beta}$ is reducible and is characterized by the eigenvalue of $\hat{G}^{(3)}_7 = diag(n, n-2,\cdots,-n+2-n)$ which is a generalized chirality matrix. We consider a subspace which is labelled by $\hat{G}^{(3)}_7 = n1$, which corresponds to the North pole of a fuzzy six-sphere. It can be shown that the size of the unit matrix

[a] A dual description of a fuzzy four-sphere and a fuzzy six-sphere is discussed in [3,4] in the context of a fuzzy funnel solution. In our paper, an emphasis is placed on a relationship between noncommutative geometry and (nonabelian) gauge fields. Some parts of the calculation in these papers overlap with those in our paper.

1 is N_2. This $\hat{G}^{(3)}_{\alpha\beta}$ which is labelled by $\hat{G}^{(3)}_7 = n1$ is equivalent to $\hat{\Sigma}^N_{\alpha\beta}$. This is the reason why we added the index N.

The field strength at the north pole ($x_\alpha = 0$, $x_7 = r$) is calculated as

$$F_{\alpha\beta} = -\frac{i}{2r^2}\hat{\Sigma}^N_{\alpha\beta}, \quad F_{7\alpha} = 0. \tag{23}$$

We note that it satisfies

$$\epsilon_{\alpha_1\alpha_2\alpha_3\alpha_4\alpha_5\alpha_6} F^{\alpha_3\alpha_4} F^{\alpha_5\alpha_6} = \frac{4}{r^2}(n+2)F_{\alpha_1\alpha_2}, \tag{24}$$

which is a generalization of a self-dual equation of the instanton. The calculation of the Chern numbers is easily done by making full use of some formulae in Appendix A. The first and second Chern numbers are calculated as $c_1 = TrF \simeq Tr\Sigma_{\alpha\beta} = 0$ and

$$\begin{aligned} c_2 &\sim Tr\,(F \wedge F) \\ &\sim \epsilon_{\alpha_1\alpha_2\alpha_3\alpha_4} Tr\,(F_{\alpha_1\alpha_2}F_{\alpha_3\alpha_4}) \\ &\sim \epsilon_{i_1i_2i_3i_4} Tr\left(\hat{G}^{(2)}_{i_1i_2}\hat{G}^{(2)}_{i_3i_4}\right) + \epsilon_{i_1i_2i_3} Tr\left(\hat{G}^{(2)}_{i_1i_2}\hat{G}^{(2)}_{i_3}\right) \\ &\sim Tr\left(\hat{G}^{(2)}_i\right) + Tr\left(\hat{G}^{(2)}_{ij}\right) = 0. \end{aligned} \tag{25}$$

These imply that the net charge of D2-branes and that of D4-branes vanish. The third Chern number is calculated as

$$\begin{aligned} c_3 &= \frac{1}{48\pi^3}\int_{S^6} Tr\,(F \wedge F \wedge F) \\ &= \frac{1}{48\pi^3}\frac{1}{8}\Omega_6 r^6 \sum_{\alpha_i=1}^{6} Tr\,(F_{\alpha_1\alpha_2}F_{\alpha_3\alpha_4}F_{\alpha_5\alpha_6})\,\epsilon_{\alpha_1\alpha_2\alpha_3\alpha_4\alpha_5\alpha_6} \\ &= \frac{1}{360}n(n+1)(n+2)^2(n+3)(n+4) \equiv \bar{c}_3. \end{aligned} \tag{26}$$

$\Omega_6 = 16\pi^3/15$ is the volume of a six-sphere with a unit radius. c_3 represents the number of D0-branes and can be compared with N_3 in (6). We find that these two values, $\bar{c}_3$ and N_3, coincide at large n.

We now consider a world-volume action for N_2 $D6$-branes with the monopole field (23). It is given by the Born-Infeld action:

$$S_{BI} = -T_6 \int d^{6+1}\sigma Str\sqrt{-\det(P[G+\lambda F]_{ab})}. \tag{27}$$

The determinant is evaluated as follows,

$$\begin{aligned} -\det(G_{ab}+\lambda F_{ab}) &= 1 + \frac{\lambda^2}{2}F_{\alpha\beta}F_{\alpha\beta} \\ &\quad + \frac{\lambda^4}{128}(F_{\alpha_1\alpha_2}F_{\alpha_3\alpha_4}\epsilon^{\alpha\beta\alpha_1\alpha_2\alpha_3\alpha_4})(F_{\beta_1\beta_2}F_{\beta_3\beta_4}\epsilon^{\alpha\beta\beta_1\beta_2\beta_3\beta_4}) \end{aligned}$$

$$+\frac{\lambda^6}{2304}\left(\epsilon^{\alpha_1\alpha_2\alpha_3\alpha_4\alpha_5\alpha_6}F_{\alpha_1\alpha_2}F_{\alpha_3\alpha_4}F_{\alpha_5\alpha_6}\right)^2$$
$$\simeq 1+\frac{3}{4}\left(\frac{\lambda n}{r^2}\right)^2+\frac{3}{16}\left(\frac{\lambda n}{r^2}\right)^4+\frac{1}{64}\left(\frac{\lambda n}{r^2}\right)^6, \quad (28)$$

where we have assumed a large n to obtain the last expression. We define $y \equiv r^2/\lambda n$. The Born-Infeld action is then evaluated as

$$S_{BI}/\int dt = -T_6\Omega_6 r^6 N_2\sqrt{1+\frac{3}{4y^2}+\frac{3}{16y^4}+\frac{1}{64y^6}}$$
$$\simeq -\bar{c}_3 T_0 - \frac{3}{4}T_6 N_2\Omega_6 r^4\lambda n. \quad (29)$$

To obtain the last expression, we have expanded the square root by assuming the condition $y \ll 1$. This assumption is valid since it satisfies (38). We next consider the coupling of D6-branes to the external R-R field. The constant R-R eight-form field strength background (14) is given from the following R-R seven-form field in a certain gauge,

$$C^{(7)}_{t123456} = \frac{1}{7}f_3 x_7 \simeq \frac{1}{7}f_3 r. \quad (30)$$

The Chern-Simons term becomes

$$S_{CS} = \mu_6\int dt\left(N_2\frac{f_3}{7}\Omega_6 r^7\right), \quad (31)$$

and the potential for the D6-branes is provided by

$$V(r) = \bar{c}_3 T_0 + T_6\Omega_6 N_2\left(\frac{3}{4}\lambda n r^4 - \frac{f_3}{7}r^7\right). \quad (32)$$

The first term represents the rest energy of $\bar{c}_3$ D0-branes. We search for extrema of this potential by regarding it as a function of r. We find two extrema; one is a trivial extremum $r=0$ and another is

$$r = \sqrt[3]{\frac{3\lambda n}{f_3}} = \sqrt[3]{\alpha^3 n(n+2)(n+4)} \equiv r_\star. \quad (33)$$

This corresponds to a fuzzy six-sphere solution in the matrix model of D0-branes. The radius of this spherical solution agrees with that of the fuzzy six sphere (9) at large n. The potential value for this extremum is

$$V(r_\star) - \bar{c}_3 T_0 \simeq T_0\frac{1}{140}\frac{\alpha^4}{\lambda^2}n^8. \quad (34)$$

We recall the D0-brane calculation in the previous section:

$$V_3 = T_0\frac{1}{140}\frac{\alpha^4}{\lambda^2}n(n+1)(n+2)(n+3)^2(n+4)(n+5)(n+6)$$

$$\simeq T_0 \frac{1}{140} \frac{\alpha^4}{\lambda^2} n^8. \tag{35}$$

$V(r_\star)$ and V_3 give the same value including the coefficient when n is large. This result manifests the fact that a fuzzy six-sphere of the matrix size N_3 is the same object as N_2 spherical D6-branes with a nonabelian $SO(6)$ monopole gauge field.

Let us consider the validity of these two descriptions [3,13]. We first analyzed the world-volume theory of D0-branes. The assumption $\lambda[\Phi_i, \Phi_j] \ll 1$ was used to derive the low energy effective action of D0-branes (12). Since this condition is rewritten as

$$l_{nc}^2 \simeq \frac{r^2}{n} \ll l_s^2, \tag{36}$$

the noncommutative scale l_{nc} has to be much smaller than the string scale l_s. On the other hand, the computations in the world-volume action of D6-branes can be trusted as long as the field strength is slowly varying $| l_s \partial F | \ll | F |$. This condition is satisfied when the radius is much larger than the string scale:

$$l_s \ll r. \tag{37}$$

If r satisfies the following inequalities

$$l_s \ll r \ll \sqrt{n} l_s, \tag{38}$$

both of (36) and (37) are satisfied. It is sometimes convenient to rewrite (38) as

$$\frac{l_s}{\sqrt{n}} \ll l_{nc} \ll l_s. \tag{39}$$

We can expect the agreement of two descriptions in a large region by taking a large n limit. This is the reason why we could obtain the agreement of two descriptions in a large n limit. We can provide another explanation in Section 5.

5. Noncommutativity and nonabelian gauge field

In the previous sections, we considered two world-volume theories. Since they provide the same values for various quantities at large n, they are supposed to be same. In this section, we discuss a concrete correspondence by noticing a relationship between the noncommutativity of the fuzzy sphere and the nonabelian gauge fields. An interesting fact is that the nonabelian gauge field on D6-branes is expressed by a matrix which are related to the fuzzy 4-sphere. We recall the gauge field strength at the north pole ($x_\alpha = 0$, $x_7 = r$):

$$F_{\alpha\beta} = -\frac{i}{2r^2}\hat{\Sigma}^N_{\alpha\beta}, \tag{40}$$

where $\hat{\Sigma}^N_{\alpha\beta} = (\hat{\Sigma}^N_{ij}, \hat{\Sigma}^N_{2k,i}) = (\hat{G}^{(2)}_{ij}, i\hat{G}^{(2)}_i)$ $(i = 1, \cdots, 5)$. We note that the matrix $\hat{G}^{(2)}_{ij}$ is realized by the N_2-dimensional irreducible representation of $SO(5)$ and forms the fuzzy 4-sphere algebra.

A commutation relation for coordinates of a fuzzy 6-sphere is given by the first equation in (7). This relation reduces to the following relation at the north pole of a fuzzy 6-sphere ($\hat{G}_7 = n$),

$$[\hat{G}^N_\alpha, \hat{G}^N_\beta] = 2\hat{G}^N_{\alpha\beta}. \tag{41}$$

As is explained in the previous section, $\hat{G}^N_{\alpha\beta}$ is the same matrix as $\Sigma^N_{\alpha\beta}$. Since the two descriptions agree in the large n limit, we may combine two relations (40) and (41) in the large n limit. Therefore we are led to the following noncommutative relationship,

$$[X_\alpha, X_\beta] = 4i\alpha^2 r^2 F_{\alpha\beta}, \tag{42}$$

where $X_\alpha = \alpha\hat{G}^N_\alpha$. We have identified the noncommutativity of the fuzzy six-sphere with the field strength of the monopole.

The commutation relation (42) is valid at the north pole. It is natural to expect that such a relation holds without restricting to the north pole because a fuzzy 6-sphere has the $SO(7)$ symmetry. Therefore we suppose

$$[X_\mu, X_\nu] = 2\alpha^2\hat{G}_{\mu\nu} = 4i\alpha^2 r^2 F_{\mu\nu}, \tag{43}$$

where the size of the matrices is N_3. $F_{\mu\nu}$ has been expressed by the N_3-dimensional irreducible representation of $SO(7)$. This relationship is interesting in the following sense. In the fuzzy sphere algebra (5) and (7), $\hat{G}_{\mu\nu}$ and $\hat{G}_\mu$ are treated on the same footing. If we regard $\hat{G}_\mu$ as a coordinate of a fuzzy sphere, $\hat{G}_{\mu\nu}$ acts on it as the $SO(7)$ rotation generator (or we may regard $\hat{G}_{\mu\nu}$ as a coordinate of an extra space). On the other hand, the role of coordinates and that of a field strength in the dual description are clearly different. The relation (43) suggests that the internal degrees of freedom $F_{\mu\nu}$ should be identified with the rotation generator in the large n limit. This is a characteristic of noncommutative geometry. This identification can be also seen in the lowest Landau level physics of a higher dimensional quantum Hall system [16,1,6]. Considering the analogy to the quantum Hall system can provide an intuitive explanation for the agreement of two descriptions in the large n limit. In this system, the angular momentum operator $\Lambda_{\mu\nu} = -i(x_\mu D_\nu - x_\nu D_\mu)$ and the monopole field strength are related as

$$G_{\mu\nu} = \Lambda_{\mu\nu} + 2ir^2 F_{\mu\nu}. \tag{44}$$

$\Lambda_{\mu\nu}$ does not satisfy the $SO(7)$ algebra due to the existence of the monopole background, and it is $G_{\mu\nu}$ that satisfies the $SO(7)$ algebra. The angular momentum generated by $\Lambda_{\mu\nu}$ characterizes the Landau level and therefore the representation of $G_{\mu\nu}$ depends on the Landau level. Since the magnitude of the field strength F in (40) is given by $O(n/r^2)$, the contribution of the second term in the right hand side of (44) becomes large compared to the first term in a large n limit. Therefore the field strength $F_{\mu\nu}$ is identified with the rotation generator $G_{\mu\nu}$. It can be also shown that $G_{\mu\nu}$ is given by the spinor representation of $SO(7)$ [16,6]. This limit is just the lowest Landau level limit. As is well known in the two-dimensional quantum Hall system, coordinates of electrons are described by noncommutative geometry in the lowest Landau level. Fuzzy spheres are actually realized in the higher dimensional quantum Hall system after we take the lowest Landau level limit. This is an intuitive explanation for the agreement of two descriptions in the large n limit.

6. Summary

We have considered two descriptions for a bound state of D6-branes and D0-branes:

- D0-brane description:
  a fuzzy six-sphere
- D6-brane description:
  spherical D6-branes with an $SO(6)$ monopole gauge field

These provide the same descriptions when n is large. We make some comments about this correspondence:

- Coordinates and monopole field strength are treated on the same footing when we take the strong magnetic field limit.
- Due to the existence of a nonabelian gauge field, a fuzzy sphere has some extra spaces compared to a usual sphere.

Acknowledgments

I would like to thank the organizers of *Noncommutative Geometry and Physics 2004* for inviting me to give this talk and for their hospitality.

Appendix A. Some Formulae of Fuzzy Sphere

In this appendix, we summarize several formulae involving $\hat{G}_\mu$ and $\hat{G}_{\mu\nu}$ in diverse dimensions. The dimension N_k is given by

$$\begin{aligned} N_1 &= n+1, \quad N_2 = \frac{1}{6}(n+1)(n+2)(n+3), \\ N_3 &= \frac{1}{360}(n+1)(n+2)(n+3)^2(n+4)(n+5), \\ N_4 &= \frac{1}{302400}(n+1)(n+2)(n+3)^2(n+4)^2(n+5)^2(n+6)(n+7), \end{aligned} \tag{B.1}$$

where n is a positive integer. We have the following relations

$$\hat{G}_\mu \hat{G}_\mu = n(n+2k) \tag{B.2}$$

and

$$\hat{G}_{\mu\nu}\hat{G}_{\nu\mu} = 2kn(n+2k). \tag{B.3}$$

The following relations are also satisfied

$$\hat{G}_{\mu\nu}\hat{G}_\nu = 2k\hat{G}_\mu \tag{B.4}$$

and

$$\hat{G}_{\mu\nu}\hat{G}_{\nu\lambda} = n(n+2k)\delta_{\mu\lambda} + (k-1)\hat{G}_\mu\hat{G}_\lambda - k\hat{G}_\lambda\hat{G}_\mu. \tag{B.5}$$

G_μ satisfy the following relation

$$\epsilon^{\mu_1\cdots\mu_{2k}\mu_{2k+1}}\hat{G}_{\mu_1}\cdots\hat{G}_{\mu_{2k}} = C_k\hat{G}_{\mu_{2k+1}} \tag{B.6}$$

where $\epsilon^{\mu_1\cdots\mu_{2k}\mu_{2k+1}}$ is the $SO(2k+1)$ invariant tensor. C_k is a constant which depends on n,

$$\begin{aligned} &C_1 = 2i, \quad C_2 = 8(n+2), \quad C_3 = -48i(n+2)(n+4), \\ &C_4 = -384(n+2)(n+4)(n+6). \end{aligned} \tag{B.7}$$

References

1. B.A. Bernevig, J.P. Hu, N. Toumbas and S.C. Zhang, *The Eight Dimensional Quantum Hall Effect and the Octonions.* **cond-mat/0306045**.
2. J. Castelino, S. Lee and W. Taylor, *Longitudinal 5-branes as 4-spheres in Matrix theory.* Nucl.Phys. **B526** (1998), 334.

3. N.R. Constable, R.C. Myers and O. Tafjord, *Non-abelian Brane Intersections.* JHEP **0106** (2001), 023.
4. P. Cook, R. de Mello Koch, and J. Murugan, *Non-Abelian BIonic Brane Intersections.* Phys. Rev. **D68** (2003), 126007.
5. M.R. Douglas, *Branes within Branes.* **hep-th/9512077**.
6. K. Hasebe and Y. Kimura, *Dimensional Hierarchy in Quantum Hall Effects on Fuzzy Spheres.* Phys. Lett. **B602** (2004), 255.
7. P.M. Ho and S. Ramgoolam, *Higher Dimensional Geometries from Matrix Brane constructions.* Nucl.Phys. **B627** (2002), 266.
8. Y. Kimura, *On Higher Dimensional Fuzzy Spherical Branes.* Nucl.Phys. **B664** (2003), 512.
9. Y. Kimura, *Nonabelian gauge field and dual description of fuzzy sphere.* JHEP **0404** (2004), 058.
10. R.G. Leigh, *Dirac-Born-Infeld Action from Dirichlet Sigma Model.* Mod. Phys. Lett. **A4** (1989), 2767.
11. M. Li, *Boundary States of D-Branes and Dy-Strings.* Nucl.Phys. **B460** (1996), 351.
12. R. Myers, *Dielectric-Branes.* JHEP **9912** (1999), 022.
13. R.C. Myers, *Nonabelian Phenomena on D-branes.* Class. Quant. Grav. **20** (2003), S347.
14. S. Ramgoolam, *On spherical harmonics for fuzzy spheres in diverse dimensions.* Nucl.Phys. **B610** (2001), 461.
15. W. Taylor and M.V. Raamsdonk, *Multiple Dp-branes in Weak Background Fields.* Nucl.Phys. **B573** (2000), 703.
16. S. Zhang and J. Hu, *A Four Dimensional Generalization of the Quantum Hall Effect.* Science **294** (2001), 823.

INSTANTONS OF σ-MODELS IN NONCOMMUTATIVE GEOMETRY

GIOVANNI LANDI

Dipartimento di Matematica e Informatica, Università di Trieste
Via A. Valerio 12/1, I-34127, Trieste, Italia
and INFN, Sezione di Napoli, Napoli, Italia
landi@univ.trieste.it

We review some recent work on harmonic maps and non-linear σ-models in noncommutative geometry. After a general discussion we concentrate mainly on models on noncommutative tori.

1. Introduction

Recent applications of noncommutative geometry started with the papers [7] and [28] on matrix and string theories. These developments involve gauge theories defined on noncommutative spaces, a first model of which was constructed almost twenty years ago [8]. One has a vast formalism for gauge fields in noncommutative geometry which allows one to define connections, their curvatures, the associated Yang-Mills action and most of the classical concepts [4]. Methods of differential topology are also available within the realm of cyclic and Hochschild homology and cohomology and this leads, via the coupling of the former with K-theory, to quantities that are stable under deformation and that generalize topological invariants like, for instance, winding numbers and topological charges. A notably results is the fact that one can prove a topological bound for the Yang-Mills action in four dimension [4]. Also, one can construct a Chern-Simons type theory and interpret its behavior under large gauge transformations as a coupling between cyclic cohomology and K-theory [12].

In [9] and [10] we have constructed noncommutative analogues of two dimensional non-linear σ-models. In a more mathematical parlance, these are examples of noncommutative harmonic maps. We have proposed three different models: a continuous analogue of the Ising model which admits instantonic solutions, the analogue of the principal chiral model together with its infinite number of conserved currents and a noncommutative Wess-

Zumino-Witten model together with its modified conformal invariance. In particular, we constructed instantonic solutions carrying a nontrivial topological charge q and satisfying a Belavin-Polyakov bound [1], an example of the use of cyclic and Hochschild homology and cohomology mentioned above. The moduli space of these instantons can be identified with an ordinary torus endowed with a complex structure times a projective space $\mathbb{CP}^{q-1}$. This work have relations with recent work on quantized theta-functions [16,17,18,19,30] as well as work on complex and holomorphic structures on noncommutative spaces, notably noncommutative tori [4,6,11,23,29].

2. Noncommutative harmonic maps

Ordinary non-linear σ-models are field theories whose configuration space consists of maps X from the *source* space, a Riemannian manifold (Σ, g) which we assume to be compact and orientable, to a *target* space, an other Riemannian manifold $(\mathcal{M}, G)$. From a more mathematical point of view they are examples of a theory of harmonic maps (see [33] and [13] for reviews).

In local coordinates one has an action functional given by

$$S[X] = \frac{1}{2\pi} \int_\Sigma \sqrt{g}\, g^{\mu\nu}\, G_{ij}(X) \partial_\mu X^i\, \partial_\nu X^j, \tag{1}$$

where $g = \det(g_{\mu\nu})$ and $g^{\mu\nu}$ is the inverse of $g_{\mu\nu}$; moreover $\mu, \nu = 1, \ldots, \dim \Sigma$, and $i, j = 1, \ldots, \dim \mathcal{M}$. Here and in the following we use the convention of summing over repeated indexes, and indices are lowered and raised by means of a metric.

The stationary points of the function (1) are harmonic maps from Σ to $\mathcal{M}$ and describe minimal surfaces embedded in $\mathcal{M}$.

Different choices of the source and target spaces lead to different field theories, some of them playing a major role in physics. Especially interesting are their applications to conformal field theory and statistical field theory (see for instance [14]). Furthermore, in supersymmetric generalizations, they are the basic building blocks of superstring theory (see for instance [15] and [21]).

When Σ is two dimensional the action S is conformally invariant, namely it is left invariant by any rescaling of the metric $g \to g e^\sigma$, where σ is any map from Σ to $\mathbb{R}$. As a consequence, the action only depends on the conformal class of the metric and may be rewritten using a complex structure on Σ as

$$S[X] = \frac{i}{\pi} \int_\Sigma G_{ij}(X)\, \partial X^i \wedge \overline{\partial} X^j, \tag{2}$$

where $\partial = \partial_z dz$ and $\overline{\partial} = \partial_{\overline{z}} d\overline{z}$, z being a suitable local complex coordinate.

To construct a noncommutative generalization of the previous construction, we must first dualize the picture and reformulate it in terms of the $*$-algebras $\mathcal{A}$ and $\mathcal{B}$ of complex valued smooth functions defined respectively on Σ and $\mathcal{M}$. Embeddings X of Σ into $\mathcal{M}$ are then in one to one correspondence with $*$-algebra morphisms π from $\mathcal{B}$ to $\mathcal{A}$, the correspondence being simply given by pullback, $\pi_X(f) = f \circ X$.

All this makes perfectly sense for (fixed) not necessarily commutative algebras $\mathcal{A}$ and $\mathcal{B}$; both algebras are over $\mathbb{C}$ and for simplicity we take them to be unital. Then, we take as configuration space the space of all $*$-algebra morphisms from $\mathcal{B}$ (the target algebra) to $\mathcal{A}$ (the source algebra). To define an action functional we need noncommutative generalizations of the conformal and Riemannian geometries. Following an idea of Connes [4,6] conformal structures can be understood within the framework of positive Hochschild cohomology. In the commutative case the following tri-linear map $\Phi : \mathcal{A} \otimes \mathcal{A} \otimes \mathcal{A} \to \mathbb{R}$

$$\Phi(f_0, f_1, f_2) = \frac{i}{\pi} \int_\Sigma f_0 \partial f_1 \wedge \overline{\partial} f_2 \tag{3}$$

is an extremal element of the space of positive Hochschild cocycles that belongs to the Hochschild cohomology class of the cyclic cocycle Ψ given by

$$\Psi(f_0, f_1, f_2) = \frac{i}{2\pi} \int_\Sigma f_0 df_1 \wedge df_2. \tag{4}$$

Clearly, both (3) and (4) still make perfectly sense for a general noncommutative algebra $\mathcal{A}$. One can say that Ψ allows to integrate 2-forms in dimension 2,

$$\frac{i}{2\pi} \int a_0 da_1 da_2 = \Psi(a_0, a_1, a_2) \tag{5}$$

so that it is a metric independent object, whereas Φ defines a suitable scalar product

$$\langle a_0 da_1, b_0 db_1 \rangle = \Phi(b_0^* a_0, a_1, b_1^*) \tag{6}$$

on the space of 1-forms and thus depends on the conformal class of the metric. Furthermore, this scalar product is positive and invariant with respect to the action of the unitary elements of $\mathcal{A}$ on 1-forms, and its relation to the cyclic cocycle Ψ allows to prove various inequalities involving topological quantities. In particular we shall get a topological bound for the action which is a two dimensional analogue of the inequality in four dimensional Yang-Mills theory. In analogy with this theory, the configurations giving (absolute) minima will be called σ-model instantons.

Having such a cocycle Φ we can compose it with a morphism $\pi : \mathcal{B} \to \mathcal{A}$ to obtain a positive cocycle on $\mathcal{B}$ defined by

$$\Phi_\pi = \Phi \circ (\pi \otimes \pi \otimes \pi). \tag{7}$$

Since our goal is to build an action functional, which assigns a number to any morphism π, we have to evaluate the cocycle Φ_π on a suitably chosen element of $\mathcal{B} \otimes \mathcal{B} \otimes \mathcal{B}$. Such an element is provided by the noncommutative analogue of the metric on the target, which we take simply as a positive element

$$G = \sum_i b_0^i \delta b_1^i \delta b_2^i \tag{8}$$

of the space of universal 2-forms $\Omega^2(\mathcal{B})$. Thus, the quantity

$$S[\pi] = \Phi_\pi(G) \tag{9}$$

is well defined and positive. We take it as a noncommutative analogue of the action functional (2) for non-linear σ-models.

Clearly, we consider π as the dynamical variable (the embedding) whereas Φ (the conformal structure on the source) and G (the metric on the target) are background structures that have been fixed. Alternatively, one could take only the metric G on the target as a background field and use the morphism $\pi : \mathcal{B} \to \mathcal{A}$ to define the induced metric $\pi_* G$ on the source as

$$\pi_* G = \sum_i \pi(b_0^i) \delta\pi(b_1^i) \delta\pi(b_2^i), \tag{10}$$

which is obviously a positive universal 2-form on $\mathcal{A}$. To such an object one can associate, by means of a variational problem (see [4] and [6]), a positive Hochschild cocycle that stands for the conformal class of the induced metric.

The critical points of the σ-model corresponding to the action functional (9) are noncommutative generalization of harmonic maps and describe 'minimally embedded surfaces' in the noncommutative space associated with $\mathcal{B}$.

3. Models on the noncommutative torus

In this section we shall work out explicitly the previous construction for the noncommutative torus as the source space and a two point space as a target. The cocycle Φ and the metric G will be replaced by their explicit simple expressions.

3.1. *Two points as a target space*

The simplest example of a target space one can think of is that of a finite space made of two points $\mathcal{M} = \{1, 2\}$, like in the Ising model. Of course, any continuous map from a connected surface to a discrete space is constant and the resulting (commutative) theory would be trivial. However, this is no longer true if the source is a noncommutative space and one has, in general, lots of such maps (i.e. algebra morphisms). Now, the algebra of functions over $\mathcal{M} = \{1, 2\}$ is just $\mathcal{B} = \mathbb{C}^2$ and any element $f \in \mathcal{M}$ is a couple of complex numbers (f_1, f_2) with $f_a = f(a)$, the value of f at the point a. As a vector space $\mathcal{B}$ is generated by the function e defined by $e(1) = 1$, $e(2) = 0$. Clearly e is a hermitian idempotent (a projection), $e^2 = e^* = e$, and $\mathcal{B}$ can be thought of as the unital *-algebra generated by such a projection e. As a consequence, any $*$-algebra morphism π from $\mathcal{B}$ to $\mathcal{A}$ is given by a projection $p = \pi(e)$ in $\mathcal{A}$.

Choosing the metric $G = \delta e \delta e$ on the space $\mathcal{M}$ of two points, the action functional (9) simply becomes

$$S[p] = \Phi(1, p, p), \tag{11}$$

where Φ is a given Hochschild cocycle standing for the conformal structure.

As we have already mentioned, from general consideration of positivity in Hochschild cohomology this action is bounded by a topological term [4]. In the following we shall explicitly prove this fact when taking the noncommutative torus as source space.

3.2. *The Noncommutative Torus*

We recall the very basic aspects of the noncommutative torus that we shall need in the following and refer the reader to [27] for a survey.

Consider an ordinary square two-torus $\mathbb{T}^2$ with coordinate functions $U_1 = \mathrm{e}^{2\pi i x}$ and $U_2 = \mathrm{e}^{2\pi i y}$, where $x, y \in [0, 1]$. By Fourier expansion the algebra $C^\infty(\mathbb{T}^2)$ of complex-valued smooth functions on the torus is made up of all power series of the form

$$a = \sum_{(m,n)\in\mathbb{Z}^2} a_{m,n}\, U_1^m\, U_2^n\ , \tag{12}$$

with $\{a_{m,n}\} \in S(\mathbb{Z}^2)$ a complex-valued Schwartz function on $\mathbb{Z}^2$. This means that the sequence of complex numbers $\{a_{m,n} \in \mathbb{C} \mid (m, n) \in \mathbb{Z}^2\}$ decreases rapidly at 'infinity', i.e. for any $k \in \mathbb{N}_0$ one has bounded seminorms

$$||a||_k = \sup_{(m,n)\in\mathbb{Z}^2} |a_{m,r}| \left(1 + |m| + |r|\right)^k < \infty \, . \tag{13}$$

Let us now fix a real number θ. The algebra $\mathcal{A}_\theta = C^\infty(\mathbb{T}^2_\theta)$ of smooth functions on the noncommutative torus is the associative algebra made up of all elements of the form (12), but now the two generators U_1 and U_2 satisfy

$$U_2 \, U_1 = \mathrm{e}^{\,2\pi i\theta} \, U_1 \, U_2 \, . \tag{14}$$

The algebra $\mathcal{A}_\theta$ can be made into a $*$-algebra by defining a involution by

$$U_1^* := U_1^{-1} \, , \quad U_2^* := U_2^{-1} \, . \tag{15}$$

From (13) with $k = 0$ one gets a C^*-norm and the corresponding closure of $\mathcal{A}_\theta$ in this norm is the universal C^*-algebra A_θ generated by two unitaries with the relation (14); the algebra $\mathcal{A}_\theta$ is dense in A_θ and is thus a pre-C^*-algebra.

There is a one-to-one correspondence between elements of the noncommutative torus algebra $\mathcal{A}_\theta$ and the commutative torus algebra $C^\infty(\mathbb{T}^2)$ given by the Weyl map Ω and its inverse, the Wigner map. As is usual for a Weyl map there are operator ordering ambiguities, and so we will take the prescription

$$\Omega\left(\sum_{(m,r)\in\mathbb{Z}^2} f_{m,r} \; \mathrm{e}^{\,2\pi i(m\,x + r\,y)}\right) := \sum_{(m,r)\in\mathbb{Z}^2} f_{m,r} \; \mathrm{e}^{\,\pi i m\, r\,\theta} \; U_1^m \, U_2^r \, . \tag{16}$$

This choice (called Weyl or symmetric ordering) maps real-valued functions on $\mathbb{T}^2$ into Hermitian elements of $\mathcal{A}_\theta$. The inverse map is given by

$$\Omega^{-1}\left(\sum_{(m,r)\in\mathbb{Z}^2} a_{m,r} \; U_1^m \, U_2^r\right) = \sum_{(m,r)\in\mathbb{Z}^2} a_{m,r} \; \mathrm{e}^{\,-\pi i m\, r\,\theta} \; \mathrm{e}^{\,2\pi i(m\,x + r\,y)} \, . \tag{17}$$

Clearly, the map $\Omega : C^\infty(\mathbb{T}^2) \rightarrow \mathcal{A}_\theta$ is not an algebra homomorphism. It can be used to deform the commutative product on the algebra $C^\infty(\mathbb{T}^2)$ into a noncommutative star-product by defining

$$f \star g := \Omega^{-1}\big(\Omega(f)\,\Omega(g)\big) \, , \quad f, g \in C^\infty\left(\mathbb{T}^2\right) \, . \tag{18}$$

A straightforward computation gives

$$f \star g = \sum_{(r_1,r_2)\in\mathbb{Z}^2} (f \star g)_{r_1,r_2} \, \mathrm{e}^{\,2\pi i(r_1 x + r_2 y)} \, , \tag{19}$$

with the coefficients of the expansion of the star-product given by a twisted convolution

$$(f \star g)_{r_1,r_2} = \sum_{(s_1,s_2)\in\mathbb{Z}^2} f_{s_1,s_2}\, g_{r_1-s_1,r_2-s_2}\ \mathrm{e}^{\,\pi i(r_1 s_2 - r_2 s_1)\,\theta} \tag{20}$$

which reduces to the usual Fourier convolution product in the limit $\theta = 0$. Up to isomorphism, the deformed product depends only on the cohomology class in the group cohomology $\mathsf{H}^2(\mathbb{Z}^2, U(1))$ of the $U(1)$-valued two-cocycle on $\mathbb{Z}^2$ given by

$$\lambda(r,s) := \mathrm{e}^{\,\pi i(r_1 s_2 - r_2 s_1)\,\theta} \tag{21}$$

with $r = (r_1, r_2), s = (s_1, s_2) \in \mathbb{Z}^2$.

Heuristically, the noncommutative structure (14) of the torus is the exponential of the Heisenberg commutation relation $[y, x] = i\theta/2\pi$. Acting on functions of x alone, the operator U_1 is represented as multiplication by $\mathrm{e}^{\,2\pi i x}$ while conjugation by U_2 yields the shift $x \mapsto x + \theta$,

$$\begin{aligned} \Omega^{-1}\Big(U_1\,\Omega\big(f(x)\big)\Big) &= \mathrm{e}^{\,2\pi i x}\, f(x)\ , \\ \Omega^{-1}\Big(U_2\,\Omega\big(f(x)\big)\,U_2^{-1}\Big) &= f(x+\theta)\ . \end{aligned} \tag{22}$$

Analogously, on functions of y alone we have

$$\begin{aligned} \Omega^{-1}\Big(U_1\,\Omega\big(g(y)\big)\,U_1^{-1}\Big) &= g(y-\theta)\ , \\ \Omega^{-1}\Big(U_2\,\Omega\big(g(y)\big)\Big) &= \mathrm{e}^{\,2\pi i y}\, g(y)\ . \end{aligned} \tag{23}$$

From (14) it follows that $\mathcal{A}_\theta$ is commutative if and only if θ is an integer, and one identifies $\mathcal{A}_0$ with the algebra $C^\infty(\mathbb{T}^2)$. Also, for any $n \in \mathbb{Z}$ there is an isomorphism $\mathcal{A}_\theta \cong \mathcal{A}_{\theta+n}$ since (14) does not change under integer shifts $\theta \mapsto \theta + n$. Thus we may restrict the noncommutativity parameter to the interval $0 \le \theta < 1$. Furthermore, since $U_1U_2 = \mathrm{e}^{\,-2\pi i\theta}\, U_2U_1 = \mathrm{e}^{\,2\pi i(1-\theta)}\, U_2U_1$, the correspondence $U_2 \mapsto U_1, U_1 \mapsto U_2$ yields an isomorphism $\mathcal{A}_\theta \cong \mathcal{A}_{1-\theta}$. These are the only possible isomorphisms and the interval $\theta \in [0, \frac{1}{2}]$ parametrizes a family of non-isomorphic algebras.

When the deformation parameter θ is a rational number the corresponding algebra is related to the commutative torus algebra $C^\infty(\mathbb{T}^2)$, i.e. $\mathcal{A}_\theta$ is Morita equivalent to it in this case [24]. Let $\theta = M/N$, with M and N integers which we take to be relatively prime with $N > 0$. Then $\mathcal{A}_{M/N}$ is isomorphic to the algebra of all smooth sections of an algebra bundle $\mathcal{B}_{M/N} \to \mathbb{T}^2$ whose typical fiber is the algebra $\mathbb{M}_N(\mathbb{C})$ of $N \times N$ complex matrices. Moreover, there is a smooth vector bundle $E_{M/N} \to \mathbb{T}^2$ with typical fiber $\mathbb{C}^N$ such that $\mathcal{B}_{M/N}$ is the endomorphism bundle $\mathrm{End}(E_{M/N})$.

With $\omega = \mathrm{e}^{2\pi i M/N}$, one introduces the $N \times N$ clock and shift matrices

$$\mathcal{C}_N = \begin{pmatrix} 1 & & & & \\ & \omega & & & \\ & & \omega^2 & & \\ & & & \ddots & \\ & & & & \omega^{N-1} \end{pmatrix}, \qquad \mathcal{S}_N = \begin{pmatrix} 0 & 1 & & & 0 \\ & 0 & 1 & & \\ & & \ddots & \ddots & \\ & & & \ddots & 1 \\ 1 & & & & 0 \end{pmatrix}, \tag{24}$$

which are unitary and traceless (since $\sum_{k=0}^{N-1} \omega^k = 0$), satisfy

$$(\mathcal{C}_N)^N = (\mathcal{S}_N)^N = \mathbb{I}_N \ , \tag{25}$$

and obey the commutation relation

$$\mathcal{S}_N \, \mathcal{C}_N = \omega \; \mathcal{C}_N \, \mathcal{S}_N \ . \tag{26}$$

Since M and N are relatively prime the matrices (24) generate the finite dimensional algebra $\mathbb{M}_N(\mathbb{C})$: they generate a C^*-subalgebra which commutes only with multiples of the identity matrix $\mathbb{I}_N$ which thus has to be the full matrix algebra. Were M and N not coprime the generated algebra would be a proper subalgebra of $\mathbb{M}_N(\mathbb{C})$.

The algebra $\mathcal{A}_{M/N}$ has a 'huge' center $\mathcal{C}(\mathcal{A}_{M/N})$ which is generated by the elements U_1^q and U_2^q, and one identifies $\mathcal{C}(\mathcal{A}_{M/N})$ with the commutative algebra $C^\infty(\mathbb{T}^2)$ of smooth functions on an ordinary torus $\mathbb{T}^2$ which is 'wrapped' N times onto itself. There is a surjective algebra homomorphism

$$\pi_N \, : \, \mathcal{A}_{M/N} \longrightarrow \mathbb{M}_N(\mathbb{C}) \tag{27}$$

given by

$$\pi_N \left(\sum_{(m,r) \in \mathbb{Z}^2} a_{m,r} \; U_1^m \, U_2^r \right) = \sum_{(m,r) \in \mathbb{Z}^2} a_{m,r} \; (\mathcal{C}_N)^m \; (\mathcal{S}_N)^r \ . \tag{28}$$

Under this homomorphism the whole center $\mathcal{C}(\mathcal{A}_{M/N})$ is mapped to $\mathbb{C}$.

From now on we will assume that θ is an irrational number unless otherwise explicitly stated. On $\mathcal{A}_\theta$ there is a unique normalized positive definite trace, which we shall denote by the symbol $⨍ : \mathcal{A}_\theta \to \mathbb{C}$, given by

$$⨍ \sum_{(m,r)\in\mathbb{Z}^2} a_{m,r}\, U_1^m\, U_2^r := a_{0,0}$$

$$= \int_{\mathbb{T}^2} dx\; dy\; \Omega^{-1} \left(\sum_{(m,r)\in\mathbb{Z}^2} a_{m,r}\, U_1^m\, U_2^r \right) (x,y)\ . \tag{29}$$

Then, for any $a, b \in \mathcal{A}_\theta$, one readily checks the properties

$$⨍ a\, b = ⨍ b\, a\ , \quad ⨍ \mathbb{I} = 1\ , \quad ⨍ a^* a > 0\ , \quad a \neq 0\ , \tag{30}$$

with $⨍ a^* a = 0$ if and only if $a = 0$ (i.e. the trace is faithful). This trace is invariant under the natural action of the commutative torus $\mathbb{T}^2$ on $\mathcal{A}_\theta$ whose infinitesimal form is by two commuting derivations ∂_1, ∂_2 acting by

$$\partial_\mu(U_\nu) = 2\pi i\ \delta_\mu^\nu U_\nu\ , \quad \mu, \nu = 1, 2\ . \tag{31}$$

Invariance is just the statement that $⨍ \partial_\mu(a) = 0\ ,\ \mu = 1, 2$, for any $a \in \mathcal{A}_\theta$.

The cyclic 2-cocycle allowing the integration of two forms is simply given by

$$\Psi(a_0, a_1, a_2) = \frac{i}{2\pi} ⨍ a_0 \left(\partial_1 a_1 \partial_2 a_2 - \partial_2 a_1 \partial_1 a_2\right). \tag{32}$$

Its normalization ensures that for any projection $p \in \mathcal{A}_\theta$ the quantity $\Psi(p, p, p)$ is an integer: it is indeed the index of a Fredholm operator.

In two dimensions, the conformal class of a general constant metric is parametrized by a complex number $\tau \in \mathbb{C}$, $\Im\tau > 0$. Up to a conformal factor, the metric is given by

$$g = (g_{\mu\nu}) = \begin{pmatrix} 1 & \Re\tau \\ \Re\tau & |\tau|^2 \end{pmatrix}\ . \tag{33}$$

Clearly $\sqrt{det g} = \Im\tau$, and the inverse metric is found to be

$$g^{-1} = (g^{\mu\nu}) = \frac{1}{(\Im\tau)^2} \begin{pmatrix} |\tau|^2 & -\Re\tau \\ -\Re\tau & 1 \end{pmatrix}\ . \tag{34}$$

By using the two derivations ∂_1, ∂_2 defined in (31) we may think of 'the complex torus' $\mathbb{T}^2$ as acting on the noncommutative torus $\mathcal{A}_\theta$ and construct two associated derivations of $\mathcal{A}_\theta$ given by

$$\partial_{(\tau)} = \frac{1}{(\tau - \overline{\tau})} \left(-\overline{\tau}\partial_1 + \partial_2\right)\ , \quad \overline{\partial}_{(\tau)} = \frac{1}{(\tau - \overline{\tau})} \left(\tau\partial_1 - \partial_2\right)\ . \tag{35}$$

In the terminology of [30,11] we are endowing the noncommutative torus $\mathcal{A}_\theta$ with a complex structure. Then, with the two complex derivatives (35) one easily finds that

$$\partial_{(\tau)}\overline{\partial}_{(\tau)} = \overline{\partial}_{(\tau)}\partial_{(\tau)} = \frac{1}{4}g^{\mu\nu}\partial_\mu\partial_\nu = \frac{1}{4}\Delta \,, \tag{36}$$

the operator $\Delta = g^{\mu\nu}\partial_\mu\partial_\nu$ being just the Laplacian of the metric (33).

By working with the metric (33) the positive Hochschild cocycle Φ associated with the cyclic one (32) will be given by

$$\Phi(a_0, a_1, a_2) = \frac{2}{\pi}\int\!\!\!\!\!- \sqrt{detg}\; a_0\partial_{(\tau)}a_1\overline{\partial}_{(\tau)}a_2 \,. \tag{37}$$

A construction of the cocycle (37) as the conformal class of a general constant metric on the torus can be found in [4] and [6].

Before we procede, we briefly mention how to construct a spectral geometry in the sense of [4,5] for the noncommutative torus.

The algebra $\mathcal{A}_\theta$ can be represented faithfully as operators acting on a separable Hilbert space $\mathcal{H}_0$, the GNS representation space $\mathcal{H}_0 = L^2(\mathcal{A}_\theta\,, \int\!\!\!\!\!-)$, defined as the completion of $\mathcal{A}_\theta$ itself in the Hilbert norm

$$||a||_{\mathrm{GNS}} := \left(\int\!\!\!\!\!- a^*\, a\right)^{1/2} \,, \tag{38}$$

with $a \in \mathcal{A}_\theta$. Since the trace is faithful the map $\mathcal{A}_\theta \ni a \mapsto \widehat{a} \in \mathcal{H}_0$ is injective and the faithful GNS representation $\pi : \mathcal{A}_\theta \to \mathcal{B}(\mathcal{H})$ is simply given by

$$\pi(a)\widehat{b} = \widehat{a\,b}\,, \tag{39}$$

for any $a, b \in \mathcal{A}_\theta$. The vector $1 = \widehat{\mathbb{I}}$ of $\mathcal{H}_0$ is cyclic (i.e. $\pi(\mathcal{A}_\theta)1$ is dense in $\mathcal{H}_0$) and separating (i.e. $\pi(a)1 = 0$ implies $a = 0$) so that the Tomita involution is just

$$J(\widehat{a}) = \widehat{a^*}\,, \qquad \forall \widehat{a} \in \mathcal{H}_0\,. \tag{40}$$

It is worth mentioning that the C^*-algebra norm on $\mathcal{A}_\theta$ given in (13) with $k = 0$ coincides with the operator norm in (38) when $\mathcal{A}_\theta$ is represented on the Hilbert space $\mathcal{H}_0$, and also with the L^∞-norm in the Wigner representation. For ease of notation, in what follows we will not distinguish between elements of the algebra $\mathcal{A}_\theta$ and their corresponding operators in the GNS representation.

A two dimensional noncommutative geometry $(\mathcal{A}_\theta, \mathcal{H}, D, \gamma, J)$ for the torus $\mathcal{A}_\theta$ is constructed as follows. The Hilbert space $\mathcal{H}$ is just $\mathcal{H} = \mathcal{H}_0 \oplus \mathcal{H}_0$

on which $\mathcal{A}_\theta$ acts diagonally with two copies of the representation π in (39). Moreover,

$$\gamma = \begin{pmatrix} 1 & 0 \\ 0 & -1 \end{pmatrix} , \qquad J = \begin{pmatrix} 0 & -J_0 \\ J_0 & 0 \end{pmatrix} . \tag{41}$$

The Dirac operator D is

$$D = \begin{pmatrix} 0 & \bar{\partial}_{(\tau)} \\ \partial_{(\tau)} & 0 \end{pmatrix} . \tag{42}$$

For the particular choice $\tau = i$ this reduces to $D = \partial_1 \sigma_1 + \partial_2 \sigma_2$ with σ_1, σ_2, two of the Pauli matrices.

The Hochschild 2-cycle c giving the orientation and volume form is

$$c = \frac{1}{(2i\pi)^2} \frac{1}{(\tau - \bar{\tau})} \left(U_2^{-1} U_1^{-1} \otimes U_1 \otimes U_2 - U_1^{-1} U_2^{-1} \otimes U_2 \otimes U_1 \right) . \tag{43}$$

Detail on this geometry can be found in [5] (see also [31]) where one shows that all properties of a real spectral geometry are satisfied. Here we only mention that the 'area' $⨍ D^{-2}$ of the torus depends on the complex parameter τ, as it should be. Indeed, with Φ the cyclic 2-cocycle in (32), after some computations one finds that

$$⨍ D^{-2} = \langle \Phi, c \rangle = \frac{1}{2\pi \, \Im\tau} . \tag{44}$$

Here $\langle \cdot, \cdot \rangle$ indicates the pairing between cocycles and cycles.

3.3. *The Action and the Field Equations*

With $\mathcal{P}_\theta = Proj(\mathcal{A}_\theta)$ denoting the collection of all projections in the algebra $\mathcal{A}_\theta$ we construct an action functional $S : \mathcal{P}_\theta \to \mathbb{R}^+$ by

$$S_{(\tau)}(p) = \Phi(1, p, p) = \frac{2}{\pi} ⨍ \sqrt{detg}\; \partial_{(\tau)} p \bar{\partial}_{(\tau)} p . \tag{45}$$

By using (35) this action functional can also be written as

$$S_{(\tau)}(p) = \frac{1}{2\pi} ⨍ \sqrt{detg}\; g^{\mu\nu} \partial_\mu p \partial_\nu p = \frac{1}{\pi} ⨍ \sqrt{detg}\; g^{\mu\nu} p \partial_\mu p \partial_\nu p . \tag{46}$$

Here the two derivations ∂_μ are the ones defined in (31) while the metric g is the one in (33)-(34) which carries also the dependence on the complex parameter τ. The equality follows from the constraint $p^2 = p$ and the use of Leibniz rule. That the value of the action is a positive real number follows from the properties of the trace $⨍$.

Then, we shall look for critical points of the action functional (46) in a given connected component of the space $\mathcal{P}_\theta$. It is well known [24,20] that these are parametrized by two integers $r, q \in \mathbb{Z}$ such that $r + \theta q \geq 0$. When θ is irrational the corresponding projections are of trace $r + \theta q$ and of K_0-class equal to (r, q).

In order to derive field equations from the action functional (46), we need to have a look at the tangent space to $\mathcal{P}_\theta$ at any of its 'point' p. An element $\delta p \in T_p(\mathcal{P}_\theta)$ is not arbitrary but needs to fulfill two requirements. First of all it must be hermitian, $(\delta p)^* = \delta p$, and this implies that it is written as

$$\delta p = pxp + (1-p)y(1-p) + (1-p)zp + pz^*(1-p) , \tag{47}$$

with $x = x^*, y = y^*, z$ three arbitrary elements in $\mathcal{A}_\theta$. Furthermore, it must be such that $(p + \delta p)^2 = p + \delta p + O(\delta p)$ which implies that $(1-p)\delta p = \delta pp$. When using (47) we get the most general tangent vector as

$$\delta p = (1-p)zp + pz^*(1-p) , \tag{48}$$

with z an arbitrary elements in $\mathcal{A}_\theta$.

As usual, the equation of motion are obtained by equating to zero the first variation of the action functional (46),

$$\begin{aligned}
0 &= \delta S_{(\tau)}(p) \\
&= -\frac{1}{2\pi} \int \sqrt{detg}\; g^{\mu\nu} (\partial_\mu \partial_\nu p)\; \delta p \\
&= -\frac{1}{2\pi} \int \sqrt{detg}\; g^{\mu\nu} (\partial_\mu \partial_\nu p) \Big[(1-p)zp + pz^*(1-p)\Big] \\
&= -\frac{1}{2\pi} \int \Big[\sqrt{detg}\; p\; \Delta(p)\; (1-p)\Big] z + \Big[\sqrt{detg}\; (1-p)\; \Delta(p)\; p\Big] z^* . \qquad (49)
\end{aligned}$$

With $\Delta = g^{\mu\nu}\partial_\mu\partial_\nu$ the Laplacian of the metric g. We have 'integrated by part' and used the invariance of the integral to get rid of the 'boundary terms', and we have used ciclicity of the integral. Since z is arbitrary we get the field equations

$$p\; \Delta(p)\; (1-p) = 0 , \quad \text{and} \quad (1-p)\; \Delta(p)\; p = 0 , \tag{50}$$

or, equivalently

$$p\; \Delta(p) \;-\; \Delta(p)\; p = 0 . \tag{51}$$

The previous equations are non-linear equations of the second order and it is rather difficult to explicit their solutions in closed form. Presently, we shall show that the absolute minima of (46) in a given connected component of $\mathcal{P}_\theta$ actually fulfill first order equations which are easier to solve.

3.4. *Topological Charges and Self-Duality Equations*

Given a projection $p \in \mathcal{A}_\theta$, its 'topological charge' (the first Chern number) is computed by using the cyclic 2-cocycle in (32),

$$\Psi(p) := \frac{i}{2\pi} ⨍ p\Big[\partial_1(p)\partial_2(p) - \partial_2(p)\partial_1(p)\Big] \in \mathbb{Z} \tag{52}$$

(we refer to [3] for a detailed discussion). Then we have the following

Proposition 3.1. *For any $p \in \mathcal{P}_\theta$ there is the inequality*

$$S_{(\tau)}(p) \geq 2\|\Psi(p)\| \,. \tag{53}$$

Proof. From positivity of the trace ⨍ and its cyclic properties, we have that

$$\begin{aligned}
0 &\leq ⨍ \sqrt{detg}\; g^{\mu\nu}\Big[\partial_\mu(p)\; p \pm i\epsilon_\mu{}^\alpha \partial_\alpha(p)\; p\Big]^* \Big[\partial_\nu(p)\; p \pm i\epsilon_\nu{}^\theta \partial_\theta(p)\; p\Big] \\
&\leq ⨍ \sqrt{detg}\; \Big[g^{\mu\nu} p\partial_\mu(p)\partial_\nu(p) + g^{\mu\nu}\epsilon_\mu{}^\alpha\epsilon_\nu{}^\theta p\partial_\alpha(p)\partial_\theta(p)\Big] \\
&\qquad \pm i ⨍ \sqrt{detg}\; \Big[\epsilon^{\mu\theta} p\partial_\mu(p)\partial_\theta(p) - \epsilon^{\nu\alpha} p\partial_\alpha(p)\partial_\nu(p)\Big] \\
&\leq 2 ⨍ \sqrt{detg}\; g^{\mu\nu} p\partial_\mu(p)\partial_\nu(p) \pm 2i ⨍ \epsilon^{\mu\nu}_{flat}\; p\partial_\mu(p)\partial_\nu(p) \,.
\end{aligned} \tag{54}$$

By comparing (54) with (46) and (52) we get the inequality (53). □

In the derivation of (54) we have used the following equalities which are valid for any metric g,

$$g^{\mu\nu}\epsilon_\mu{}^\alpha\epsilon_\nu{}^\theta = g^{\alpha\theta} \,, \qquad \epsilon^{\mu\nu} = \frac{1}{\sqrt{detg}}\, \epsilon^{\mu\nu}_{flat} \,, \tag{55}$$

with $\epsilon^{\mu\nu}_{flat}$ the antisymmetric two-tensor determined by $\epsilon^{1,1}_{flat} = 1$.

From (54) it is clear that the equality in (53) occurs when the projection p satisfies the following *self-duality* or *anti-self duality* equations

$$\Big[\partial_\mu p \pm i\epsilon_\mu{}^\alpha \partial_\alpha p\Big]\; p = 0 \qquad \text{and/or} \qquad p\Big[\partial_\mu p \mp i\epsilon_\mu{}^\alpha \partial_\alpha p\Big] = 0. \tag{56}$$

By using the equality

$$\epsilon_\mu{}^\alpha = \sqrt{detg}\; g^{\alpha\nu}\epsilon^{flat}_{\mu\nu} \tag{57}$$

and the derivations $\partial_{(\tau)}$ and $\overline{\partial}_{(\tau)}$ in (35), the self-duality equations (56) reduce to

$$\overline{\partial}_{(\tau)}(p)\; p = 0 \qquad \text{and/or} \qquad p\; \partial_{(\tau)}(p) = 0 \,, \tag{58}$$

while the anti-self-duality one is

$$\partial_{(\tau)}(p)\, p = 0 \qquad \text{and/or} \qquad p\, \overline{\partial}_{(\tau)}(p) = 0\ . \tag{59}$$

Simple manipulations show directly that each of the equations (58) and (59) implies the field equations (51) as it should be.

3.5. *The σ-model instantons*

From now on we shall concentrate on the self-dual equations (58), a solution of which we shall call σ-model instanton. A similar analysis would be possible for the anti self-dual equations (59), whose solutions would be σ-model anti-instantons .

As we have mentioned already, the connected components of $\mathcal{P}_\theta$ are parametrized by two integers r and q and when θ is irrational, the corresponding projections are exactly the projections of trace $r + q\theta$ and the topological charge $\Psi(p)$ appearing in (52) is just given by q. Thus we have to find projections that belongs to the previous homotopy classes and satisfy the self-duality equation $\overline{\partial}_{(\tau)}(p)\ p = 0$ or, equivalently, $p\partial_{(\tau)}p = 0$. Although these equations look very simple, they are far from being easy to solve because of their non-linear nature; the next step will be to reduce them to a linear problem. The key point is to identify the algebra $\mathcal{A}_\theta$ as the endomorphism algebra of a suitable bundle and to think of any projection in it as an operator on such a bundle. The bundle in question will be a finite projective module on a different copy $\mathcal{A}_\alpha$ of the noncommutative torus, the two algebras $\mathcal{A}_\theta$ and $\mathcal{A}_\alpha$ being related by the fact that they are Morita equivalent to each other.

3.6. *The modules*

Let us then consider another copy $\mathcal{A}_\alpha$ of the noncommutative torus with generators Z_1, Z_2 obeying the relation

$$Z_2\ Z_1 = e^{2\pi i\alpha} Z_1\ Z_2\ . \tag{60}$$

When α is not rational, every finitely generated projective module over the algebra $\mathcal{A}_\alpha$ which is not free is isomorphic to a Heisenberg module [3,25]. Any such a module $\mathcal{E}_{r,q}$ is characterized by two integers r, q. If $q = 0$, $\mathcal{E}_{r,q} = \mathcal{A}_\alpha^{|r|}$. Otherwise, they can be taken to be relatively coprime with $q > 0$ (a similar construction being possible for $q < 0$), or $r = 0$ and $q = 1$. We shall briefly describe them. As a vector space

$$\mathcal{E}_{r,q} = \mathcal{S}(\mathbb{R} \times \mathbb{Z}_q) \simeq \mathcal{S}(\mathbb{R}) \otimes \mathbb{C}^q\ , \tag{61}$$

the space of Schwartz functions of one continuous variable $s \in \mathbb{R}$ and a discrete one $k \in \mathbb{Z}_q$ (we shall implicitly understand that such a variable is defined modulo q). By introducing the shorthand

$$\varepsilon = r/q - \alpha \ , \tag{62}$$

the space $\mathcal{E}_{r,q}$ is made a *right* module over $\mathcal{A}_\alpha$ by defining

$$\begin{aligned}(\xi \ Z_1)(s,k) &:= \xi(s-\varepsilon, k-r) \ , \\ (\xi \ Z_2)(s,k) &:= e^{2\pi i(s-k/q)}\xi(s,k) \ ,\end{aligned} \tag{63}$$

for any $\xi \in \mathcal{E}_{r,q}$, the relations (60) being easily verified.

On the module $\mathcal{E}_{r,q}$ one defines an $\mathcal{A}_\alpha$-valued hermitian structure, namely a sesquilinear form $\langle\cdot,\cdot\rangle_\alpha : \mathcal{E}_{r,q} \times \mathcal{E}_{r,q} \to \mathcal{A}_\alpha$, which is antilinear in the first variable and such that

$$\begin{aligned}&\langle\xi, \eta a\rangle_\alpha = \langle\xi,\eta\rangle_\alpha \, a \ , \\ &(\langle\xi,\eta\rangle_\alpha)^* = \langle\eta,\xi\rangle_\alpha \ , \\ &\langle\xi,\xi\rangle_\alpha \geq 0 \ , \quad \langle\xi,\xi\rangle_\alpha = 0 \iff \xi = 0 \ ,\end{aligned} \tag{64}$$

for all $\xi, \eta \in \mathcal{E}_{r,q}$, $a \in \mathcal{A}_\alpha$.
This hermitian structure is explicitly given by

$$\langle\xi,\eta\rangle_\alpha := \sum_{m,n}\sum_{k=0}^{q-1}\int_{-\infty}^{+\infty} ds \ \overline{\xi(s-m\varepsilon, k-mp)} \ \eta(s,k)e^{-2\pi i n(s-k/q)} \ Z_1^m \ Z_2^n \ . \tag{65}$$

It is proven in [8] that the endomorphism algebra $End_{\mathcal{A}_\alpha}(\mathcal{E}_{r,q})$, which acts on the left on $\mathcal{E}_{r,q}$, can be identified with another copy of the noncommutative torus $\mathcal{A}_\theta$ where the parameter θ is 'uniquely' determined by α in the following way. Since r and q are coprime, there exist integer numbers $a, b \in \mathbb{Z}$ such that $ar + bq = 1$. Then, the transformed parameter is

$$\theta = \frac{a\alpha + b}{-q\alpha + r} \ . \tag{66}$$

Given any two other integers $a', b' \in \mathbb{Z}$ such that $a'r + b'q = 1$, one would find that $\theta' - \theta \in \mathbb{Z}$ so that $\mathcal{A}_{\theta'} \simeq \mathcal{A}_\theta$. Thus we are saying that the algebra $End_{\mathcal{A}_\alpha}(\mathcal{E}_{r,q})$ is generated by two operators U_1, U_2 acting on the *left* on $\mathcal{E}_{r,q}$ by

$$\begin{aligned}(U_1\xi)(s,k) &:= \xi(s-1/q, k-1) \ , \\ (U_2\xi)(s,k) &:= e^{2\pi i \ (s/\varepsilon - ak)/q} \ \xi(s,k) \ ,\end{aligned} \tag{67}$$

and one easily verify that

$$U_2\, U_1 = e^{2\pi i\theta} U_1\, U_2 \ , \tag{68}$$

namely the defining relations of the algebra $\mathcal{A}_\theta$.

Furthermore, the $\mathcal{A}_\theta$-$\mathcal{A}_\alpha$-bimodule $\mathcal{E}_{r,q}$ is a *Morita equivalence* between the two algebras $\mathcal{A}_\theta$ and $\mathcal{A}_\alpha$. That is, there exists also a $\mathcal{A}_\theta$-valued hermitian structure on $\mathcal{E}_{r,q}$,

$$\langle\cdot,\cdot\rangle_\theta : \mathcal{E}_{r,q} \times \mathcal{E}_{r,q} \to \mathcal{A}_\theta \ , \tag{69}$$

which is compatible with the $\mathcal{A}_\alpha$-valued one $\langle\cdot,\cdot\rangle_\alpha$,

$$\langle\xi,\eta\rangle_\theta \ \zeta = \xi \ \langle\eta,\zeta\rangle_\alpha \ , \tag{70}$$

for all $\xi,\eta,\zeta \in \mathcal{E}_{r,q}$.

The second hermitian structure is explicitly given by

$$\langle\xi,\eta\rangle_\theta := \frac{1}{|q\varepsilon|} \sum_{m,n}\sum_{k=0}^{q-1} \int_{-\infty}^{+\infty} ds\ \xi(s,k)\ \overline{\eta(s-m/q,k-m)} \\ \times e^{-2\pi i n/q[(s-m/q)/\varepsilon - ak]}\ Z_1^m\ Z_2^n \ . \tag{71}$$

Notice that now the antilinearity is in the second variable. The hermitian structure $\langle\cdot,\cdot\rangle_\theta$ will obey properties analogous to the ones in (64) but now $\langle a\xi,\eta\rangle_\theta = a\,\langle\xi,\eta\rangle_\theta$, with $a \in \mathcal{A}_\theta$.

The proof of the compatibility condition (70) goes as follows. By using the explicit formula (65) for the right hermitian structure, the right hand side of (70) is given by

$$(\xi\ \langle\eta,\zeta\rangle_\alpha)(s,k) = \sum_{m,n}\sum_{l=0}^{q-1} \int_{-\infty}^{+\infty} dt\ \ \xi(s-m\varepsilon,k-mp) \\ \times\ \overline{\eta}(t-m\varepsilon,l-mp)\zeta(t,l)\ e^{2\pi i n[s-t-1/q(k-l)]}$$

$$= \sum_{m,n}\sum_{l=0}^{q-1}\ \xi(s-m\varepsilon,k-mp)\ \overline{\eta}(s+\frac{l-k}{q}+n-m\varepsilon,l-mp) \\ \times\ \zeta(s+\frac{l-k}{q}+n,l) \ ,$$

where we have used the Poisson resummation formula for the periodic delta function, $\sum_{n\in\mathbb{Z}} e^{2\pi i n x} = \sum_{n\in\mathbb{Z}} \delta(x+n))$, and integrated over the variable

t. On the other hand, by using similar techniques, we find that the left hand side of (70) is given by

$$(\langle\xi,\eta\rangle_\theta\ \zeta)(s,k) = \sum_{x,y}\sum_{d=0}^{q-1}\ \xi(s+\varepsilon[a(d-k)+qy],d)$$
$$\times\ \overline{\eta}(s+\varepsilon[a(d-k)+qy]-\frac{x}{q},d-x)\zeta(s-\frac{x}{q},k-x)\ ,$$

By comparing these two expressions we see that they coincide provide the indices on which we sum are related by

$$m = qy - a(d-k), \quad n = (k-l-x)/q, \quad l = k - x \bmod q;$$

notice that the last relation assures that n is an integer.

By using the previous construction one can construct projections in the algebra $\mathcal{A}_\theta$ by picking suitable vectors $\xi \in \mathcal{E}_{p,q}$ with $\langle\xi,\xi\rangle_\alpha = \mathbb{I}$. Then, the bimodule property (70) implies that $p = \langle\xi,\xi\rangle_\theta$ is a non-trivial projection in $\mathcal{A}_\theta$,

$$p^2 = \langle\xi,\xi\rangle_\theta\,\langle\xi,\xi\rangle_\theta = \langle\langle\xi,\xi\rangle_\theta\,\xi,\xi\rangle_\theta = \langle\xi\,\langle\xi,\xi\rangle_\alpha\,,\xi\rangle_\theta = \langle\xi,\xi\rangle_\theta = p\ . \qquad (72)$$

By using the identification $\mathcal{A}_\theta \simeq \mathrm{End}_{\mathcal{A}_\alpha}(\mathcal{E}_{p,q})$, any such a projection may be equivalently written in the more suggestive form

$$p_\psi = \left\langle \psi\,(\langle\psi,\psi\rangle_\alpha)^{-1/2}\ ,\ \psi\,(\langle\psi,\psi\rangle_\alpha)^{-1/2}\right\rangle_\theta = |\psi\rangle\,\frac{1}{\langle\psi,\psi\rangle_\alpha}\,\langle\psi|\,, \qquad (73)$$

where for each vector $|\psi\rangle \in \mathcal{E}_{p,q}$ the corresponding dual vector $\langle\psi| \in (\mathcal{E}_{p,q})^*$ is defined via the $\mathcal{A}_\alpha$-valued Hermitian structure as $\langle\psi|\,(\eta) := \langle\psi,\eta\rangle_\alpha \in \mathcal{A}_\alpha$. We are only assuming that $\langle\psi,\psi\rangle_\alpha$ is an invertible element of $\mathcal{A}_\alpha$.

In order to translate the self-duality equations (58) for p_ψ to equations for ψ, we need a gauge connection on the right $\mathcal{A}_\alpha$-module $\mathcal{E}_{p,q}$.

3.7. *The constant curvature connections*

Again the theory of gauge connection on noncommutative tori is worked out in [8]. A gauge connection on the right $\mathcal{A}_\alpha$-module $\mathcal{E}_{r,q}$ is given by two covariant derivatives

$$\nabla_\mu : \mathcal{E}_{r,q} \to \mathcal{E}_{r,q}\ , \quad \mu = 1,2\ , \qquad (74)$$

which satisfy a right Leibniz rule

$$\nabla_\mu(\xi\ a) = (\nabla_\mu\xi)a + \xi(\partial_\mu a)\ , \quad \mu = 1,2\ , \qquad (75)$$

for any $\xi \in \mathcal{E}_{r,q}$ and any $a \in \mathcal{A}_\alpha$. One also requires compatibility with the $\mathcal{A}_\alpha$-valued hermitian structure

$$\partial_\mu(\langle \xi, \eta \rangle_\alpha) = \langle \nabla_\mu \xi, \eta \rangle_\alpha + \langle \xi, \nabla_\mu \eta \rangle_\alpha \ , \quad \mu = 1, 2 \ . \tag{76}$$

It is not difficult to see that any other compatible connection $\widetilde{\nabla}_\mu$ must be of the form $\widetilde{\nabla}_\mu = \nabla_\nu + \eta_\mu$ with the η_μ's (skew-adjoint) elements in $End_{\mathcal{A}_\alpha}(\mathcal{E}_{r,q})$.

Given any compatible connection, the operators ∇_μ can be used to define derivations on the endomorphism algebra $End_{\mathcal{A}_\alpha}(\mathcal{E}_{r,q})$ by

$$\hat{\delta}_\mu(T) := \nabla_\mu \circ T - T \circ \nabla_\mu \ , \quad \mu = 1, 2, \tag{77}$$

for any $T \in End_{\mathcal{A}_\alpha}(\mathcal{E}_{r,q})$. By using the compatibility (76) and the right Leibniz rule (75), one easily prove that the connection ∇_μ is compatible for the derivations $\hat{\delta}_\mu$ and the hermitian structure $\langle \cdot, \cdot \rangle_\theta$, that is,

$$\hat{\delta}_\mu(\langle \xi, \eta \rangle_\theta) = \langle \nabla_\mu \xi, \eta \rangle_\theta + \langle \xi, \nabla_\mu \eta \rangle_\theta \ , \quad \mu = 1, 2 \ . \tag{78}$$

Moreover, there is also a left Leibniz rule,

$$\nabla_\mu(T\xi) = T(\nabla_\mu \xi) + (\hat{\delta}_\mu T)\xi \ , \quad \mu = 1, 2 \ , \tag{79}$$

for any $\xi \in \mathcal{E}_{r,q}$ and any $T \in End_{\mathcal{A}_\alpha}(\mathcal{E}_{r,q})$.

A particular connection on the right $\mathcal{A}_\alpha$-module $\mathcal{E}_{r,q}$ is given by the operators

$$(\nabla_1 \xi)(s, k) := \frac{2\pi i}{\varepsilon} s \ \xi(s, k) \ , \quad (\nabla_2 \xi)(s, k) := \frac{d\xi}{ds}(s, k) \ . \tag{80}$$

Notice that the discrete index k is not touched. The previous connection is of constant curvature,

$$F_{1,2} := [\nabla_1, \nabla_2] - \nabla_{[\partial_1, \partial_2]} = -\frac{2\pi i}{\varepsilon} \ \mathbb{I}_{\mathcal{E}_{r,q}} \ , \tag{81}$$

where $\mathbb{I}_{\mathcal{E}_{r,q}}$ is the identity operator on $\mathcal{E}_{r,q}$.

Finally, by remembering that $End_{\mathcal{A}_\alpha}(\mathcal{E}_{r,q}) \simeq \mathcal{A}_\theta$, one also proves that the derivations $\hat{\delta}_\mu$ on $End_{\mathcal{A}_\alpha}(\mathcal{E}_{r,q})$ determined by the constant curvature connection (80) are proportional to the generators of the infinitesimal action of the commutative torus $\mathbb{T}^2$ on $\mathcal{A}_\theta$, that is,

$$\hat{\delta}_\mu(U_\nu) = \frac{2\pi i}{q\varepsilon} \ \delta_\mu^\nu U_\nu \ , \quad \mu, \nu = 1, 2 \ . \tag{82}$$

That the induced derivations are the canonical derivations on $\mathcal{A}_\theta$ is crucial for our analysis in the following (see the proof of Prop. 3.2). Any other connection with this same property will have the form $\widetilde{\nabla}_\mu = \nabla_\nu + \lambda_\mu$ where λ_1 and λ_2 are (pure-imaginary) constants. Then, the corresponding curvature $\widetilde{F}_{1,2}$ will be constant and in fact will coincide with the constant curvature $F_{1,2}$ in (81). It is known [8] that constant curvature connections are exactly the connections that minimize a suitable Yang-Mills functional and that all this minimizing connections must have the same curvature. Moreover, the moduli spaces of these connections is just (in the simplest possible case) an ordinary two-torus $\mathbb{T}^2$.

In the following we shall also need the covariant derivatives $\nabla_{(\tau)}$ and $\overline{\nabla}_{(\tau)}$ which lift the complex derivations $\partial_{(\tau)}$ and $\overline{\partial}_{(\tau)}$ in (35). By using (80) they are given by

$$\nabla_{(\tau)} = \frac{1}{(\tau-\overline{\tau})}\left(-\overline{\tau}\nabla_1 + \nabla_2\right), \quad \overline{\nabla}_{(\tau)} = \frac{1}{(\tau-\overline{\tau})}\left(\tau\nabla_1 - \nabla_2\right) \tag{83}$$

These connections will endow the module $\mathcal{E}_{r,q}$ with a complex (or better holomorphic) structure (see also [11,30,23,29]).

3.8. *The Instantons*

We are now ready to look for solutions of the self-dual equations (58) of the form

$$p_\psi := |\psi\rangle \left(\langle\psi,\psi\rangle_\alpha\right)^{-1} \langle\psi| \,, \tag{84}$$

with the vector $|\psi\rangle \in \mathcal{E}_{p,q}$ assumed to be such that $\langle\psi,\psi\rangle_\alpha$ is invertible in $\mathcal{A}_\alpha$. One has the following

Proposition 3.2.
Let the vector $|\psi\rangle \in \mathcal{E}_{r,q}$ be such that $\langle\psi,\psi\rangle_\alpha \in \mathcal{A}_\theta$ is invertible. And consider the projection $p_\psi := |\psi\rangle \left(\langle\psi,\psi\rangle_\alpha\right)^{-1} \langle\psi|$.
Then,
1.) If there exists an element $\lambda \in \mathcal{A}_\alpha$ such that $\psi = |\psi\rangle$ is a solution of the equations

$$\overline{\nabla}\psi - \psi\lambda = 0 \,, \tag{85}$$

where $\overline{\nabla}$ is the anti-holomorphic connection (83), then the projection p_ψ is a solution of the self-duality equations,

$$\overline{\partial}_{(\tau)}(p_\psi)\, p_\psi = 0 \,. \tag{86}$$

2.) Conversely, if p_ψ is a solution of the equations (86), then $\psi = |\psi\rangle$ obeys the equations (85) with

$$\lambda = (\langle\psi,\psi\rangle_\alpha)^{-1} \langle\psi, \overline{\nabla}\psi\rangle_\alpha . \tag{87}$$

Proof. A straightforward computation using the properties of the connection, notably the fact that the induced derivations on the endomorphism algebra are the canonical derivations on $\mathcal{A}_\theta$, and left and right Leibniz rules and metric compatibilities. □

For a generic element $\lambda \in \mathcal{A}_\alpha$, equations (85) still are horribly complicated. A very important class of examples is obtained by considering constant parameters $\lambda \in \mathbb{C}$.

Proposition 3.3. *Let $\lambda \in \mathbb{C}$. Then the Gaussian*

$$\psi_\lambda(s,k) = A_k \, \mathrm{e}^{\,i\tau\pi s^2/\varepsilon + \lambda(\overline{\tau}-\tau)s} \quad , \; k = 1, \ldots, q \, , \tag{88}$$

is a solution of the equation (85) such that the element $\langle\psi_\lambda,\psi_\lambda\rangle_\alpha \in \mathcal{A}_\alpha$ is invertible. The vector $A = (A_1, \ldots, A_q) \in \mathbb{C}^q$ can be taken to lie in the complex projective space $\mathbb{CP}^{q-1}$ by removing an inessential normalization.

Proof. Recall first that we are taking $\Im\tau > 0$, hence the generalized gaussian $\psi_\lambda(s,k)$ in (88) is an element of $\mathcal{E}_{r,q}$. That it is also a solution of the equations (85) (with the constant curvature connection (80)) is a straightforward computation. As for invertibility of the corresponding element $\langle\psi_\lambda,\psi_\lambda\rangle_\alpha \in \mathcal{A}_\alpha$ this is more difficult. For the lowest values of the parameters, i.e. $q = 1$, $r = 0$, (and with $\tau = i$, $\lambda = 0$), invertibility of

$$\langle\psi_{\lambda=0},\psi_{\lambda=0}\rangle_{-1/\theta} = \frac{1}{\sqrt{2\theta}} \sum_{(m,n)\in\mathbb{Z}^2} \mathrm{e}^{\,i\pi mn/\theta - \pi(m^2+n^2)/2\theta} \, Z_1^m \, Z_2^n \tag{89}$$

(now $\alpha = -1/\theta$) was proved in [2] for a restricted range of the deformation parameter θ. The invertibility was extended in [32] for all values $0 < \theta < 1$. As for the general case, a careful extension of the techniques of [2] and [32] provides invertibility. □

Furthermore, a lengthy computation gives that the projection $p_\lambda \in \mathcal{A}_\theta$ corresponding to the Gaussian (88) has dimension (= rank) $r + q\theta$, namely $⨍(p_\lambda) = r + q\theta$, and topological charge q: $\Psi(p_\lambda) = q$.

Before we pass to the general case of nonconstant 'parameter' λ in the equations (85) we need to introduce gauge transformations.

3.9. *Gauge Transformations*

Having lifted our duality equations to equations on the bundle $\mathcal{E}_{r,q}$ has introduced what we could call 'gauge degrees of freedom'. Then, there are analogous of (complex) gauge transformations. These are given by invertible elements in $\mathcal{A}_\alpha$ acting on the right on $\mathcal{E}_{r,q}$,

$$\mathcal{E}_{r,q} \ni |\psi\rangle \to |\psi^g\rangle = |\psi\rangle\, g \in \mathcal{E}_{r,q}\ , \quad \forall\, g \in \mathrm{GL}(\mathcal{A}_\alpha)\ . \tag{90}$$

Notice that we do not require g to be unitary. Then, it is straightforward to prove that projections of the form (84) are invariant under gauge transformations. Indeed,

$$\begin{aligned} p_\psi \to p_\psi^g &= |\psi^g\rangle \left(\langle\psi^g, \psi^g\rangle_\alpha\right)^{-1} \langle\psi^g| = |\psi\rangle\, g(g^* \langle\psi,\psi\rangle_\alpha\, g)^{-1} g^* \langle\psi| \\ &= |\psi\rangle\, g g^{-1}(\langle\psi,\psi\rangle_\alpha)^{-1}(g^*)^{-1} g^* \langle\psi| = |\psi\rangle \left(\langle\psi,\psi\rangle_\alpha\right)^{-1} \langle\psi| \\ &= p_\psi. \end{aligned}$$

Equally straightforward is to find the gauge transformed of the self duality equations (85). Indeed, let $|\psi\rangle \in \mathcal{E}_{r,q}$ be a solution of the equations (85): $\overline{\nabla}\psi - \psi\lambda = 0$; and let $g \in \mathrm{GL}(\mathcal{A}_\alpha)$. Then by using the Leibniz rule for the connection, one finds that the gauge transformed vector $|\psi^g\rangle$ will be a solution of an equation of the same form: $\overline{\nabla}\psi^g - \psi^g\lambda_g = 0$ with λ_g given by

$$\lambda_g = g^{-1}\lambda g + g^{-1}\overline{\partial}_{(\tau)} g\ . \tag{91}$$

Let us then analyze how the gaussian projections transform under the action of the gauge transformations (90). We have the following

Proposition 3.4. *Let $|\psi\rangle$ be a solution of equation (85), $\lambda \in \mathbb{C}$; and let $g \in \mathrm{GL}(\mathcal{A}_\alpha)$. Then the transformed λ_g will again be constant if an only if there exists a couple of integer $(m,n) \in \mathbb{Z}^2$ such that*

$$g = g_{mn} U^m V^n \qquad \text{no sum.} \tag{92}$$

Furthermore,

$$\lambda_g - \lambda = \frac{2\pi i \tau}{\tau - \overline{\tau}}\,(m - \frac{1}{\tau} n). \tag{93}$$

Proof. The first statement in the direction right to left of the statement is obvious. On the other hand, assume that $\lambda_g \in \mathbb{C}$. Any $g \in \mathrm{GL}(\mathcal{A}_\alpha)$ is written as $g = \sum_{(m,n)\in\mathbb{Z}^2} g_{mn} U^m V^n$. When substituting in (91) one gets an equation which requires that only one term in the sum does not vanish. The relation (93) is straightforwardly worked out. □

One could also act directly on gaussian functions, which are elements of the module $\mathcal{E}_{r,q}$ on which $\mathrm{GL}(\mathcal{A}_\alpha)$ acts on the left. Indeed, let the solution corresponding to the parameter $\lambda \in \mathbb{C}$ be the Gaussian ψ_λ as in (88). When acting on ψ_λ with the invertible element $g = g_{mn}U^mV^n \in \mathrm{GL}(\mathcal{A}_\alpha)$ one produces a new Gaussian ψ_{λ_g} with the constant parameter λ_g given exactly as in (93). By putting together these results with Proposition 3.3, we have the following result which parallel an analogous one for connections minimizing a Yang-Mill functional (for which the connections are forced to be of constant curvature) [8].

Corollary 3.1. *The moduli space of 'gaussian projection' is given by an ordinary complex torus times an ordinary projective space*

$$\mathbb{C}/(\tau\mathbb{Z} + \mathbb{Z}) \times \mathbb{CP}^{q-1} . \tag{94}$$

As for the generic case of the equations (85) for a nonconstant element $\lambda \in \mathcal{A}_\alpha$, at the moment we are unable to state a general result. In [9,10] it was suggested that any solution of the these self-duality equations could be gauged away to a gaussian solution. This would be equivalent to the statement that given any element $\rho \in \mathcal{A}_\alpha$, there exists an invertible element $g \in \mathrm{GL}(\mathcal{A}_\alpha)$ such that $\rho = \lambda + g^{-1}\overline{\partial}_{(\tau)}g$, with $\lambda \in \mathbb{C}$. That this fact is not true follows from recent work in [22].

Acknowledgments

This is a review of work done with Thomas Krajewski and Ludwik Dąbrowski most of which is published in [9,10]; I am most grateful to them for the collaboration. I thank Yoshiaki Maeda, Nobuyuki Tose and Satoshi Watamura for their kind invitation to Keio University and for the fantastic hospitality there. I also thank all participants of the conference for the great time we had together. Jacopo was a great travel companion and I enjoyed his fifteenth birthday in Tokyo.

References

1. A.A. Belavin, A.M. Polyakov, *Metastable states of two-dimensional isotropic ferromagnets*, JETP Lett. **22** (1975) 245-247.
2. F.-P. Boca, *Projections in Rotation Algebras and Theta Functions*, Commun. Math. Phys. **202** (1999) 325–357.
3. A. Connes, *C^*-algèbres et géométrie différentielle*, C.R. Acad. Sci. Paris Sér. A **290** (1980) 599-604.
4. A. Connes, *Noncommutative Geometry*, Academic Press, 1994.

5. A. Connes, *Gravity coupled with matter and the foundation of noncommutative geometry*, Commun. Math. Phys. **182** (1996) 155-176.
6. A. Connes, *A short survey of noncommutative geometry*, J. Math. Phys. **41** (2000) 3832-3866.
7. A. Connes, M. Douglas, A. Schwarz, *Matrix theory compactification on tori*, J. High Energy Phys. **02** (1998) 003.
8. A. Connes, M. Rieffel, *Yang-Mills for Non-commutative Two-Tori*, in *Operator Algebras and Mathematical Physics*, Contemp. Math. **62** (1987) 237-266.
9. L. Dabrowski, T. Krajewski, G. Landi, *Some Properties of Non-linear σ-models in Noncommutative Geometry*; Int. J. Mod. Phys. **B14** (2000) 2367-2382.
10. L. Dabrowski, T. Krajewski, G. Landi, *Non-linear σ-models in noncommutative geometry: fields with values in finite spaces*, Mod. Phys. Lett. **A18** (2003) 2371-2380.
11. M. Dieng, A. Schwarz, *Differential and complex geometry of two-dimensional noncommutative tori*, Lett. Math. Phys. **61** (2002) 263-270.
12. T. Krajewski, *Gauge invariance of the Chern-Simons action in noncommutative geometry*, ISI GUCCIA Conference 'Quantum Groups, Noncommutative Geometry and Fundamental Physical Interactions', Palermo December 1997, `math-ph/9810015`.
13. J. Eells, *Harmonic maps: Selected papers of James Eells and collaborators*, World Scientific, 1992.
14. K. Gawedzki, *Lectures on conformal field theory*, in 'Quantum Fields and Strings: a Course for Mathematicians', P. Deligne et al. editors, American Mathematical Society 1999; pp 727-805.
15. M. Green, J.H. Schwarz, E. Witten, *Superstring theory*, Cambridge University Press, 1987.
16. Yu. I. Manin, *Quantized theta-function*, Prog. Theor. Phys. Suppl. **102** (1990) 219-228.
17. Yu. I. Manin, *Theta functions, quantum tori and Heisenberg group*, Lett. Math. Phys. **56** (2001) 295-320.
18. Yu. I. Manin, *Real multiplication and noncommutative geometry*, `math.QA/0202109`.
19. Yu. I. Manin. *Functional equations for quantum theta functions*, `math.QA/0307393`.
20. M. Pimsner, D. Voiculescu, *Exact Sequences for K-Groups and Ext-Groups of Certain Cross-Product C^*-Algebras*, J. Oper. Theory **4** (1980) 93-118.
21. J. Polchinski, *String theory*, Cambridge University Press, 1998.
22. A. Polishchuk, *Analogues of the exponential map associated with complex structures on noncommutative two-tori*, `math.QA/0404056`.
23. A. Polishchuk, A. Schwarz, *Categories of holomorphic vector bundles on noncommutative two-tori*, Commun. Math. Phys. **236** (2003) 135-159.
24. M. Rieffel, *C^*-algebras associated with irrational rotations*, Pacific J. Math. **93** (1981) 415-429.
25. M. Rieffel, *The Cancellation Theorem for projective Modules over irrational C^*-algebras*, Proc. London Math. Soc. **47** (1983) 1285-302.

26. M. Rieffel, *Projective modules over higher-dimensional noncommutative tori*, Can. J. Math. **40** (1988) 257-338.
27. M. Rieffel, *Non-commutative Tori -A case study of Non-commutative Differentiable Manifolds* Contemp. Math. **105** (1991) 191-211.
28. N. Seiberg, E. Witten, *String Theory and Noncommutative Geometry*, JHEP **09** (1999) 032.
29. M. Spera, *Yang-Mills theory in non commutative differential geometry*, Rend. Sem. Fac. Scienze Univ. Cagliari, Suppl. **58** (1988) 409-421.
30. A. Schwarz, *Theta-functions on noncommutative tori*, Lett. Math. Phys. **58** (2001) 81-90.
31. J.C. Varilly, *An Introduction to Noncommutative Geometry*, Lectures at EMS Summer School on NCG and Applications, Sept 1997, physics/9709045.
32. S. Walters, *The AF Structure of Noncommutative Toroidal $\mathbb{Z}/4\mathbb{Z}$ Orbifolds*, J. Reine Angew. Math. **568** (2004) 139-196.
33. W.J. Zakrzewski, *Low dimensional sigma models*, Adam Hilger, Bristol 1989.

ON VECTORIAL GERBES AND POINCARÉ–CARTAN CLASSES *

NAOYA MIYAZAKI

Department of Mathematics,
Faculty of Economics, Keio University
Yokohama, 223-8521, Japan
miyazaki@math.hc.keio.ac.jp

1. Introduction

In recent years notable progress has been made in the theory of deformation quantization ([2], [6], [7], [9], [10], [14], [15], [15], [19]. See also Definition 2.1 for precise definition.). It has brought to light some interesting connections between deformation quantization and noncommutative differential geometry [5].

In the present paper, we give an elementary proof of Mehler's formula in deformation quantization, which is closely related to index theorem for deformation quantization developed by Fedosov [6], [7] under topological K-theoretical setting [a]. Our approach is based on applying conformal rescaling to the heat kernel of Dirac-Laplacian (the square of Dirac operator) coupled with a twisted quantum vectorial gerbe (cf. Theorem 2.2).

In order to construct an appropriate Dirac operator $\not{D}_Q$, we need a twisted quantum vectorial gerbe which is an infinite dimensional vectorial gerbe equipped with an algebra module structure and a twisted quantum connection whose 1st Chern class coincides with the Poincaré-Cartan class[b] $c_\nu(\mathcal{A}_M)$ which is a complete invariant of deformation quantization of symplectic manifolds (cf. [15]). Thanks to the twisted quantum connection, we

---

*This research is partially supported by Grant-in-Aid for Scientific Research (#15740045, #15540094), Ministry of Education, Culture, Sports, Science and Technology, Japan, and is also partially supported by Keio Gijuku Academic Funds.

[a] More precisely his argument is based on L-functor which is equivalent to topological K-thoery

[b] See Theorem 2.1 for precise definition. This class correspondes to the Deligne Čech-cohomology class [9].

can see that the Poincaré-Cartan form appears as a relative curvature of the twisting deformation quantum vectorial gerbe in Mehler's formula. Then we first have the following.

Theorem 1.1. *Assume that (M,ω) is an $n = 2m$-dimensional symplectic manifold, and (Q_M, ∇^Q) is a twisted quantum vectorial gerbe (see Theorem 2.2). Let $\not{D}_Q$ be a Dirac operator coupled with (Q_M, ∇^Q). Then the heat equation $(\partial_t + (\not{D}_Q)^2)p_t = 0$ with respect to the Dirac-Laplacian operator $(\not{D}_Q)^2$ has an analytic solution in a certain sense (cf. Proposition 4.1).*

In terms of a good local trivialization and the rescaling transform with rescaling parameter u in the sense of Getzler calculus (cf. [4]), we also define the rescaled Dirac (resp. Dirac-Laplacian) $\not{D}_{Q,u}$ (resp. $(\not{D}_{Q,u})^2$) (cf. (20)). As for this operator, we have the following.

Theorem 1.2. *Under the same assumpution in Theorem 1.1, the heat kernel of $(\not{D}_{Q,u})^2$ has the rescaling limit as $u \to 0$ in the following form:*

$$(4\pi t\hbar)^{\frac{-n}{2}} \det{}^{1/2}\left(\frac{t\hbar R/2}{\sinh(t\hbar R/2)}\right) \times \exp\left(-\tfrac{1}{4t\hbar}\left\langle x \,\middle|\, \tfrac{t\hbar R}{2}\coth\left(\tfrac{t\hbar R}{2}\right) \middle|\, x\right\rangle\right) \\ \times \exp\left(-t\,\Omega_\nu(\mathcal{A}_M)\right) \tag{1}$$

where R is the curvature of base manifold M, and $\Omega_\nu(\mathcal{A}_M)$ is the Poincaré-Cartan form in the sense of Theorems 2.1 and 2.2 (see also [15]).

The proof of Theorem 1.2 will be given in subsection 4.2. We remark that Theorems 1.1 and 1.2 are already given in [12]. In this paper, we improve the description of Dirac operator $\not{D}_Q$ and the proofs of Theorems.

2. Twisted quantum vectrial gerbes

We begin with definition of deformation quantization [2].

Definition 2.1. A deformation quantization of Poisson manifold (M,π) is a family of product $* = *_\hbar$ (depending on the Planck constant $\hbar$) on the space of formal power series of parameter $\hbar$ with coefficients in $C^\infty(M)$, $C^\infty(M)[[\hbar]]$ defined by

$$f *_\hbar g = fg + \hbar\pi_1(f,g) + \cdots + \hbar^n\pi_n(f,g) + \cdots, \quad \forall f,g \in C^\infty(M)[[\hbar]]$$

satisfying 1. $*$ is associative, 2. $\pi_1(f,g) = \frac{1}{2\sqrt{-1}}\{f,g\}$, 3. each π_n $(n \geq 1)$ is a $\mathbb{C}[[\hbar]]$-bilinear and bidifferential operator, where $\{,\}$ is the Poisson bracket defined by the Poisson structure π.

Deformed algebra (resp. deformed algebra structures) is called a *star algebra* (resp. *star-products*). As to classification problem we have the following [15]:

Theorem 2.1. *The equivalence classes of the star algebras (or deformation quantization)* $(\mathcal{A}_M = (C^\infty(M)[[\nu]], *)$ *over a symplectic manifold* (M, ω) *have a bijection to the set of all formal power series in* ν^2 *with coefficients in* $\check{H}^2(M)$ *having the following form:*

$$c_\nu(\mathcal{A}_M) = [\omega] + \nu^2 c_2 + \cdots + \nu^{2k} c_{2k} + \cdots \in \check{H}^2(M)[[\nu^2]], \quad (\nu = i\hbar). \quad (2)$$

The element $c_\nu(\mathcal{A}_M)$ corresponding to $\mathcal{A}_M$ is called a *Poincaré-Cartan* class[c] of $\mathcal{A}_M$.

In the theory of geometric quantization [18], it is well-known that *for any symplectic manifold endowed with a symplectic form belonging to an integral cohomology class, there exists a line bundle equipped with a connection whose curvature coinsides with the symplectic form, and resulting bundle is called a prequantum bundle.* By a similar manner, for a given Poincaré-Cartan class, we can construct an infinite dimensional vectorial gerbe and Poincaré-Cartarn form by mimic of Čech-de Rham zig-zag construction. In order to explain it, we need the presice definition of gerbes and vectorial gerbes following [8] and [1] (see also [3]).

Let M be a smooth manifold and $\mathcal{U} = \{U_i\}_{i \in I}$ an open covering of M. Set

$$\pi : Y = Y^{[0]} = \coprod U_i \to M \text{ (projection)}, \quad (3)$$

$$Y^{[p]} = \{(y_0, y_1, \cdots, y_p) | \pi(y_0) = \pi(y_1) = \cdots = \pi(y_p)\}, \quad (4)$$

$$\pi_i : Y^{[p]} \to Y^{[p-1]}; \text{ omit the i-th argument} \quad (5)$$

$$\delta = \sum_{i=0}^{p} (-1)^i \pi_i^* : \Lambda^q(Y^{[p]}) \to \Lambda^q(Y^{[p+1]}) \quad (6)$$

Definition 2.2. Under the notations above, $\mathcal{G} = (Y, P, s)$ is called a (H-)*gerbe* (where H is an abelian group) over M, if the following conditions are satisfied

(1) π is a submersion equipped with a local section,
(2) $P \to Y^{[1]}$ is a H-principal bundle,

---

[c]Strictly speaking, in [15], the Čech cohomology class corresponding to $\Omega_\nu(\mathcal{A}_M)$ is called a Poincaré-Cartan class.

(3) $s : Y^{[2]} \to \delta P$ is a section (it will be referred to as "discordance"), where $\delta := \pi_1^* P \otimes (\pi_2^* P)^{\otimes -1} \otimes \pi_3^* P$.
(4) $\delta s = 1$.

It is well-known that for any G-principal bundle and any representation (ρ, V) of G, there exists an associated vector bundle. In the same way, for any H-gerbe, we can define *vectorial gerbe* associated with a representation (V, ρ) of H. It is known that for any vectorial gerbe we can define global sections, inner product, completion of space of smooth global sections, connection, etc. [1], [13]. Especially, there exists a connection over $Y^{[0]}$ which is compatible with the discordance. Moreover, we can define Chern class, Chern character (these notion is well-defined since $\mathrm{tr}(gAg^{-1}) = \mathrm{tr}(A)$, $\mathrm{Det}(gAg^{-1}) = \mathrm{Det}(A)$).

The following example is fundamental.

Example (Spin gerbe [13]) Consider the Bockstein's exact sequence:

$$\cdots \to H^1(M, Spin(n)) \to H^1(M, SO(n)) \xrightarrow{b} H^2(M, \mathbb{Z}_2) \to \cdots . \tag{7}$$

Then Stiefel-Whitney class w_2 defined as an image of (cocycle) $(g_{ij}) \in C^\infty(U_{ij}, SO(M))$ by b can be regarded as a discordance, and then we can define a gerbe. It will be referred to as a *spin gerbe*.

For our purpose, we also need the following example (cf. [12]):

Theorem 2.2. *For any Poincaré-Cartan class $c_\nu(\mathcal{A}_M)$, there exists a vectorial gerbe with a connection (Q_M, ∇^Q) whose 1st Chern class coincides with a Poincaré-Cartan form $\Omega_\nu(\mathcal{A}_M)$ belonging to the Poincaré-cartan class. Furthermore, each fiber has a natural noncommutative associative algebra right-module structure. The resulting right-module bundle (resp. connection) is referred to as a twisted quantum vectrial gerbe (resp. a twisted quantum connection).*

Proof. Fix a good covering $\mathcal{U} = \{U_\alpha\}$ of M. Using this covering, we introduce a double complex which is obtained by replacing Čech-de Rham double complex $C^q(\mathcal{U}, \Lambda^p_M)$ (resp. de Rham differential operator) by $C^q(\mathcal{U}, \Lambda^p_M \otimes W_M))$ (resp. Fedosov connection ∇^F in [6] and [7]). Then we obtain the desired vectorial gerbe by the standard method. Since ∇^F is compatible with Moyal product on each fiber, the vectorial gerbe admits right module structure. See [12] for details. □

3. Dirac operator coupled with a twisted quantum vectorial gerbe

In this section we modify the Clifford algebra to fit in with our situation.

Definition 3.1. Let (V^*, g) be an $n = 2m$ dimensional Euclidean space. Clifford algebra $\mathbb{C}\ell_\hbar(V^*)$ is an algebra formally generated by an orthonomal basis $e^1, \cdots, e^n$ with respect to g over $\mathbb{C}_\hbar = \mathbb{C}[\hbar^{-1}, \hbar]]$ satisfying

$$e^i e^j + e^j e^i = -2\delta^{ij}\hbar. \tag{8}$$

There is a natural *degree $C\ell$*-deg defined in the following way:

$$C\ell\text{-deg}(e^{i_1} \cdots e^{i_k} \hbar^l) = k + 2l. \tag{9}$$

We can also define the spinor space $S_\hbar$ and Clifford action $c_\hbar$ in the following ways: Set

$$P_\hbar = \Big\langle \{w^j = \frac{1}{2}\{e^{2j-1} - \sqrt{-1}e^{2j}\} : 1 \leq j \leq \frac{n}{2}\} \Big\rangle_{\mathbb{C}[\hbar^{-1},\hbar]}, \tag{10}$$

$$S_\hbar = \Lambda P_\hbar, \quad S_\hbar^\pm = \Lambda^\pm P_\hbar, \tag{11}$$

and for any element $s = \sum_i s_i \hbar^i \in S_\hbar$, set

$$c_\hbar(w)s = 2^{1/2} w \wedge s \quad (w \in P_\hbar), \tag{12}$$

$$c_\hbar(\bar{w})s = -2^{1/2}\iota_\hbar(\bar{w})s = -2^{1/2}\hbar g(\bar{w}, s) \quad (\bar{w} \in \bar{P}_\hbar), \tag{13}$$

$$c_\hbar(\hbar)s = \hbar s. \tag{14}$$

Let ∇^Q be a twisted quantum connection constructed in the previous section, and ∇^S a Clifford (spin) connection which is obtained by lifting a metric connection. Then we also obtain a desired vectorial gerbe $\mathcal{E} = Q_M \otimes_{\mathbb{C}[\hbar^{-1},\hbar]]} S_\hbar$ equipped with a twisted connection $\nabla = \nabla^{\mathcal{E}}$ and coupled Dirac $\not{D}_Q$ operator by the standard manner.

We remark that by rescaling $g \longrightarrow \frac{1}{\hbar}g = g_\hbar$ the volume form (density), the Levi-Civita connection, the connection Laplacian vary. Especially, as for the Lichnerowicz formula we have the following.

Proposition 3.1. *Under the rescaling of Riemannian metric as above, we have* $(\not{D}_Q)^2 = \hbar\nabla^*\nabla + V$, *where* $V = \frac{\hbar r_M}{4} + c_\hbar(\Omega_\nu(\mathcal{A}_M))$, *and* r_M *is a scalar curvature.*

4. Proofs of Theorems 1.1 and 1.2

The proofs of Theorems 1.1 and 1.2 will be given in the present section by showing how the twisted quantum vectorial gerbe framework fits with the more standard formalism of Dirac bundle.

4.1. *Construction of an asymptotic solution for Dirac-Laplacian*

First we show existence of asymptotic solution of Dirac-Laplacian $(\not{D}_Q)^2$ However, since the rank of our vectorial gerbe is of infinite, thus it is impossible to directly apply usual arguments using C^l-norm of the space of all smooth sections to construct an asymptotic solution of the heat equation with respect to the square of Dirac operator $\not{D}_Q$.

To overcome this difficluty, we introduce a filtration derived from the total degree which is the summation of Weyl degree and Clifford degree: $deg = W\ell\text{-deg} + C\ell\text{-deg}$ of $\hbar$-differential form, where $W\ell\text{-deg}(\nu) = 2$ and $C\ell\text{-deg}(\hbar) = 2$. According to this degree, we define the projective system in the following way:

$$\mathcal{E}^n := \{a \in \mathcal{E} : \deg a = n\}, \qquad \mathcal{E}_n := \mathcal{E}/\sum_{l>n} \mathcal{E}^l.$$

Then we can construct a strict solution by the standard asymptotic method in the following way ([12]):

Proposition 4.1. *As to the heat equation*

$$(\partial_t + (\not{D}_Q)^2)p_t = 0, \tag{15}$$

for each $\mathcal{E}_n$ there exists a fundamental solution having the following asymptotic expansion:

$$p_t(x, y) = q_t(x, y) \sum_{i=0}^{\infty} t^i \Phi_i(x, y) |vol_{M_y}|^{\frac{1}{2}} \Big|_{\mathcal{E}_n}, \tag{16}$$

which is compatible with the forgetting map, where $q_t(x, y) = (4\pi\hbar t)^{\frac{-n}{2}} e^{\frac{||x-y||^2}{4\hbar t}} |dx|^{\frac{1}{2}}$ is the Euclidean heat kernel. Thus, it defines a projective system of fundamental solution.

4.2. *Local trivializations and conformal rescalings*

In this subsection, we derive Mehler's formula. With a slight modification, we can adapt the method used in the proof of Atiyah-Singer index theorem based on the heat kernel method in such a way that we can apply it to the present case.

1. Fix a normal coordinate system $(U, \mathbf{x})$ around a point q. Let $\{e_i\}$ be a local orthonormal frame obtained by the parallel transformtion of a orthonormal basis $\partial_{x^1}, \cdots, \partial_{x^n}$ at q with respect to the metric g_q along geodesic curves starting from q. We denote its dual basis by $\{e^i\}$. Note that under this trivialization, the Clifford action becomes a constant action. Using the above trivialization, we represent $\nabla = \nabla^{\mathcal{E}} = d+\omega$, where ω is the connection form of a twisted quantum connection. Let Ω be the curvature tensor of ∇. Then the following is well-known (cf. [17]):

Proposition 4.2.

$$\sum(|\alpha|+1)\partial^\alpha\omega_i(0)\frac{x^\alpha}{\alpha!} = \sum \partial^\alpha\Omega(\partial_k,\partial_i)(0)x^k\frac{x^\alpha}{\alpha!} \tag{17}$$

Comparing coefficients of each degree in the both sides, we have

$$\partial_j\omega_i(0) = -\frac{1}{2}\Omega(\partial_i,\partial_j)(0), \qquad \partial_\ell\partial_j\omega_i = \frac{1}{3}\partial_\ell\Omega(\partial_j,\partial_i)(0). \tag{18}$$

Hence thanks to the Taylor expansion, we have

$$\begin{aligned}\nabla_i(x) = \partial_i &- \frac{1}{8}\sum_{k\ell} g\Big(R^{LC}(\partial_k,\partial_\ell)\partial_i, \sum_j x^j\partial_j\Big)e_k e_\ell \\ &-\frac{1}{2}\sum_j F(\partial_i,\partial_j)x^j + \sum O(|x|^2)e_k e_\ell + \sum O(|x|),\end{aligned} \tag{19}$$

where R^{spin} (resp. R^{LC}) denotes the curvature of the spin (resp. Levi-Civita) connection, $F = \Omega_\nu(\mathcal{A}_M)$, and $O(r)$ denotes Landau's symbol.

2. Next, we give a conformal rescaling on the evolution equation obtained in Proposition 4.1. Roughly speaking, conformal rescaling means multiplication u (resp.u^{-1}) to covariant (resp. contravariant) time direction, and $u^{\frac{1}{2}}$ (resp. $u^{\frac{-1}{2}}$) to covariant (resp. contravariant) space direction. Since the heat kernel $k_\hbar(t,\mathbf{x})$ should be regarded as a $\Lambda V^*[\hbar^{-1},\hbar]]\otimes_{\mathbb{C}[\hbar^{-1},\hbar]]} End_{C\ell_\hbar}(C_M\otimes S_\hbar)\otimes$ "density bundle"-valued section (where V^* denotes a vector bundle appeared in Definition 3.1), in order to define conformal rescaling, we took account of not only t, $\mathbf{x}$ and dx, but also "density bundle". Here we give the precise definition ([4]):

Definition 4.1. For

$$\alpha \in C^\infty\big(\mathbb{R}_+\times U\ ;\ \Lambda V^*[\hbar^{-1},\hbar]]\otimes_{\mathbb{C}[\hbar^{-1},\hbar]]} End_{C\ell_\hbar}(C_M\otimes S_\hbar\otimes|\Omega|^{\frac{1}{2}})\big)$$

we define the conformal rescaling $\delta_u(\alpha)$ in the folloiwng:

$$r_\hbar(t,u,\mathbf{x}) = (\delta_u(\alpha))(t,\mathbf{x}) = u^{n/2}\sum_{i=0}^{n} u^{-i/2}\alpha(ut, u^{1/2}\mathbf{x})_{[i]}$$

where $\alpha_{[i]}$ denotes ith-differential forms of α. Note that the factor $u^{n/2}$ included in this definition comes from the effect of "density bundle", and α is regarded as a $\Lambda V^*[\hbar^{-1},\hbar]] \otimes_{\mathbb{C}[\hbar^{-1},\hbar]]} End_{C\ell_\hbar}(Q_M \otimes S_\hbar)$-valued section. We also remark that the Euclidean heat kernel $q_t(\mathbf{x}) = (4\pi\hbar t)^{-n/2}e^{-||\mathbf{x}||^2/4\hbar t}|d\mathbf{x}|^{1/2}$ is invariant under the conformal rescaling.

The rescaled heat equation in Proposition 4.1 becomes

$$\begin{aligned}(\partial_t + u\delta_u(\not{D}_Q)^2\delta_u^{-1})r(u,t,\mathbf{x}) &= u^{\frac{n}{2}}\big(u\delta_u\partial_t\delta_u^{-1} + u\delta_u(\not{D}_Q)^2\delta_u^{-1}\big)(\delta_u k)(t,x)\\ &= u^{\frac{n}{2}+1}\big(\delta_u\partial_t + \delta_u(\not{D}_Q)^2\big)k(t,x)\\ &= 0. \end{aligned} \tag{20}$$

Thus we would like to know the rescaling limit of $(\not{D}_{Q,u})^2 := u\delta_u(\not{D}_Q)^2\delta_u^{-1}$ as $u \longrightarrow 0$. A direct computation gives

Proposition 4.3. *Set* $\nabla_i^u := u^{\frac{1}{2}}\delta_u\nabla_i\delta_u^{-1}$

$$\nabla_i^u \overset{u\to 0}{\longrightarrow} \partial_i - \frac{1}{8}\sum_{k\ell} g(R^{LC}(\partial_k,\partial_\ell)\partial_i,\partial_j)e_k\wedge e_\ell\wedge = \partial_i - \frac{1}{4}\sum_{ij} R_{ij}x^j.$$

According to Propositions 3.1 and 4.3 above, we easily have the following.

Proposition 4.4. *The operator* $u\delta_u\not{D}_Q^2\delta_u^{-1}$ *acting on* $C^\infty(U,\Lambda^*T_qM \otimes End_{Cl}(Q_q\otimes S_\hbar))$ *has the following limit as* $u \to 0$*:*

$$K = -\hbar\sum_i\Big(\partial_i - \frac{1}{4}\sum_j R_{ij}\mathbf{x}^j\Big)^2 + \Omega_\nu(\mathcal{A}_M).$$

See [12] for detailed proofs of Propositions 4.3 and 4.4

Thus, multiplying $1/\hbar$, we obtain the following equation.

$$\Big(\frac{\partial}{\hbar\partial t} - \sum_i\Big(\partial_i - \frac{1}{4}\sum_j R_{ij}\mathbf{x}_j\Big)^2 + \frac{\Omega_\nu(\mathcal{A}_M)}{\hbar}\Big)p_t(\mathbf{x}) = 0 \tag{21}$$

Solving this equation, we have the following formula:

$$(4\pi t)^{\frac{-n}{2}}\det{}^{1/2}\Big(\frac{tR/2}{\sinh(tR/2)}\Big)\times\exp\Big(-\frac{1}{4t}\langle\mathbf{x}|\frac{tR}{2}\coth(\frac{tR}{2})|\mathbf{x}\rangle - t\,\Omega_\nu(\mathcal{A}_M)\Big).$$

Summing up what mentioned above, we complete the proof of Theorem 1.2.

Remark.
This remark is devoted to some speculations. First we remark that our method can be applied to ths study of relation between local index formula and RK-theory for generalized operator algebras. Especially we can also derive Fedosov index theorem via Mehler's formula under certain topological conditions. See [12]. Secondary we remark that our method can be applied to local index theorem with a certain group action and deformation quantization for regular Poisson manifolds.

Acknowledgements
The author is greatful to Professors Hideki Omori and Yoshiaki Maeda for their encouragements. He also expresses his thanks to Professor Akira Yoshioka, Doctors Yasushi Homma, Shingo Kamimura and Kiyonori Gomi for helpful discussions and valuable suggestions.

References

1. T. Aristide,*An Atiyah-Singer theorem for gerbes.* math.DG/0302050v1.
2. F. Bayen, M. Flato, C. Fronsdal, A. Lichnerowicz and D. Sternheimer, *Deformation theory and quantization I.* Ann. of Phys. 111 (1978), 61-110.
3. K. Behrend and P. Xu, *Differentiable stacks and Gerbes.* preprint.
4. N. Berline, E. Getzler and M. Vergne, *Heat Kernels and Dirac operators.* Springer-Verlag, 1996.
5. A. Connes, *Noncommutative Geometry.* Academic Press, 1994.
6. B. V. Fedosov, *A simple geometrical construction of deformation quantization.* J. Differential Geom. **40** (1994), 213-238.
7. B. V. Fedosov, *Deformation quantization and Index theory.* Akademie Verlag, 1996.
8. K. Gomi, *Connections and curvings on lifting bundle gerbes.* J. London Math. Soc. **(2) 67**, no.2 (2003), 510-529.
9. S. Gutt and J. Rawnsley, *Equivalence of star products on a symplectic manifold; an introduction of Deligne's Cech cohomology classes.* J. Geom. Phys. **29** (1999), 347-392.
10. M. Kontsevich, *Deformation quantization of Poisson manifolds.* **q-alg/9709040**.
11. Y. Maeda and H. Kajiura, *Introduction to deformation quantization.* Lectures in Math. Sci. The Univ. of Tokyo, **20** (2002), Yurinsya.
12. N. Miyazaki, *Contact Weyl manifold, Index theorem and representable K-theory.* to appear in Kôkyûroku, RIMS. **1408**, Kyoto Univ.
13. M. Murray and M. Singer, *Gerbes, Clifford modules and the index theorem.* **math.DG/0302096v2**.
14. H. Omori, Y. Maeda and A. Yoshioka, *Deformation quantization and Weyl manifolds.* Advances in Mathematics **85** (1991), 224-255.

15. H. Omori, Y. Maeda, N. Miyazaki and A. Yoshioka, *Poincaré-Cartan class and deformation quantization of Kähler manifolds.* Commun. Math. Phys. **194** (1998), 207-230.
16. H. Omori, Y. Maeda, N. Miyazaki and A. Yoshioka, *Strange phenomena related to ordering problems in quantizations.* Jour. Lie Theory vol. **13**, no 2, (2003), 481-510.
17. T. Sakai, *Riemannian geometry.* Shokabo, 1992.
18. N. Woodhouse, *Geometric quantization.* Clarendon Press, Oxford, 1980.
19. A. Yoshioka, *Contact Weyl manifold over a symplectic manifold.* in "Lie groups, Geometric structures and Differential equations", Adv. St. in Pure Math. **37** (2002), 459-493.

A SHORT NOTE ON SYMPLECTIC FLOER THEORY

KAORU ONO*

Department of Mathematics
Hokkaido University
Sapporo, 060-0810, Japan
ono@math.sci.hokudai.ac.jp

In the middle of 1980's, Andreas Floer initiated the $\infty/2$-dimensional (co)homology thoery, which is now called the Floer theory. He seems to be motivated by the so-called Arnold conjecture for fixed points of Hamiltonian diffeomorphisms. The first series of papers [12] dealt with Lagrangian intersections under Hamiltonian deformations. For technical reasons, he assumed that the second relative homotopy group of the pair of the ambient symplectic manifold and its Lagrangian submanifold is trivial. In particular, he gave lower bounds for the number of Lagrangian intersections under this assumption. This framework was soon adapted in the Donaldson theory and gave rise to the instanton homology theory [13]. He also constructed the Floer cohomology for periodic Hamiltonian systems under the assumption that the symplectic manifold is monotone and verified the homological version of the Arnold conjecture for Hamiltonian diffeomorphisms [14]. These works manifested a discovery of new fruitful theories. There followed other works, e.g., generalizations of his construction, applications, etc.

In this short note, we would like to give a brief review on Floer theory in symplectic geometry for non-specialists. This is by no means a thorough survey in this rich field and we apologize people whose works are not mentioned. There are also good introduction and survey articles, e.g., [40], [30], [43], [54].

1. Periodic Hamiltonian systems

Let (M, ω) be a closed symplectic manifold of dimension $2n$ and $H : M \times \mathbf{R} \to \mathbf{R}$ be a smooth function. The (time-dependent) Hamiltonian vector

*The author is partly supported by the Grant-in-Aid for Scientific Research, JSPS.

field X_t is defined by $i(X_t)\omega = dH(\cdot, t)$ for each $t \in \mathbf{R}$. By integrating X_t, we get an isotopy ϕ_t, i.e., $\frac{d}{dt}\phi_t = X_t \circ \phi_t$. We call the time-one map ϕ_1 a Hamiltonian diffeomorphism or an exact symplectomorphism. We recall the Arnold conjecture for fixed points of Hamiltonian diffeomorphisms.

Arnold's conjecture.

$$\sharp\mathrm{Fix}(\phi) \geq \min\{\sharp\mathrm{Crit}(f) | f : M \to \mathbf{R} \text{ smooth}\}.$$

If, in addition, all the fixed points are non-degenerate,

$$\sharp\mathrm{Fix}(\phi) \geq \min\{\sharp\mathrm{Crit}(f) | f : M \to \mathbf{R} \text{ Morse function}\}.$$

The version we will discuss is the following weaker one.

homological Arnold's conjecture.

$$\sharp\mathrm{Fix}(\phi) \geq \textit{cup-length}(M) + 1.$$

If, in addition, all the fixed points are non-degenerate,

$$\sharp\mathrm{Fix}(\phi) \geq \sum_p b_p(M),$$

where $b_p(M)$ denotes the p-th Betti number of M.

We can arrange H so that $H(\cdot, t) = H(\cdot, t+1)$ without changing the time-one map. Then the fixed points of $\phi = \phi_1$ correspond in one-to-one way to 1-periodic orbits of X_t, i.e., $\ell : S^1 = \mathbf{R}/\mathbf{Z} \to M$ satifying

$$\dot{\ell}(t) = X_t(\ell(t)).$$

We define a 1-form α_H, in a formal sense, on the space LM of smooth free loops by

$$\alpha_H(\xi) = \int_0^1 \omega(\xi(t), \dot{\ell}(t) - X_t(\ell(t)))dt,$$

where $\xi \in T_\ell LM$, i.e., a section of ℓ^*TM. Then the set of 1-periodic orbits is exactly the zero set of α_H. We can see that α_H is "closed". For instance, if $\pi_2(M) = 0$, the restriction of α_H to the space of contractible loops admits a primitive function given by

$$\mathcal{A}_H(\ell) = \int_{D^2} w^*\omega + \int_0^1 H(\ell(t), t)dt,$$

where $w : D^2 \to M$ is a bounding disk for ℓ, i.e., $w|_{\partial D^2} = \ell$. Note that the first term does not depend on the choice of w, since $\pi_2(M) = 0$. Hence if there is some way to find critical points of $\mathcal{A}_H$, we obtain existence of 1-periodic orbits, or the fixed points of ϕ. However, the functional $\mathcal{A}_H$ is not bounded in either positive nor negative direction. Moreover, even if there is critical points, the Hessian of $\mathcal{A}_H$ has infinitely many positive, resp. negative, eigenvalues. Thus the usual Morse theory on infinite dimensional spaces cannot be applied to this situation.

The first breakthrough was made by Conley and Zehnder, who used finite dimensional approximation of $\mathcal{A}_H : LM \to \mathbf{R}$ in the case that $M = \mathbf{R}^{2n}/\mathbf{Z}^{2n}$, where $\mathbf{R}^{2n}$ here stands for the symplectic vector space [7]. Roughly, they considered the finite dimensional subspace in LM spanned by finitely many Fourier modes and reduced the infinite dimensional variational problem to the one on this finite dimensional space, i.e., a trivial vector bundle over $\mathbf{R}^{2n}/\mathbf{Z}^{2n}$. The induced function is asymptotic at infinity, in each fiber, to a non-degenerate quadratic function and the "stable Morse theory" can be applied to this situation. As a consequence, they proved the Arnold conjecture for $\mathbf{R}^{2n}/\mathbf{Z}^{2n}$, namely, the number of fixed points of ϕ is at least $2n+1$, i.e., the cup-length of the $2n$-torus. If, in addition, all the fixed points are non-degenerate, the number of fixed points is at least 2^{2n}, i.e., the sum of Betti number of $2n$-torus. This method was generalized to wider classes including closed oriented surfaces of genus > 1 by Floer [11] and Sikorav [61]. (Note also that the Arnold conjecture for S^2 was proved by Nikishin and Simon.)

Finite dimensional approximation method was also explored by Chaperon [2], Laudenbach-Sikorav [35] for Hamiltonian deformations of the zero section in the cotangent bundle to verify a similar estimate for their intersections with the zero section. Later Viterbo used this technique to invent symplectic invariants and established some rigidity results in sympletic geometry [62]. Finite dimensional approximation method was also applied to existence of periodic orbits of the Reeb flow for a contact form. In particular, Rabinowitz established existence for the boundary of a star-shaped domain in a symplectic vector space [51]. Viterbo extended the existence result to closed hypersurfaces of contact type [a] [62]. Although finite dimensional approximation is a powerful technique, the cases, for which this method is applicable, are restricted. In fact, the constructions of finite dimensional approximation in [11] and [61] are already complicated. Therefore, it was desired to find a variational method similar to Conley-Zehnder's work

[a] Hofer and Zehnder gave an alternative proof by the linking argument. This technology was used by Ekeland and Hofer to construct symplectic capacities.

without using finite dimensional approximation. It was Floer who realized this approach by combining the Conley-Zehnder variational set-up and the theory of pseudoholomorphic curves due to Gromov [25]. We also note that there were also Hofer's works [26] around the same time about Lagrangian intersection problems.

Let J be an almost complex structure J calibrated by ω, i.e, $g_J(v_1, v_2) = \omega(v_1, Jv_2)$ gives a Riemannian metric. Then we set

$$\langle \xi_1, \xi_2 \rangle = \int_0^1 g_J(\xi_1(t), \xi_2(t))dt,$$

for $\xi_1, \xi_2 \in T_\ell LM$. With respect to this Riemannian metric on LM, the gradient vector field of $\mathcal{A}_H$, in a formal sense, is

$$\text{grad } \mathcal{A}_H(\ell)(t) = -J\dot{\ell}(t) + \nabla H_t(\ell(t)),$$

where ∇H_t is the gradient of $H_t(\cdot) = H(\cdot, t)$ with respect to g_J. Then the gradient flow line $\gamma : \mathbf{R} \to LM$, i.e., solutions of $\frac{d}{d\tau}\gamma(\tau) = \text{grad } \mathcal{A}_H(\gamma(\tau))$ is interpreted as $u = u_\gamma : \mathbf{R} \times S^1 \to M$ satisfying the following equation[b]

$$\frac{\partial}{\partial \tau}u(\tau, t) + J\frac{\partial}{\partial t}u(\tau, t) - \nabla H_t(u(\tau, t)) = 0.$$

We call solutions of this equation with $\ell^\pm(t) = \lim_{\tau \to \pm\infty} u(\tau.t)$ connecting orbits joining $\ell^\pm$, which are necessarily 1-periodic orbits. Note that, if $H = 0$, the gradient flow lines of $\mathcal{A}_0$ are J-holomorphic cylinders.

If $\pi_2(M) \neq 0$, $\mathcal{A}_H$ is not, in general, well-defined on LM and we have to work with a certain covering space $\widetilde{L}M$ of LM. Here is an additional issue about the index or grading of critical points. In the case of finite dimensional Morse theory, the index is defined by the number of negative eigenvalues of the Hessian. Although the number of negative, resp. positive, eigenvalues are infinitely many in our case, there is a notion of the Conley-Zehnder index of 1-periodic orbits. Namely, the linearization of ϕ_t along a 1-periodic orbit ℓ gives a one-parameter family of symplectic transformations once TM is trivialized, as a symplectic vector bundle, along ℓ. If w is a bounding disk of ℓ, $TM|_\ell$ is trivialized by the trivialization of $w^*TM \to D^2$, which is unique up to homotopy. Thus we assign the Conley-Zehnder index for each pair (ℓ, w) of a 1-periodic solution and its bounding disk by a kind of the Maslov index of the 1-parameter family of symplectic matrices. The Conley-Zehnder index is also interpreted as the Atiyah-Patodi-Singer type index of a certain elliptic operator on D^2.

[b]There may not exist any solutions of this equation for given initial condition $u(0, t)$.

The space $\widetilde{L}M$ consists of equivalence classes of pairs (ℓ, w) of a loop ℓ in M and its bounding disk w by $(\ell, w) \sim (\ell', w')$ iff $\ell = \ell'$, $\langle \omega, w\sharp(-w')\rangle = \langle c_1(M), w\sharp(-w')\rangle = 0$.

Let us consider the graded free module generated by critical points of $\mathcal{A}_H$. If $\widetilde{L}M$ is a non-trivial covering of LM, we need to take its completion with respect to the filtration induced by $\mathcal{A}_H$[c], which is denoted by $\mathrm{CF}^*(H)$. The covering transformation group G of $\widetilde{L}M \to LM$ acts naturally on $\mathrm{CF}^*(H)$, and it is finitely generated free module over the Novikov ring Λ_ω of (M, ω), which is a certain completion of the group ring of G.

The coboundary operator is defined by counting with signs the connecting orbits:

$$\delta[\ell, w] = \sum \langle [\ell, w], [\ell', w'] \rangle [\ell', w'],$$

where $[\ell' w']$ runs over critical points of $\mathcal{A}_H$, the Conley-Zehnder index of which is greater, by 1, than that of $[\ell, w]$ and $\langle [\ell, w], [\ell', w'] \rangle$ is the signed number of connecting orbits joining them modulo translation in the τ-variable. In order to show that $(\mathrm{CF}^*(H), \delta)$ is a cochain complex, we need to study the moduli space of connecting orbits. Transversality of the connecting orbit equation, i.e., the surjectivity of its linearized operator, is achieved by generic choice of J and H. However compactness may fail. The lack of compactness is caused by bubbling-off of J-holomorphic spheres and splitting into several connecting orbits. The splitting phenomenon is not troublesome at all, but the key to establish $\delta \circ \delta = 0$. If $\pi_2(M) = 0$, there are no J-holomorphic spheres and we can construct the Floer cochain complex, hence its resulting cohomology, i.e., the Floer cohomology $\mathrm{HF}^*(H)$. Floer showed that this construction works for monotone symplectic manifold, i.e., $c_1(M)$ is positively proportional to $[\omega]$. Note also that the moduli spaces of connecting orbits are canonically oriented[d]. It was shown that Floer cohomology groups are canonically isomorphic one another. By choosing H as a C^2-small Morse function, $\mathrm{HF}^*(H)$ is isomorphic to $\mathrm{H}^{*+n}(M) \otimes \Lambda_\omega$. The homological version of Arnold's conjecture (non-degenerate case) is a direct consequence of these results. Hofer-Salamon and the author extended these results to so-called weakly monotone symplectic manifolds [28], [45].

In order to generalize the construction to arbitrary closed symplectic manifolds, we need to find appropriate compactification of the moduli spaces of connecting orbits and establish transversality. Note that transver-

[c]The completion is necessary for the coboundary operator is well-defined.
[d]Orientation is a subtle issue in Lagrangian case, as we will see later.

sality fails at infinity when there occur multiply covered bubbles of negative Chern numbers (negative multiple cover problem). This difficulty was also in common with the theory of Gromov-Witten invariants. Kontsevich [32], [34] found an appropriate notion of stable maps for compactification of moduli spaces of holomorphic curves. Based on the framework of stable maps or stable connecting orbits, K. Fukaya and the author, J. Li-G.Tian, Y. Ruan, B. Siebert, G. Liu-G.Tian introduced new techniques (Kuranishi structures, normal cones, virtual neighborhod system, etc) to define the virtual fundamental cycles, with coefficients in $\mathbf{Q}$, of the stable compactification of those moduli spaces (for details, see [22], [38], [52], [60], [39]). Here the fact that the coefficient is $\mathbf{Q}$ is caused by existence of stable connecting orbits with non-trivial automorphisms. To achieve transversality, we use multi-valued perturbation of the Kuranishi map, which provides a local model describing the moduli space of stable maps/connecting orbits. Each branch of the zero locus of perturbed multi-section is given a multiplicity, which is a rational number. K. Fukaya and the author announced a modified construction to obtain Floer cohomology with integer coefficients [23].

For Hamiltonian diffeomorphisms with degenerate fixed points, the cup-length estimate is known in $2n \geq 4$ only for special cases. When $\pi_2(M) = 0$, Floer and Hofer, independently, proved the cup-length estimate[e]. The case of complex projective spaces was first proved by Fortune-Weinstein and later by Floer. Floer's argument was slightly extended by H.-V. Lê and the author [37] and M. Schwarz [56]. At present, we do not know the cup-length estimate holds or not in a wider class. It may be worth mentioning that there is a closed 1-form representing a given non-trivial cohomology class $c \in \mathrm{H}^1(X)$ with at most 1 zero. Hence there are no significant estimate, like the cup-length estimate or the Lusternik-Schnirelman estimate for critical points of a smooth function, for zeros of closed 1-forms.

We also note that there are, so far, no general estimate for the number of fixed point of Hamiltonian diffeomorphisms in terms of the fundamental group, although the original Arnold conjecture (non-degenerate case) guarantees fixed points at least as many as the minimal numbers of generators and relations of the fundamental group.

There are also other nice applications of Floer theory for periodic Hamiltonian systems and we would like to mention some of them. Floer cohomology for periodic Hamiltonian systems carries a product structure, the

[e]Later the estimate by Lusternik-Schnirelman category was established by Oprea and Rudyakk under the same hypothesis.

pair-of-pants product. The Floer cohomology is isomorphic to the quantum cohomology as rings [53], [49].

For a given H, the topologically essential critical values of $\mathcal{A}_H$, like the min-max values in finite dimensional case, give invariants. Schwarz found a way of selecting a canonical critical value for each Hamiltonian diffeomorphism when (M, ω) is symplectically aspherical, e.g., $\pi_2(M) = 0$ [57]. This direction was further explored by Y. G. Oh, who also found applications to so-called Hofer's geometry, see [44]. For a domain U in a symplectic vector space, find an appropriate class of Hamiltonian functions and then define the invariants of U. This is carried out by Cieliebak, Floer, Hofer, Wysocki [15], [16], [4], [5].

H.-V. Lê and the author extended the construction to symplectic isotopies which are not necessarily Hamiltonian and obtained an estimate for number of fixed points by the sum of Novikov numbers of the flux provided the symplectic manifold is $\pm$-monotone [36]. This Floer theory, the Floer-Novikov cohomology, was recently used to establish the flux conjecture by the author [47]. Here the flux conjecture states that the group of Hamiltonian diffeomorphisms is C^1-closed in the group of symplectomorphisms.

2. Lagrangian intersections

For a closed Lagrangian submanifold L in (M, ω), we can consider a relative version of the construction in the previous section. Denote by $\mathcal{P}(L, L)$ the space of paths with end points in L and define the action 1-form α_H on $\mathcal{P}(L, L)$ by

$$\alpha_H^L(\xi) = \int \omega(\xi, \dot{\sigma} - X_{H_t}(\sigma))dt,$$

where $\sigma \in \mathcal{P}(L, L)$ and ξ is a section of σ^*TM such that $\xi(0)$ and $\xi(1)$ are tangent to L. Note that this set-up is more general than the one in the previous section. Let ϕ_t be a Hamiltonian isotopy ϕ_t generated by H. Then $(\phi_{(1-t)/2}, \phi_{(1+t)/2})$ is a Hamiltonian isotopy of $(M \times M, -\omega \oplus \omega)$. Denote by $\mathbf{H}$ the Hamiltonian generating this Hamiltonian isotopy. Note that the diagonal Δ_M is a Lagrangian submanifold. The mapping $\Psi : (\ell(t)) \in LM \mapsto \sigma = (\ell((1-t)/2), \ell((1+t)/2)) \in \mathcal{P}(\Delta, \Delta)$ is bijective and satisfies that $\Psi^*\alpha_{\mathbf{H}}^L = \alpha_H$.

We can make it a more general set-up. Denote by $\mathcal{P}(L_0, L_1)$ the space of paths from L_0 to L_1 and consider the action 1-form

$$\alpha^{L_0,L_1}(\xi) = \int \omega(\xi, \dot{\sigma})dt.$$

For a Hamiltonian isotopy ϕ_t generated by H, we have an isomorphism Φ : $(\sigma(t)) \in \mathcal{P}(L, L) \mapsto (\phi_t(\sigma(t)) \in \mathcal{P}(L, \phi_1(L))$ such that $\Phi^* \alpha^{L,\phi_1(L)} = \alpha_H^L$. From now on, we discuss the set-up with $\mathcal{P} = \mathcal{P}(L_0, L_1)$ and $\alpha = \alpha^{L_0, L_1}$, unless otherwise stated.

Under the assumption that $\pi_1(M, L_i) = 0$, $i = 0, 1$, Floer constructed the Floer cochain complex. (He discussed in the literature the case of $L_1 = \phi_1(L_0)$ and a certain connected component of $\mathcal{P}(L_0, L_1)$.) In this case, α^{L_0, L_1} admits a primitive functional $\mathcal{A}$. Pick a base point σ_0 in each connected component of $\mathcal{P}$. For σ in the same component, there is a mapping $w : [0,1] \times [0,t] \to M$ such that $w(\tau, i) \in L_i$ for $i = 0, 1$, $w(0,t) = \sigma_0(t)$ and $w(1,t) = \sigma(t)$. Then we define

$$\mathcal{A}(\sigma) = \int_0^1 \int_0^1 w^* \omega.$$

The critical points of $\mathcal{A}$, or the zeros of α, are exactly the constant path at intersection points $L_0 \cap L_1$. The gradient flow lines of $\mathcal{A}$ is interpreted as $u : \mathbf{R} \times [0,1] \to M$ satisfying the equation of pseudoholomorphic mappings:

$$\frac{\partial}{\partial \tau} u(\tau, t) + J(u(\tau, t)) \frac{\partial}{\partial t} u(\tau, t) = 0$$

with the boundary condition $u(\tau, i) \in L_i$, $i = 0, 1$. Note that in the set-up with Lagrangian submanifolds the moduli space of connecting orbits may not be orientable. (We will come back this point later.) Hence they use $\mathbf{Z}/2$ as (the ground field of) the coefficient ring for the cochain complexes.

The construction is carried out in the same way as in the case of periodic Hamiltonian systems. The cochain modules $\mathrm{CF}^*(L_0, L_1)$[f] are free modules generated by $L_0 \cap L_1$ over $\mathbf{Z}/2$, where the grading is induced by the so-called Maslov-Viterbo index. Note that the Maslov-Viterbo index is well-defined under the assumption that $\pi_1(M, L_i) = 0$, $i = 0, 1$. The coboundary operator δ is defined by counting, modulo 2, connecting orbits joining critical points of $\mathcal{A}$. The lack of compactness of the moduli spaces of connecting orbits is caused by bubbling-off of J-holomorphic spheres or J-holomorphic disks with boundary on L_i as well as splitting into several connecting orbits. Since $\pi_1(M, L_i) = 0$, $i = 0, 1$, there is no possibilities of bubbling-off of J-holomorphic spheres and J-holomorphic disks with boundary on L_i. Therefore we can achieve transversality of the moduli spaces by perturbing J and L_i and prove $\delta \circ \delta = 0$. Thus we obtain the cochain complex $(\mathrm{CF}^*(L_0, L_1), \delta)$, whose resulting cohomology is the Floer

[f] In a recent version of [21], we use a different notation $\mathrm{CF}^*(L_1, L_0)$.

cohomology $\mathrm{HF}^*(L_0, L_1)$. The invariance under Hamiltonian deformations of L_i as well as the choice of a generic compatible almost complex structure were also established. The computation is a bit subtle issue here. Recall that Weinstein's neighborhood theorem states existence of a tubular neighborhood of $L = L_0$, which is symplectomorphic to a tubular neighborhood of the zero section in the cotangent bundle T^*L with the standard symplectic structure. A C^1-small Hamiltonian deformation of the zero section in T^*L is regarded as the graph of an exact 1-form. We choose L_1 as the corresponding Lagrangian submanifold, which is sufficiently C^1-close to L_0. We may assume that the exact 1-form is transvesal to the zero section, i.e., the exterior derivative of a Morse function f. Then the connecting orbits in Floer thoery with "small energy" correspond to gradient flow lines of f. A connecting orbit is either with "small energy", i.e., approximated by a gradient flow line of f or with "big energy", i.e., approximated by non-trivial J-holomorphic disk with boundary on L. Again the assumption that $\pi_1(M, L_i) = 0$, $i = 0, 1$, there are no connecting with "big energy". Therefore the Floer cohomology is isomorphic to the ordinary cohomology $\mathrm{H}^*(L; \mathbf{Z}/2)$. As a consequence, he established an estimate of $\sharp(L \cap \phi_1(L))$ by the sum of $\mathrm{rank}_{\mathbf{Z}/2}\mathrm{H}^p(L; \mathbf{Z}/2)$ in this case.

A similar estimate for Lagrangian intersection was studied in the case of the zero section in the cotangent bundles. The finite dimensional approximation method was used by Chaperon, Laudenbach-Sikorav. Persistence of intersection was shown by Gromov using the theory of pseudoholomorphic curves. Cup-length estimate was obtained by Hofer[g]. He used both the theory of pseudoholomorphic curves and the variational approach.

Floer's construction was extended by Y. G. Oh to the case of monotone Lagrangian submanifolds with the minimal Maslov number being at least 3 [41]. (He weakend the assumption to that the minimal Maslov number is at least 2 when L_1 is Hamiltonian isotopic to L_0.) He carefully analyzed the situation to exclude the bubbling-off of pseudoholomorphic disks from moduli spaces of connecting orbits of dimension 0 or 1, which are used to define the Floer coboundary operator. Computation is a subtle matter, since there may appear connecting orbits with "big" energy even when L_1 is a sufficiently C^1-small Hamiltonian deformation of L_0. Oh computed the Floer cohomology in the case that L_0 is a real form in a Hermitian symmetric space M of compact type and verified so-called Arnold-Givental's

[g]Floer also established the cup-length estimate using moduli spaces of connecting orbits in the case that $\pi_1(M, L) = 0$.

conjecture[h] for them. Here Arnold-Givental's conjecture states the following. Let ι be an anti-symplectic involution of (M,ω), i.e., $\iota^*\omega = -\omega$, and L the fixed point set of ι. Then the number of intersection points of L and its image under a Hamiltonian deformation is expected to satisfy a similar estimate to the one in Arnold's conjecture for fixed points.

Note that Lagrangian intersection may not persist without certain conditions. For instance a loop on the 2-sphere is Lagrangian. For a loop C contained in the interior of a hemisphere, there is a rotation of the 2-sphere, which moves C away from itself. Hence, in such a case, some of the above argument (exclusion of bubbling-off phenomena, invariance under Hamiltonian deformation, computation, etc) fail. It sounds like a bad news. On the other hand, the failure is caused by exisitence of certain non-trivial holomorphic disks, which is also a strong device for applications. For instance, non-existence of closed exact Lagrangian submanifolds in a symplectic vector space (Gromov's theorem) is a consequence of existence of non-trivial holomorphic disks [25], [50].

Although the bubbling-off phenomena cannot be excluded, Chekanov noticed a variant of Floer cohomology can be constructed for a Hamiltonian diffeomorphism ϕ, whose Hofer norm is less than the least symplectic area A of non-constant pseudoholomorphic disks with boundary on L (and non-constant pseudoholomorphic spheres in M) [3]. Namely, he consider only the critical points of the action functional in the "window" $c < \mathcal{A}^L < c + A$ and connecting orbits between them. Since the bubbling-off can only happen in the moduli space of connecting orbits with energy $> A$, we can forget bubbling-off phenomena. Computation can be also done by comparing this construction with Morse cohomology on M. As a consequence, he gave a new proof for non-degeneracy of Hofer's distance.

Associated to a contact manifold (X,ξ) is the symplectization $\mathrm{Symp}(X,\xi)$, which is diffeomorphic to $X \times \mathbf{R}$ when ξ is co-oriented. For a closed Legendrian submanifold Λ and a closed pre-Lagrangian submanifold L, i.e., $L \times \{0\} \subset X \times \mathbf{R}$ is Lagrangian in the symplectization, we can consider a variant of Floer theory for the pair (Λ, L). Under certain technical assumptions, Floer theory in symplectizations was studied by Eliashberg-Hofer-Salamon [10]. This provides a tool to distinguish contact structures. The case of the prequantum bundle of a symplectic manifold (M,ω) and the

[h]The machinery in [21] as well as [23] provides an approach to Arnold-Givental's conjecture in the case that the fix point set of the anti-symplectic involution is a semi-positive Lagrangian submanifold.

Legendrian lift of a Lagrangian submanifold in M was studied by the author [46] (see also a related work [24] by Givental) in order to show a Legendrian version of Arnold's conjecture.

Encouraged by the success in periodic Hamiltonian systems and Gromov-Witten invariants, we started, with K. Fukaya, Y. G. Oh and H. Ohta, the study of obstruction theory in defining Floer cohomology for pairs of Lagrangian submanifolds. As we mentioned, the troubles are the bubbling-off of pseudoholomorphic disks. More concretely, the bubbling-off of pseudoholomorphic disks occurs in real codimension 1, while the bubbling-off of pseudoholomorphic spheres occurs in real dimension 2 (complex dimension 1). In the construction of cycles, the real codimension 2 effect is negligible but the real codimension 1 contribution is essentially the boundary of the fundamental chain of the moduli space. We briefly explain why the bubbling-off of pseudoholomorphic disks is a real codimension 1 phenomenon for readers' convenience.

Let u_j be a sequence of J-holomorphic disks with boundary on L, which tends to split into w^1 and w^2. Denote by $\beta \in \pi_2(M, L)$ the class represented by u_j and by β^i represented by w^i, $i = 1, 2$. For simplicity, we assume transversality. The dimension of the moduli space around u_j is $n+\mu_L(\beta)-3$, where μ_L is the Maslov class of L, by the index formula and the fact that the biholomorphic automorphism group of D^2 is 3-dimensional. Meanwhile, the dimension of the moduli space of pairs of holomorphic disks with a common point on the boundary and representing β^i, $i = 1, 2$, is equal to

$$(n + \mu_L(\beta^1)) + (n + \mu_L(\beta^2)) - n - 4 = n + \mu_L(\beta) - 4,$$

where 4 is the dimension of the biholomorphic automorphism of one-point union of two copies of the unit disk and $-n$ comes from the constraint having a common point. Thus the bubbling-off of J-holomorphic disks occurs in real codimension 1.

In cotrast to the case of J-holomorphic curves from closed Riemann surfaces, the moduli space of J-holomorphic curves from compact Riemann surfaces with the totally real boundary condition along non-empty boundaries is not necessarily orientable. (The simplest example is the moduli space of constant maps to a non-orientable Lagrangian submanifold.) In the case of closed J-holomorphic curves, the linearization of the J-holomorphic curve equation has the same principal symbol as the Dolbeault operator twisted by a certain holomorphic vector bundle. Thus the linearlization operator can be deformed to the twisted Dolbeault operator, which is a complex linear Fredholm operator. The (virtual) tangent bundle of the moduli space of J-holomorphic curves is described by the index of a family of complex linear Fredholm operators, hence carries a natural orientation, modulo the automorphism group of the domain Riemann surface. A similar argument works for the moduli space of J-holomorphic stable curves without boundary, resp., stable connecting orbits in Floer theory of periodic Hamiltonian systems (for details, see, e.g., [22]). Although the linearization operator for J-holomorphic curves with boundary is the same as partial differential operators but with boundary condition. Note that the totally real boundary condition is not preserved under the multiplication by J, hence, the above argument cannot be applied to the case with Lagrangian boundary conditions.

The basic observation is the following. Let E be a holomorphic vector bundle over a compact Riemann surface Σ and F a totally real subbundle of $E|_{\partial\Sigma}$. Then the index of the Dolbeault operator twisted by E with boundary condition F is canonically oriented, provided F is trivialized (and the components of $\partial\Sigma$ are ordered). In the case of J-holomorphic disk $w : D^2 \to M$, the above argument works, if $(w|_{\partial D^2})^*TL$ is canonically trivialized up to homotopy. This condition is guranteed when a spin structure[i] is specified for L. Note that the manifold is already oriented, when we discuss the spin structure. For a pair of spin Lagrangian submanifold, we can give orientations for moduli spaces of stable connecting orbits. For details, refer to [21]. Orientability of these moduli spaces were also independently studied by de Silva [8]. From now on, we work with (weakly) spin Lagrangian submanifolds.

The obstruction to defining Floer cochain complex for (L_0, L_1) is the bubbling-off of J-holomorphic disks with boundary on L_i. Note that the bubbling-off of J-holomorphic spheres is negligible for our purpose. We study all the J-holomorphic disks systematically to formulate obstruction classes. Here is a rough idea of constructing obstruction classes. For details, see [21]. By Gromov's weak compactness, we can introduce an order among elements in $\pi_2(M, L)$ represented by stable J-holomorphic disks as $\beta_0 = 0, \beta_1, \beta_2, \ldots$ so that $\langle \omega, \beta_i \rangle \leq \langle \omega, \beta_{i+1} \rangle$ and $\langle \omega, \beta_i \rangle$ tends to $+\infty$ as $i \to +\infty$. Since the class β_1 has the least symplectic area among those represented by stable J-holomorphic disks, the moduli space $\mathcal{M}_1(\beta_1)$ of

[i] A weaker condition that L is relatively spin, i.e., there exists a cohomology class $a \in H^2(M; \mathbf{Z}/2)$ such that $a|_L = w_2(L)$, is actually enough. For relatively spin structures, see [21]. For instance, the diagonal set in $M \times M$ is relatively spin. When we deal with a pair of Lagrangian submanifolds, we need a cohomology class $a \in H^2(M; \mathbf{Z}/2)$ such that $a|_{L_i} = w_2(L_i)$ for both $i = 0, 1$ in order to carry out the Floer theory.

J-holomorphic disks representing β_1 with one marked point on the boundary is compact without boundary. The evaluation map $\mathrm{ev}_0 : \mathcal{M}_1(\beta_1) \to L$ at the marked point is a cycle in L, which is the first obstruction cycle $o_1 = o(\beta_1)$, which defines a class in $\mathrm{H}_*(L;\mathbf{Q})$. If the obstruction cycles $o_i = o(\beta_i)$ are defined for $i = 1, \ldots, k$ and $o_i = (-1)^n \partial \mathcal{B}_i$ holds for $i = 1, \ldots, k$, we proceed to define the next obstruction cycle $o_{k+1} = o(\beta_{k+1})$ as follows. The moduli space $\mathcal{M}_1(\beta_{k+1})$ of stable J-holomorphic disks representing β_{k+1} with one boundary marked point may have non-empty boundary. Using the "bounding chains" $\mathcal{B}_i$, $i = 1, \ldots, k$, we construct other moduli spaces, whose boundary compensate $\partial \mathcal{M}_1(\beta_{k+1})$. Namely, we consider the moduli space of J-holomorphic disks w, with one boundary marked point, representing β' such that $w(z_j) \in \mathcal{B}_{i_j}$ for some $z_j \in \partial D^2$ and $\beta_{k+1} = \beta' + \sum_j \beta_{i_j}$. Gluing these moduli spaces and $\mathcal{M}_1(\beta_{k+1})$ we obtain a rational cycle[j], whose image under the evaluation map is the obstruction class o_{k+1}.

If all obstruction cycles $o_i(L_0)$ and $o_j(L_1)$ are defined and there are chains $\mathcal{B}_i(L_0)$ in L_0 and $\mathcal{B}_j(L_1)$ in L_1 such that $o_i(L_0) = (-1)^n \mathcal{B}_i(L_0)$ and $o_j(L_1) = (-1)^n \mathcal{B}_j(L_1)$, then we also count the connecting orbits intersecting some of $\mathcal{B}_i(L_0)$ and/or $\mathcal{B}_j(L_1)$ in addition to the usual connecting orbits in order to define the coboundary operator of the Floer cochain complex. In this way, we obtain Floer cohomology for the pair of L_0 and L_1, provided all the obstruction cycles are defined and vanish for both L_i $(i = 1, 2)$. In such a case, we say that L is unobstructed.

In fact, Floer cochain complex can be constructed under a weaker condition, the weak unobstructedness for each Lagrangian submanifold and coincidence of the "potential functions" of L_i, $i = 1, 2$. A typical case is the pair of a spin Lagrangian submanifold L and its image under a Hamiltonian perturbation such that the only non-trivial obstruction cycles are of the top dimension. This is an extension of Oh's result for monotone Lagrangian submanifolds with minimal Maslov number 2 under Hamiltonian perturbations. There is another condition, which allows us to define Floer cohomology. Namely, if the inclusion $L \subset M$ induces an isomorphism $\mathrm{H}_*(L_i;\mathbf{Q}) \to \mathrm{H}_*(M;\mathbf{Q})$, then we can define Floer cohomology for (L_0, L_1)[k].

Note that the above construction depends on several data, e.g. the

[j]The multi-valued perturbation techinque is used in order to achieve transversality. Hence we work with rational coefficients.

[k]There was a flaw in the 2000 preprint version of [21]. We need a new kind of modification in the construction, which will appear in the revised version of it.

bounding disks, on which the obstruction cycles are also dependent, as well as bunch of perturbations to achieve transversality. In order to clarify dependence, we adopt the framework of filtered A_∞-algebra associated to the Lagrangian submanifold $L = L_i$. In contrast to the theory of Gromov-Witten invariants, we cannot deal only with cycles, since the moduli space of J-holomorphic disk with boundary on L_i may have non-empty codimension 1 boundary as we saw. Once we are forced to work with chains rather than cycles, we have to overcome failure of transversality concerning intersection of chains. This leads us to introduce A_∞-algebra even without presence of J-holomorphic disks. This is already a bit involved. The filtered A_∞-algebra associated to L is a "quantization" of the (classical) A_∞-algebra associated to L by including contributions of J-holomorphic disks with boundary on L. Although we do not explain, there are notions of morphisms, homotopy equivalence, etc, and the homotopy equivalence class of the filtered A_∞-algebra is well-defined for $L \subset M$. For the pair (L_0, L_1), we can associate the filtered A_∞-bimodules over the filtered A_∞-algebras associated to L_0 and L_1, respectively. The system of bounding chains $\mathcal{B}_i(L_0)$, resp. $\mathcal{B}_j(L_1)$, gives rise to a solution of the A_∞-version of the Maurer-Cartan equation. Homotopies of perturbations in the construction derives homotopy equivalence of filtered A_∞-algebras, which induces an identification between the moduli spaces of solutions of the Maurer-Cartan equiations. These machinery controls the behavior of bounding chains, Floer cohomologies, etc, under deformations, e.g., Hamiltonian deformation of Lagrangian submanifolds.

For practical purposes, we need to compute Floer cohomology, or at least show non-vanishing result. Consider the case that L_1 is a Hamiltonian deformation of L_0. We have a spectral sequence, whose E_2-term is the ordinary cohomology of $L = L_0$, converging to $\mathrm{HF}^*(L_0, L_1)$. When $L_0 = L_1$, the Bott-Morse situation, the cohomology with respect to the classical coboundary operator is $\mathrm{H}^*(L) \otimes \Lambda_{nov}$. Contribution from J-holomorphic disks with the next energy level is responsible for the coboundary operator on the E_2-level. We successively continue this process to obtain the spectral sequence. Y. G. Oh is the first who used this spectral sequence in the case of monotone Lagrangian submanifold [42]. As a consequence, he obtained restrictions on the Maslov numbers for monotone Lagrangian submanifolds in a symplectic vector space.

As we mentioned before, if the Lagrangian submanifold L can be moved away from itself by a Hamiltonian deformation, we get non-trivial J-holomorphic disks. There are some results for such Lagrangian submanifolds. For instance, we can show that the Maslov class of a spin Lagrangian submanifold $L \subset \mathbf{R}^{2n}$ is non-zero, provided $H_2(L; \mathbf{Q}) = 0$. There are also important contributions by P. Biran in this direction [1]. For other recent applications to Lagrangian submanifolds, see [20]. There are also other important results, e.g., [59] in Floer theory for Lagrangian submanifolds.

3. Some related topics

We mentioned the existence problem of Reeb periodic orbits on hypersurface of contact type in a symplectic manifold known as Weinstein's conjecture. Hofer and Viterbo were the first to apply holomorphic curve technique toward Weinstein's conjecture [29] (see also [17]). They realized that Reeb periodic orbits exist if existence[l] of holomorphic curves intersecting both sides of the hypersurface is guaranteed, e.g., by a certain variant of Gromov-Witten invariants. Hofer proved Weinstein's conjecture for contact forms of overtwisted contact 3-manifolds. Combined with results of Rabinowitz, Viterbo, Weinstein's conjecture was settled in 3-dimensional case. This direction was much further developed in a series of works by Hofer-Wysocki-Zehnder. If a Legendrian submanifold Λ is given in the contact manifold, one may ask existence question for Reeb chords with distinct end points on Λ (Arnold's chord conjecture). There are works by Abbas, Cieliebak and Mohnke in this direction. The so-called symplectic field theory [9] is also one of important framework in theory of pseudoholomorphic curves and Floer theories.

It is now widely known that Floer theory (more precisely, Fukaya's A_∞-category [18]) appears in the A-model side of homological Mirror Symmetry conjecture proposed by M. Kontsevich [33]. For this topic, see [19], [58].

When a non-singular complex projective algebraic variety is defined over $\mathbf{R}$, the real part is a Lagrangian submanifold in the complex manifold. Viterbo proved that closed manifolds carrying metrics of negative curvature cannot be realized as the real part of "strongly Fano manifolds" using Floer theory [m] This direction is very interesting.

There are also some successful Floer theories, such as Ozsvath-Szabo's Heegaard Floer theory [48] and Floer theory based on the symplectic vortex equation ([6]) [n]. In particular, Heegaard Floer theory provides an efficient

[l]Here holomorphic curve should persist under perturbations of the holomorphic curve equation.
[m]For more information, see [31].
[n]U. Frauenfelder adapted this framework in the Lagrangian intersection problem.

machinery in three dimensional topology and has brought great progress in this field. The symplectic vortex equation has an advantage of compactness of the moduli space, provided, say, the target symplectic manifold has trivial second homotopy group. Thus, for the corresponding symplectic reduced phase space, this theory sometimes offers a theory, which is easier handled than the theory of pseudoholomorphic curves. The same is true for Floer theories. There are certainly much more works to be mentioned.

References

1. P. Biran, *Geometry of symplectic intersections.* Proceeding of ICM, 2002, Beijing, Vol. 2, 241-255, 2002.
2. M. Chaperon, *Une idée du type géodesiques brisées.* Comptes Rendues, Paris, **298** (1984), 293-296.
3. Y. Chekanov, *Hofer's symplectic energy and lagrangian intersections.* In: "Contact and Symplectic Geometry" Edited by C. B. Thomas, Cambridge University Press, 1996. *Lagrangian intersections, symplectic energy and areas of holomorphic curves.* Duke Math. J., **95** (1998), 213-226.
4. K. Cieliebak, A. Floer and H. Hofer, *Symplectic homology II: A general construction.* Math. Z., **218** (1995), 103-122.
5. K. Cielibak, A. Floer, H. Hofer and K. Wysocki, *Applications of symplectic homology II. Stability of the action spectrum.* Math. Z., **223** (1996), 27-45.
6. K. Cieliebak, R. Gaio, I. Mundet-Riera, D. Salamon, *The symplectic vortex equations and invariants of Hamiltonian group actions.* J. Symplectic Geom. **1** (2002), 543-645.
7. C. Conley and E. Zehnder, *The Birkhoff-Lewis fixed point theorem and a conjecture of V. I. Arnold.* Invent. Math., **73** (1983) 33-49, *Morse type index theory for flows and periodic solutions for Hamiltonian systems.* Comm. Pure Appl. Math., **37** (1984) 207-253.
8. V. de Silva, thesis, Oxford University.
9. Y. Eliashberg, A. Givental and H. Hofer, *Introduction to symplectic field theory.* Geom. Funct. Anal. special volume 2000, 560-673.
10. Y. Eliashberg, H. Hofer and D. Salamon, *Lagrangian intersection in contact geometry.* Geom. Funct. Anal. **5** (1995) 244-269.
11. A. Floer, *Proof of the Arnold conjecture for surfaces and generalizations to certain Kahler manifolds.* Duke Math. J., **53** (1986), 1-32.
12. A. Floer, *Morse theory for lagrangian intersections.* Journ. Differ. Geom. **28** (1988), 513-547, *The unregularized gradient flow of the symplectic action.* Comm. Pure Appl. Math. **41** (1988), 775-813, *A relative Morse index for the symplectic action.* Comm. Pure Appl. Math. **41** (1988), 393-407, *Witten's complex and infinite dimensional Morse theory.* Journ. Differential Geom. **30** (1989), 207-221, *Cup length estimate on lagrangian intersections.* Comm. Pure Appl. Math. **42** (1989), 335-357.

13. A. Floer, *An instanton invariant for homology 3-spheres.* Commun. Math. Phys., **118** (1998), 215-240.
14. A. Floer, *Holomorphic spheres and symplectic fixed points.* Comm. Math. Phys. **120** (1989), 575-61.
15. A. Floer and H. Hofer, *Symplectic homology I: Open sets in* $\mathbf{C}^n$. Math. Z., **203** (1989), 355-378.
16. A. Floer, H. Hofer and K. Wysocki, *Applications of symplectic homology I.* Math. Z., **217** (1994), 577-606.
17. A. Floer, H. Hofer and C. Viterbo, *The Weinstein conjecture in* $\mathbf{C}P^n$. Math. Z., **203** (1990), 469-482.
18. K. Fukaya, *Morse homotopy,* A_∞*-category and Floer homologies.* Proc. Garc Workshop, Seoul, 1993, 1-102.
19. K. Fukaya, *Floer homology and mirror symmetry, I.* AMS/IP Stud. Adv. Math.,**23** (2001), 15-43, Amer. Math. Soc., II, Advanced Studies in Pure Math., **34** (2002), 31-127.
20. K. Fukaya, Lecture notes for NATO summer school, 2004.
21. K. Fukaya, Y.G. Oh, H. Ohta and K. Ono, *Lagrangian intersection Floer theory -obstruction and anomaly-.* preprint 2000.
22. K. Fukaya and K. Ono, *Arnold conjecture and Gromov-Witten invariant for general symplectic manifolds.* Fields Institute Commun., **21** (1999), 173-190, Amer. Math. Soc. *Arnold conjecture and Gromov-Witten invariant.* Topology **38** (1999), 933-1048.
23. K. Fukaya and K. Ono, *Floer homology and Gromov-Witten invariant over integer for general symplectic manifolds.* Advanced Studies in Pure Math. **31** (2001), 75-91.
24. A. Givental, *Nonlinear generalization of the Maslov index.* In: Theory of singularities and its applications, Edited by V. Arnold, 71-103, Adv. Soviet Math. **1**, Amer. Mah. Soc. 1990.
25. M. Gromov, *Pseudoholomorphic curves in symplectic manifolds.* Invent. Math. **82** (1985), 307-347.
26. H. Hofer, *Lagrangian embeddings and critical point theory.* Ann. Inst. H. Poincaré, Anal. Non Linéaire **2** (1985) 407-462, *Lusternik-Schnirelmann theory for Lagrangian intersections.* ibid. **5**(1988) 465-499.
27. H. Hofer, *Pseudoholomorphic curves in symplectization with application to Weinstein conjecture in dimension three.* Invent. Math. **114**(1993) 515-563.
28. H. Hofer and D. Salamon, *Floer homology and Novikov rings.* The Floer memorial volume, Edited by H. Hofer, C. Taubes, A. Weinstein and E. Zehnder, 483-524, Progr. Math. 133 Birkhäuser 1995.
29. H. Hofer and C. Viterbo, *The Weinstein conjecture in cotangent bundles and related results.* Annali Sc. Norm. Sup. Pisa, **15** (1988), 411-445, *The Weinstein conjecture in the presence of holomorphic spheres.* Comm. Pure Appl. Math., **45** (1992), 583-622.
30. H. Hofer and E. Zehnder, *Symplectic Invariants and Hamiltonian Dynamics.* Birkhäuser, 1994.
31. V. Kharlamov, *Variétés de Fano réels (d'après C. Viterbo).* Séminaire Bourbaki, 1999/2000, Astérisque No. **276** (2002), 189-206.

32. M. Kontsevich, *Enumeration of rational curves by torus action.* In: Moduli space of curves, Edited by H. Dijkgraaf, C. Faber, G. v. d. Geer, 335-368, Progr. Math. 129 Birkhäuser 1995.
33. M. Kontsevich, *Homological algebra in mirror symmetry.* Proceeding of ICM, Zürich, 1994, 120-139, Birkhäuser, 1995.
34. M. Kontsevich and Y. Manin, *Gromov-Witten classes, quantum cohomology and enumerative geometry.* Comm. Math. Phys., **164** (1994) 525-562.
35. F. Laudenbach and J.-C. Sikorav, *Persistence d'intersection avec la section nulle au cours d'une isotopie hamiltonienne dans un fibré cotangent.* Invent. Math., **82** (1985), 349-358.
36. H. V. Lê and K. Ono, *Symplectic fixed points, the Calabi invariant and Novikov homology.* Topology, **34** (1995), 155-176.
37. H. V. Lê and K. Ono, *Cup-length estimate for symplectic fixed points.* In: Contact and Symplectic Geometry, 268-295, Edited by C. B. Thomas, Publication of the Newton Institute, Cambridge Univ. Press, 1996.
38. J. Li and G. Tian, *Virtual moduli cycles and Gromov-Witten invariants of general symplectic manifolds.* Topics in symplectic 4-manifolds, 47-83, International Press, 1998.
39. G. Liu and G. Tian, *Floer homology and Arnold conjecture.* J. Differential Geom., **49** (1998), 1-74.
40. D. McDuff, *Elliptic methods in symplectic geometry.* Bull Amer. Math. Soc. **23** (1990), 311-358.
41. Y. G. Oh, *Floer cohomology of Lagrangian intersections and pseudo-holomorphic disks, I.* Comm. Pure Appl. Math. **46** (1993), 949-994, *II* ibid. **46** (1993), 995-1012, *III* In :Floer Memorial Volume, Birkhäuser 1995.
42. Y. G. Oh, *Floer cohomology, spectral sequences and the Maslov class of Lagrangian embeddings.* Intern. Math. Res. Notices **7** (1996), 305-346.
43. Y. G. Oh, *Relative Floer and quantum cohomology and the symplectic topology of Lagrangian submanifolds.* In: Contact and Symplectic Geometry, Edited by C. B. Thomas, 201-267, Cambridge Univ. Press 1996.
44. Y. G. Oh, Lecture notes for NATO summer school, 2004.
45. K. Ono, *On the Arnold conjecture for weakly monotone symplectic manifolds.* Invent. Math. **119** (1995) 519-537.
46. K. Ono, *Lagrangian intersection under legendrian deformations.* Duke Math. Journ. **85** (1996) 209-225.
47. K. Ono, *Floer-Novikov cohomology and the flux conjecture.* preprint.
48. P. Ozsvath and Z. Szabo, to appear in Annals of Math.
49. S. Piunikhin, D. Salamon and M. Schwarz, *Symplectic Floer-Donaldson theory and quantum cohomology.* In: Contact and Symplectic Geometry, Edited by C. B. Thomas, 201-267, Cambridge Univ. Press 1996.
50. L. Polterovich, *Monotone Lagrangian submanifolds of linear spaces and the Maslov class in cotangent bundles.* Math. Z. **207** (1991), 217-222.
51. P. Rabinowitz, *Periodic solutions of Hamiltonian systems.* Comm. Pure Appl. Math., **31** (1978), 157-184.
52. Y. Ruan, *Virtual neighborhoods and pseudoholomorphic curves.* Turkish J. Math.

53. Y. Ruan and G. Tian, *A mathematical theory of quantum cohomology.* Journ. Differential Geom. **42** (1995), 259-367, *Higher genus symplectic invariants and sigma model coupled with gravity.* Invent. Math. **130** (1997), 455-516.
54. D. Salamon, *Morse thoery, the Conley index and Floer homology.* Bull London Math. Soc. **22** (1990), 113-140.
55. D. Salamon and E. Zehnder, *Morse theory for periodic solutions of Hamiltonian systems and the Maslov index.* Comm. Pure Appl. Math., **45** (1992), 1303-1360.
56. M. Schwarz, *A quantum cup-length estimate for symplectic fixed points.* Invent. Math., **133** (1998), 353-397.
57. M. Schwarz, *On the action spectrum for closed symplectically aspherical manifolds.* Pacific J. Math., **193** (2000), 419-461.
58. P. Seidel, *Fukaya categories and deformations.* Proceeding of ICM 2002, Beijing, Vol. 2, 351-360, 2002.
59. P. Seidel, *A long exact sequence for symplectic Floer cohomology.* Topology **42** (2003), 1003-1063.
60. B. Siebert, *Gromov-Witten invariants for general symplectic manifolds.* preprint 1996.
61. J.-C. Sikorav, *Points fixes d'une application symplectique homologue à l'identité.* J. Differential Geom., **22** (1985), 49-79.
62. C. Viterbo, *A proof of the Weinstein conjecture in* $\mathbf{R}^{2n}$. Ann. Inst. H. Poincaré, Annal. non linéaire, **4** (1987), 337-357.
63. C. Viterbo, *Symplectic topology as the geometry of generating functions.* Math. Ann., **292** (1992), 685-710.

NONCOMMUTATIVE COHOMOLOGICAL FIELD THEORIES AND TOPOLOGICAL ASPECTS OF MATRIX MODELS

AKIFUMI SAKO

Department of Mathematics,
Faculty of Science and Technology,
Keio University
3-14-1 Hiyoshi, Kohoku-ku,
Yokohama 223-8522, Japan
sako@math.keio.ac.jp

We study topological aspects of matrix models and noncommutative cohomological field theories (N.C.CohFT). N.C.CohFT have symmetry under an arbitrary infinitesimal deformation with noncommutative parameter θ. This fact implies that N.C.CohFT are topologically less sensitive than K-theory, but the classification of manifolds by N.C.CohFT opens the possibility to get a new view point for global characterization of noncommutative manifolds. To investigate the properties of N.C.CohFT, we construct some models whose fixed point loci are given by sets of projection operators. In particular, the partition function on the Moyal plane is calculated by using a matrix model. The moduli space of the matrix model is a union of Grassman manifolds. The partition function of the matrix model is calculated using the Euler number of the Grassman manifold. Identifying the N.C.CohFT with the matrix model, we obtain the partition function of the N.C.CohFT. To check the independence of the noncommutative parameters, we also study the moduli space in the large θ limit and for finite θ, for the case of the Moyal plane. If the partition function of N.C.CohFT is topological in the sense of noncommutative geometry, then this should reveal some relation with K-theory. Therefore we investigate certain models of CohFT and N.C.CohFT from the point of view of K-theory. Our observations give us an analogy between CohFT and N.C.CohFT in connection with K-theory. Furthermore, we verify for the Moyal plane and noncommutative torus cases that our partition functions are invariant under those deformations which do not change the K-theory. Finally, we discuss the noncommutative cohomological Yang-Mills theory.

1. Introduction

Recent developments in string theory provide a fruitful framework and motivation for physicists to study noncommutative field theories. From the viewpoint of physics, much progress has been achieved by using noncommutative geometry. On the other hand, from the point of view of noncommutative space geometry and topology investigated by physical techniques, there are some successful cases, like for example Kontsevich's deformation quantization. It is known that we can construct Kontsevich's deformation quantization by using some kind of topological string theory [1,2]. Another example are the investigations of topological charges. The role of these charges in the noncommutative case is not completely clear yet, however they are topological in commutative space, and thus these charges have some kind of topological nature [3,4,5,6,7,8,9,10,11,12].

Topology and geometry of "commutative" space are studied by many methods. One important way to investigate them is to use quantum (or classical) field theories and string theories. For example, Donaldson theory, Seiberg-Witten theory, Gromov-Witten theory are constructed by cohomological field theories(CohFT). Therefore, it is natural to ask: " Can Noncommutative Cohomological field theories (N.C.CohFT) be used for the investigation of noncommutative geometry or topology?" Here we call N.C.CohFT the CohFT which is naively extended to noncommutative space. One of the aims of this article is to give the circumstantial evidence for a positive answer to this question.

Noncommutative spaces are often defined by using an algebraic formulation, for example by using C^* algebras. So their topological discussions are usually based on algebraic K-theory. For example, the rank of K_0 identifies each noncommutative torus T^2_θ that is characterized by the noncommutative parameter θ. In this sense, even if $\theta - \theta'$ is arbitrary small, T^2_θ is distinguished from $T^2_{\theta'}$ unless they are Morita equivalent. Meanwhile, some topological charges in commutative space seem to remain "topological" on the noncommutative space, and some do not depend on θ. ("Topological" is used in a slightly different sense than usual and its definition is given below.) For example, the Euler number of a noncommutative torus is independent of the noncommutative parameter θ and it is defined as a topological invariant by the difference of K_0 and K_1. Another example is the possibility to define the instanton number (the integral of the first Pontrjagin class) as an integer for Moyal space [3,4], and this fact implies that the instanton number has some kind of "topological" nature even for the case that the base manifold is noncommutative space. (Here, we call Moyal space noncommutative Euclidian space whose commutation relations of the coordinates are given by $[x^\mu, x^\nu] = i\theta^{\mu\nu}$, where $\theta_{\mu\nu}$ is an anti-symmetric constant matrix.) The instanton number does not depend on θ, at least for Moyal space. Also the partition functions of CohFT are examples of such "topological" invariants [13].

These observations show that "topological" charges defined by noncommutative field theory have a tendency to be independence of θ. Therefore it is natural to expect the existence of a topological class less sensitive than K-theory but nontrivial. Here, we define an "insensitive topological invariant" as follows: If noncommutative manifolds A and B give the same K-group, then the topological invariant defined on both A and B takes the same value. However, the reverse of this statement is not always true. In short, if K-theory does not distinguish A from B, then the "insensitive topological invariant" does not classify them. To express thess insensitive topology classes we use the word "topological" in the explanations given above. One may think that such an insensitive topology is not useful for geometrical classification. Possibly, the expression "topological" in the above sense might not be suitable for the instanton number or the partition function of N.C.CohFT, since there is some circumstantial evidence, but this is not proved. Nevertheless, even if they are not "topological", they have an indisputable value from field theoretical point of view, since it is possible to classify manifolds by global characters whose equivalent relations are defined by field theories. In this sense, this classification is similar to the mirror of Calabi-Yau manifolds or to duality in a physical sense. Therefore, one of the aims of this article is to investigate the partition functions of some models of N.C.CohFT as examples of such "topological" invariants.

As mentioned above, N.C.CohFT have the property of θ-shift invariance and the proof of θ-shift invariance is based on the smoothness for θ [13,14]. In some cases, at the commutative point ($\theta = 0$) theories have singularities, as we know e.g. from U(1) instantons. So, we have to keep in mind that there are difficulties to connect a noncommutative theory to a commutative theory and the smoothness of θ for the proof should be checked whenever we consider new models.

On the other hand, there are interesting phenomena caused by the θ-shift. For examples, when we consider Moyal spaces, derivative terms in the action functional become irrelevant in the large θ limit. Then the theory is determined by the potential of the action and the calculation of the partition function sometimes becomes easy. If we can compare the moduli space topology in the large θ limit with the one for finite θ, the θ invariance of the partition function may be checked. We shall verify this for one model in this article.

Here, we comment on the relation between [13] and this article. As an example of N.C.CohFT, a scalar field theory was investigated and its partition function was calculated in [13]. This model is essentially equivalent to the model that is studied in this article. We found that the partition function was given as the "Euler number" of a moduli space by using the method of the fundamental theorem of Morse theory extended to the operator space. This fact implies that the partition function is still the sum of the Euler numbers even if the base manifold is a noncommutative space. But it is not enough to verify the equivalence of the above "Euler number" and usual the Euler number defined for commutative manifolds, because we do not know the connection between the usual Euler number and the extension of the fundamental theorem of Morse theory to the operator formalism, in the sense of local geometry. The calculation in [13] is done by choosing some representation of Hilbert space caused from noncommutativity, and choosing a representation can be understood as a gauge fixing. The computation of [13] lacks of the view point of a local differential geometry of moduli space. On the other hand, when the moduli spaces are defined as spread commutative manifolds, their Euler number is given by the Chern-Weil theorem, and we expect that the partition function is obtained by the Chern-Weil theorem. In other words, we will find that the fundamental theorem of Morse theory extended to the operator formulation connects to the usual local geometry or the usual Euler number on commutative space. It is worth verifying this statement. In this article, we demonstrate this for one example.

We remark that the operator representation of N.C.field theories can be interpreted as an infinite dimensional matrix model. The partition function of N.C.CohFT is determined by the geometry of the moduli space of the matrix model. In particular, when the noncommutative space is a Moyal space, the matrix model does not include kinetic terms like the IKKT matrix model in the $\theta \to \infty$ limit, because terms with differential operator in the Lagrangian like kinetic terms become infinitesimal. Then, we can calculate the partition functions from potential terms for the Moyal space by using the matrix models. This relation between N.C.CohFT and matrix models is also important to the matrix models, since this relation allows us to investigate the topology of their moduli spaces by the N.C.CohFT. Besides that, this correspondence is not just for particular cases. The connection between noncommutative cohomological Yang-Mills theory and the IKKT matrix model is discussed in this article. One of our aims is to observe these relations between the matrix models and N.C.CohFT.

The plan of the article is the following: In the next section N.C.CohFT is reviewed. We see that the partition function of N.C.CohFT is independent of the deformation parameter of the $*$ product. In section 3, we introduce a finite size Hermitian matrix model (finite matrix model) as a 0 dimensional cohomological field theory and we calculate its partition function. This partition function is determined by merely topological information. In section 4, we construct some models of noncommutative cohomological field theory whose moduli spaces are defined by projection operators. Projection operators play an important role in the topology of noncommutative space, because K_0 is obtained by the Grothendieck construction of equivalent classes of projection operators. The partition function of one of the models is given by the sum of the Euler numbers of moduli space of the projectors spaces. In particular, using the result of finite matrix model in section 3, the partition function of the noncommutative cohomological scalar field theory on the Moyal plane is obtained in section 4. Independence from noncommutative parameters is also discussed. The models that contain derivative terms are investigated for the finite noncommutative parameter case and the large limit case. We see that the topology of the moduli spaces of both cases are equivalent. In section 5, a model mirrored by N.C.CohFT in section 4 is constructed on COMMUTATIVE space and this model gives the model in section 4 by large N dimensional reduction. We see the connection between the model and the homotopy classification of vector bundles or topological K-theory. Furthermore, from the point of view of K_0 we see that our partition function of N.C.CohFT is "topological" for the Moyal plane and noncommutative torus cases. In section 6, the correspondence between matrix models and N.C.CohFT is investigated for the case of N.C.cohomological Yang-Mills theories. In the last section, we summarize this article.

2. Brief Review of N.C.CohFT

In this section, we give a brief review of cohomological field theory (CohFT) and the nature of its noncommutative version. The CohFT is formulated in several ways [15] [16] however we only use the Mathai-Quillen formalism in this article.

2.1. *Review of Mathai-Quillen Formalism*

Atiyah and Jeffrey gave a very elegant approach to CohFT [17]. The Atiyah-Jeffrey approach is an infinite dimensional generalization of the Mathai-Quillen formalism that is Gaussian shaped Thom forms [18]. We recall some well known facts here. Details can be found in several lecture notes [19], [20] and [21].

For simplicity we only consider the finite dimensional case in this subsection. Let X be an orientable compact finite dimensional manifold. For a local coordinate x and Grassmann odd variable ψ corresponding to dx, we introduce the BRS operator $\hat{\delta}$:

$$\hat{\delta}x_\mu = \psi_\mu, \quad \hat{\delta}\psi_\mu = 0. \tag{1}$$

Let us consider a vector bundle V with $2n$ dimensional fiber and Grassman-odd variables χ_a and Grassmann-even variables H_a, $a = 1, \cdots 2n$. For these variables, we define the BRS operator $\hat{\delta}$ transformations:

$$\hat{\delta}\chi_a = H_a, \quad \hat{\delta}H_a = 0. \tag{2}$$

Note that $\hat{\delta}$ is a nilpotent operator. Using some section s and a connection A of the vector bundle, the action of the CohFT is defined by the BRS-exact form:

$$\begin{aligned} S &= \hat{\delta}\left\{\frac{1}{2}\chi_a(2is^a + A^{ab}_\mu \psi^\mu \chi_b + H^a)\right\} \\ &= \frac{1}{2}|s^a|^2 - \frac{1}{2}\chi_a \Omega^{ab}_{\mu\nu}\psi^\mu\psi^\nu\chi_b - i\nabla_\mu s^a(\psi)^\mu \chi_a. \end{aligned} \tag{3}$$

To get the second equality, we integrate out the auxiliary field H_a. The partition function is defined by

$$Z = \int \mathcal{D}x\mathcal{D}\psi\mathcal{D}\chi\mathcal{D}H \exp(-S). \tag{4}$$

In the commutative space, the Mathai-Quillen formalism tells us that the partition function is a sum of Euler numbers of the vector bundle on the space $\mathcal{M} = \{s_a^{-1}(0)\}$ with sign. We can see this fact as follows. We expand the bosonic part $|s^a|^2$ around the zero section $s^a = 0$ as

$$|s^a|^2 = (\nabla_\mu s^a x^\mu)^2 + \cdots \ . \tag{5}$$

In general, the CohFT is invariant under rescaling of the BRS-exact terms, then the exact expectation value is given by a Gaussian integral. The Gaussian integral of the bosonic parts gives

$$1/\sqrt{det|\nabla_\mu s^a|^2}. \tag{6}$$

Note that if the connected submanifolds $\mathcal{M}_k$ defined by

$$\bigcup_k \mathcal{M}_k := \{x|s = 0\}, \tag{7}$$

$$\mathcal{M}_i \cap \mathcal{M}_j = \emptyset \text{ for } i \neq j$$

have finite dimension, the Gaussian integral is performed over $X\backslash\{x|s=0\}$. The fermionic non-zero mode ψ, χ integral is

$$det(\nabla_\mu s^a) \tag{8}$$

from the fermionic action $\nabla_\mu s^a(\psi)^\mu \chi_a$. From (6) and (8), the sign $\epsilon_k = \pm$ is given. Here, the remaining zero modes of ψ are tangent to $\mathcal{M}_k$ and the zero modes of χ are understood as a section of the vector bundle over $\mathcal{M}_k$. Let ψ_0 and χ_0 dentote these zero-modes and V_k be the vector bundle over $\mathcal{M}_k$. Then the remaining integral over $\mathcal{M}_k$ is expressed as

$$\int_{\mathcal{M}_k} \mathcal{D}\psi_0 \mathcal{D}\chi_0 e^{-\frac{1}{2}\chi_{a0}\Omega^{ab}_{\mu\nu}\psi^\mu_0\psi^\nu_0\chi_{b0}} = \chi(V_k). \tag{9}$$

Here $\Omega_{\mu\nu}$ is the curvature. After using Chern-Weil theorem the right hand side is given by the Euler number of the vector bundle V_k. Finally we obtain the partition function

$$Z = \sum_k \epsilon_k \chi(V_k). \tag{10}$$

The Cohomological field theories are naive extensions of this Mathai-Quillen formalism to the infinitesimal dimensional cases. The transition to the N.C.CohFT is trivially achieved by going over to operator valued objects everywhere or by replacing the product by the $*$ product everywhere.

2.2. *Some Aspects of N.C.CohFT*

In this subsection we review some aspects of N.C.CohFT that are investigated in [13,14].

In this article we use both, the $*$ product formulation and the operator formulation [22]. We define the $*$ product of noncommutative deformation by using the Poisson bracket $\{\ ,\ \}_\theta$ as follows

$$\phi_1 * \phi_2 = \phi_1\phi_2 + \frac{1}{2}\{\phi_1, \phi_2\}_\theta + (\text{higher order of } \theta), \tag{11}$$

where ϕ_i (i=1,2) are sections of vector bundles whose base manifold is a Poisson manifold. Note that the Poisson brackets are defined on Poisson manifolds. The $*$ product is frequently expressed by $\hbar$ expansion and this $\hbar$ is distinguished from the symplectic form used for the definition of the Poisson bracket. However, we make no distinction between $\hbar$ and the symplectic form and collectively call them noncommutative parameters θ, hereinafter, for simplicity. The index θ of $\{\ ,\ \}_\theta$ denotes the set of noncommutative parameters. For example, we will use Moyal product for $\mathbb{R}^{2n}$ and T^2 when we perform concrete calculations in section 4. In these cases, the following Poisson brackets are used,

$$\{\phi_1, \phi_2\}_\theta = \frac{i}{2}\theta^{\mu\nu}(\partial_\mu\phi_1\partial_\nu\phi_2 - \partial_\mu\phi_2\partial_\nu\phi_1), \tag{12}$$

where the noncommutative parameter $\theta^{\mu\nu}$ is a constant anti-symmetric matrix. Then the $*$ product, called the "Moyal product" [23], for $\mathbb{R}^2$ or T^2 is given by

$$\phi_1 * \phi_2(x) = e^{\frac{i}{2}\theta^{\mu\nu}\partial_\mu\partial'_\nu}\phi_1(x)\phi_2(x')|_{x=x'}. \tag{13}$$

In the following, $*$ is used for both general Poisson manifolds and $\mathbb{R}^2$ or T^2. So, when we consider $\mathbb{R}^2$ or T^2, we write "Moyal product" , "Moyal plane" etc. to make a distinction from the general $*$ product.

Let us consider the CohFT on some Poisson manifolds deformed by the $*$ product. Take the Lagrangian and the partition function as in the previous subsection but with infinite dimensions. Naively, replacing x, χ etc. by some fields $\phi^i(x)$, $\chi^a(x)$ etc. gives the infinite dimensional extension of the Mathai-Quillen formalism. Since the action functional is defined by an BRS-exact functional like $\hat{\delta}V$, its partition function is invariant under any infinitesimal transformation δ' which commutes (or anti-commutes) with the BRS transformation:

$$\begin{aligned} &\hat{\delta}\delta' = \pm\delta'\hat{\delta}, \\ &\delta'\, Z_\theta = \int \mathcal{D}\phi\mathcal{D}\psi\mathcal{D}\chi\mathcal{D}H \;\; \delta'\left(-\int dx^D\hat{\delta}V\right)\; \exp(-S_\theta) \\ &\qquad = \pm\int \mathcal{D}\phi\mathcal{D}\psi\mathcal{D}\chi\mathcal{D}H \;\; \hat{\delta}\left(-\int dx^D\delta'V\right)\; \exp(-S_\theta) = 0. \end{aligned} \tag{14}$$

Let δ_θ be the infinitesimal deformation operator of the noncommutative parameter θ which operates as

$$\delta_\theta\, \theta^{\mu\nu} = \delta\theta^{\mu\nu}, \tag{15}$$

where $\delta\theta^{\mu\nu}$ are some infinitesimal anti-symmetric two form elements. To express the dependence on θ, we use $*_\theta$ as the $*$ product defined by (11) with noncommutative parameter θ in the following discussion. For $*_\theta$, the δ_θ operation is represented as

$$\delta_\theta *_\theta = *_{\theta+\delta\theta} - *_\theta. \tag{16}$$

Then we see that $\hat{\delta}$ commute with δ_θ as follows,

$$\begin{aligned}\hat{\delta}\delta_\theta(\phi_1 *_\theta \phi_2) &= \hat{\delta}(\phi_1 *_{\theta+\delta\theta} \phi_2 - \phi_1 *_\theta \phi_2) \\ &= (\psi_1 *_{\theta+\delta\theta} \phi_2 + (-1)^{P_{\phi_1}} \phi_1 *_{\theta+\delta\theta} \psi_2) \\ &\quad -(\psi_1 *_\theta \phi_2 + (-1)^{P_{\phi_1}} \phi_1 *_\theta \psi_2) \\ &= \delta_\theta(\psi_1 *_\theta \phi_2 + (-1)^{P_{\phi_1}} \phi_1 *_\theta \psi_2) \\ &= \delta_\theta \hat{\delta}(\phi_1 *_\theta \phi_2), \end{aligned} \tag{17}$$

where $\psi_i = \hat{\delta}\phi_i$ and P_{ϕ_i} is the parity of ϕ_i. This fact shows that the partition function of the N.C.CohFT is invariant under the θ deformation.

Note that this proof of invariance under the θ deformation is available when the definition of the BRS transformation does not include the $*$ product explicitly. If the $*$ product is used in the definition of the BRS transformation, the δ_θ and $\hat{\delta}$ do not commute, i.e. we can not apply the above proof. Cohomological Yang-Mills theory is one of such theories, and this issue will be disscussed in section 6.

If we restrict the models to Moyal spaces, more concrete and interesting properties appear from θ shifting. To clarify their character, we introduce the rescaling operator δ_s that satisfies

$$x'^\mu = x^\mu - \delta_s x^\mu, \tag{18}$$

$$\delta_s x^\mu = (\frac{1}{2}\delta\theta^{\mu\nu}(\theta^{-1})_{\nu\rho})x^\rho \tag{19}$$

and

$$(1-\delta_s)[x^\mu, x^\nu] = [x'^\mu, x'^\nu] = i(\theta^{\mu\nu} - \delta\theta^{\mu\nu}). \tag{20}$$

The transformation matrix is given as

$$J^\mu_\rho \equiv \delta^\mu{}_\rho + \frac{1}{2}\delta\theta^{\mu\nu}(\theta^{-1})_{\nu\rho}, \tag{21}$$

and the integral measure is expressed as

$$dx^D = \det \mathbf{J} dx'^D, \qquad \frac{\partial}{\partial x^\mu} = (J^{-1})_{\mu\nu}\frac{\partial}{\partial x'^\nu}, \tag{22}$$

where $\det \mathbf{J}$ is the Jacobian.
Using these new variables the Moyal product is rewritten as

$$(1-\delta_s)(*_\theta) = \delta_s(\exp(\frac{i}{2}\overleftarrow{\partial}_\mu(\theta-\delta\theta)^{\mu\nu}\overrightarrow{\partial}_\nu)) = *_{\theta-\delta\theta}. \tag{23}$$

The above processes are simply changing variables, so the theory itself is not changed. The action before and after this variable change is written as follows.

$$S_\theta = \int dx^D \mathcal{L}(*_\theta, \partial_\mu)$$
$$= \int \det \mathbf{J} dx'^D \mathcal{L}(*_{\theta-\delta\theta}, (J^{-1})^{\mu\nu} \frac{\partial}{\partial x'^\nu}), \qquad (24)$$

where $\mathcal{L}(*_\theta, \partial_\mu)$ is an explicit description to emphasise that the product of fields is the Moyal product and the Lagrangian contains derivative terms.

As the next step, we shift the noncommutative parameter θ as follows

$$\theta \to \theta' = \theta + \delta\theta. \qquad (25)$$

This deformation changes theories in general. However, the partition function of the N.C.CohFT does not change under this shift as we have seen. After a change of variables (18) and a deformation θ (25), the action is expressed as follows.

$$S_{\theta'} = \int \det \mathbf{J} dx'^D \mathcal{L}(*_\theta, (J^{-1})^{\mu\nu} \frac{\partial}{\partial x'^\nu}). \qquad (26)$$

Here $\mathcal{L}(*_\theta, (J^{-1})^{\mu\nu} \frac{\partial}{\partial x'^\nu})$ is a Lagrangian in which the multiplication of fields is defined by $*_\theta$ and all differential operators $\frac{\partial}{\partial x^\mu}$ in the original Lagrangian are replaced by $(J^{-1})^{\mu\nu} \frac{\partial}{\partial x'^\nu}$ without derivations in $*_\theta$. This action (26) shows that the θ deformation is equivalent to rescaling of x by δ_s, but the Moyal product $*_\theta$ is fixed. Note that the $\theta \to \infty$ limit is given by omitting kinetic terms in the action, because the limit $\theta^{\mu\nu} \to \infty$ means $\det \mathbf{J} \to \infty$ in Eq.(26) (see also [24] and [25]). Using this property we investigate both, the large θ limit case and the finite θ case for some N.C.CohFT model on the Moyal plane in section 4.

3. Finite Matrix model with Connections

In this subsection, we study a matrix model and its partition function. Finite size or infinite size Hermitian matrix models are important in physics, even for one-matrix models (see for example [26], [27] and [28]). The model considered here is different from these models, however the methods of the analysis done here is also applicable to them when the geometry of their moduli spaces is studied. The matrix model of this section is regarded as an operator representation of the N.C. cohomological scalar model of section 4, with a cut-off taken in the Hilbert space. From this fact, the calculations performed in this section make it possible to determine the partition function of the N.C.CohFT on the Moyal plane in section 4. (This model is also obtained by a reduction to dim=0 of the model in section 5.)

Let M be the set of all $N \times N$ Hermitian matrices, then this is a N^2 dim Euclidian manifold $\mathbb{R}^{N^2}$. Let V be a rank N^2 (trivial) vector bundle over M. Let $s : M \to V$ denote some given section of a trivial bundle. We adopt the Killing form as a positive-definite inner product.

We construct the finite matrix model as the 0 dimensional CohFT. Take some orthonormal basis of $N \times N$ Hermitian matrices as a canonical coordinate of M, and write $\phi = (\phi^{ab}) \in M$. The other fields (matrices) are introduced by the method of general CohFT. H^{ab} is a bosonic auxiliary field that is a $N \times N$ Hermitian matrix. Fermionic matrices are ψ^{ab} and χ^{ab}, i.e. the BRS partners of ϕ and H, and they are $N \times N$ Hermitian matrices, too. Their BRS transformation is given as

$$\hat{\delta}\phi = \psi,\ \hat{\delta}\psi = 0,\ \hat{\delta}\chi = H,\ \hat{\delta}H = 0. \tag{27}$$

Let ∇ be a connection $\Gamma(V) \to \Gamma(T^*M \otimes V) = V$, where $\Gamma(V)$ is a set of all sections. Let $A_{ji;mn}^{\ \ kl}(\phi)$ be a component of a connection 1-form in the vector bundle V, and e_{ij} be a component of a local frame field of V. Using e_{ij}, the relation between A and ∇ is written as $\nabla_{ij} e_{kl} = \sum A_{ij;kl}^{\ \ mn} e_{mn}$. In the following, we take the section of the trivial bundle as $s(\phi) = \phi(1-\phi)$. Then the CohFT action is given by

$$S = \sum_{i,j} \hat{\delta}\{\chi^{ij}(2[\phi(1-\phi)]_{ji} + i \sum_{m,n,k,l} \chi^{mn} A_{ji,mn}^{\ \ kl}(\phi)\psi_{kl} - iH_{ij})\}. \tag{28}$$

After performing the Gaussian integral of H_{ij}, the bosonic part of the action becomes

$$\mathrm{tr}(\phi(1-\phi))^2, \tag{29}$$

and the fermionic part of the action is

$$\mathcal{L}_F = \mathrm{tr}i\chi\Big\{2(\psi(1-\phi) - \phi\psi) - \sum_{ijklmn} \psi_{ij}\psi_{kl} F(ij,kl;ab,mn)\chi_{mn}\Big\}. \tag{30}$$

Here $F(ij,kl;ab,mn)$ is the curvature defined by

$$F(ij,kl;ab,mn) \equiv \tag{31}$$
$$\frac{\delta}{\delta\phi_{ij}} A_{kl;ab}^{\ \ mn} - \frac{\delta}{\delta\phi_{kl}} A_{ij;ab}^{\ \ mn} + i \sum_{(c,d)} [A_{ij;ab}^{\ \ cd} A_{kl;cd}^{\ \ mn} - A_{kl;ab}^{\ \ cd} A_{ij;cd}^{\ \ mn}]$$

The fixed points of this action are determined by

$$(\phi(1-\phi)) = 0. \tag{32}$$

Non-zero solutions of ϕ are the projection operators P defined by $P^2 = P$. We denote by P_k the projector that restricts a rank N vector space to a dimension k vector space. The set of all P_k is connected and

$$\mathcal{M}_{k,N} \equiv \{P_k\} = G_k(N), \tag{33}$$

where $G_k(N)$ is a Grassman manifold $(\frac{U(N)}{U(k)U(N-k)})$ whose dimension is $2k(N-k)$.

Let us investigate the $\mathcal{M}_k$ from a local geometric aspect. At first, we prove the non-degeneracy of s in the normal directions to $\mathcal{M}_k$. The definition of non-degeneracy is as follows. Locally one can pick a coordinate e_{ij} (number of combination (i,j) is $N^2 - 2k(N-k)$) in the directions normal to $\mathcal{M}_k$ and a trivialization of V such that

$$s^{ab} = \sum_{i,j} f_{ij}^{ab} e^{ij}, \quad \text{for} \quad (i,j),(a,b) \in \mathbf{N} \tag{34}$$

$$s^{ab} = 0\ , \quad \text{for } (a,b) \in \mathbf{T}. \tag{35}$$

Here, $\mathbf{N}$ and $\mathbf{T}$ are sets of indices (i,j) and the numbers of their elements are $N^2 - 2k(N-k)$ and $2k(N-k)$, respectively. Let us prove this non-degeneracy of $\mathcal{M}_k$. After an appropriate coordinate choice, we can take a rank k solution $P_k \in \mathcal{M}_k$ as

$$P_k = \left(\begin{array}{c|c} \mathbf{1}_k & 0 \\ \hline 0 & 0 \end{array}\right), \tag{36}$$

where P is a $N \times N$ matrix valued projection operator and $\mathbf{1}_k$ is the $k \times k$ unit matrix. The (co)tangent vectors at this point are determined by variation of ϕ equation around this solution;

$$\delta\phi(1-P_k) - P_k\delta\phi = 0. \tag{37}$$

Its solutions are given by

$$\delta\phi_{ij} = 0,\ \delta\phi_{mn} = 0,\ \delta\phi_{in} = \delta\bar{\phi}_{ni}, \quad \text{for} \quad i,j \in \{1,2,\cdots,k\}$$
$$m,n \in \{k+1,\cdots,N\}. \tag{38}$$

Here $\bar{\phi}$ is the complex conjugate of ϕ. We can chose a $2(N-k)k$ dim orthonormal basis of $N \times N$ matrices $\delta\phi$:

$$(\phi_R^{(in)}) = \left(\begin{array}{c|c} O & (\delta_{in}) \\ \hline (\delta_{ni}) & O \end{array}\right),\ (\phi_I^{(in)}) = \left(\begin{array}{c|c} O & i(\delta_{in}) \\ \hline -i(\delta_{ni}) & O \end{array}\right) \tag{39}$$

where $i \in \{1, 2, \cdots, k\}$ and $n \in \{k+1, \cdots, N\}$. Let us not confuse "i" of $\sqrt{-1}$ and index in this article. On the other hand, it is possible to choose a basis of the normal direction e_{normal} as a Lie algebra of $U(k) \times U(N-k)$ whose non-zero elements lie only in the block diagonal part i.e. $(e_{normal})_{in} = 0$ for $i \in \{1, 2, \cdots, k\}$ and $n \in \{k+1, \cdots, N\}$. (Note that $\mathrm{tr}\phi_I^{(in)} e_{normal} = \mathrm{tr}\phi_R^{(in)} e_{normal} = 0$ shows that the direction of e_{normal} is normal to $\delta\phi$.) This basis of the Lie algebra may be chosen to be non-degenerate. For example, we can take this non-degenerate $N^2 - 2k(N-k)$ dim basis e_{normal} as,

$$(e^{(ij)}_{normal}) = \begin{cases} \left(\begin{array}{c|c} U^k_{i,j} & O \\ \hline O & O \end{array}\right), & \text{for } i \text{ and } j \in \{1, \cdots, k\} \\ \\ \left(\begin{array}{c|c} O & O \\ \hline O & U^{N-k}_{i,j} \end{array}\right), & i \text{ and } j \in \{k+1, \cdots, N\} \end{cases} \tag{40}$$

where $\{U^{N-k}_{i,j;a,b}\}$ is a orthonormal basis of $u(k)$ and $\{U^{N-k}_{i,j;a,b}\}$ is one of $u(N-k)$. We found the local coordinate e_{normal} in the directions normal to $\mathcal{M}_k$ such that (34) holds. This shows non-degeneracy. This discussion for non-degeneracy is parallel to the one in [29].

Let us investigate the mass matrix of fermions near the $\mathcal{M}_k$ and the fermionic zero-modes. The χ equation and the ψ equation are

$$\psi(1-P) - P\psi = 0, \text{ and } \chi(1-P) - P\chi = 0, \tag{41}$$

where we neglect nonlinear terms. Note that f^{ab}_{ij} in (34) is the mass matrix of χ and ψ near $\mathcal{M}_k$. Using the χ equation, we see that the massless components of ψ are those ones that are tangent to $\mathcal{M}_k$. There are massless components of χ^{ab} that are regarded as the above trivialization i.e. $(a, b) \in \mathbf{T}$. Furthermore we can understand from the ψ equation that the χ zero-modes are sections of the (co)tangent bundle of $\mathcal{M}_{k,N}$.

Now we evaluate the integral for Z. The mass components integral gives an overall factor $(-1)^{k^2} = (-1)^k$ (see [13]). Recall that the moduli space $\{\phi | s = 0\} = \bigcup_k \{P_k\}$ and $\{P_k\} = G_k(N)$. The Poincare polynomial of the Grassman manifold is given as

$$P_t(G_k(N)) = \frac{(1-t^2)\cdots(1-t^{2N})}{(1-t^2)\cdots(1-t^{2(N-k)})(1-t^2)\cdots(1-t^{2k})}.$$

(See for example [30].) Using these results and (10), the partition function is written as

$$Z = \sum_{k=0}^{N} (-1)^k P_{-1}(G_k(N)). \tag{42}$$

When we take $t = \pm 1$, the Poincare polynomial becomes the number of combinations,

$$P_{\pm 1}(G_k(N)) = \frac{N!}{k!(N-k)!} \equiv \binom{N}{k}. \tag{43}$$

The proof of (43) is given as follows.

$$\begin{aligned} P_{\pm 1}(G_k(N)) &= \left. \frac{(1-t^2)\cdots(1-t^{2N})}{(1-t^2)\cdots(1-t^{2(N-k)})(1-t^2)\cdots(1-t^{2k})} \right|_{t=1} \\ &= \left. \frac{(1-t^{2(N-k+1)})\cdots(1-t^{2N})}{(1-t^2)\cdots(1-t^{2k})} \right|_{t=1}. \end{aligned} \tag{44}$$

After replacing t^2 by a positive real number x,

$$\begin{aligned} P_{\pm 1}(G_k(N)) &= \left. \frac{(1-x^{(N-k+1)})\cdots(1-x^N)}{(1-x)\cdots(1-x^k)} \right|_{x=1} \\ &= \left. \frac{\{(1-x)(1+x+\cdots+x^{N-k})\}\cdots\{(1-x)(1+x+\cdots+x^{N-1})\}}{\{(1-x)\}\{(1-x)(1+x)\}\cdots\{(1-x)(1+x+\cdots+x^{k-1})\}} \right|_{x=1} \\ &= \frac{(N-k+1)(N-k+2)\cdots N}{1\cdot 2\cdots k} = \binom{N}{k}. \end{aligned} \tag{45}$$

This is what we were aiming for. From (42), (43) and the binomial theorem, the final result is then

$$Z = \sum_{k=0}^{N} (-1)^k 1^{N-k} P_{-1}(G_k(N)) = (1-1)^N = 0. \tag{46}$$

The calculation of the finite matrix model in this section will be used directly in the noncommutative cohomological scalar model in the next section.

4. N.C.Cohomological Scalar Model

In this section, we study some N.C.cohomological scalar models and evaluate their partition functions for Moyal space by using the matrix model partition function of the previous section. We also check the θ-shift invariance of Z.

4.1. *N.C cohomological scalar model*

Let M be a $2n$ dimensional Poisson manifold with Riemannian metric. Let ϕ and H be real scalar fields on M and, ψ and χ be the BRS partner fermionic scalar fields of ϕ and H, respectively. In other words, (ϕ, H, ψ, χ) are elements of $\Omega^0(M)$ with ghost number $(0, 0, 1, -1)$ and parity (even, even, odd, odd).

We introduce a nilpotent operator $\hat{\delta}$, i.e.

$$\hat{\delta}^2 = 0, \tag{47}$$

as a BRS operator whose transformation is given by

$$\hat{\delta}\phi = \psi, \ \hat{\delta}\chi = H, \ \hat{\delta}\psi = \hat{\delta}H = 0. \tag{48}$$

We consider the deformation quantization defined by some $*$ product. ($*$ product exist on arbitrary Poisson manifolds [1].)

We consider two actions :

$$S_1 = \int_M dx^D \sqrt{g}\mathcal{L} \tag{49}$$

$$S_2 = S_1 + S_{top}, \tag{50}$$

where the Lagrangian $\mathcal{L}$ is given by

$$\mathcal{L} = \hat{\delta}\left(\frac{1}{2}\chi * \left(2(\phi * (1-\phi)) + \frac{2i}{g}\int d^{2n}zd^{2n}y\psi(z)A(z;x,y)\chi(y) - iH\right)\right). \tag{51}$$

Here, g is a coupling constant, $x, y, z \in M$, and $A(z; x, y)$ is some functional of ϕ that should be defined as a connection on the trivial bundle over the set of all ϕ. $A(z; x, y)$ is an anti-symmetric matrix with respect to x and y, and the multiplication between $A(z; x, y)$, $\psi(z)$ and $\chi(y)$ is not $*$ multiplication since the trace operation (integral) over z and y has been performed. (However, we can also express their products by $*$ product in the integral.) This may look like some strange kind of non-local interaction, but it is possible to regard this as an integral kernel. In many cases, deformation quantization is introduced by an integral kernel, therefore such kind of non-local interactions are not unusual in noncommutative field theory. The precise definition of $A(z; x, y)$ depends on M and the deformation by $*$, the connection is formally introduced, here. When we consider the $\mathbb{R}^2$ case in the following subsection, we will verify that $A(z; x, y)$ is a connection and particularly that it becomes a nontrivial connection on a submanifold of $\{\phi\}$. Especially in conjunction with the matrix model in previous section, after using the Weyl correspondence, we can regard $A(z;x,y)$ as the usual connection of the (co)tangent vector bundle over some Grassman manifold that appears as a moduli space of ϕ.

The topological action in S_2 is

$$S_{top} = g'\tau_{2n}(\mathcal{F},\cdots,\mathcal{F}), \tag{52}$$

where g' is a coupling constant and $\mathcal{F}$ is defined by

$$\mathcal{F}_{ij} = [\phi\partial_i\phi, \phi\partial_j\phi]. \tag{53}$$

This action itself is not topological, but in our case ϕ is replaced by projection operators. In such a case, we can regard S_{top} as Connes's Chern character. Connes's Chern character homomorphism is:

$$\begin{aligned} ch_{2n} &: K_0(\mathcal{A}) \to HC_{2n}(\mathcal{A}) \\ ch_{2n}(p) &= \sum_{n=0}^{\infty} \tau_{2n}(f,\cdots,f) \end{aligned} \tag{54}$$

where $f_{ij} = [p\partial_i p, p\partial_j p]$. It is worth emphasizing that S_{top} is not invariant in general under changing the noncommutative parameter θ, since it is not a BRS exact action. Indeed $ch_{2n}(p)$ apparently depends on θ for the noncommutative torus example. Therefore S_2 is not suitable if we are interested in only constructing the θ-shift invariant theory. However, there is another motivation to construct the N.C.CohFT, that is to construct some "topological" invariant. In the commutative case, we often add a topological action to the BRS exact one, and the topological terms play important roles. In analogy with commutative CohFT, it seems useful to consider both the S_1 and the S_2 case.

The Lagrangian $\mathcal{L}$ without the S_{top} part is divided into a bosonic part $\mathcal{L}_B$ and a fermionic part $\mathcal{L}_F$:

$$\mathcal{L} = \mathcal{L}_B + \mathcal{L}_F, \tag{55}$$

$$\mathcal{L}_B = |\phi * (1-\phi)|^2\,, \tag{56}$$

$$\begin{aligned} \mathcal{L}_F = i\chi * \Big\{ & 2(\psi * (1-\phi) - \phi * \psi) \\ & - \frac{i}{2g}\int d^n z d^n w d^n y \psi(z)\psi(w)F(z,w;x,y)\chi(y)\Big\}. \end{aligned}$$

Here $F(z,w;x,y)$ is defined by

$$\begin{aligned} &\frac{\delta A(z;x,y)}{\delta\phi(w)} - \frac{\delta A(w;x,y)}{\delta\phi(z)} \\ &\qquad + \frac{i}{g}\int d^n u(\, A(z;x,u)A(w;u,y) - \, A(w;x,u)A(z;u,y)), \end{aligned} \tag{57}$$

and it corresponds to the curvature.

From the general argument of the Mathai-Quillen formalism and the similar analysis as in the previous section, the partition function of this theory is given by the sum of the Euler numbers of the solution space of ϕ. From Eq.(56), the fixed point loci of ϕ are given by the set of all projection operators P, i.e. $P * P = P$, and they are called GMS solitons [24]. We denote by $\mathcal{M}_k$ the set of projections distinguished by index k. An example of the index k is the rank of projections for the case that we can define the rank by a discrete number. If there is ghost number anomaly, the partition function vanishes in general. In our case, there is no ghost number anomaly as we saw in section 3, thus we get some nontrivial partition functions for S_1 and S_2:

$$Z_1 = \sum_k \epsilon_k \chi(\mathcal{M}_k), \tag{58}$$

$$Z_2 = \sum_k \epsilon_k \chi(\mathcal{M}_k) e^{g' \tau_{2n}(k)} \tag{59}$$

where $\chi(\mathcal{M}_k)$ is the Euler number of $\mathcal{M}_k$ and ϵ_k gives a sign $\pm$.

When we consider the noncommutative theory from the topological view point, the most important operators are projectors and unitary operators since they define K_0 and K_1. This partition function is a sum of integer valued Euler numbers of the sets of all projections that construct the K_0 elements when the moduli space is a manifold. So it is natural to expect the partition function to be "topological".

Concrete calculation of the partition function will be done for the Moyal plane case below. We are interested in whether the "topological" quantity is invariant under the continuous change of the noncommutative parameter. For the S_1 model of this section, it is clear that the partition function is invariant under the θ change as far as there is no singularity. A more interesting case is when the Lagrangian has kinetic terms. To investigate the behavior of the partition function whose Lagrangian contains kinetic terms, we slightly deform our models in the following.

4.2. *N.C. cohomological scalar model with kinetic terms*

Let M be a $2n$ dimensional Poisson manifold with a Riemannian metric. Let ϕ and H be real scalar fields on M, and ψ and χ be the BRS partner fermionic scalar fields of ϕ and H, respectively. Let B_μ and H_μ be real vector fields and ψ_μ and χ_μ be BRS partner fermionic vector fields of B_μ

and H_μ. In other words, (ϕ, H, ψ, χ) are elements of $\Omega^0(M)$ with ghost number $(0,0,1,-1)$ and parity (even, even, odd, odd). $(B_\mu, H_\mu, \psi_\mu, \chi_\mu)$ are elements of $\Omega^1(M)$ with ghost number $(0,0,1,-1)$ and parity (even, even, odd, odd). The BRS operator transformation is given by

$$\hat{\delta}\phi = \psi \ , \ \hat{\delta}\chi = H, \ \hat{\delta}\psi = \hat{\delta}H = 0, \ \hat{\delta}B^\mu = \psi^\mu, \\ \hat{\delta}\chi^\mu = H^\mu, \ \hat{\delta}\psi^\mu = \hat{\delta}H^\mu = 0. \tag{60}$$

One of our interests is to investigate the behavior of the partition function of N.C.CohFT under a change of the noncommutative parameter. It is difficult to study the general case of deformation quantization. Therefore, we put an assumption in this subsection such that terms including derivatives like kinetic terms become irrelevant in the large noncommutative parameter limit ($\theta \to \infty$) as far as evaluation of perturbative contribution is concerned. For example, when we consider the deformation of $\mathbb{R}^d$ by the Moyal product, only the potential terms become relevant in the $\theta \to \infty$ limit [13,24]. Note that we make this assumption only for simplicity of the calculation. The invariance under a change of θ is essential for us, and this is not affected by our assumption.

Similar to the previous subsection, we consider two types of action:

$$S_1 = \int_M dx^D \sqrt{g}\mathcal{L} \tag{61}$$

$$S_2 = S_1 + S_{top}, \tag{62}$$

where the Lagrangian is slightly different from (51),

$$\mathcal{L} = \\ \hat{\delta}\left(\chi * \left((\phi * (1-\phi)) - i\partial_\mu B^\mu) + \frac{i}{g}\int d^n z d^n y \psi(z) A(z;x,y)\chi(y) - iH\right)\right) \\ +\hat{\delta}\left(\frac{1}{2}\chi^\mu * (2(i\partial_\mu \phi + B_\mu) - iH_\mu)\right). \tag{63}$$

As noted in the previous subsection, the topological term S_{top} has noncommutative parameter θ dependence in general. One such example is the noncommutative torus. On the other hand, the Moyal plane theory does not depend on θ. When we construct a θ independent "topological" invariant Z, we find whether we can add S_{top} to the action S_1 obtained from the K-theory (cyclic cohomology) information of the base manifold.

The Lagrangian $\mathcal{L}$ without S_{top} term is divided into bosonic part and fermionic part:

$$\mathcal{L} = \mathcal{L}_B + \mathcal{L}_F, \tag{64}$$

$$\mathcal{L}_B = |\phi * (1-\phi) - i\partial_\mu B^\mu|^2 + |i\partial_\mu \phi + B_\mu|^2, \tag{65}$$

$$\mathcal{L}_F = i\chi * \Big\{ 2(\psi * (1-\phi) - \phi * \psi - i\partial_\mu \psi^\mu) \tag{66}$$

$$-\frac{i}{2g}\int d^n z d^n w d^n y \psi(z)\psi(w) F(z,w;x,y)\chi(y) \Big\} + i\chi^\mu * \{2i\partial_\mu \psi + 2\psi_\mu\}.$$

Note that this theory is invariant under arbitrary A deformation ($A \to A + \delta A$) and deformation of the coupling constant g. In the following subsections, we investigate the deformation of moduli space and the invariance of partition function under change of θ. If we observe $\theta \to \infty$ in the Moyal plane case by using the scaling method discussed in section 2, the $F(z,w;x,y)$ contribution to the partition function becomes dominating since each integral measure d^2z, d^2w and d^2y is of order θ. Then, the surviving terms in the limit are not the BRS exact terms. Therefore, we have to tune the other parameters in such a limit if we want to use the convenient methods of CohFT. From the fact that the partition function has symmetry under arbitrary g and A variation, we can fit g without changing Z for surviving terms which are BRS exact terms in the limit $\theta \to \infty$.

4.3. *Moyal plane case in* $\theta \to \infty$

In this subsection, the partition function of the N.C.cohomological scalar model is calculated. To compute it explicitly, we consider the two dimensional Moyal plane. There are two reasons for choosing the Moyal plane here. First, the Moyal plane satisfies the assumption of the previous subsection that derivative terms like kinetic terms in the Lagrangian become irrelevant in $\theta \to \infty$. The other reason is that the rank of a projection operator is defined by an integer. From this, the solution space of ϕ is given by a Grassmann manifold the properties of which are well known. In particular, if we represent our theory by operators, this theory can be regarded as an infinite dimensional matrix model. It is possible to represent noncommutative Euclidian plane by a Hilbert space and we can chose some set of eigenvectors with discrete eigenvalues as the basis of this Hilbert space, like for example Fock states. So if we introduce a cut-off in this Hilbert space, we can regard our model as the finite matrix model which we discussed in section 3.

So far we have used the $*$ product representation of noncommutative field theory, however, in this subsection the operator representation is used since it is convenient to see the relation between the finite matrix model and the large θ N.C. cohomological scalar model.

In $\theta \to \infty$, we can ignore the terms including derivative as we saw in section 2. Then, the remaining part of the action in the operator formalism is

$$S_\infty = \sum_{i,j} \hat{\delta}\{\chi^{ij}(2[\phi(1-\phi)]_{ji} + i(\sum_{m,n,k,l} \chi^{mn} A_{ji,mn}^{kl}(\hat{\phi})\psi_{kl}) - iH_{ij})\}$$
$$+Tr\hat{\delta}\{\hat{\chi}^\mu(2\hat{B}_\mu - i\hat{H}_\mu)\}, \tag{67}$$

where $\hat{\phi}$, $\hat{\psi}$, $\cdots$ are operator representation of ϕ, ψ, $\cdots$ that have infinite dimensional matrix representation $\hat{\phi} = \sum_{ij} |i\rangle\phi_{ij}\langle j|$ in some complete system $\{|i\rangle\}$.

We introduce a cut-off to restrict the Hilbert space to a finite N dimensional vector space. Let $\{|i\rangle | i = 1, \cdots, N\}$ be a set of orthonormal basis. Using this representation, the operators $\hat{\phi}$, $\hat{H}$, $\hat{\psi}$, $\cdots$ are expressed by $N \times N$ Hermitian matrices, i.e. $\hat{\phi} \to (\phi_{ij})$ etc. After integration of B_μ, H_μ, χ_μ and ψ_μ, we find that this model is equivalent to the finite dimensional matrix model discussed in section 3.

The bosonic part of the action is

$$Tr\{(\hat{\phi}(1-\hat{\phi}))^2 + \hat{B}_\mu \hat{B}^\mu\}. \tag{68}$$

The fixed point locus is determined by $(\hat{\phi}(1-\hat{\phi})) = 0$ and $\hat{B}_\mu = 0$. The solution is given by $\phi = P$, where P is an arbitrary projection operator, which is called GMS soliton. The moduli space is obtained as a set of Grassman manifolds $\{G_k(N)\} := \{\frac{U(N)}{U(k)\times U(N-k)}\}$ since the rank k projection operator determines a subspace the codimension of which is $N-k$. This solution of rank k projector is interpreted as symmetry breaking from $U(N)$ to $U(k) \times U(N-k)$.

On the other hand, the integration of the fermionic part generates the Euler numbers of the Grassmann manifolds that are given in section 3. For the Moyal plane, the topological term S_{top} (ch_2) is $g'k$ when the solution of $\hat{\phi}$ is given by a rank k projection operator. Note that the value of ch_2 is independent of θ for the Moyal plane (see for example [32]).

Using the Euler number of the Grassman manifolds and the contribution from the topological term, the partition function is then

$$\begin{aligned} Z_2 &= \lim_{N\to\infty} \sum_{k=0}^{N} P_{-1}(G_k(N)) e^{g'k}(-1)^{N-k} \\ &= \lim_{N\to\infty} (1-e^{g'})^N, \end{aligned} \tag{69}$$

where we take $N \to \infty$ after using the result from the finite matrix model.

If we take S_1 as the total action of the theory, the partition function is given by (69) with the condition $g' = 0$, then

$$Z_1 = 0. \tag{70}$$

It is worth to comment here on the introduction of the cut-off in the above analysis. It is a well known fact that some properties of noncommutative field theories are originated in the characteristic nature of infinite dimensional Hilbert space. For example, the trace of a commutation $\mathrm{tr}(AB - BA)$ does not vanish in noncommutative theories in general. This phenomenom does not exist in the finite matrix model. So one might think that we have to add some correction in order to account for this effect from infinite dimension to the above partition functions. However it turns out that we do not have to correct the partition function.

First, we consider the real scalar field ϕ, its fixed point being given by a projector in this case. If the solution is given by a shift operator like the complex scalar field case in [33,34], then the calculation does not close for the case of finite size matrices, even though the trace operation is done. On the other hand, our solutions are given by projection operators in this case, thus there is a possibility that the calculation closes in the finite Hilbert space. Additionally, even if we treat the shift operator, there is a way to take the infinite dimension effect into account. The method is to put a cut-off only for the initial and final states, to define the trace operation for finite matrices, i.e. the intermediate states are not restricted by the cut-off, (see [3,4] for details). Using such methods we can estimate the effect of infinite dimension, like the shift operator, by finite size computation. The other reason is that we should discuss the partition function in the terms of the weak topology since the trace operation is performed in the partition function calculation. Thus, it is difficult to distinguish $U(\mathcal{H})$ from $U(\infty) = \lim_{N\to\infty} U(N)$ by our calculation. From these facts, it is reasonable to evaluate the partition function by using the finite matrix model.

4.4. *finite* θ

One of our aims is to confirm that the partition function does not change under a change of the noncommutative parameter. The proof of the invariance under the θ-shift is based on the smoothness of θ. So, we have to check the smoothness for each model. In the previous subsection, we considered the $\theta \to \infty$ case and we calculated the partition function of the N.C.cohomological scalar model on the 2-dimensional Moyal space by using the result of the finite matrix model. Obeying the general property of N.C.CohFT, for finite θ, we expect that the partition function takes the same value as Eq.(69). This statement is realized when the moduli space deforms smoothly and its topology does not change under the variation of θ. Therefore, let us compare the moduli space of the large θ limit with the one of finite θ in this subsection.

It is difficult to analyze the arbitrary finite θ case because derivative terms and the nonlinear terms are intertwining. Therefore our strategy is to analyze the moduli space deformation from the large θ limit perturbatively. Let ϕ_0 and $B_{\mu 0}$ be large θ limit solutions of ϕ and B_μ, i.e. $\phi_0 = P$, $B_{\mu 0} = 0$. We consider that the fields belong to $C^\infty(\mathbb{R}^2)[[1/\sqrt{\theta}]]$. ϕ and B_μ are expanded as

$$\phi = \phi_0 + \frac{1}{\sqrt{\theta}}\phi_1 + \cdots, \quad B_\mu = B_{\mu 0} + \frac{1}{\sqrt{\theta}}B_{\mu 1} + \cdots, \tag{71}$$

and we substitute them into the action. The leading order bosonic action is then

$$\frac{1}{\theta}\mathrm{tr}|\phi_1(\phi_0 - 1) + \phi_0\phi_1|^2 + \frac{1}{\theta}\mathrm{tr}|\partial_\mu\phi_0 + B_{\mu 1}|^2 \tag{72}$$

$$= \frac{1}{\theta}\mathrm{tr}\left\{|\phi_1(P-1)|^2 + |P\phi_1|^2 + |\partial_\mu P + B_{\mu 1}|^2\right\} \tag{73}$$

Let $|P, i\rangle$ be an eigenvector of the projector P with eigenvalue 1, i.e. $P|P,i\rangle = |P,i\rangle$. Using this vector, $\sum_{i,j} |1-P, i\rangle a_{ij}\langle P, j| + h.c.$ is a solution of ϕ_1, where (a_{ij}) is a Hermitian matrix. However, the deformation of the moduli space from $\{a_{1,ij}\}$ is trivial and retractable. On the other hand, $B_{\mu 1} = -\partial_\mu P$. B_μ is deformed but it is determined completely by the given P. Therefore the moduli space topology is not changed at all. In other words, we can deform the moduli space smoothly. This result is consistent with our expectation, hence the partition function is invariant under θ deformation.

5. K-theory and Cohomological Scalar model

We discuss the relation between our theory and K-theory in this section.

5.1. *Commutative CohFT and Homotopy of Vector bundle*

The relation between a model of CohFT on a COMMUTATIVE space and the homotopy of the classifying map of a vector bundle are studied in this subsection. The model is deeply related to the N.C.CohFT models which are treated in section 4. By means of this model we will find an analogy of the correspondence between our N.C.cohomological scalar model and algebraic K-theory, namely the correspondence between CohFT and topological K-theory.

Let M be a n dim Riemannian manifold, V be a rank N trivial vector bundle.

$$\begin{aligned} \phi : M &\to H \\ x &\mapsto \phi^{ab}(x) \in H,\ a,b \in \{1,\cdots,N\} \end{aligned} \tag{74}$$

where H is set of all $N\times N$ Hermitian matrices, i.e. $H \equiv \{h|h^{ab} = \bar{h}^{ba}\}$. In other words, ϕ is a $N\times N$ Hermitian matrix valued scalar field on M. $N\times N$ Hermitian matrix valued scalar fields $\phi^{ab}(x)$ and $H^{ab}(x)$ have ghost number 0 and the fermionic BRS partners $\psi^{ab}(x)$ and $\chi^{ab}(x)$ have ghost number 1 and -1, respectively. The BRS transformation is similar to the previous one, however there is a difference caused by the $U(N)$ gauge symmetry. [a] The BRS operator is nilpotent up to gauge transformation δ_g, i.e. $\hat{\delta}^2 = \delta_g$. We denote by $c(x)$ the scalar field corresponding to a local gauge parameter with ghost number 2, then the explicit BRS transformation is given by

$$\begin{aligned} &\hat{\delta}\phi(x) = \psi(x),\ \hat{\delta}\chi(x) = H(x)\ ,\ \hat{\delta}c(x) = 0 \\ &\hat{\delta}\psi(x) = \delta_g\phi(x) = i[c(x),\phi(x)]\ ,\ \hat{\delta}H(x) = \delta_g\chi(x) = i[c(x),\chi(x)]. \end{aligned} \tag{75}$$

We introduce the following action;

$$S = S_0 + S_p + S_g \tag{76}$$

$$S_0 = \int_M tr\hat{\delta}\{\frac{1}{2}\chi(2\phi(1-\phi) - iH)\}, \tag{77}$$

[a] The theory of this subsection has $U(N)$ gauge symmetry. Since gauge symmetry is not our main subject here, we do not discuss the technical problems related with gauge symmetry.

where S_0 has $U(N)$ gauge symmetry and we have to project out the pure gauge degrees of freedom. Therefore, we introduce S_p for the projection to the gauge horizontal part and S_g for the gauge fixing.

After performing the Gaussian integral, the bosonic part of S_0 is

$$(\phi(1-\phi))^2, \tag{78}$$

and the fermionic action is

$$\chi(-2\psi(1-\phi)+2\phi\psi-[c,\chi]). \tag{79}$$

The fixed point is determined by $(\phi(1-\phi))=0$. If this ϕ is not matrix valued, then the only nontrivial solution which is a smooth function is $\phi=1$. However, if $N>1$ then a projection operator P which restricts the rank N vector space to dimension k for each point in M, is a solution. In other words, the solution of ϕ is a classification map to $G_k(N)$ whose homotopy class classifies the vector bundle.

We can construct S_{pro} by following the general method of cohomological gauge theory [21]. Let us introduce an anti-ghost $\bar{c}$ whose ghost number is -2 and its BRS partner η. Then, S_{pro} is given as

$$S_{pro}=i\int Tr\hat{\delta}((C^{\dagger}\psi)\bar{c}). \tag{80}$$

Here $C^{\dagger}$ is the adjoint operator of C. C is defined by $\delta_g\phi(x)=i[c,\phi]=Cc(x)$ i.e. $C=i[\ ,\phi]$. (Precisely speaking, we define a group action of $U(N)$ for some point p in the principal bundle P over the base manifold M. Then we can define C as the differential of the group action on the point p; $C:u(N)\to T_pP$. The image of C is the vertical tangent space of p.)

$$S_{pro}=\int Tr\{i[\phi,[c,\phi]]\bar{c}+[\psi,\psi]\bar{c}-[\psi,c]\eta\} \tag{81}$$

When we consider the theory near the rank k solution, the gauge symmetry $U(N)$ is broken to $U(k)\times U(N-k)$. Note that for a rank k projection operator ϕ there are objects c satisfying $C^{\dagger}Cc=[\phi,[c,\phi]]=\phi c(\phi-1)-(1-\phi)c\phi=0$, i.e. if c is a generator of the gauge group of $U(N-k)\times U(k)$, then the first term of the right hand side of (81) vanishes. This zero mode causes other types of problems that should be solved by inserting observables and choosing a good gauge. However, to inquire further into this matter would lead us into that specialized area, and such a digression would obscure the outline of our argument. Therefore, we do not go into these details here.

In the following discussion, $1/C^\dagger C$ operates non-zeromodes and we assume here that there are methods to take care of the zero modes. It is a well known fact of Cohomological gauge theory, that from the $\bar{c}$ equation of motion c is given as the curvature of the moduli space. But this discussion cannot be adopted to our case, since the non-trivial solution of ϕ causes symmetry breaking. The moduli space is the coset space whose equivalent relation is given by a left gauge symmetry.

$$\mathcal{M}_{k,N} = \{\phi \mid M \to G_k(N)\}/\mathcal{G}_{k,N}, \tag{82}$$

where $\mathcal{G}_{k,N}$ is the group of gauge transformations with gauge group of $U(N-k) \times U(k)$. From the $\bar{c}$ equation of motion we get

$$c = -\frac{1}{C^\dagger C}[\psi, \psi]. \tag{83}$$

Unlike the usual case, we can not regard c as the curvature on the principal bundle whose base manifold is the moduli space.

Let us consider fermionic zermo-modes of χ and ψ. Similar to N.C.CohFT and the finite matrix model, the equations of motion of ψ and χ without nonlinear terms are

$$\psi(1-P) - P\psi = 0, \text{ and } \chi(1-P) - P\chi = 0. \tag{84}$$

Note that the solution of both equations represents the cotangent vector of the solution space of ϕ. As far as these equations are concerned, the number of zero modes of ψ is equal to that of χ and there is no ghost number anomaly. After the integration of the nonzeromodes that produces some sign factor $\epsilon_{k,N} = \pm 1$, the integral of the zero-modes is obtained as

$$\mathcal{E}_{k,N} := \int_{\mathcal{M}_{k,N}} \mathcal{D}\phi_0 \mathcal{D}\chi_0 e^{-\int \frac{1}{C^\dagger C}[\psi_0,\psi_0][\chi_0,\chi_0]}. \tag{85}$$

Now, recall that our theory has a symmetry that allows arbitrary infinitesimal ϕ deformation i.e. $\phi \to \phi + \delta\phi$, where $\delta\phi$ is arbitrary infinitesimal $N \times N$ Hermite matrix valued scalar field. This due to the fact that we can regard the BRS exact action as a gauge fixing action of this local symmetry. This symmetry means that the partition function is homotopy invariant. Therefore, the equivalence class of this symmetry corresponds to homotopy equivalent class of ϕ. So the zero-mode integral (85) is summed up by the homotopy class $[M, G_k(N)]$.

Finally, the partition function is given as

$$Z \sim \sum_{[M,G_k(N)]} \epsilon_{k,N}\, \mathcal{E}_{k,N} \tag{86}$$

To interpret this partition function from the point of view of classifying homotopy of vector bundles, we note that ϕ is a classifying map for complex vector bundles when N is large enough (see [30]). (Note that there are no non-trivial vector bundles with dimension of fiber space larger than $n+2$.)

We introduce the homotopy class $Vect_k(M) = [M, BU(k)]$, where

$$BU(k) \equiv \bigcup_{m=k+n+1}^{\infty} Gr_k(m) \; ; \; m > k+n,$$

and consider the case where N is sufficiently large. Using this, the partition function is represented as

$$Z \sim \sum_{Vect_k(M)} \epsilon_{k,N} \; \mathcal{E}_{k,N}. \tag{87}$$

Note that this homotopy class is related to the $K'(M)$ group with virtual dimmension 0, where $K'(M) = [M, BU(\infty)]$ (see for example [31]). In particular, when M is connected $K(M) = \mathbb{Z} \oplus \tilde{K}$ and $K' = \tilde{K}$. For stable range $k > \frac{1}{2}\dim M$, we can put the relation between the homotopy class and $K'(M)$ as $K'(M) = [M, BU(k)]$. Therefore the partition function is proportional to the sum of $\epsilon_{k,N}$ $\mathcal{E}_{k,N}$ over the $K'(M)$ elements for sufficiently large N. This is analogous to the N.C.CohFT partition function which is given by a sum over the elements of the algebraic K-group. (See also the next subsection.)

To compare with the noncommutative theory with kinetic terms, we consider the model (77) with kinetic terms and investigate its large scale limit and finite scale case. The Lagrangian is similar to the N.C.CohFT given in section 4.3;

$$\mathcal{L} = \hat{\delta}\left(\frac{1}{2}\chi\left(2(\phi(1-\phi) - \partial_\mu B^\mu) - iH\right)\right)$$
$$+\hat{\delta}\left(\frac{1}{2}\chi^\mu(\partial_\mu\phi + B_\mu - iH_\mu)\right). \tag{88}$$

Since the $U(N)$ gauge symmetry is not our main concern, we break gauge symmetry here, i.e. we do not introduce gauge fields and gauge covariant derivatives. In the N.C.CohFT case, we took the large θ limit. We can perform a similar discussion by scaling

$$g_{\mu\nu} \to (1+\epsilon^2)g_{\mu\nu}, \;\; g^{\mu\nu} \to (1-\epsilon^2)g^{\mu\nu}. \tag{89}$$

Since the partition function is invariant under this transformation, the kinetic terms become irrelevant in the large scale limit and B_μ becomes an auxiliary field. After integration, the theory is equivalent to the one with the above action (77). This observation is similar to the case of N.C.CohFT in the limit $\theta \to \infty$.

The N.C.CohFT in the previous section is the naive extension of the model dealt with in this section. Considering the noncommutative deformation of the model in this subsection, we can identify it with the N.C.CohoFT model of section 4, after renumbering the U(N) indices and Hilbert space indices such that we do not have to distinguish them. Alternatively, the N.C.CohFT model is obtained by dimensional reduction to dimension zero and taking the large N limit.

5.2. K_0 and N.C.CohFT

In this subsection, we disscuss the correspondence with K_0-theory. As mentioned in section 1, one of our purposes is to construct a less sensitive topology than K-theory, where the term "topology" means that the vacuum expectation value of the field theory which is invariant under continuous deformation. It is natural to expect that our partition function is invariant under deformations which do not change the K-theory. In a sense, θ independence of the partition function implies this fact. To see this, we consider the examples of the Moyal plane and the noncommutative torus.

For the Moyal plane, as we saw in the previous section, the partition function (69) is expressed as a sum over the projection operators that are identified by their rank. The rank of the projection can be identified as $\tau_0(P_k) = k$ or $\tau_2(P_k) = k$ (see for example [35] and [32]). Furthermore, the Euler numbers of the Grassmann manifolds are determined essentially by k since we take $N \to \infty$ in the end. Therefore, the partition function is determined by the K_0 data only.

Next, we consider the N.C.torus T^2_θ. The classification by Morita equivalence corresponds to the one by K-theory and the equivalence is determined by a noncommutative parameter θ up to SL(2,$\mathbb{Z}$) transformation. If T^2_θ and $T^2_{\theta'}$ are Morita equivalent, θ' should be written as

$$\theta' = \frac{a\theta + b}{c\theta + d} \ , \ ad - bc = 1, \quad a, b, c, d \in \mathbb{Z}. \tag{90}$$

For arbitrary θ we can transform T^2_θ to a non-Morita equivalent noncommutative torus by infinitesimal θ deformation. So, the θ shift changes the K-group. On the other hand, the model whose action is given by (49) or

(61) is invariant under the θ shift when there is no singular point. (Note that the models with action (50) or (62) are not invariant under an arbitrary θ deformation, but they are invariant under SL(2,$\mathbb{Z}$) transformation.) At least, if some deformation of noncommutative manifolds does not change the K-theory, we may expect that the partition function of N.C.CohFT will not change. This fact implies that the partition function satisfies the condition we are interested in, that is, less sensitive topological invariant than K-theory.

6. N.C. Cohomological Yang-Mills Theory

In this section, Cohomological Yang-Mills theories on noncommutative manifolds are discussed. If there is gauge symmetry, the BRS-like symmetry is slightly different from (48), i.e., it is not nilpotent but

$$\delta^2 = \delta_{g,\theta}, \tag{91}$$

where $\delta_{g,\theta}$ is gauge transformation operator deformed by the star product $*_\theta$. The partition function of the N.C.CohFT is invariant under a change of the noncommutative parameter when the BRS transformation is nilpotent, since the BRS transformation δ and the θ deformation δ_θ commute. Conversely, if the definition of the BRS-like operator (91) depends on the noncommutative parameter θ, then δ and δ_θ do not commute;

$$\delta_\theta \delta \neq \delta \delta_\theta \Rightarrow \delta_\theta \delta = \delta' \delta_\theta, \tag{92}$$

where δ' is the BRS-like operator that generates the same transformations as the original BRS-like operator δ without the square

$$\delta'^2 = \delta_{g,\theta+\delta\theta}. \tag{93}$$

This fact causes some complexity when we want to prove the θ-shift invariance of N.C.cohomological Yang-Mills theory as mentioned in section 2.

Note that the important point of this problem is not the change of the nilpotency, but the dependence on the $*$ product in the definition of the BRS operator. (In fact, we can construct the BRS operator for the cohomological Yang-Mills theory as a nilpotent operator [36].)

However, we can prove the invariance of the partition function of cohomological Yang-Mills theory in noncommutative $\mathbb{R}^4$ under the noncommutative parameter deformation

$$\theta \to \theta + \delta\theta \quad , (\theta^{\mu\nu}) = \left(\begin{array}{cc|cc} 0 & \theta & 0 & 0 \\ -\theta & 0 & 0 & 0 \\ \hline 0 & 0 & 0 & \theta \\ 0 & 0 & -\theta & 0 \end{array}\right) .$$

When we consider only the case of noncommutative $\mathbb{R}^4$, field theories are expressed by the Fock space formalism. Then the differential operator ∂_μ is expressed by using commutation bracket $-i\theta^{-1}_{\mu\nu}[x^\nu, \;]$ and $\int d^Dx$ is replaced by $det(\theta)^{1/2}Tr$. Therefore the noncommutative parameter deformation is equivalent to replacing $-i\theta^{-1}_{\mu\nu}[x^\nu, \;]$ and $det(\theta)^{1/2}Tr$ by $-i(\theta+\delta\theta)^{-1}_{\mu\nu}[x^\nu, \;]$ and $det(\theta+\delta\theta)^{1/2}Tr$, respectively. Let us consider Donaldson-Witten theory (topological twisted $\mathcal{N} = 2$ Yang-Mills theory) on noncommutative $\mathbb{R}^4$ [15]. This theory is constructed by bosonic fields $(A_\mu, H_{\mu\nu}, \bar{\phi}, \phi)$ and fermionic fields $(\psi_\mu, \eta, \chi_{\mu\nu})$, where $(A_\mu, H_{\mu\nu}, \bar{\phi})$ and $(\psi_\mu, \eta, \chi_{\mu\nu})$ are supersymmetric (BRS) pairs. Using the BRS symmetry, it has been proved that partition function and expectation values have symmetry under the following reparametrization :

$$A_\mu \to \frac{1}{g}A_\mu, \quad \psi_\mu \to \frac{1}{g}\psi_\mu \;, \bar{\phi} \to \frac{1}{g^2}\bar{\phi}, \quad \eta \to \frac{1}{g^2}\eta$$
$$\chi_{\mu\nu} \to \frac{1}{g^2}\chi_{\mu\nu} \;, H_{\mu\nu} \to \frac{1}{g^2}H_{\mu\nu}, \tag{94}$$

where $g \in \mathbb{R}$. Taking $g = \theta$, the invariance of Donaldson-Witten theory under $\theta \to \theta + \delta\theta$ is proved.

Similarly, we can discuss the Vafa-Witten theory (topological twisted $\mathcal{N} = 4$ Yang-Mills theory) on noncommutative $\mathbb{R}^4$ [29]. In this case, there are additional fields $(B_{\mu\nu}, c, H_\mu)$ and $(\psi_{\mu\nu}, \bar{\eta}, \chi_\mu)$, where $(B_{\mu\nu}, c, H_\mu)$ are bosonic fields, $(\psi_{\mu\nu}, \bar{\eta}, \chi_\mu)$ are fermionic, and they are supersymmetric partners. For these fields, we assign the weight of the above transformation as

$$B_{\mu\nu} \to \frac{1}{g}B_{\mu\nu} \quad , \quad \psi_{\mu\nu} \to \frac{1}{g}\psi_{\mu\nu} \;, c \to \frac{1}{g}c \quad , \quad \bar{\eta} \to \frac{1}{g}\bar{\eta},$$
$$H_\mu \to \frac{1}{g^2}H_\mu \quad , \quad \chi_\mu \to \frac{1}{g^2}\chi_\mu \tag{95}$$

with (94), then the invariance of Vafa-Witten theory under $\theta \to \theta + \delta\theta$ is proved in the same way as Donaldson-Witten theory.

By applying these facts to several physical models, interesting information can be found. For example, the partition function of the N.C. Cohomological Yang-Mills Theory on 10-dim Moyal space and the partition function of the IKKT matrix model have a correspondence, since the IKKT matrix model is constructed as a dimensional reduction of super $U(N)$ Yang-Mills theory with large N limit [37] [38]. This dimensional reduction is regarded as a large limit of the noncommutative parameter ($\theta \to \infty$ in section 4). Taking the large N limit of the matrix model is equivalent to considering the Yang-Mills theories on noncommutative Moyal space, i.e. matrices are regarded as linear transformation of the Hilbert space caused by the noncommutativity in a similar manner as in the case of the N.C.CohFT on the Moyal plane. In particular, the Noncommutative Cohomological Yang-Mills model on Moyal space in the large θ limit is almost the same as the model of Moore, Nekrasov and Shatashvili [39]. Moore et al. showed that the partition function is calculated by a cohomological matrix model in [39] and related works can be seen in [40,41,42]. We can be fairly certain that we can reproduce their results by using N.C.cohomological Yang-Mills theories.

Another example is an application to N=4 d=4 Vafa-Witten theory [29]. The theory is constructed as a balanced CohFT (see [43] and [44]). The partition function of the Vafa-Witten theory is given by a sum of the Euler numbers of the instanton moduli space over all instanton numbers, provided the vanishing theorem holds. Here the vanishing theorem guaranties that the fixed point locus of the theory is the instanton moduli space. On commutative manifolds, one of the conditions for the vanishing theorem to hold is that there is no $U(1)$ instanton. On the other hand, existence of $U(1)$ instantons is well-known in noncommutative Moyal space [45,46], so it is likely that the $U(1)$ instantons also exist on the other noncommutative manifolds even if these manifolds do not have $U(1)$ instantons before the noncommutative deformation. Therefore, if we consider the Vafa-Witten theory on noncommutative manifolds, the $U(1)$ instanton effect appears as the difference to the commutative manifold case. The result is a sum of Euler numbers of the instanton moduli spaces and the moduli spaces are deformed by the $U(1)$ instanton effect. In this case, we expect that its partition function on a commutative manifold is computed by a matrix theory calculation like [39]. By comparing this partition function, it is reasonable to suppose that the Euler number of the deformed moduli space is given, and we obtain a partition function on a manifold that does not satisfy the vanishing theorem. Such differences from the CohFT on commutative manifolds emphasize that N.C.CohFT is non-trivial, though it is less sensitive than K-theory.

With these considerations we find many interesting subjects to be studied by using N.C.chomological Yang-Mills theory, and many of them will be left for our future work.

7. Summary

Let us summarize this article. We have studied topological aspects of N.C.CohFT and matrix models. At first, we reviewed N.C.CohFT and its properties. In particular, through this article we have used the property that the N.C.CohFT has symmetry under an arbitrary infinitesimal deformation of the noncommutative parameter. This symmetry implies that the partition function of N.C.CohFT is an insensitive "topological" invariant. In section 3, we introduced a Hermitian finite size matrix model of CohFT and calculated its partition function. The calculation was done by using only topological information of its moduli space. The partition function was given as a sum of Euler numbers of Grassman manifolds with sign, and we showed that the partition function vanishes. This calculation is the first example of determining its partition function by only moduli space topology of a matrix model. The scalar field models of N.C.CohFT were discussed in section 4. The variations of the models are caused by adding kinetic terms or topological action that correspond to Connes's Chern character. The fixed point loci of the scalar fields were given by the set of all projection operators on the noncommutative manifold. From the analogy of the finite size matrix model, we introduced a connection functional in these N.C.CohFT models. Using the curvature obtained from these connections, the partition functions were represented as a sum of Euler numbers of the set of all projection operators. This partition function is an example of a new "topological" invariant, and the fundamental theorem of Morse theory extended to the operator formalism [13] is connected to the usual local geometry by this model. As a concrete example, we calculated the partition function of a model including kinetic terms on the Moyal plane. Through the operator formulation, this calculation boiled down to the calculation of the Hermitian finite size matrix model of CohFT in section 3. Additionally, to confirm the independence of the noncommutative parameter of the N.C.CohFT we studied the moduli space for finite θ. If the partition function of CohFT is "topological", then it should have some relation with K-theory and the partition function should not change under a deformation that does not change the K-group. Therefore we investigated the models of CohFT and N.C.CohFT from the point of view of K-theory. At first, one CohFT was constructed. This model and the N.C.CohFT model in section 4 are related by dimensional reduction or noncommutative deformation. The partition function is invariant under scaling and this scaling is similar to the θ-shift. In the large scaling limit, the kinetic terms become irrelevant and the fixed point loci are given by a classifying map. The partition function was given by sum of topological invariants with sign. This sum is taken over all the homotopy equivalent classes of the classifying map of the vector bundle. The corresponding homotopy class is regarded as K'. Comparing the connection between the CohFT model and K-theory with the relation between the N.C.cohomological scalar model and algebraic K-theory, respectively, we found an analogy. Furthermore, we studied the correspondence with the K_0-theory for the Moyal plane and the noncommutative torus. It was verified that our partition function is invariant under deformations which do not change the K_0, at least for the Moyal plane and noncommutative torus. These fact implies the existence of a new classification by the "topological" invariants. To clarify the nature of this classification, it is a future work to collect examples for which similar calulations like the case of the Moyal plane are possible.

Furthermore, we considered the noncommutative cohomological Yang-Mills theory. The noncommutative parameter independence is non-trivial for noncommutative gauge theory, and it is possible to be proved. Therefore, we can remove kinetic terms in the large θ limit on the Moyal spaces, similar as in the case of the N.C.CohFT studied in section 4. The observations of the N.C.CohFT of scalar models give us a general correspondence between N.C.CohFT and matrix models. An example is the connection between the IKKT matrix model and noncommutative cohomological Yang-Mills theory. As another example, we considered the Vafa-Witten theory. The contribution from noncommutative solitons like $U(1)$ instantons may make the expectation value of N.C.CohFT different from the expectation value of CohFT on a commutative manifold. In such case, N.C.CohFT lead to a topological invariant which is different from the topological invariant of the commutative case, and which is less sensitive than algebraic K-theory. In other words, there will be a new nontrivial global characterization of the geometry, though its classification is less sensitive than K-theory. It is likely that the Vafa-Witten theory is one such example. A detailed analysis of similar variations for noncommutative cohomological Yang-Mills theory corresponding to matrix model will be carried out in future work.

The unsettled questions are the following. As we have seen, there is evidence to suggest the partition function of N.C.CohFT is an insensitive but nontrivial topological invariant. To clarify this, a more strict topological discussion about N.C.CohFT for the general case should be done, since there are many ambiguous problems concerning the relation to K-theory. This subject is left as a future issue.

Acknowledgments

We are grateful to Y.Maeda and H.Moriyoshi for helpful suggestions and observations. We also would like to thank Y.Matsuo, M.Furuta, Y.Kametani for valuable discussions and useful comments. This work is supported by the 21st Century COE Program at Keio University (Integrative Mathematical Sciences: Progress in Mathematics Motivated by Natural and Social Phenomena). The Yukawa Institute for Theoretical Physics at Kyoto University is also greatly acknowledged. Discussions during the YITP workshop YITP-W-03-07 on "Quantum Field Theory 2003" were useful to complete this work. Thanks also to U.Carow-Watamura for reading the manuscript and correcting the proofs.

References

1. M. Kontsevich, *Deformation Quantization of Poisson Manifolds. 1.* `math.QA/9709040`.
2. A.S. Cattaneo, G. Felder, *A Path Integral Approach to the Kontsevich quantization formula*, Commun.Math.Phys.212(2000)591-611, `math.QA/9902090`.
3. T. Ishikawa, S. Kuroki and A. Sako, *Instanton number on Noncommutative* R^4, `hep-th/0201196`. (The title of this paper was changed to "*Calculation of the Pontrjagin class for U(1) instantons on noncommutative* $\mathbf{R}^4$" JHEP 0208(2002)028.
4. A. Sako,*Instanton number of Noncommutative U(N) Gauge Theory*, JHEP 0304(2003)023, `hep-th/0209139`.
5. K. Furuuchi, *Instantons on Noncommutative* R^4 *and Projection Operators*, Prog.Theor.Phys.103(2000)1043-1068, `hep-th/9912047`.
6. K. Furuuchi, *Topological Charge of U(1) Instantons*, `hep-th/0010006`.
7. K. Furuuchi, *Dp-D(p+4) in noncommutative Yang-Mills*, JHEP 0103(2001) 033, `hep-th/0010119`.
8. K. Furuuchi, *Equivalence of Projections as Gauge Equivalence on Noncommutative Space*, Comm.Math.Phys.217(2001)579-593, `hep-th/0005199`
9. Y. Tian, C. Zhu and X. Song, *Topological Charge of Noncommutative ADHM Instanton*, `hep-th/0211225`.
10. K. Kim, B. Lee and H.S. Yang, *Zero Modes and the Atiyah-Singer Index in Noncommutative Instantons*, Phys.Rev.B.66(2002)025034, `hep-th/0205010`.
11. D. Bak, K. Lee, J. Park, *Noncommutative Vortex Solitons* Phys.Rev.D63 (2001)125010, `hep-th/0011099`.
12. D.P. Jatkar, G. Mandal, S. R. Wadia, *Nielsen-Olesen Vortices in Noncommutative Abelian Higgs Model*, JHEP 0009(2000)018, `hep-th/0007078`.

13. A. Sako, S-I. Kuroki and T. Ishikawa, *Noncommutative Cohomological Field Theory and GMS soliton*, J.Math.Phys.43(2002)872-896, hep-th/0107033.
14. A. Sako, S-I. Kuroki and T. Ishikawa, *Noncommutative-shift invariant field theory*, proceeding of 10th Tohwa International Symposium on String Theory, (AIP conference proceedings 607, 340).
15. E. Witten, *Topological quantum field theory*, Commun.Math.Phys.117(1988) 353;
Introduction to cohomological field theories, Int.J.Mod.Phys.A.6(1991)2273.
16. E. Witten, *SuperSymmetry and Morse Theory*, J.Diff.Geom.**17**(1982)661-692.
17. M.F. Atiyah, L. Jeffrey, *Topological Lagrangians and Cohomology*, J.Geom. Phys.7(1990)119-136.
18. V. Mathai, D. Quillen, *Superconnections, Thom Classes and Equivariant Differential Forms* Topology 25(1986)85-110.
19. J.M.F. Labastida, C. Lozano, *Lectures in Topological Quantum Field Theory* hep-th/9709192.
20. M. Blau, *The Mathai-Quillen Formalism and Topological Field Theory* J.Geom.Phys.11(1993)95-127, hep-th/9203026.
21. S. Cordes, G.W. Moore, S. Ramgoolam, *Lectures on 2-D Yang-Mills Theory, Equivariant Cohomology and Topological Field Theories* Nucl.Phys. Proc.Suppl.41:184-244,1995 (Part 1 of the lectures is published in the Nucl. Phys. B, Proc. Suppl., and part 2 in the Les Houches proceedings. Also in Trieste Spring School 1994:0184-244 (QCD161:T732:1994), hep-th/9411210.
22. F. Bayen, M. Flato, C. Fronsdal, A. Lichnerowicz, D. Sternheimer, *Deformation Theory and Quantization. 1. Deformations of Symplectic Structures*, Annals Phys.111(1978)61.
F. Bayen, M. Flato, C. Fronsdal, A. Lichnerowicz, D. Sternheimer, *Deformation Theory and Quantization. 2.Physical Applications*, Annals Phys.111(1978)111.
23. J.E. Moyal, *Quantum Mechanics as a Statistical Theory*, Proc.Cambridge Phil.Soc.45(1949)99-124.
24. R. Gopakumar, S. Minwalla and A. Strominger, *Noncommutative Solitons*, JHEP 05(2000)020, hep-th/0003160.
25. R. Gopakumar, M. Headrick and M. Spradlin, *On Noncommutative Multisolitons*, Commun.Math.Phys.233(2003)355-381, hep-th/0103256.
26. A. Alexandrov, A. Mironov, A. Morozov, *Partition Functions of Matrix Models as the First Special Functions of String Theory. 1. Finite Size Hermitean One Matrix Model*, hep-th/0310113.
27. W. Bietenholz, F. Hofheinz, J. Nishimura, *The Noncommutative $\lambda\phi^4$ Model*, Acta Phys.Polon.B34(2003)4711-4726, hep-th/0309216.
28. J. Nishimura, T. Okubo, F. Sugino, *Testing the Gaussian Expansion Method in Exactly Solvable Matrix Models*, JHEP 0310(2003)057, *hep-th/0309262.*
29. C. Vafa and E. Witten, *A Strong coupling test of S-duality*, Nucl.Phys.B431 (1994)3-77, hep-th/9408074.
30. R. Bott, L. Tu, *Differential Forms in algebraic Topology*, Springer -Verlag, New York, 1982.

31. K. Olsen, R.J. Szabo, *Constructing D-branes from K Theory*, Adv.Theor. Math.Phys.3(1999)889-1025, hep-th/9907140.
32. H. Kajiura, Y. Matsuo, T. Takayanagi, *Exact Tachyon Condensation on Noncommutative Torus*, JHEP 0106(2001)041, hep-th/0104143.
33. J A. Harvey, P. Kraus, F. Larsen, E J. Martinec *D-branes and String as Noncommutative Solitons* JHEP 0007(2000)042, hep-th/0005031 .
34. E. Witten, *Noncommutative tachyons and String Field Theory* hep-th/0006071.
35. A. Connes, *C^* algebras and Differential Geometry*, hep-th/0101093.
36. D. Birmingham, M. Blau, M. Rakowski, and G. Thompson, Phys.Report. 209(1991)129.
37. N. Ishibashi, H. Kawai, Y. Kitazawa, A. Tsuchiya, *A Large N Reduction Model as Superstring*, Nucl.Phys.B498(1997)467-491, hep-th/9612115.
38. H. Aoki, S. Iso, H. Kawai, Y. Kitazawa, A. Tsuchiya, T. Tada, *IIB Matrix Model*, Prog.Theor.Phys.Suppl.134(1999)47-83, hep-th/9908038.
39. G.W. Moore, N. Nekrasov, S. Shatashvili, *D-particle bound states and generalized instantons*, Commun.Math.Phys.209(2000)77-95, hep-th/9803265.
40. S. Hirano, M. Kato, *Topological Matrix Model*, Prog.Theor.Phys.98(1997) 1371-1384, hep-th/9708039.
41. F. Sugino, *Cohomological Field Theory Approach To Matrix Strings*, Int.J. Mod.Phys.A14(1999)3979-4002, hep-th/9904122.
42. U. Bruzzo, F. Fucito, J.F. Morales, A. Tanzini, *Multi-Instanton Calculus and Equivalent Cohomology*, JHEP 0305(2003)054, hep-th/0211108.
43. R. Dijkgraaf and G. Moore, *Balanced Topological Field Theories*, Commun. Math.Phys. 185(1997)411-440, hep-th/9608169;
J.M.F. Labastida and Carlos Lozano, *Mathai-Quillen Formulation of Twisted $N = 4$ Supersymmetric Gauge Theories in Four Dimensions*, Nucl.Phys. B502(1997)741-790, hep-th/9702106.
44. R. Dijkgraaf, J. Park and B. Schroers, *N=4 Supersymmetric Yang-Mills Theory on a Kähler Surface*, hep-th/9801066.
45. N. Nekrasov and A. Schwarz, *Instantons on noncommutative R^4 and (2,0) superconformal six dimensional theory*, Comm.Math.Phys.198(1998)689, hep-th/9802068.
46. N.A. Nekrasov, *Trieste lectures on solitons in noncommutative gauge theories*, hep-th/0011095.

GAUGE THEORY ON FUZZY S^2 AND $\mathbb{C}P^2$ AND RANDOM MATRICES

HAROLD STEINACKER
Department für Physik
Ludwig–Maximilians–Universität München
Theresienstr. 37, D-80333 München, Germany
hsteinac@theorie.physik.uni-muenchen.de

We review the formulation of gauge theory on the fuzzy sphere as a matrix model, and show how the path integral can be reduced to an integral over eigenvalues. This integral can be evaluated explicitly for large N, recovering the partition function of $U(n)$ Yang-Mills theory on the classical sphere as a sum over instanton contributions. The monopole solutions are briefly discussed. We also briefly discuss a generalization to fuzzy $\mathbb{C}P^2$.

1. Introduction

Gauge theories provide the best known description of the fundamental forces in nature. At very short distances however, physics is not known, and it seems unlikely that spacetime is a perfect continuum down to arbitrarily small scales. Indeed, physicists have started to learn in recent years how to formulate field theory on quantized, or noncommutative (NC) spaces, see [4,11] for reviews. However, the quantization of non-commutative field theories by straightforward generalization of the conventional, perturbative methods turns out to be very difficult in general. The reason is a new phenomenon called UV/IR mixing, which appears to be very generic in NC field theories, both for scalar and for gauge field theories. In essence it means that the UV divergences not only lead to the usual infinite renormalization of the mass and couplings, but also to new divergences in the infrared behaviour of the propagator (hence the name), which are likely to signal new physics. It is therefore important to develop new techniques for the quantization of NC field theories, which take into account these features of the non-commutative case. Gauge theories are of particular interest, not only because of their obvious importance in physics, but also because here the differences to the commutative case are particularly striking already at the level of the action: they can be formulated as *(multi-) matrix models*, which from a conventional point of view would be interpreted as *zero-dimensional* theories with a large "internal" space.

This formulation leads to a very simple formal definition of the quantization in terms of a "path" integral over the matrices. The difficult task to estimate the effects of the quantization can then be attacked using new methods. I will explain here the main ideas of [10], where such a matrix formulation of pure $U(n)$ gauge theory on the fuzzy sphere was used to calculate its partition function in the commutative limit. This is done using matrix techniques which *cannot* be applied in the commutative case. While this 2-dimensional case is essentially very simple, it suggests that more sophisticated versions of this strategy may lead to entirely new techniques for gauge field theory.

2. The model

Consider the matrix model for 3 hermitian $N \times N$ matrices B_i, with action

$$S(B) = \frac{2}{g^2 N} Tr\Big((B_i B^i - \frac{N^2-1}{4})^2 + (B_i + i\varepsilon_{ijk} B^j B^k)(B^i + i\varepsilon^{irs} B_r B_s)\Big) \tag{1}$$

where g is the coupling constant and N is a (large) integer. This action describes pure gauge theory on the fuzzy sphere S^2_N, cf.[8,12]. It is invariant under the $U(N)$ gauge symmetry acting as

$$B_i \to U^{-1} B_i U.$$

To see that this corresponds to the usual $U(1)$ local gauge symmetry in the classical limit, we first note that the absolute minimum (the "vacuum") of the action is given by

$$B_i = \lambda_i = \pi_N(J_i)$$

up to gauge transformation, where π_N is the N-dimensional representation of $su(2)$ with generators J_i. Upon rescaling $\lambda_i = x_i \sqrt{\frac{N^2-1}{4}}$, on finds the generators x_i of the fuzzy sphere [8] which satisfy

$$\sum x_i x_i = 1, \quad [x_i, x_j] = i\sqrt{\frac{4}{N^2-1}}\, \epsilon_{ijk} x_k.$$

This means that the vacuum of this matrix model is the fuzzy sphere. We can now write any field ("covariant coordinate") as

$$B_i = \lambda_i + A_i. \tag{2}$$

Then

$$B^i + i\varepsilon^{ikl} B_k B_l = \frac{1}{2}\varepsilon^{ikl} F_{kl}, \quad F_{kl} := i[\lambda_k, A_l] - i[\lambda_l, A_k] + i[A_k, A_l] + \varepsilon_{klm} A^m.$$

Notice that the kinetic terms in the field strength F_{kl} arise automatically due to the shift (2). The $U(N)$ gauge symmetry acts on A_i as

$$A_i \to U^{-1} A_i U + U^{-1}[\lambda_i, U] \tag{3}$$

which for $U = \exp(ih(x))$ and $N \to \infty$ becomes the usual (abelian!) gauge transformation for a gauge field. One can furthermore show that the "radial" field

$$\varphi := \lambda^i A_i$$

decouples in the large N limit, and $1NTr \to \int$. Hence the model reduces to the usual $U(1)$ Yang-Mills action

$$S = \frac{1}{g^2} \int F_{mn} F^{mn} \tag{4}$$

on S^2 in the commutative (=large N) limit.

Monopoles. One can easily find new, non-trivial solutions of this model using the ansatz [10]

$$B_i = \alpha_m \lambda_i^{(M)} \tag{5}$$

for suitable normalization constant α_m. Here $\lambda_i^{(M)} = \pi_M(J_i)$ is the generator of the M-dimensional irrep of $su(2)$, which can be embedded in the configuration space of $N \times N$ matrices if $m = N - M > 0$. It turns out that (5) describes monopole solutions with monopole charge m, and the corresponding gauge potential can be calculated explicitly [10]. Notice that negative monopole charges $m < 0$ can also be obtained by admitting matrices B_i of size $N' \times N'$ with $N < N' \ll 2N$, while keeping the action (1) as it is.

As a subtlety, it should be noted that the correct action $S = \frac{m^2}{2g^2}$ of the above monopoles is recovered only upon a slight modification of the constraint term in the action (1) as indicated in [10], which does not affect the classical limit. The reason is that the "empty" blocks in (5) if embedded in $N \times N$ matrices give a large contribution due to the first term in (1). This is certainly unphysical (it could be interpreted as action of a Dirac string), and can be avoided by the slightly modified action (78) in [10]. Then the energy of all monopoles is correctly reproduced in the commutative limit $N \to \infty$. All this extends immediately to the non-abelian case:

Non-abelian case This model is readily extended to the nonabelian case by using matrices of size nN, i.e. $B_i = B_{i,\alpha} t^\alpha = \lambda_i\, t^0 + A_{i,0}\, t^0 + A_{i,a}\, t^a$ where t^a denote the Gell-Mann matrices of $su(n)$. The action then reduces to the usual $U(n)$ Yang-Mills action

$$S = \frac{1}{g^2} \int (F_{mn,0} F^{mn,0} + F_{mn,a} F^{mn,a}) \tag{6}$$

in the commutative limit. Again, all "instanton" sectors are recovered if one admits matrices of arbitrary size $M \approx nN$ for the above action.

3. Quantization

The quantization of $U(n)$ Yang-Mills gauge theory on the usual 2-sphere is well known, see e.g. [9,14]. In particular, the partition function and correlation functions of Wilson loops have been calculated. Our goal is to calculate the partition function for the YM action (1) on the fuzzy sphere, taking advantage of the formulation as matrix model. This can be achieved by collecting the 3 matrices B_i into a single $2M \times 2M$ matrix

$$C = C_0 + B_i \sigma^i \tag{7}$$

The main observation is that the above action (1) can be rewritten simply as

$$S(B) = TrV(C)$$

imposing the constraint $C_0 = \frac{1}{2}$, for the potential

$$V(C) = \frac{1}{g^2 N}(C^2 - (\frac{N}{2})^2)^2$$

Then we proceed as

$$\begin{aligned} Z &= \int dB_i \ \exp(-S(B))) \\ &= \int dC\ \delta(C_0 - \frac{1}{2})\ \exp(-TrV(C)) \\ &= \int d\Lambda_i \Delta^2(\Lambda_i) \exp(-TrV(\Lambda)) \int dU \delta((U^{-1}\Lambda U)_0 - \frac{1}{2}) \end{aligned}$$

where dU is the integral over $2M \times 2M$ unitary matrices, $C = U^{-1}\Lambda U$, and $\Delta(\Lambda_i)$ is the Vandermonde-determinant of the eigenvalues Λ_i. Here

$\delta(C_0 - \frac{1}{2})$ is a product over M^2 delta functions, which can be calculated by introducing $J = \begin{pmatrix} K & 0 \\ 0 & K \end{pmatrix} = K\,\sigma^0$ where K is a $N \times N$ matrix. Then

$$\delta((U^{-1}CU)_0 - \frac{1}{2}) = \int dK \exp(iTr(U^{-1}(C - \frac{1}{2})UJ)).$$

By gauge invariance, the r.h.s. depends only on the eigenvalues Λ_i of C. Hence

$$\begin{aligned} Z &= \int dK \int d\Lambda_i \Delta^2(\Lambda_i) \exp(-TrV(\Lambda)) \int dU \exp(iTr(U^{-1}\Lambda UJ - \frac{1}{2}J)) \\ &= \int dK\; Z[J]\; e^{-\frac{i}{2}TrJ} \end{aligned} \tag{8}$$

where

$$Z[J] := \int dC\; \exp(-TrV(C) + iTr(CJ)) \tag{9}$$

depends only on the eigenvalues J_i of J. Diagonalizing $K = V^{-1}kV$, we get

$$Z = \int dk_i \Delta^2(k) \int d\Lambda_i \Delta^2(\Lambda_i) \exp(-TrV(\Lambda)) \int dU \exp(iTr(U^{-1}(\Lambda - \frac{1}{2})UJ))$$

where $\int dV$ was absorbed in $\int dU$. The integral over $\int dU$ can now be done using the Itzykson-Zuber-Harish-Chandra formula [7]

$$\int dU \exp(iTr(U^{-1}CUJ)) = const \frac{\det(e^{i\Lambda_i J_j})}{\Delta(\Lambda_i)\Delta(J_j)},$$

which also depends only on the eigenvalues of J and C.

In this step the number of integrals is reduced from N^2 to $2N$. This basically means that the integral over fields on S^2_N is reduced to the integral over functions in one variable. This is a huge step, just like in the usual matrix models. The constraint however forces us to evaluate in addition the integral over k_i, which is quite complicated due to the rapid oscillations in $\det(e^{i\Lambda_i J_j})$; note that $\Lambda_i \approx \pm\frac{N}{2}$. Nevertheless, the integrals can be evaluated for large N [10], with the result

$$Z_m = \sum_{m_1+...+m_n=m} \int_{-\infty}^{\infty} d\kappa_1 ... d\kappa_n\, \Delta^2(\kappa)\, e^{i\kappa_i m_i}\, \exp(-\frac{g^2}{2}\sum \kappa_i^2).$$

Here we consider matrices of size $M = nN - m$, which corresponds to the monopole sectors with total $U(1)$ charge $m = m_1 + ... + m_n$. This can be rewritten in the "localized" form as a weighted sum of saddle-point contributions, as advocated by Witten [14]:

$$Z_m = \sum_{m_1+\ldots+m_n=m} P(m_i, g) \exp(-\frac{1}{2g^2}\sum m_i^2)$$

where $P(m_i, g)$ is a totally symmetric polynomial in the m_i which can be given explicitly. In order to include all monopole configurations, we should simply sum over matrices of different sizes $M = nN - m$, for the same action given by $V(C)$. One can indeed find corresponding saddle-points of the action (1) which have the form [10]

$$A_i = \begin{pmatrix} m_1 A_i & 0 & \ldots & 0 \\ 0 & m_2 A_i & \ldots & 0 \\ \vdots & \vdots & \ddots & \vdots \\ 0 & 0 & \ldots & m_n A_i \end{pmatrix}$$

where

$$\vec{A}_i^{(c)} = \vec{r} \times \vec{A} \approx \frac{m}{2}\frac{1}{1+x_3}\begin{pmatrix} x_2 \\ -x_1 \\ 0 \end{pmatrix}$$

becomes the usual monopole field for large N, and action becomes

$$S(C^{(m_1,\ldots,m_n)}) = \frac{1}{2g^2}\sum_i m_i^2$$

for large N. This is a standard result on the classical sphere.

Hence the full partition function is obtained by summing[a] over all Z_m,

$$Z = \sum_m Z_m = \sum_{m_1,\ldots,m_n=-\infty}^{\infty} \int d\kappa_i\, \Delta^2(\kappa)\, e^{i\kappa_i m_i} \exp(-\frac{g^2}{2}\sum \kappa_i^2).$$

Using a Poisson resummation, this can be rewritten in the form

$$Z = \sum_{p_1,\ldots,p_n \in Z} \Delta^2(p) \exp(-2\pi^2 g^2 \sum_i p_i^2)$$

or equivalently

$$Z = \sum_R (d_R)^2 \exp(-4\pi^2 g^2 C_{2R}).$$

---

[a]the relative weights of Z_m for different m is not determined here. However, it could be calculated in principle

Here the sum is over all representations of $U(n)$, d_R is the dimension of the representation and C_{2R} the quadratic casimir. This form was found in [9] for the partition function of a $U(n)$ Yang-Mills theory on the ordinary 2-sphere.

We see that the limit $N \to \infty$ of the partition function for $U(n)$ YM on the fuzzy sphere is well-defined, and reproduces the result for YM on the classical sphere. This strongly suggests that the same holds for the full YM theory on the fuzzy sphere, and that there is no UV/IR mixing for pure gauge theory on S^2_N. This is unlike the case of a scalar field, which exhibits a "non-commutative anomaly" [3] related to UV/IR mixing.

4. Multi-Matrix Models for Yang-Mills on fuzzy $\mathbb{C}P^2$

4.1. *Fuzzy* $\mathbb{C}P^2$

We briefly describe how the above formulation of gauge theory can be generalized to 4 dimensions, in the example of $\mathbb{C}P^2$ following [6]. From a physical point of view, $\mathbb{C}P^2$ should be considered here as compactified $\mathbb{R}^4$. Fuzzy $\mathbb{C}P^2$ is a quantization of $\mathbb{C}P^2$ as (co)adjoint orbit; for the basic properties of this space we refer to [5,1,2,13]. We only recall here that this space is defined in terms of non-commutative coordinates x_a, $a = 1, ..., 8$ which describe the embedding of $\mathbb{C}P^2 \subset \mathbb{R}^8$, and satisfy [1]

$$i f_c^{ab} x_a x_b = -3\Lambda_N \, x_c = -3 \frac{R}{\sqrt{\frac{1}{3}N^2 + N}} \, x_c, \tag{10}$$

$$g^{ab} x_a x_b = R^2, \tag{11}$$

$$d_c^{ab} x_a x_b = R \, \frac{2N/3 + 1}{\sqrt{\frac{1}{3}N^2 + N}} \, x_c. \tag{12}$$

Here R is an arbitrary radius, which will usually be 1 here. This generates the algebra of functions on fuzzy $\mathbb{C}P^2_N$, which is simply $Mat(D_N, \mathbb{C})$ for D_N being the dimension of a certain irrep of $su(3)$.

4.2. *Degrees of freedom and field strength*

Our basic degrees of freedom are 8 hermitian matrices $C_a \in Mat(D_N, \mathbb{C})$ transforming in the adjoint of $su(3)$, which are naturally arranged as a single $3D_N \times 3D_N$ matrix

$$C = C_a \tau^a + C_0 \mathbb{1} \tag{13}$$

where $C_0 = 0$ in much of the following. Here τ^a are the conjugated Gell-mann matrices. The size D_N of these matrices will be relaxed later. We want to find a multi-matrix model in terms these C_a, which for large N reduces to Yang-Mills gauge theory on $\mathbb{C}P^2$. The idea is again to interpret the C_a as suitably rescaled "covariant coordinates" on fuzzy $\mathbb{C}P^2_N$, with the gauge transformation

$$C_a \to U^{-1} C_a U \tag{14}$$

for unitary matrices U of the same size. The C_a can also be interpreted as components of a one-form if desired [6]. Following the above model for the fuzzy sphere, we look for an action which has the "vacuum" solution

$$C_a = \xi_a \tag{15}$$

corresponding to $\mathbb{C}P^2_N$, and forces C_a to be at least approximately the corresponding representation $V_{N\Lambda_2}$ of $su(3)$. Then the fluctuations

$$C_a = \xi_a + A_a \tag{16}$$

are small, and describe the gauge fields. By inspection, these gauge fields A_a transform as

$$\delta A_a = i[\xi_a + A_a, \Lambda] = iL_a\Lambda + i[A_a, \Lambda] \tag{17}$$

for $U = e^{i\Lambda}$, which is the appropriate formula for a gauge transformation. Since the C_a resp. ξ_a correspond to "global" coordinates in the embedding space $\mathbb{R}^8$, we can hope that nontrivial solutions such as instantons can also be described in this way.

A suitable definition for the field strength is then given by

$$F_{ab} = i[C_a, C_b] + \frac{1}{2}\, f_{abc}C_c = i(L_aA_b - L_bA_a + [A_a, A_b]) + \frac{1}{2}\, f_{abc}A_c. \tag{18}$$

We will also need

$$\begin{aligned} F_a &= if_{abc}C_bC_c + 3C_a = \tfrac{1}{2} f_{abc}F_{ab}, \\ D_a &= d_c^{ab}C_aC_b - \left(\tfrac{2N}{3} + 1\right) C_c. \end{aligned} \tag{19}$$

Under gauge transformations, the field strength transforms as

$$F_{ab} \to U^{-1} F_{ab} U. \tag{20}$$

F can also be interpreted as 2-form

$$F = dA + AA \tag{21}$$

if one considers the fields C_a as one-forms $C = C_a\theta_a = \Theta + A$, using the differential calculus introduced in [6]. Furthermore, one can show that F_{ab} is (approximately) tangential if C_a satisfies (approximately) the constraints of $\mathbb{C}P^2$. Assuming that A_a tend to well-defined functions on $\mathbb{C}P^2$ in the large N limit, this implies that F_{ab} are the components of the usual field strength 2-form in the commutative (large N) limit. This justifies the above definition of F_{ab}, and it is a matter of taste whether one works with the components or with forms.

4.3. *Constraints*

In order to describe fuzzy $\mathbb{C}P^2$, the fields C_a should satisfy at least approximately the constraints (11), (12) of $\mathbb{C}P^2_N$,

$$D_a = 0, \tag{22}$$

$$g_{ab}C_aC_b = \frac{1}{3}N^2 + N \tag{23}$$

which are gauge invariant. These constraints ensure that C_a can be interpreted as describing a ("dynamical" or fluctuating) $\mathbb{C}P^2_N$. These constraints are analyzed in considerable detail in [6] in the non-commutative case.

4.4. *The Yang-Mills action*

Assume that the C_a satisfy the constraints (22), (23) of $\mathbb{C}P^2_N$ exactly or approximately. This implies that F_{ab} is tangential in the commutative limit, as shown in [6]. Then one can define the "Yang-Mills" action as

$$S_{YM} = \frac{1}{g}\int F_{ab}F_{ab} = \frac{1}{gD_N}\,Tr(-[C_a, C_b]^2 + 2if_{abc}C_aC_bC_c + 3C_aC_a). \tag{24}$$

It reduces to the classical Yang-Mills action on $\mathbb{C}P^2$, because only the tangential indices contribute in the commutative limit. The corresponding equation of motion is

$$2[C_b, F_{ab}] - iF_a = 0 \tag{25}$$

We now have to impose the constraints (22), (23) either exactly or approximately, and there are several possibilities how to proceed. Imposing both of them exactly seems too restrictive, since they are not independent even classically. One can hence either

(1) consider all 8 fields C_a as dynamical and add something like

$$S_D = \frac{1}{gD_N}\, Tr\left(\mu_1(dCC - (\frac{2N}{3}+1)C)^2 + \mu_2(C\cdot C - (\frac{N^2}{3}+N))^2\right) \tag{26}$$

to the action. This will impose the constraint dynamically for suitable $\mu_1 > 0$ and $\mu_2 \geq 0$, by giving the 4 transversal fields a large mass $m \to \infty$. Or,

(2) impose the constraint $D = dCC - (\frac{2N}{3}+1)C = 0$ exactly, or a slightly modified version.

In the second approach, it is not clear whether there are sufficiently many solutions of $D = 0$ in the noncommutative case to admit 4 tangential gauge fields. This concern could be circumvented by modifying the constraint, which is discussed in [6]. However we have not been able to find instanton-like solutions in this case (which may just be a technical problem). Therefore we concentrate on the first approach here, where we do find topologically nontrivial instanton solutions. It offers the additional possibility to give physical meaning to the non-tangential degrees of freedom.

Therefore our action is

$$\boxed{S = S_{YM} + S_D.} \tag{27}$$

It is shown in [6] that this reproduces the classical Yang-Mills action on $\mathbb{C}P^2$ in the large N limit provided

$$\mu_1 = o(\frac{1}{N}), \quad \text{and} \quad \mu_2 \leq o(\frac{1}{N^3}). \tag{28}$$

Here $o(\frac{1}{N})$ stands for a function which scales exactly like $\frac{1}{N}$. These constraints on the scaling of $\mu_{1,2}$ ensure that the vacuum which defines the geometry of $\mathbb{C}P^2_N$ is stable (note that the geometry is determined dynamically in noncommutative gauge theories!), and the monopole-and instanton solutions which will be discussed below survive. Imposing e.g. $\mu_1 = 0$ strictly would suppress the instanton solutions, hence in some sense fix the topology of the gauge fields. These issues certainly need further investigations; similarly, one may or may not fix the size of the matrices to be exactly D_N, which also has some influence on the existence of certain non-trivial solutions. Our choices are such that the conventional Yang-Mills theories emerge in the large N limit. These issues are discussed in more detail in [6].

We proceed to find the "vacuum", i.e. the minimum of the action. Assume first that the size of the matrices is D_N. Then the absolute minima of the action are given by solutions of $F_{ab} = 0$ and $D_a = 0$, which means that C_a is a representation of $su(3)$ with $D_a = 0$. The latter implies that the only allowed irreps are $V_{N\Lambda_2}$ or the trivial representation. Ignoring the latter (it has a smaller "phase space" of fluctuations), the vacuum solution is therefore

$$(C_{vac})_a = \xi_a \tag{29}$$

in a suitable basis. These arguments go through if we allow the size of the matrices C_a to be somewhat bigger that D_N, say

$$C_a \in Mat(D_N + N, \mathbb{C}) \tag{30}$$

(anything much smaller that $2D_N$ will do), which is needed to accomodate all the nontrivial solutions found in [6]. Any configuration with finite action is therefore close to (29), and can hence be written as

$$C_a = \xi_a \; + A_a \tag{31}$$

in a suitable basis, with "small" A_a. This justifies the assumptions made in the beginning of Section 4.4

It is quite straightforward to include scalar fields in this construction. Assume that we have an additional complex scalar field ϕ. Without gauge coupling, a natural action would be $\int([\xi_a, \phi])^\dagger[\xi_a, \phi] = -\int \phi^\dagger \Delta\phi$. If we assume that ϕ is charged and transforms under gauge transformations as

$$\phi \to U\phi, \tag{32}$$

then a natural gauge-invariant action would be

$$S[\phi] = \int (C_a\phi - \phi\xi_a)^\dagger(C_a\phi - \phi\xi_a). \tag{33}$$

This reduces to $\int(D_a\phi)^\dagger D_a\phi$ where $D_a = [\xi_a, .] + A_a$. Fermions are much more difficult to handle since $\mathbb{C}P^2$ is a spin^c manifold but not spin, and a fully satisfactory treatment in the fuzzy case is still lacking; see [1] for a possible approach.

Acknowledgments

The author is grateful for the invitation to the International Workshop on Noncommutative Geometry and Physics February 26- March 3, 2004, Keio Univ., Yokohama, and in particular for collaboration and useful discussions with S. and U. Watamura while preparing [13].

References

1. G. Alexanian, A.P. Balachandran, G. Immirzi and B. Ydri, *Fuzzy* $\mathbb{C}P^2$. J.Geom.Phys.**42** (2002), 28-53.
2. A. Balachandran, B. Dolan, J. Lee, X. Martin, and D. O'Connor, *Fuzzy Complex Projective Spaces and their Star-products.* J.Geom.Phys. **43** (2002), 184-204.
3. C. S. Chu, J. Madore and H. Steinacker, *Scaling Limits of the Fuzzy Sphere at one Loop.* JHEP **0108** (2001), 038.
4. M. R. Douglas and N. A. Nekrasov, *Noncommutative field theory.* Rev.Mod.Phys. **73** (2001), 977 **hep-th/0106048**.
5. H. Grosse and A. Strohmaier, *Towards a Nonperturbative Covariant Regularization in 4D Quantum Field Theory.* Lett.Math.Phys. **48** (1999), 163-179.
6. H. Grosse and H. Steinacker, *Finite Gauge Theory on Fuzzy* CP^2. Nucl. Phys. **B707** (2005), 145-198,**hep-th/0407089**
7. C. Itzykson and J.B. Zuber, *The planar approximation.* Journ. Math. Phys. **21** (1980), 411.
8. J. Madore, *The Fuzzy Sphere.* Class. Quant. Grav. **9** (1992), 69.
9. A.A. Migdal, *Recursion equations in gauge theories.* Sov.Phys.JETP **42** (1976), 413. B.E. Rusakov, *Loop averages and partition functions in U(N) gauge theory on two-dimensional manifolds* Mod.Phys.Lett. **A5** (1990), 693.
10. H. Steinacker, *Quantized Gauge Theory on the Fuzzy Sphere as Random Matrix Model.* Nucl. Phys. **B679**, vol. 1-2 (2004), 66-98.
11. R. Szabo, *Quantum Field Theory on Noncommutative Spaces.* Phys.Rept. bf 378 (2003), 207-299, **hep-th/0109162**.
12. U. Carow-Watamura and S. Watamura, *Noncommutative Geometry and Gauge Theory on Fuzzy Sphere.* Commun.Math.Phys. **212** (2000), 395.
13. U. Carow-Watamura, H. Steinacker and S. Watamura, *Monopole Bundles over Fuzzy Complex Projective Spaces.* To appear in J. Geom. Phys, **hep-th/0404130**.
14. E. Witten, *Two-dimensional gauge theories revisited.* J. Geom. Phys. **9** (1992), 303.

A RELATION ON SPIN BUNDLE GERBES AND MAYER'S DIRAC OPERATORS

ATSUSHI TOMODA

Department of Mathematics, Keio University,
3-14-1 Hiyoshi, Kohoku-ku, Yokohama, Japan
ton@math.keio.ac.jp

Exploiting the notion of bundle gerbe due to Murray, Murray and Singer constructed a generalization of Dirac operators for possibly non-spin manifolds. We shall provide an alternative proof for their index formula, and clarify the relation between a generalized Dirac operators due to Mayer and their operators. Furthermore, we determine the twisted Chern character of some bundle gerbe modules.

Introduction

Murray and Singer introduced the Dirac operators associated with the spin bundle gerbe modules, which are , roughly speaking, roots of complex vector bundles, by the use of the spin bundle gerbes.

The purpose of this paper is to describe the index formula of the Dirac operators associated with spin bundle gerbe modules from a different point of view. Mayer introduced Mayer's Dirac operator associated with a $G(2l, n, 2)$-structure σ. We also relate their operators with the Mayer's Dirac operators and describe the twisted Chern character for some bundle gerbe modules explicitly:

Theorem 0.1. *Let σ be a $G(2l, n, 2)$-structure of X induced by a real vector bundle E with $W_3(TX \oplus E) = 0$ and let F be the associated line bundle of σ. Then we have*

$$\mathrm{ch}^{\tau}(W_\sigma) = 2^{[n/2]} e^{d/2} \prod_{i=1}^{[n/2]} \cosh(x_i/2),$$

where $d = c_1(F)$ and $p(E) = \prod_i (1 + x_i^2)$.

1. Bundle gerbes

1.1. *Bundle gerbes*

We shall give an exposition of the bundle gerbes. We refer to Murray[5].

Let X be a compact oriented smooth manifold and let $\pi : Y \to X$ be a fiber bundle. Then we can consider the fiber product $Y \times_\pi Y$ with itself. For simplicity, we denote this by $Y^{[2]}$.

Definition 1.1. Let X and Y be as above and let $L \to Y^{[2]}$ be a hermitian line bundle. A triple (X, Y, L) is said to be *a bundle gerbe* over X if L is equipped with a product which is an isomorphism

$$L_{(y_1,y_2)} \otimes L_{(y_2,y_3)} \xrightarrow{\cong} L_{(y_1,y_3)} \quad \text{for every } (y_1, y_2), (y_2, y_3) \in Y^{[2]},$$

and which is associative.

As line bundles have Euler classes, bundle gerbes have a kind of characteristic classes:

Definition 1.2. Fix a good cover $\{U_\alpha\}$ of X and local sections $\{s_\alpha : U_\alpha \to P\}$. Then we obtain line bundles $L_{\alpha\beta} = (s_\alpha, s_\beta)^* L$ over $U_{\alpha\beta}$. A choice of sections $z_{\alpha\beta} \in \Gamma(L_{\alpha\beta})$ and the induced product gives a unique $\{\varepsilon_{\alpha\beta\gamma} : U_{\alpha\beta\gamma} \to U(1)\}$ such that

$$L_{\alpha\beta} \otimes L_{\beta\gamma} \to L_{\alpha\gamma} \ ; \ z_{\alpha\beta} \otimes z_{\beta\gamma} \mapsto \varepsilon_{\alpha\beta\gamma} z_{\alpha\gamma}.$$

Then $\{\varepsilon_{\alpha\beta\gamma}\}$ gives a cohomology class $[\{\varepsilon_{\alpha\beta\gamma}\}] \in \check{H}^2(X; U(1)) \cong \check{H}^3(X; \mathbb{Z})$ and we denote this by $\mathrm{dd}(X, Y, L)$ and we call this the Dixmier-Douady class for the bundle gerbe (X, Y, L).

Remark 1.1. The Dixmier-Douady class $\mathrm{dd}(X, Y, L)$ of (X, Y, L) does not depend on the choice of open covering of X and local sections $\{s_\alpha\}$ and $\{z_{\alpha\beta}\}$.

Example 1.1. Consider a principal $SO(n)$-bundle P and a central extension:

$$1 \to \mathbb{Z}_2 \to Spin(n) \xrightarrow{p} SO(n) \to 1.$$

We define a $\mathbb{Z}_2$-bundle Q over $P^{[2]} \to X$ by

$$Q = \left\{ ((y_1, y_2), \alpha) \in P^{[2]} \times Spin(n) \middle| y_1 p(\alpha) = y_2 \right\}$$

so we obtain the complex line bundle $L \to P^{[2]}$ associated with Q. Then the triple (X, P, L) is a bundle gerbe over X and we call this the spin bundle gerbe of P. Especially every oriented closed n-dimensional manifold with a Riemannian metric yields a $SO(n)$-frame bundle P of TX and hence a spin bundle gerbe (X, P, L) of P. We simply call this the spin bundle gerbe of X.

We note that the Dixmier-Douady class $\mathrm{dd}(X, P, L)$of the spin bundle gerbe (X, P, L) of X coincides with the third integral Stiefel-Whitney class $W_3(X)$.

In the subsequent discussion we deal with only spin bundle gerbes.

1.2. *Bundle gerbe modules and the twisted Chern character*

We introduce the notion of bundle gerbe modules. This is introduced in Bouwknegt, Carey, Mathai, Murray, and Stevenson[3].

Definition 1.3. Let (X, P, L) be a spin bundle gerbe. Then a hermitian vector bundle W over P is called a *bundle gerbe module for* (X, P, L) if it is endowed with the multiplication of L:

$$L_{(y_1,y_2)} \otimes W_{y_2} \xrightarrow{\cong} W_{y_1} \quad \text{for every } (y_1, y_2) \in P^{[2]}$$

which satisfies the commutative diagram:

$$\begin{array}{ccccc} (L_{(y_1,y_2)} \otimes L_{(y_2,y_3)}) \otimes W_{y_3} & \longrightarrow & L_{(y_1,y_3)} \otimes W_{y_3} & \longrightarrow & W_{y_1} \\ \| & & & & \| \\ L_{(y_1,y_2)} \otimes (L_{(y_2,y_3)} \otimes W_{y_3}) & \longrightarrow & L_{(y_1,y_2)} \otimes W_{y_2} & \longrightarrow & W_{y_1}. \end{array}$$

We denotes by $\mathrm{Mod}(X, P, L)$ the isomorphism classes of bundle gerbe modules for (X, P, L).

Spin(n) is naturally included in L. Hence *Spin*(n) acts on a bundle gerbe module W. The action of $\mathbb{Z}_2 \subset Spin(n)$ on each fiber W_y is multiplication by ± 1. Consider two bundle gerbe modules V and W. Then *Spin*(n) acts on $V \otimes W$ and the action induces $SO(n)$-action. Moreover, it gives rise to a complex vector bundle E over X satisfying $\pi^* E = V \otimes W$. We denote this by $(V \otimes W)_0$.

Let (X, P, L) be the spin bundle gerbe for X. Then we have a canonical flat connection ∇ on L induced by $\mathbb{Z}_2$-bundle Q. ∇ is a bundle gerbe connection. That is, the product of L preserves the connection.

Definition 1.4. Let $W \to P$ be a bundle gerbe module for (X, P, L). Then a hermitian connection ∇^W on W is called *a bundle gerbe module* *connection on W compatible with* (X, P, L, ∇) if the multiplication

$$\varphi : L \otimes \pi_1^* W \xrightarrow{\cong} \pi_2^* W$$

preserves the connections, where $L \otimes \pi_1^* W$ ($\pi_2^* W$ resp.) is endowed with the induced connection $\nabla \otimes 1 + 1 \otimes \pi_1^* \nabla^W$ ($\pi_2^* \nabla^W$ resp.). Here $\pi_i : Y^{[2]} \to Y$ denotes the i-th element forgetting map for $i = 1, 2$. If there is no confusion, we simply call ∇^W a bundle gerbe module connection on W.

Let ∇^W be a bundle gerbe module connection. Then we have

$$1 \otimes \pi_1^* F(\nabla^W)^k = \varphi \circ \pi_2^* F(\nabla^W)^k \circ \varphi^{-1} \text{ for every } k > 0.$$

So every invariant polynomial P defines $P(F(\nabla^W)) \in \Omega^{2*}(P)$ and $(\pi_1^* - \pi_2^*)P(F(\nabla^W)) = 0$. Therefore we obtain $[\eta] \in H^{2*}(X)$ such that $\pi^* \eta = P(F(\nabla^W))$. The cohomology class $[\eta]$ does not depend on a choice of bundle gerbe module connections ∇^W. Especially tr is invariant polynomial and there is unique $[\eta_k] \in H^{2k}(X)$ such that

$$\pi^* [\eta_k] = \mathrm{tr} \left(\left(\frac{-1}{2\pi i} \left(F(\nabla^W) \right) \right)^k \right)$$

Definition 1.5. Analogously to Chern character of complex vector bundles, we define the twisted Chern character $\mathrm{ch}^\tau(W)$ of a bundle gerbe module W by

$$\mathrm{ch}^\tau(W) = [\eta_0] + [\eta_1] + \cdots + [\eta_k] + \cdots \in H^{2*}(X; \mathbb{R}).$$

2. Dirac operators associated with bundle gerbe modules

Let X be a $2l$-dimensional smooth oriented closed manifold and let (X, P, L) be the lifting bundle gerbe of X. In this section we shall introduce the Dirac operator associated with a bundle gerbe module for a probably non-spin manifold.

Definition 2.1. Let (X, P, L) be as above. Then we define $\mathbb{S}(P)$ as

$$\mathbb{S}(P) = P \times \mathbb{S}_{2l}, \text{ where } \mathbb{S}_{2l} \text{ denotes the spinor space of } \mathbb{R}^{2l}.$$

It is easy to see that $\mathbb{S}(P) \to P$ is a bundle gerbe module for (X, P, L), and we call $\mathbb{S}(P)$ the spinor bundle gerbe module of P.

Endowed with some Riemannian metric g, X has the Levi-Civita connection ∇^{LC} on TX. The connection ∇^{LC} induces a connection 1-form

$\omega \in \Omega^1(P;\mathfrak{so}(2l))$ on P. Here we note that there is the natural isomorphism $\varphi_* : \mathfrak{spin}(2l) \xrightarrow{\cong} \mathfrak{so}(2l)$ induced by $\varphi : Spin(2l) \to SO(2l)$. The spinor representation $\rho : Spin(2l) \to \mathrm{Aut}(\mathbb{S}_{2l})$ gives rise to $\rho_* : \mathfrak{spin}(2l) \to \mathrm{End}_+(\mathbb{S}_{2l})$. The composition $\rho_* \circ \varphi_*^{-1} : \mathfrak{so}(2l) \to \mathrm{End}_+(\mathbb{S}_{2l})$ defines $\widetilde{\omega} \in \Omega^1(P;\mathrm{End}_+\mathbb{S}(P))$ by $\widetilde{\omega} = \rho_* \circ \varphi_*^{-1}(\omega)$. We define a canonical connection $\widetilde{\nabla}$ on $\mathbb{S}(P)$ by $\widetilde{\nabla} = d + \widetilde{\omega}$.

We can easily show that $\widetilde{\nabla}$ is a bundle gerbe module connection on $\mathbb{S}(P)$.

Next, we shall define Clifford multiplication $\widetilde{c}$ on P. The connection 1-form $\omega \in \Omega^1(P;\mathrm{End}_-(\mathbb{S}(P)))$ induces

$$T^*P \to \pi^*T^*X.$$

Moreover,

$$P \times \mathbb{R}^{2l} \times \mathbb{S}_{2l} \to P \times \mathbb{S}_{2l}\ ;\ (y, v, \xi) \mapsto (y, v\xi)$$

defines $c : \pi^*TX \to \mathrm{End}_-(\mathbb{S}(P))$ since $P \times_{SO(2l)} \mathbb{R}^{2l}$ is isomorphic to TX. So we have

$$\widetilde{c} : T^*P \to \pi^*T^*X \xrightarrow{\cong} \pi^*TX \xrightarrow{c} \mathrm{End}(\mathbb{S}^{\pm}(P), \mathbb{S}^{\mp}(P)).$$

We have prepared some notions so far. Now we define the Dirac operator $\not{D}_W$ associated with a bundle gerbe module W.

Choose and fix a bundle gerbe module W for (X, P, L) and an arbitrary bundle gerbe module connection ∇^A compatible with (X, P, L, ∇^L). Then we obtain $\overline{\not{D}}_{W,A}$ as the composition:

$$\overline{\not{D}}_{W,A} = (\widetilde{c} \otimes \mathrm{Id}) \circ (\widetilde{\nabla} \otimes 1 + 1 \otimes \nabla^A).$$

We have to note that $\overline{\not{D}}_{W,A}$ is a first order **non-elliptic** differential operator from $\mathbb{S}^+(P) \otimes W$ to $\mathbb{S}^-(P) \otimes W$ over P because the symbol $\sigma(\overline{\not{D}}_{W,A})$ vanishes in direction of the fiber of P. However, $\overline{\not{D}}_{W,A}$ is $Spin(2l)$-equivariant operator and it gives rise to an elliptic operator

$$\not{D}_{W,A} : \Gamma\left(X, (\mathbb{S}^+(P) \otimes W)_0\right) \to \Gamma\left(X, (\mathbb{S}^-(P) \otimes W)_0\right)$$

over X. We call this the Dirac operator associated with a bundle gerbe module W. The next theorem is due to Murray and Singer[6]:

Theorem 2.1.

$$\mathrm{Index}\,(\not{D}_{W,A}) = \int_X \widehat{\mathcal{A}}(X)\mathrm{ch}^{\tau}(W)$$

They proved this theorem by the use of the method of Getzler, Berline and Vergne[2]. Here we prove in an alternative way.

Proof. The property of the twisted Chern character implies

$$\mathrm{ch}((\mathbb{S}^{\pm} \otimes W)_0) = \mathrm{ch}^{\tau}(\mathbb{S}^{\pm})\mathrm{ch}^{\tau}(W). \tag{1}$$

Atiyah-Singer index theorem[1] says:

$$\begin{aligned}\mathrm{Index}(\not{D}_{W,A}) &= (-1)^l \frac{\mathrm{ch}\,((\mathbb{S}^+ \otimes W)_0 - (\mathbb{S}^- \otimes W)_0)\,\mathrm{td}(TX \otimes \mathbb{C})}{e(X)}[X] \\ &= (-1)^l \frac{\left(\prod_{i=1}^l e^{x_i/2} - e^{-x_i/2}\right)}{x_1 \cdots x_l} \prod_{i=1}^l \frac{x_i}{1-e^{-x_i}} \frac{-x_i}{1-e^{x_i}} \mathrm{ch}^{\tau}(W)[X] \\ &= \widehat{A}(X)\mathrm{ch}^{\tau}(W)[X].\end{aligned}$$

The second equality is derived from (1) and the next theorem. □

The next theorem is essential.

Theorem 2.2. *For every $SO(2l)$-bundle $P \to X$, we have the spinor bundle gerbe module $\mathbb{S}^{\pm}(P)$, which is a bundle gerbe module for (X, P, L). Then we have*

$$\mathrm{ch}^{\tau}(\mathbb{S}^+(P) - \mathbb{S}^-(P)) = \prod_{i=1}^{l} \left(e^{x_i/2} - e^{-x_i/2}\right),$$

where the associated rank m vector bundle $E = P \times_{SO(2l)} \mathbb{R}^{2l}$ virtually splits into l complex line bundles $L_1 \oplus \cdots \oplus L_l$ and we put $c_1(L_i) = x_i$ for every i.

Proof. Consider a vector bundle $E = P \times_{SO(2l)} \mathbb{R}^{2l} \to X$. Then we have $E = L_1 \oplus \cdots L_l$ virtually. Since $w_2(E \oplus E) = 0$, we have a complex vector bundle $\mathbb{S}(E \oplus E)$ over X, and

$$\mathbb{S}^+(E \oplus E) - \mathbb{S}^-(E \oplus E) = \bigotimes_{i=1}^{l} (\mathbb{S}^+(L_i \oplus L_i) - \mathbb{S}^-(L_i \oplus L_i))$$

Easy calculation gives us the explicit description of Chern character:

$$\mathrm{ch}\left(\mathbb{S}^+(L_i \oplus L_i) - \mathbb{S}^-(L_i \oplus L_i)\right) = e^{x_i} + e^{-x_i} - 2 \text{ for } i = 1, \ldots, l,$$

where $x_i = c_1(L_i)$. So we have

$$\mathrm{ch}\left(\mathbb{S}^+(E\oplus E) - \mathbb{S}^-(E\oplus E)\right) = \prod_{i=1}^{l}\left(e^{x_i}+e^{-x_i}-2\right) = \left(\prod_{i=1}^{l}\left(e^{x_i/2}-e^{-x_i/2}\right)\right)^2.$$

On the other hand,

$$\mathrm{ch}^\tau(\mathbb{S}^+(P_E) - \mathbb{S}^-(P_E))^2 = \mathrm{ch}(\mathbb{S}^+(E\oplus E) - \mathbb{S}^-(E\oplus E)).$$

By considering the universal bundle $ESO(2l) \to BSO(2l)$ and the spin bundle gerbe $(BSO(2l), ESO(2l), \widetilde{L})$ of $ESO(2l)$, we obtain

$$\mathrm{ch}^\tau(\mathbb{S}^+(ESO(2l)) - \mathbb{S}^-(ESO(2l))) = \pm\prod_{i=1}^{l}\left(e^{x_i/2}-e^{-x_i/2}\right) \tag{2}$$

since the ring $H^*(BSO(2l);\mathbb{R})$ has no zero divisor. So far, the equation holds up to signature. For every $SO(2l)$-bundle $P \to X$, there exists $f : X \to BSO(2l)$ satisfying $f^*ESO(2l) \cong P$, and $\mathbb{S}^\pm(P)$ comes from bundle gerbe module $\mathbb{S}^\pm(ESO(2l))$ for $(BSO(2l), ESO(2l), \widetilde{L})$. Strictly speaking,

$$\mathbb{S}^\pm(P) \cong \overline{f}^*(\mathbb{S}^\pm(ESO(2l))),$$

where $\overline{f} : P \to ESO(2l)$ is a bundle map covering $f : X \to BSO(2l)$. Twisted Chern character has the naturality. Hence the equation (2) is valid for every $SO(2l)$-bundle $P \to X$.

Since the signature is universally determined, we can see which is correct by a non-trivial example. For a $SO(2l)$-bundle which admits spin structure, the signature is positive. Hence we can determine the signature in (2). □

We shall mention spin^c structures in the view point of bundle gerbe. A spin^c structure is the unit vector bundle of a bundle gerbe module G with rank 1. Then $\mathrm{ch}^\tau(G) = e^{d/2}$, where we set $d = c_1((G^{\otimes 2})_0)$. Especially a spin structure is one associated with a $\mathbb{Z}_2$-bundle over P and $(G^{\otimes 2})_0 = \underline{\mathbb{C}}$. Therefore $\mathrm{ch}^\tau(G) = 1$.

Mayer[4] defined the Dirac operator associated with some principal bundle. First, we recall the construction. Consider

$$SO(m)\times SO(n)\times SO(2) \subset SO(m+n+2)$$

and $\lambda : Spin(m+n+2) \to SO(m+n+2)$. Then we can identify $\lambda^{-1}(SO(m)\times SO(n)\times SO(2))$ with $(Spin(m)\times Spin(n)\times Spin(2))/\mathbb{Z}_2\oplus\mathbb{Z}_2$, where $\mathbb{Z}_2 \oplus \mathbb{Z}_2$ is generated by $\{(-1,-1,+1),(+1,-1,-1)\}$. We denote this by $G(m,n,2)$.

Definition 2.2. Let $\{g_{\alpha\beta}\} \in \check{C}^2(X, SO(m))$ denote the Cech cocycle defining the frame bundle P of TX. Then $\{[\widehat{g}_{\alpha\beta}, \widehat{h}_{\alpha\beta}, \widehat{z}_{\alpha\beta}]\} \in \check{C}^2(X, G(m,n,2))$ is called a $G(m,n,2)$-structure if

$$\lambda(\widehat{g}_{\alpha\beta}) = g_{\alpha\beta} \text{ and } \delta\{[\widehat{g}_{\alpha\beta}, \widehat{h}_{\alpha\beta}, \widehat{z}_{\alpha\beta}]\} = 1.$$

We have the complex spinor representation $\rho^c : G(2l,n,2) \to \mathrm{Aut}(\mathbb{S}_{2l} \otimes \mathbb{S}_n)$. Given a $G(2l,n,2)$-structure $\sigma = \{[\widehat{g}_{\alpha\beta}, \widehat{h}_{\alpha\beta}, \widehat{z}_{\alpha\beta}]\}$ over $2l$-dimensional manifold X, we obtain a spinor bundle $\mathbb{S}_\sigma$ associated with it and the Dirac operator $\not{D}_\sigma : \Gamma(\mathbb{S}_\sigma^+) \to \Gamma(\mathbb{S}_\sigma^-)$. We call this Mayer's Dirac operator.

If a real vector bundle $E \to X$ with rank n satisfies $W_3(TX \oplus E) = 0$, then we have a $G(2l,n,2)$-structure $\sigma = \{[\widehat{g}_{\alpha\beta}, \widehat{h}_{\alpha\beta}, \widehat{z}_{\alpha\beta}]\}$ such that $\{\lambda(\widehat{h}_{\alpha\beta})\}$ defines E and $\{\lambda(\widehat{z}_{\alpha\beta})\}$ defines a complex line bundle F which satisfies $w_2(TX) + w_2(E) = w_2(F)$. Mayer proved:

Theorem 2.3. *Let X, E, and F be as above. Then we have*

$$\mathrm{Index}\not{D}_\sigma = 2^{[n/2]} \int_X e^{d/2} \widehat{A}(X) \prod_{i=1} \cosh(x_i/2),$$

where $d = c_1(F)$ and $p(E) = \prod_i (1 + x_i^2)$.

Here we shall introduce twisted unitary structure of X. We define $U^\tau(n)$ as $U^\tau(n) = Spin(2l) \times_{\pm 1} U(n)$. Then $U(n) \hookrightarrow U^\tau(n)$ is a normal subgroup. *The twisted unitary structures $U^\tau(n)(X)$ of X are defined by*

$$U^\tau(n)(X) = \left\{ \widetilde{P} \to X : U^\tau(n)\text{-bundle} \;\middle|\; \widetilde{P}/U(n) \text{ is isomorphic to } P \right\}.$$

Theorem 2.4. *We obtain the isomorphism*

$$U^\tau(n)(X) \cong \mathrm{Mod}_n(X,P,L) \;;\; \widetilde{P} \mapsto \widetilde{P} \times_{U(n)} \mathbb{C}^n = W_{\widetilde{P}}.$$

Proof. For every $W \in \mathrm{Mod}_n(X,P,L)$ a $U(n)$-bundle $\widetilde{P}_W$, which consists of unitary frames of W, defines a twisted unitary structure. This is the inverse. □

The homomorphism $Spin(n) \times_{\mathbb{Z}_2} Spin(2) \to U\left(2^{[n/2]}\right)$ defines a homomorphism $G(m,n,2) \to U^\tau\left(2^{[n/2]}\right)$. Then we obtain

Theorem 2.5. *Every $G(2l,n,2)$-structure σ gives rise to a bundle gerbe module W_σ. Moreover* $\mathrm{Index}(\not{D}_\sigma) = \mathrm{Index}(\not{D}_{W_\sigma})$ *for Mayer's Dirac operator D_σ.*

Let σ be a $G(2l,n,2)$-structure over X. We denote the associated real vector bundle by E and the associated complex line bundle by F. Then we have a spin bundle gerbe (X, P_E, L_E) of E, and a bundle gerbe module $\mathbb{S}(P_E)$ for (X, P_E, L_E), where P_E denotes the $SO(n)$-frame bundle of E. Like as the proof of Theorem2.2, we have

$$\mathrm{ch}^\tau(\mathbb{S}(P_E)) = \prod_{i=1}^{[n/2]} \left(e^{x_i/2} + e^{-x_i/2}\right) = 2^{[n/2]} \prod_i \cosh(x_i/2),$$

where $p(E) = \prod_i (1+x_i^2)$. We have to remark that $\mathbb{S}(P_E)$ is a bundle gerbe module not for (X,P,L) but for (X, P_E, L_E).

We shall consider another spin bundle gerbe $(X, P\times_\pi P_E, L\otimes L_E)$. It is easy to see there is a bundle gerbe module G for this with rank 1 satisfying $(G^{\otimes 2})_0 = F$ since $W_3(TX) + W_3(E) = 0$. Then G induces an isomorphism $G\otimes : \mathrm{Mod}(X, P_E, L_E) \to \mathrm{Mod}(X,P,L)$. We will prove that $G\otimes$ maps $\mathbb{S}(P_E)$ to the bundle gerbe module W_σ for (X,P,L) and that

$$\mathrm{ch}^\tau(W_\sigma) = e^{d/2}\mathrm{ch}^\tau(\mathbb{S}(E)) = 2^{[n/2]} e^{d/2} \prod_i \cosh(x_i/2).$$

Consider the next diagram:

$$\begin{array}{ccccc}
L & & \overline{\pi}_P^* L \otimes \overline{\pi}_E^* L_E & & L_E \\
\downarrow & & \downarrow & & \downarrow \\
P^{[2]} & \xleftarrow{\overline{\pi}_P} & P^{[2]} \times \pi P_E^{[2]} & \xrightarrow{\overline{\pi}_E} & P_E^{[2]} \\
{\scriptstyle \pi_i}\downarrow & & {\scriptstyle \pi_i}\downarrow & & {\scriptstyle \pi_i}\downarrow \\
P & \xleftarrow{\pi_P} & P \times_\pi P_E & \xrightarrow{\pi_E} & P_E.
\end{array}$$

It is easy to see $\pi_1^* G \otimes \pi_2^* G^* \cong \overline{\pi}^* L \otimes \overline{\pi}_E^* L_E$. We define Q_y and Q_z by

$$Q_y = \left\{ ((y_1,y_2),(z_1,z_2)) \in P^{[2]} \times_\pi P_E^{[2]} \,\middle|\, y_1 = y_2 \right\},$$
$$Q_z = \left\{ ((y_1,y_2),(z_1,z_2)) \in P^{[2]} \times_\pi P_E^{[2]} \,\middle|\, z_1 = z_2 \right\}.$$

Then, $\pi_1^* G \otimes \pi_2^* G^*$ is isomorphic to $\overline{\pi}_E^* L_E$ over Q_y, and to $\overline{\pi}^* L$ over Q_z. We obtain the isomorphisms:

$$\begin{aligned}\pi_2^*(G \otimes \pi_E^* \mathbb{S}(P_E)) &= \pi_2^*(G) \otimes \pi_P^* \pi_E^* \mathbb{S}(P_E) = \pi_2^* G \otimes \overline{\pi}_E^* \pi_2^* \mathbb{S}(P_E) \\ &= \pi_2^* G \otimes \overline{\pi}_E^*(L_E \times \pi_1^* \mathbb{S}(P_E)) \\ &= \pi_2^* G \otimes \overline{\pi}_E^* L_E \otimes \pi_1^* \pi_E^* \mathbb{S}(P_E) \\ &= \pi_1^* G \otimes \pi_1^* \pi_E^* \mathbb{S}(P_E) = \pi_1^*(G \otimes \pi_E^* \mathbb{S}(P_E)) \text{ over } Q_y\end{aligned} \tag{3}$$

and

$$\begin{aligned}\overline{\pi}^* L \otimes \pi_1^*(G \otimes \pi_E^* \mathbb{S}(P_E)) &= \overline{\pi}^* L \otimes \pi_1^* G \otimes \pi_1^* \pi_E^* \mathbb{S}(P_E) \\ &= (\overline{\pi}^* L \otimes \overline{\pi}_E^* L_E) \otimes \pi_1^* G \otimes \pi_1^* \pi_E^* \mathbb{S}(P_E) \\ &= \pi_2^* G \otimes \pi_2^* \pi_E^* \mathbb{S}(P_E) \\ &= \pi_2^*(E \otimes \pi_E^* \mathbb{S}(P_E)) \text{ over } Q_z.\end{aligned} \tag{4}$$

The isomorphism (3) implies that there is a hermitian vector bundle W' over P such that $\pi_P^* W'$ is isomorphic to $G \otimes \pi_E^* \mathbb{S}(P_E)$, and (4) says that W' is a bundle gerbe module for (X, P, L). It is easy to see that W' is isomorphic to W_σ by the construction.

Moreover, these are isomorphisms with connections. Therefore, $\nabla^G \otimes \pi_E^* \nabla^{\mathbb{S}(P_E)}$ induces a bundle gerbe module connection ∇^{W_σ} on W_σ such that $\pi_P^* \nabla^{W_\sigma} = \nabla^G \otimes \pi_E^* \nabla^{\mathbb{S}(P_E)}$, and

$$\mathrm{ch}^\tau(W_\sigma) = \mathrm{ch}^\tau(G)\mathrm{ch}^\tau(\mathbb{S}(P_E)) = e^{d/2}\mathrm{ch}^\tau(\mathbb{S}(P_E)).$$

So we have proved Theorem 0.1.

References

1. M.F. Atiyah and I.M. Singer, *The index of elliptic operators I.* Ann. Math. **87** (1963), 484–530.
2. N. Berline, E. Getzler, and M. Vergne, *Heat kernels and Dirac operators.* Grundlehren Math. Wiss. **298** (1992), Springer-Verlag, New York.
3. P. Bouwknegt, A. Carey, V. Mathai, M.K. Murray, and D. Stevenson, *Twisted K-theory and K-theory of bundle gerbes.* Comm. Math. Phys. **228** (2002), no. 1, 17–45.
4. K.H. Mayer, *Elliptische Differentialoperatoren und Ganzzahligkeitssätze für charakteristische Zahlen.* Topology **4** (1965), 295–313.
5. M.K. Murray, *Bundle gerbes.* J. London. Math. Soc. (2) **54** (1996) no. 2, 403–416.
6. M.K. Murray and M.A. Singer, *Gerbes, Clifford Modules and the Index Theorem.* Ann. Global Anal. Geom. **26** (2004), 355–367.